INTRODUCTION TO DIGITAL SIGNAL PROCESSING

INTRODUCTION TO DIGITAL SIGNAL PROCESSING

Johnny R. Johnson

Professor Emeritus, Louisiana State University
Department of Mathematics
University of North Alabama

Prentice Hall, Englewood Cliffs, NJ 07632

Library of Congress Cataloging-in-Publication Data

JOHNSON, JOHNNY R.
Introduction to digital signal processing / Johnny R. Johnson.
p. cm.
Bibliography: p.
Includes index.
ISBN 0-13-481581-5
1. Signal processing—Digital techniques. I. Title.
TK5102.5.J62 1989
621.38'043—dc19 88-25573
CIP

Editorial/production supervision
and interior design: Debbie Young
Cover design: Edsal Enterprises
Manufacturing buyer: Mary Noonan

See page 407 for figure credits.

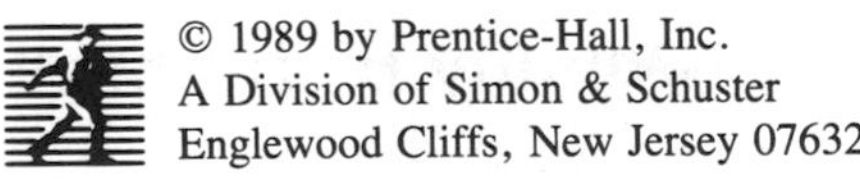

A Division of Simon & Schuster
Englewood Cliffs, New Jersey 07632

Printed in the United States of America

10 9 8 7 6 5 4 3 2 1

ISBN 0-13-481581-5

PRENTICE-HALL INTERNATIONAL (UK) LIMITED, *London*
PRENTICE-HALL OF AUSTRALIA PTY. LIMITED, *Sydney*
PRENTICE-HALL CANADA INC., *Toronto*
PRENTICE-HALL HISPANOAMERICANA, S.A., *Mexico*
PRENTICE-HALL OF INDIA PRIVATE LIMITED, *New Delhi*
PRENTICE-HALL OF JAPAN, INC., *Tokyo*
SIMON & SCHUSTER ASIA PTE. LTD., *Singapore*
EDITORA PRENTICE-HALL DO BRASIL, LTDA., *Rio de Janeiro*

CONTENTS

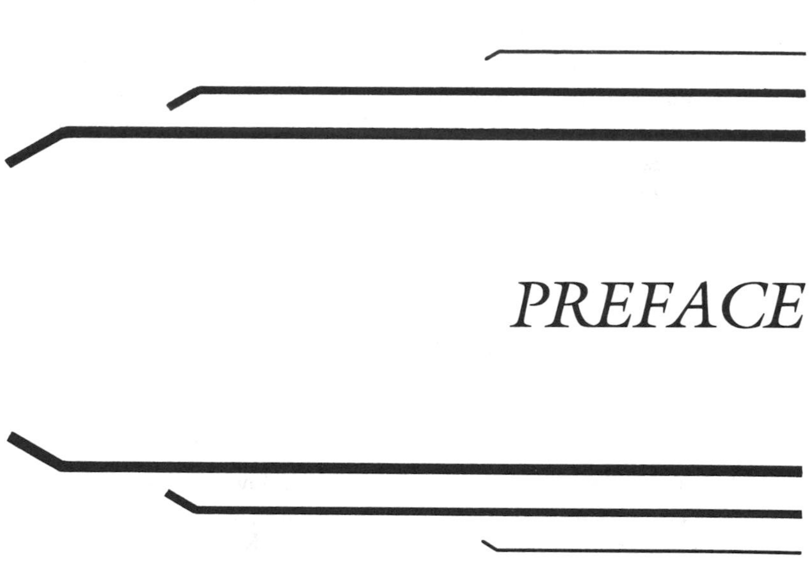

PREFACE

This book was written for a one-semester introductory course in digital signal processing. It is assumed that the reader has a background in calculus, including infinite series, and an exposure to continuous-time linear-system theory, including Laplace and Fourier transforms. Except for this background material, the book is self-contained. In many colleges, the book could be used in a course at the junior level.

In Chapter 1, we begin with the definition of discrete-time signals and give examples of important signals. We show how discrete-time signals can be represented in both the time and frequency domains. The frequency-domain representation is based on Fourier series. The discrete Fourier transform is introduced and it is shown how it can be used in approximating continuous-time Fourier transforms. Several examples of discrete-time systems are given.

Chapter 2 is devoted to a discussion of discrete-time systems. After a system is defined, several system classifications are given. Most of the discussion pertains to causal systems described by a linear difference equation with constant coefficients. Analysis techniques both in the time and frequency domains are given. The concept of stability is introduced here and a time-domain criterion is presented. The concept of finite- and infinite-duration sequences is discussed, and it is shown how linear phase can be obtained for a finite impulse-response system.

The z-transform is treated in Chapter 3, using only the concepts of infinite series that are normally included in the calculus courses. The z-transform is defined and its more important properties are given. Several methods for finding inverse z-transforms, are shown. The concept of the transfer function of a system is also given. Stability is revisited, using frequency-domain techniques.

Chapter 4 is a discussion of the realization of digital systems. Realizations using block diagrams and signal-flow graphs are given. Block diagrams provide a better analogy to actual hardware realizations, but signal-flow graphs can be more useful in computing transfer functions. The basic realizations that are discussed are the direct form, the cascade form, and the parallel form. Modifications of the direct form and the cascade form are given for finite impulse-response systems that use fewer multipliers. In addition, some of the ladder structures are discussed.

In Chapter 5, the design of infinite impulse-response filters is considered. The two methods that are emphasized are the impulse-invariance technique and the bilinear transformation. Design procedures for digital Butterworth, Chebyshev, inverse Chebyshev, and elliptic low-pass filters are given. Frequency transformations for obtraining high-pass, bandpass, and band-reject filters are discussed. Also, Bessel and all-pass filter designs are presented.

The design of finite impulse-response filters is treated in Chapter 6. The design techniques that are discussed are windowing, frequency sampling, and optimization. The rectangular, Hann, and Hamming windows are discussed and examples given. Section 6.5 is devoted entirely to the Kaiser window. The optimal filter based on a Chebyshev approximation is discussed and an example using this program is given. Finally, analytical procedures are given for some special cases of optimal filters.

In Chapter 7, the discrete Fourier transform (DFT) is again presented along with some of its properties. The circular convolution is developed and it is then shown how it can be used to obtain the linear convolution of two sequences. Various other discrete transforms, including the Hartley, Hadamard, and Walsh, are introduced in the problems.

Fast and efficient algorithms known as fast Fourier transforms (FFTs) are presented in Chapter 8 for evaluating the DFT. Decimation-in-time and decimation-in-frequency algorithms are presented for the case when N, the number of terms in the sequence being considered, is a power of 2. The algorithms are then modified and applied to the case when N is a composite number. Discrete transforms such as the number theoretic, Fermat, and Mersenne are considered in the problems. Fast transform algorithms are discussed for these transforms as well as for those discussed in the problems of Chapter 7.

Johnny R. Johnson

ACKNOWLEDGMENTS

The author is indebted to his students, colleagues, and the reviewers of the manuscript for their contributions to this text. Special thanks are due Steve Scott, Jimmy Miller, Oscar Beck, Patricia Roden, and Jean Henderson for providing computer graphics expertise, and Marie Jines, Susan Boyer, Margaret Brewer, and Ginnevere Mobley for typing the manuscript. The book, *Digital Signal Processing,* written by A. V. Oppenheim and R. W. Schafer, has greatly influenced this text. The contribution of Don Childers is greatly appreciated. Finally, the author owes a debt of gratitude to his brother, David Johnson, for his advice and continued support.

INTRODUCTION TO DIGITAL SIGNAL PROCESSING

1

DISCRETE-TIME SIGNALS AND SYSTEMS

1.1 INTRODUCTION

This is a book that is primarily concerned with digital systems and their applications in filtering and signal processing. We think of a *system* as a collection of objects interacting with each other in such a way as to achieve some objective. Systems arise in many diverse scientific fields, such as communications, power systems, control systems, speech analysis, economics, radar and sonar, seismology, telemetry and image processing. The physical phenomena that cause these systems to function can be electrical, mechanical, acoustical, optical, thermal, and so forth. These phenomena are referred to as *signals*. In order to determine how a system behaves, data are obtained by measuring these signals. For example, when a seismic signal travels through the earth, it can be picked up by various sensors and the information obtained can be used to describe the underlying strata. In the case of planetary probes, pictures are taken and transmitted back to earth in the form of electrical signals. During this transmission, the signals containing the picture information are distorted and it is desired to process the distorted signals so that the pictures can be faithfully reproduced. In this case, the signal represents light intensity as a function of two space variables. For a good overview of the field of signal processing, the reader should consult [125], a survey of digital signal-processing applications. More information concerning image analysis can be found in [5, 6, 7, 19, 53, 69, 165], whereas [153, 187] are devoted to geophysical and seismic applications. The important field of speech analysis and synthesis is the subject of [40, 94, 144, 154, 160, 166]. Recently, discrete-time system theory has been applied to switched-capacitor filters [47, 64, 182].

Until World War II, analog signal processing played a dominant role, but in the 1940s, development of the theory of sampled-data systems began. The analysis of sampled-data systems was based on the sampling theorem, which is frequently associated with Shannon [124], and the z-transform, which essentially was used much earlier by Laplace [93] and DeMoivre [32]. Hurewicz [70] was one of the early contributors in this area. The first books on sampled-data systems by Ragazzini and Franklin [150] and Jury [83] appeared in 1958. Also in this year, Blackman and Tukey [9] published articles describing how to estimate power spectra from a finite set of signal samples. Kaiser made the first major contributions to the design of digital signal-processing systems when he showed how to obtain digital filters from analog filters using the bilinear transformation [85, 86]. Other notable milestones were Kalman's paper on recursive estimation in 1960 [90] and the development of a fast Fourier-transform algorithm by Cooley and Tukey in 1965 [26]. Homomorphic signal processing was developed by Oppenheim, Schafer, and Stockham [51, 126, 128, 136].

At the present time, the field of digital signal processing is experiencing rapid expansion. This is due to the advances in integrated-circuit technology, achievements in software engineering, and improved algorithms in numerical analysis.

The first book devoted to digital signal processing was written by Gold and Rader [51] and appeared in 1969. A partial list of books that followed include Cadzow [16] in 1973; Oppenheim and Schafer [126], Rabiner and Gold [136], Childers and Durling [21], all in 1975; Tretter [180] and Peled and Liu [134] in 1976; Hamming [55] in 1977; Antoniou [8] and Chen [20] in 1979; Jong [81] in 1982; Hamming's second edition [56] in 1983; Stanley, Dougherty, and Dougherty [169] in 1984; Jackson [73] and Bose [12] in 1985; Williams [186] in 1986; Roberts and Mullis [152] in 1987; Strum and Kirk [173] in 1988, and Oppenheim and Schafer [127] in 1989. In addition to these books, many papers concerning digital signal processing have appeared in the *IEEE Transactions on Acoustics, Speech, and Signal Processing,* the *IEEE Transactions on Circuits and Systems,* and numerous other journals. Also, there are several publications [34, 35, 112, 142] that contain reproductions of important papers in digital signal processing as well as a listing of computer programs.

In recent years, considerable effort has been expended in the study of multidimensional digital systems, which includes image processing as a major topic. Much research has been done also in the use of number theory and abstract algebra to develop finite transforms similar to the discrete Fourier transform. For an excellent survey of two-dimensional signal processing, the reader should consult Mersereau and Dudgeon [110] and Rabiner and Gold [136]. Other interesting works in this area include [4, 31, 66, 67, 68, 105, 109, 111, 115, 158, 164, 168]. For an introduction to applications of number theory to digital signal processing, the reader is referred to McClellan and Rader [108] and Rader's chapter on number theoretical transforms in [136]. Additional information can be found in [2, 10, 75, 96, 104, 147, 151, 156, 174].

Another area of considerable importance is that of error analysis. When a digital filter is implemented, finite-length registers are used, and, therefore, filter coefficients and variables must be quantized. When this is done, errors result. For a discussion of these effects, the reader should consult such books as Oppenheim and Schafer [126] and Rabiner and Gold [136]. Other references include [52, 74, 98, 99, 129, 184, 185].

Signal Classifications

In our discussion of systems and signals, a *signal* is represented as a mathematical function of one or more variables. One of the most important signals is the sinusoid

$$f(t) = A \sin(\omega t + \theta) \tag{1.1}$$

a function of time that frequently arises in the study of power systems and many electronic systems. The quantity A is the amplitude, ω is the frequency in radians per second, and θ is the phase in radians.

There are two types of signals, depending on the values that the independent variables can assume. When there is a single independent variable, we refer to it as time regardless of what it actually is and this is reflected in our terminology. *Continuous-time* signals are those that are defined on a continuum of values of the independent variables. That is, they are functions of continuous variables. The sinusoid of Eq. (1.1) is an example of a continuous-time signal having a single independent variable. *Discrete-time* signals are defined at discrete values of the independent variables. For example, the set of discrete times for which the signal is defined could be the set $\{t_k\}$, $k = 0, 1, 2, \ldots$. The discrete-time signal obtained from the sinusoid of Eq. (1.1) using these time values is described by

$$f(t_k) = A \cos(\omega t_k + \theta) \qquad k = 0, 1, 2, \ldots \tag{1.2}$$

It follows that discrete-time signals are represented as *sequences of numbers*. For the most part, we shall be concerned with functions of a single variable.

If we restrict the amplitude of a discrete-time signal to a finite number of discrete values, we obtain a *digital signal*. This is, of course, the kind of signal encountered in digital computers. Digital signals are those for which time (independent variable) and amplitude are discrete. Signals that have continuous time and continuous amplitude are known as *analog signals*.

System Classifications

We can now classify systems in much the same way that we have classified signals. *Continuous-time systems* are those for which both the input and output are continuous-time signals and so are the signals within the system. *Discrete-time systems* are systems that have discrete-time input and output signals as well as discrete-time signals within the system. *Digital systems* are those that have digital signals as input and output signals and digital signals within the system.

Digital signal processing then is the transformation of digital signals into other digital signals. We can think of a digital computer performing operations on the input data as processing the input digital signal to give the output digital signal. General-purpose digital computers provide an important means for realizing digital signal processing systems. More and more, we find that microprocessors and special-purpose hardware are also being employed as digital signal processors [167].

We are concerned primarily with discrete-time systems and signals. Discrete-time signals can arise naturally from some discrete-time process, but they quite frequently arise by sampling a continuous-time signal. That is, the values of the continuous-time signal are obtained at discrete times. For example, the discrete-time signal of Eq. (1.2) is obtained by sampling the continuous-time signal of Eq. (1.1) at the discrete times t_k, $k = 0, 1, 2, \ldots$.

A Savings Account

An example of a very common discrete-time system is a savings account. This is also a digital system, since the amount in the account is quantized to the nearest cent. It is assumed that deposits are made at the end of the period after the interest has been paid. Let $y(k)$ be the amount in the account at the end of the kth period, $x(k)$ be the amount deposited at the end of the kth period [a negative $x(k)$ indicates a withdrawal], and i be the interest rate per period. Then, after any deposit at the end of the $(k + 1)$th period, the amount in the account is

$$\begin{aligned} y(k+1) &= y(k) + iy(k) + x(k+1) \\ &= (1+i)y(k) + x(k+1) \end{aligned} \tag{1.3}$$

Suppose that $x(k) = R$, a constant, and $y(0) = P$. Then the above equation, which is known as a *difference equation,* becomes

$$y(k+1) = (1+i)y(k) + R \tag{1.4}$$

There are various ways to solve difference equations, but the method we shall employ for this example is the *recursion* method. This merely involves repeatedly calculating $y(k)$ from Eq. (1.4), starting with $k = 0$. The resulting values are

$$\begin{aligned} y(1) &= (1+i)y(0) + R = (1+i)P + R \\ y(2) &= (1+i)y(1) + R = (1+i)^2P + (1+i)R + R \\ y(3) &= (1+i)y(2) + R = (1+i)^3P + (1+i)^2R + (1+i)R + R \end{aligned}$$

From the pattern that seems to be established here, it appears that

$$y(k) = (1+i)^kP + R\sum_{m=0}^{k-1}(1+i)^m$$

Mathematical induction can be used to prove that this is the correct expression. The sum in this expression is the sum of a geometric progression or sequence and appears frequently enough in what follows to merit special consideration. In many cases where it arises, we are interested in obtaining a closed-form expression.

Sum of a Geometric Progression

We shall denote the sum of the first n terms of the general geometric progression or sequence by

$$S_n = \sum_{m=0}^{n-1} a^m = 1 + a + a^2 + \cdots + a^{n-1}$$

where the quantity a is the ratio. There are two cases that must be considered separately: $a = 1$ and $a \neq 1$. If $a = 1$, then $S_n = n$. If $a \neq 1$, we first compute

$$aS_n = a + a^2 + \cdots + a^n$$

and then subtract this value from S_n. From this result,

$$S_n - aS_n = (1 - a)S_n = 1 - a^n$$

we can obtain

$$S_n = \frac{1 - a^n}{1 - a} = \frac{a^n - 1}{a - 1}$$

Summarizing the results for these two cases, we have

$$\sum_{m=0}^{n-1} a^m = \begin{cases} n & a = 1 \\ \dfrac{a^n - 1}{a - 1} & a \neq 1 \end{cases} \tag{1.5}$$

In the special case of the savings account example that we are considering, the ratio $a = 1 + i$ and $n = k$. Consequently, by using Eq. (1.5), the expression for $y(k)$ can be written

$$y(k) = (1 + i)^k P + R[(1 + i)^k - 1]/i$$

An Electric Circuit

Whenever we utilize a digital computer to analyze a continuous-time system, we are actually approximating the continuous-time system by a discrete-time system. For example, let us analyze the linear electrical circuit of Fig. 1.1 containing two resis-

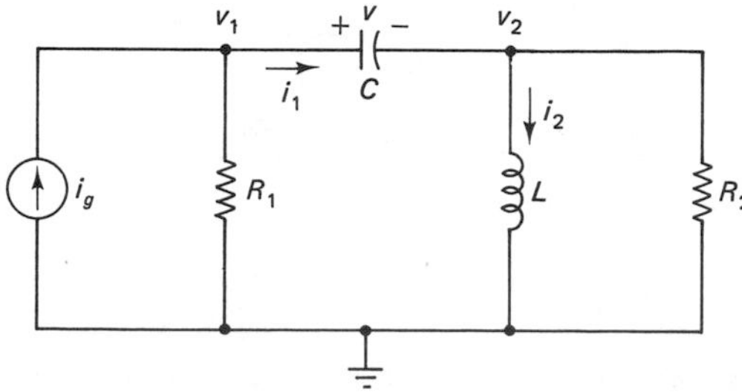

Figure 1.1 An electrical circuit.

tors, an inductor, a capacitor, and a current source. The voltage–current relations for a resistor, an inductor, and a capacitor are, respectively,

$$v = Ri$$

$$v = L\,di/dt$$

$$i = C\,dv/dt$$

where v is the voltage in volts (V), i is the current in amperes (A), R is the resistance in ohms (Ω), L is the inductance in henries (H), C is the capacitance in farads (F) and t is the time in seconds (s). The voltages and currents in the circuit are computed at discrete values t_k of time t. In order to do this on a digital computer, we must obtain approximations for the v–i relations that involve addition and multiplication only. There are numerous ways to accomplish this, but we shall integrate the equations and use the trapezoidal rule to approximate the integrals. The reader is asked to do this in Prob. 1.7 and the resulting relations are

$$v(t_k) = v(t_{k-1}) + \frac{T}{2C}[i(t_k) + i(t_{k-1})] \tag{1.6}$$

for the capacitor, and

$$i(t_k) = i(t_{k-1}) + \frac{T}{2L}[v(t_k) + v(t_{k-1})] \tag{1.7}$$

for the inductor, where $T = t_k - t_{k-1}$. These equations describe the resistive circuits shown in Fig. 1.2. When we replace the capacitor and the inductor in Fig. 1.1 by the equivalent resistive circuits of Fig. 1.2, we obtain the resistive circuit shown in Fig. 1.3. We can now apply nodal analysis to find the voltages and currents of interest. If we start with $k = 1$ and $t_0 = 0$ and assume that initial conditions on the capacitor and inductor are given, we can find voltages and currents at time t_1. From these values at t_1, we then determine values at t_2 and so forth.

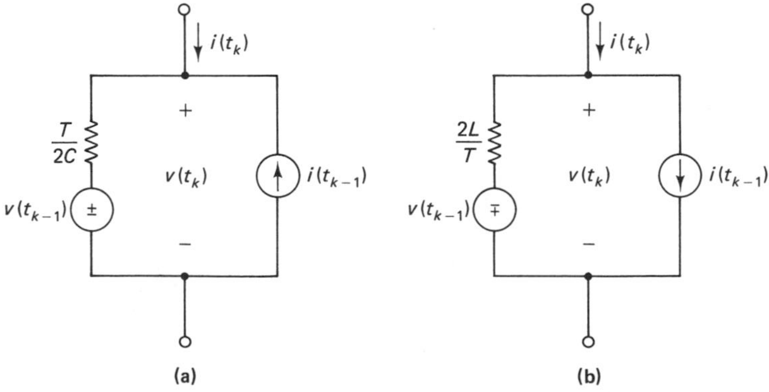

Figure 1.2 Equivalent circuits for (a) a capacitor and (b) an inductor.

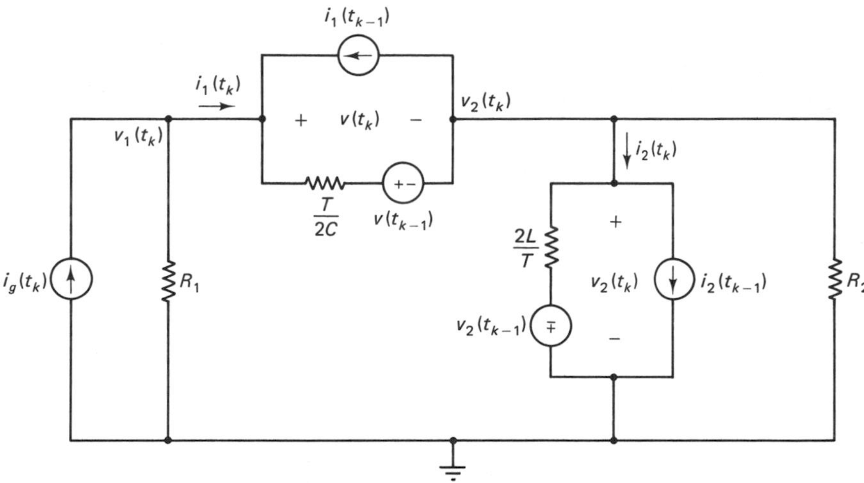

Figure 1.3 Approximate model of the circuit of Fig. 1.1.

We hope that when the circuit of Fig. 1.3 is solved, the values of voltage and current that we obtain are good approximations to the voltages and currents of Fig. 1.1 when $t = t_k$. This is a simple example of how computer programs are developed for analyzing electric circuits.

EXERCISES

1.1.1 If $P = \$500$, the interest rate is 12 percent per annum compounded monthly ($i = 0.01$), and the amount deposited at the end of each month is \$100, determine the amount in the savings account at the end of the first year.

1.1.2 If $P = R$, obtain a closed-form expression for $y(k)$ for the system of Eq. (1.4).

1.1.3 Obtain an approximate model using the circuits of Fig. 1.2 for the *RLC* series circuit shown.

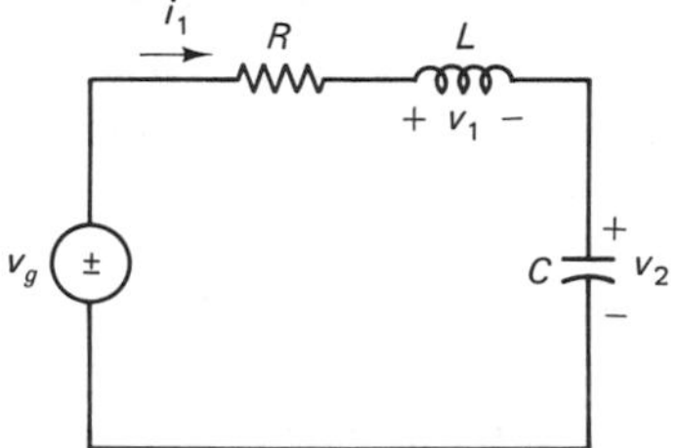

Exercise 1.1.3

1.1.4 Using Eq. (1.5), show that

$$\sum_{n=0}^{N-1} e^{jn\theta} = \frac{1 - \cos\theta + \cos(N-1)\theta - \cos N\theta}{2(1-\cos\theta)} + j\frac{\sin\theta + \sin(N-1)\theta - \sin N\theta}{2(1-\cos\theta)}$$

$$= \sum_{n=0}^{N-1} \cos n\theta + j\sum_{n=0}^{N-1} \sin n\theta$$

From this result, obtain the relations

$$\sum_{n=0}^{N-1} \cos n\theta = \frac{1 - \cos\theta + \cos(N-1)\theta - \cos N\theta}{2(1-\cos\theta)}$$

$$= \frac{\sin(N\theta/2)\cos[(N-1)\theta/2]}{\sin\theta/2}$$

$$\sum_{n=0}^{N-1} \sin n\theta = \sum_{n=1}^{N-1} \sin n\theta = \frac{\sin\theta + \sin(N-1)\theta - \sin N\theta}{2(1-\cos\theta)}$$

$$= \frac{\sin(N\theta/2)\sin[(N-1)\theta/2]}{\sin\theta/2}$$

1.2 SAMPLED-DATA SYSTEMS

Although the signals occurring in nature are analog, much of the signal processing is now being done digitally. Digital systems are used for processing because they are inexpensive, accurate, and can be readily implemented. To accomplish this feat, we must provide data-acquisition and -conversion devices that interface the real world of analog signals and the computational world of digital data processing.

Data-Conversion Devices

Most of the data-conversion devices operate on electrical signals. Consequently, if the signals of interest are not electrical, then they must be applied to *transducers* to produce electrical signals. In many cases, we want to process several signals with the same signal processor. This can be accomplished by using an analog *multiplexer,* a switch (usually electronic) that selects from among several lines the one to supply the desired signal to be converted. Usually, the multiplexer connects the input signal to a *sample-and-hold* (S/H) circuit, which acquires the signal voltage and holds its value while an *analog-to-digital* (A/D) *converter* (ADC) changes the signal value to digital form. After the digital signal is processed, the resulting digital signal can be converted into an analog signal by passing it through a *digital-to-analog* (D/A) *converter* (DAC).

Multiplexers and Sample-and-Hold Circuits

To give the reader some feel for these devices, we give some very simple circuits that perform these operations. A simple analog multiplexer is shown in Fig. 1.4 and a sample-and-hold (S/H) circuit is shown in Fig. 1.5. The latter figure is a block dia-

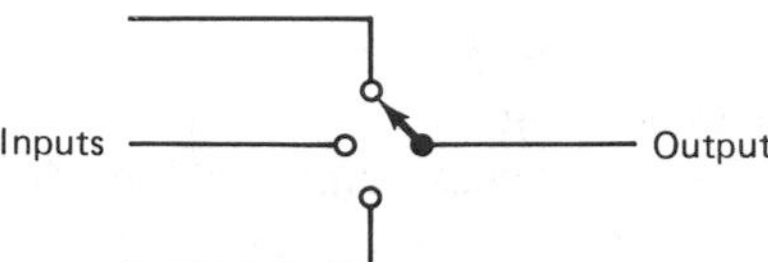

Figure 1.4 An analog multiplexer having three input signals.

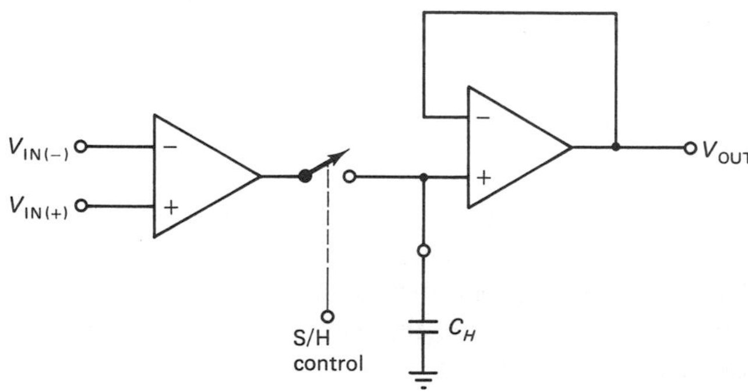

Figure 1.5 A sample-and-hold (S/H) circuit. [Reprinted, with permission, from *The Linear IC Handbook* (No. 2672), copyright 1986 by TAB BOOKS Inc., Blue Ridge Summit, PA 17294-0850. Hardbound price is $49.50. Call toll-free 1-800-233-1128. In Pennsylvania and Alaska call 717-794-2191.]

gram of the Harris HA-2420/2425 and the Analog Devices AD583 sample-and-hold IC. The IC consists of two op amps, a low-leakage bipolar analog switch, and a TTL-compatible logic circuit to control the switch. When the S/H control input is low, the switch is closed and the amplifier is in the sample mode. The voltage that is applied at the $V_{IN(+)}$ input appears across the hold capacitor C_H and at the output V_{OUT}. When the S/H control input goes high, the switch opens. The amplifier is in the hold mode and the capacitor voltage remains the same as it was just prior to the S/H control going high [118]. (Also see [18, 72, 159].)

A Digital-to-Analog Converter

A five-bit digital-to-analog converter (DAC) is shown in Fig. 1.6 [18, 72, 159]. The output voltage is given in terms of the reference voltage V_R by

$$\begin{aligned} V_{OUT} &= -\frac{R}{2}V_R\left(\frac{1}{R}b_4 + \frac{1}{2R}b_3 + \frac{1}{4R}b_2 + \frac{1}{8R}b_1 + \frac{1}{16R}b_0\right) \\ &= -\frac{1}{2}V_R\left(b_4 + \frac{1}{2}b_3 + \frac{1}{4}b_2 + \frac{1}{8}b_1 + \frac{1}{16}b_0\right) \end{aligned} \tag{1.8}$$

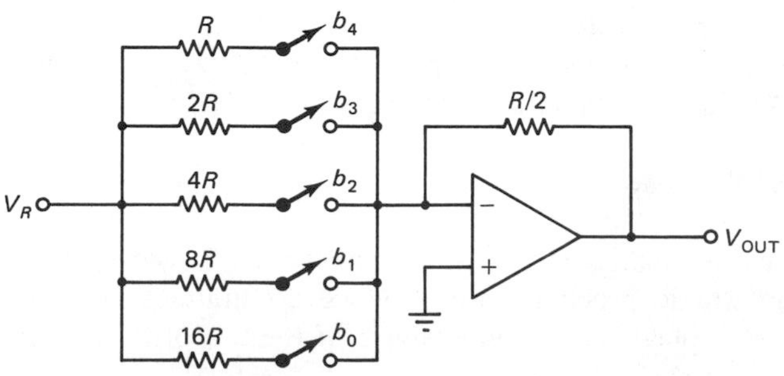

Figure 1.6 A five-bit digital-to-analog converter.

In this circuit, each switch is controlled by a bit in the digital number. If the value of the bit is one, then the corresponding switch is closed. The switch is open whenever the value of the bit is zero. The most-significant bit (MSB) is b_4 and the least-significant bit (LSB) is b_0. For example, if the digital number is $b_4b_3b_2b_1b_0 =$ 10101, then the output voltage of the DAC is

$$\begin{aligned} V_{\text{OUT}} &= -\tfrac{1}{2}V_R\left(1 + \tfrac{1}{4} + \tfrac{1}{16}\right) \\ &= -\tfrac{21}{32}V_R \end{aligned}$$

An Analog-to-Digital Converter

An analog-to-digital converter (ADC) that uses a DAC is shown in Fig. 1.7 [18, 72, 159]. In the ADC, the signal $s(t)$ clears the counter and actuates the sample-and-hold (S/H) circuit. Signal v_r is a staircase waveform that is incremented by each clock pulse in steps of $V_R/2^n$. Voltages v_r and v_{as}, the output of the S/H circuit, are fed into a comparator whose output Q is a logical signal given by

$$Q = \begin{cases} \text{high} & v_{as} > v_r \\ \text{low} & v_r > v_{as} \end{cases}$$

The pulse counting continues until $v_r > v_{as}$. Then the comparator output Q goes low, ending the conversion. The b_i's represent the digital output.

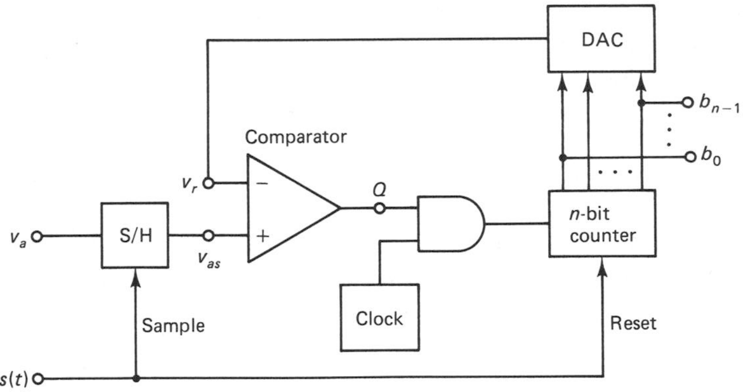

Figure 1.7 An analog-to-digital converter.

There are several other types of DACs and ADCs available besides the examples given. They are available with several other binary codes besides the natural binary and with different reference voltages that include $\pm$ 5 and $\pm$ 10 V. The reader who is interested in a more detailed description of multiplexers, sample-and-hold circuits, DACs and ADCs should consult references [18, 72, 118, 159].

Controlling an Industrial Process

Data converters have numerous applications in many industries. These applications include data telemetry, automatic process control, voice communications, digital panel meters, and computer displays. A typical example of the use of these converters is the industrial process shown in Fig. 1.8 [72]. The digital computer that controls the process can be located at a considerable distance from the plant. The physical parameters of the process are temperature, pressure, and flow. Transducers convert these parameters to voltages, which are then amplified and fed to an analog multiplexer. Each analog signal is sampled periodically. These sampled analog signals are converted to digital signals and processed by the digital computer. The computer calculates the existing "state" of the process and compares it with the "desired state," which is stored in its memory. Based on this comparison, corrections for the process parameters are determined and are fed back to the process through DACs to produce the desired state for the process.

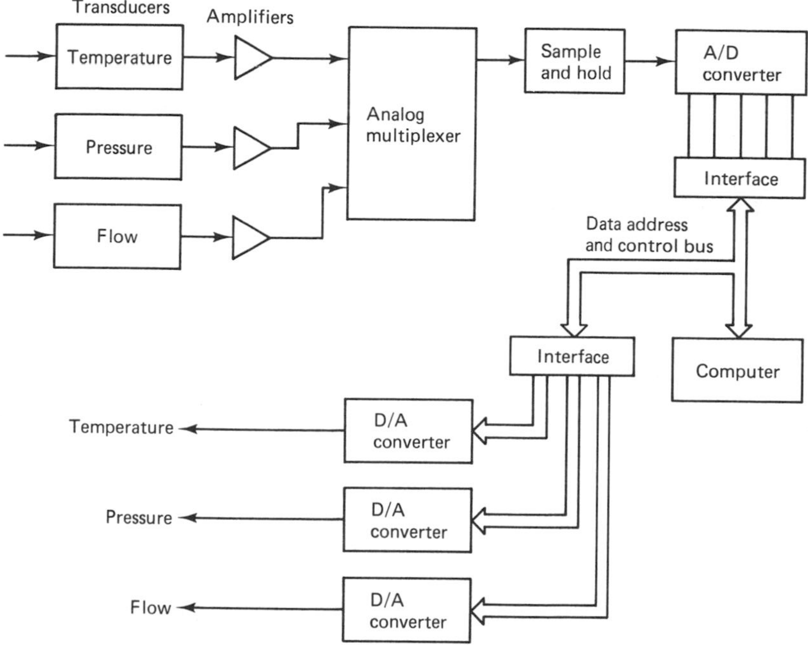

Figure 1.8 Example of process control using a digital computer.

Sampling an Analog Signal

In considering discrete-time signals and systems, we are interested in the processes of sampling an analog signal and holding this value. A discrete-time signal can be obtained by sampling an analog signal at times t_k, $k = 0, 1, 2, \ldots$. This process is illustrated in Fig. 1.9, where we have taken $t_k = kT$, which means the samples are uniformly spaced. We call this *uniform* or *conventional sampling* and T is the *sampling period*. The *sampling frequency* is given by

$$f_s = \frac{1}{T} \text{ Hz}$$

In sampled-data systems, an ideal sampler provides an output in much the same

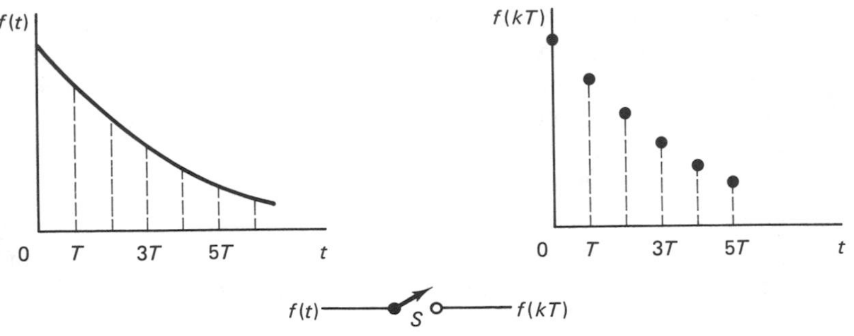

Figure 1.9 Sampling an analog signal.

way as a modulator. If $f_a(t)$ is the analog input to the sampler, then the sampled output signal $f_a^*(t)$ is the product

$$f_a^*(t) = s_a(t) f_a(t)$$

which resembles a modulated signal. (Here * does not mean complex conjugation.) The modulating function $s_a(t)$ is a train of uniformly spaced impulse functions given by

$$s_a(t) = \sum_{n=-\infty}^{\infty} \delta_a(t - nT)$$

where $\delta_a(t)$ is the analog impulse function defined by

$$\delta_a(t - \tau) = 0 \qquad t \neq \tau$$

$$\int_{-\infty}^{\infty} \delta_a(t - \tau) f_a(t)\, dt = f_a(\tau)$$

where $f_a(t)$ is continuous at $t = \tau$. The last equation is frequently referred to as the *sifting* property of the impulse function. The process of instantaneous sampling described is analogous to the process of *impulse modulation* [179, 130].

Replacing $s_a(t)$ by its series representation, we obtain

$$\begin{aligned} f_a^*(t) &= f_a(t) \sum_{n=-\infty}^{\infty} \delta_a(t - nT) \\ &= \sum_{n=-\infty}^{\infty} f_a(t)\, \delta_a(t - nT) \\ &= \sum_{n=-\infty}^{\infty} f_a(nT)\, \delta_a(t - nT) \end{aligned}$$

Consequently, the sampled signal can be represented as a train of impulses having strengths $f_a(nT)$, the value of the nth sample of $f_a(t)$.

Reconstructing an Analog Signal

In many cases, it is desirable to convert a sampled signal back into an analog signal based on the limited description provided by the sampled values. Extrapolation can be used with acceptable accuracy if the function being sampled has no rapid changes. Otherwise, large and unpredictable errors in the approximation can occur between samples. Consequently, the sampling frequency is an important factor in the amount of error that appears in the data reconstruction. Since we are dealing with physical systems, this reconstruction must be based on a preceding set of sample values.

The devices that reconstruct continuous data from a sequence of samples or numbers are generally called *data holds, extrapolators, desampling filters,* or *reconstruction filters*. In order to construct a data hold, it is necessary to assume a particular form for the function that is to be reconstructed from its samples. One form that is commonly used in control systems is a polynomial in time. The data hold must then compute the polynomial coefficients at each sample time that provide the best fit to the past data. If the extrapolating polynomial is a constant, a polynomial of degree zero, then the extrapolator is a *zero-order hold.* A *first-order* extrapolator is obtained when the extrapolating polynomial is of first degree. Practical systems very rarely use data extrapolators that are more complex than first order. For more information on data extrapolators, the reader should consult [72, 150, 159].

The result of applying a sampled signal to a zero-order hold is shown in Fig. 1.10. The effect of passing the analog signal of Fig. 1.9 through a sampler and a zero-order hold (equivalently, a sample-and-hold) is shown in Fig. 1.11.

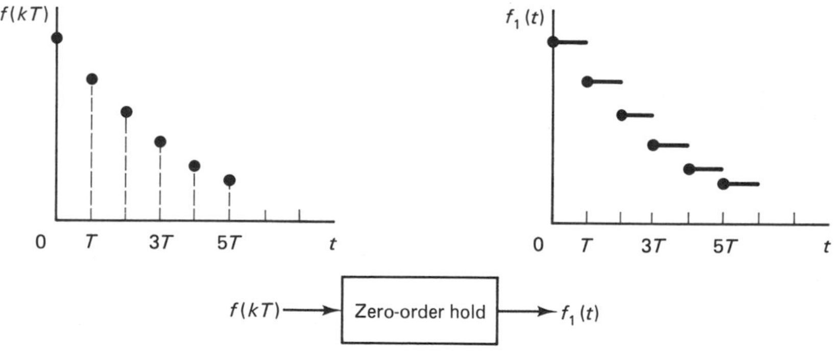

Figure 1.10 Effect of a zero-order hold.

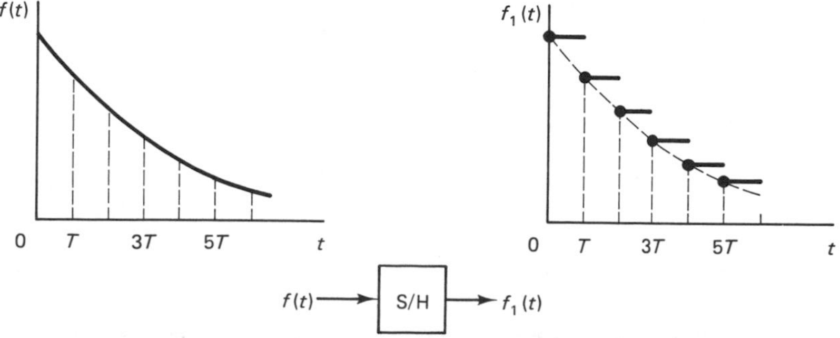

Figure 1.11 Effect of a sample-and-hold circuit on an analog signal.

Digital Processing an Analog Signal

An illustration showing how digital processing can be utilized when the signals of interest are analog signals is shown in Fig. 1.12. The analog signal $x(t)$ is passed through an A/D converter, resulting in the digital signal $\tilde{x}(kT)$. This digital signal is then processed by a digital signal processor whose output is $\tilde{y}(kT)$. The processed digital signal is then converted to the analog signal $y(t)$ by a D/A converter.

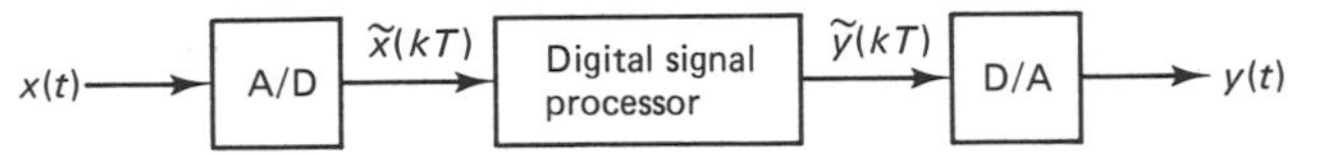

Figure 1.12 Processing an analog signal using a digital signal processor.

A Sampled-Data Control System

A typical sampled-data control system is shown in Fig. 1.13 [15, 37, 43, 91, 150]. The error signal is the difference between the input and the process output, $e(t) = r(t) - c(t)$. The error signal is sampled and the sampled signal $e^*(t)$ is fed into a data hold. The output $h(t)$ of the data hold is the controlling signal for the process.

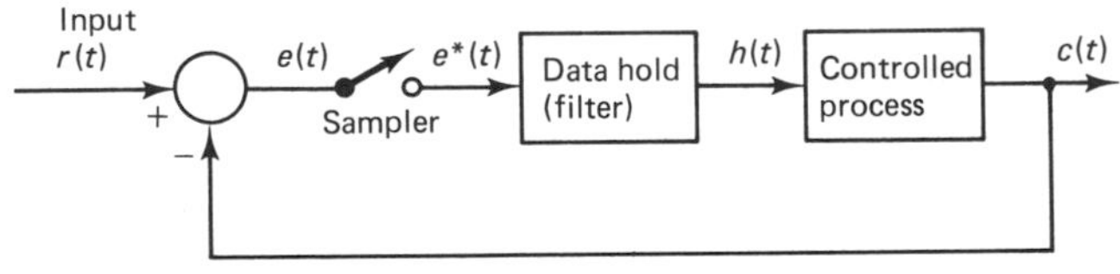

Figure 1.13 A sampled-data control system.

A Digital Phase-Locked Loop

In the implementation of coherent communications and tracking receivers, digital phase-locked loops (DPLLs) play an important role [97]. The block diagram of a basic DPLL is shown in Fig. 1.14. The input signal is sampled and compared with a reconstructed reference signal generated by an oscillator. The error signal is proportional to the phase difference of the input signals. If the loop is designed properly, the reconstructed reference is forced to resemble the input signal.

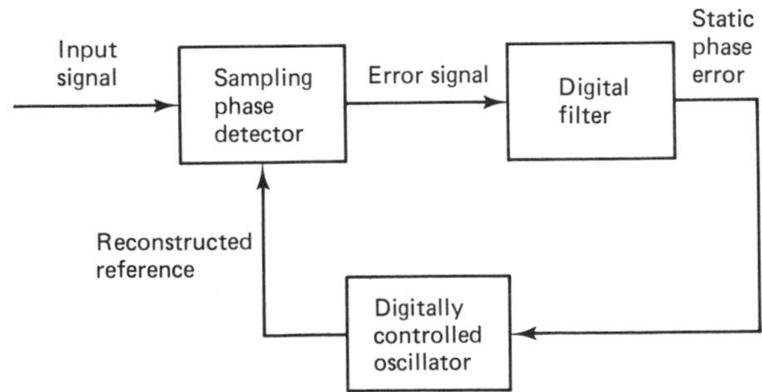

Figure 1.14 Basic DPLL Block diagram. (© 1981, IEEE)

A Switched-Capacitor Circuit

One of the most important electronic circuits is the frequency-selective filter. Active filters are usually made up of resistors, capacitors, and operational amplifiers, and it is usually desirable to fabricate a large number of these filters on a single integrated circuit. Problems arise with the integrated resistors, and in order to avoid them, analog sampled-data filters utilizing *switched capacitors* have been developed. In these circuits, a resistor is replaced by a capacitor and a switch [47, 64, 182].

A circuit that simulates the behavior of a resistor is shown in Fig. 1.15(a). The switch is initially in position 1 and capacitor C is charged to voltage v_1; that is, $v_C = v_1$. The switch is then moved to position 2 and the capacitor is discharged to voltage v_2. The amount of charge, in coulombs (C), that is transferred from the input to the output of this device is $q = C(v_1 - v_2)$. The switch is in each position for $T_C/2$ s and this process is repeated every T_C s. During this period, current i into v_2 is given by

$$i \approx \frac{\Delta q}{\Delta t} = \frac{C(v_1 - v_2)}{T_C}$$

This behavior can be simulated by the resistive circuit of Fig. 1.15(b), where it is seen that

$$i = \frac{v_1 - v_2}{R}$$

The same effect is achieved in these two circuits if

$$R = T_C/C$$

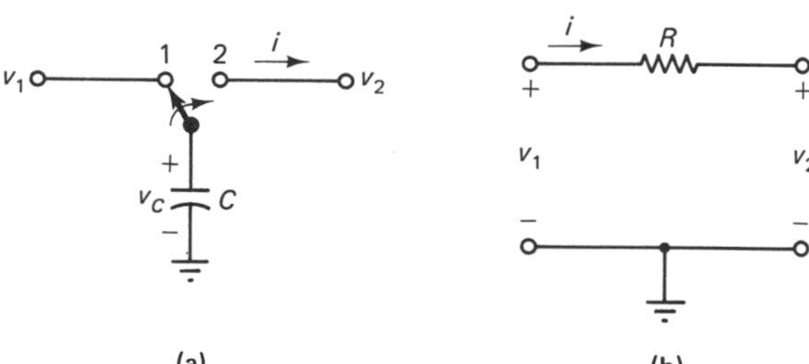

Figure 1.15 (a) Switched capacitor and (b) its resistive equivalent.

A basic circuit of active filters is the integrator shown in Fig. 1.16(a). It can be shown that the transfer function of the integrator is

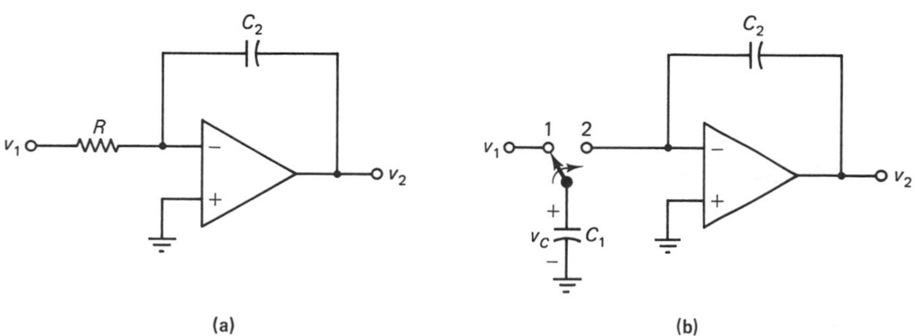

Figure 1.16 (a) An analog integrator and (b) an equivalent switched-capacitor integrator.

$$\frac{V_2(s)}{V_1(s)} = -\frac{1}{RC_2 s}$$

where $V_1(s)$ and $V_2(s)$ are the Laplace transforms of $v_1(t)$ and $v_2(t)$, respectively. If C_1 is the capacitance of the switched capacitor, then $R = T_C/C_1$, and we have

$$\frac{V_2(s)}{V_1(s)} = -\frac{1}{(T_C/C_1)C_2 s} = -f_C(C_1/C_2)/s$$

where $f_C = 1/T_C$ Hz is the switching frequency. This circuit can be implemented with very good results using MOS technology. The ratio of capacitances can be obtained with high precision and the silicon area required is less than for the resistive circuit. The switches are implemented using MOSFETs.

The relation between the input and the output voltages of the switched-capacitor integrator can be obtained using conservation of charge. If the switch is operated at $t = 0$, conservation of charge requires that

$$q_1(0+) - q_1(0-) = q_2(0+) - q_2(0-)$$

where q_1 is the charge on C_1, and q_2 is the charge on C_2. The signs in this equation are those that result if Kirchhoff's current law is applied at the inverting input terminal of the operational amplifier with the switch in position 2. Since

$$q_1(0-) = C_1 v_C(0-) = C_1 v_1(0-)$$

$$q_1(0+) = C_1 v_C(0+) = 0$$

$$q_2 = C_2 v_2$$

the conservation of charge equation becomes

$$-C_1 v_1(0-) = C_2 v_2(0+) - C_2 v_2(0-)$$

The voltage $v_C(0+) = 0$, since it is the voltage across the input terminals of the operational amplifier. During the nth switching period, $(n - 1)T_C < t < nT_C$, and the voltages before and after switching are the voltages at $(n - 1)T_C$ and nT_C. Replacing $0-$ by $(n - 1)T_C$ and $0+$ by nT_C in the last equation, we obtain

$$C_2 v_2(nT_C) = C_2 v_2[(n - 1)T_C] - C_1 v_1[(n - 1)T_C] \tag{1.9}$$

In most cases, the switching rate f_C is much larger than the signal frequencies of interest.

Even though the voltages at the switching times satisfy a difference equation, the switched-capacitor filter is an analog filter.

EXERCISES

1.2.1 Determine the output voltage of the DAC of Fig. 1.6 if the digital number is 01110.

1.2.2 Suppose that the input signal to a sample-and-hold circuit is

$$f(t) = 10(1 - e^{-t})$$

Determine the output function $f_1(t)$ for $0 \leq t \leq 4$ s if $T = 0.2$ s.

1.2.3 If the plant input in Fig. 1.13 is $x(t)$ and the plant is described by

$$dy/dt + y = x$$

obtain the difference equation for $y(nT)$. Hint: Multiply the equation through by e^t and integrate from $(n - 1)T$ to nT.

1.3 DISCRETE-TIME SIGNALS: TIME-DOMAIN REPRESENTATIONS

As was previously pointed out, a discrete-time signal is a sequence of numbers. If the nth number or sample in the sequence is $x(n)$, then the sequence is denoted by

$$\{x(n)\}, \quad -\infty < n < \infty \tag{1.10}$$

We are interested in performing certain arithmetic operations on sequences to obtain other sequences. The operations of concern are those that are carried out when a sequence is being transformed by a discrete-time system.

Multiplication by a Constant

We can multiply a given sequence $\{x(n)\}$ by a constant a, as defined by

$$a\{x(n)\} = \{ax(n)\}$$

Each sample in the new sequence is multiplied by the constant a.

Addition and Subtraction

Two sequences $\{x(n)\}$ and $\{y(n)\}$ can be added or subtracted, resulting in the new sequences

$$\{x(n)\} \pm \{y(n)\} = \{x(n) \pm y(n)\}$$

Multiplication

The product of two sequences $\{x(n)\}$ and $\{y(n)\}$ is defined as

$$\{x(n)\}\,\{y(n)\} = \{x(n)y(n)\}$$

Note that the corresponding nth samples are multiplied to form the nth sample of the product.

Shifting

If the sequence $\{x(n)\}$ is shifted to the right by k units, we obtain the new sequence $\{x(n - k)\}$.

Examples

To illustrate these operations, let $\{x(n)\} = \{(-1)^n\}$ and $\{y(n)\} = \{\cos n\pi/2\}$. Then we have

$$2\{x(n)\} = \{2(-1)^n\}$$
$$\{x(n)\} + \{y(n)\} = \{(-1)^n + \cos n\pi/2\}$$
$$\{x(n)\}\{y(n)\} = \{(-1)^n \cos n\pi/2\}$$
$$\{y(n-2)\} = \{\cos (n-2)\pi/2\}$$
$$\{x(n+3)\} = \{(-1)^{n+3}\}$$

Except in the few cases where the braces in Eq. (1.10) are needed for clarity, we shall use $x(n)$ to represent the sequence as well as the nth sample. This simplifies the notation considerably as we shall see.

Special Discrete-Time Signals

The sequence of Eq. (1.2) obtained by sampling the sinusoid of Eq. (1.1) is an important discrete-time signal and we use it quite frequently. Another very important discrete-time signal is the *unit-sample sequence*, or *discrete-time impulse* $\delta(n)$ defined by

$$\delta(n) = \begin{cases} 1 & n = 0 \\ 0 & n \neq 0 \end{cases} \tag{1.11}$$

The graphical representation of the unit-sample sequence is shown in Fig. 1.17(a). The shifted impulse sequence is given by

$$\delta(n-k) = \begin{cases} 1 & n = k \\ 0 & n \neq k \end{cases} \tag{1.12}$$

where k is an integer. The graph of $\delta(n-3)$ is shown in Fig. 1.17(b).

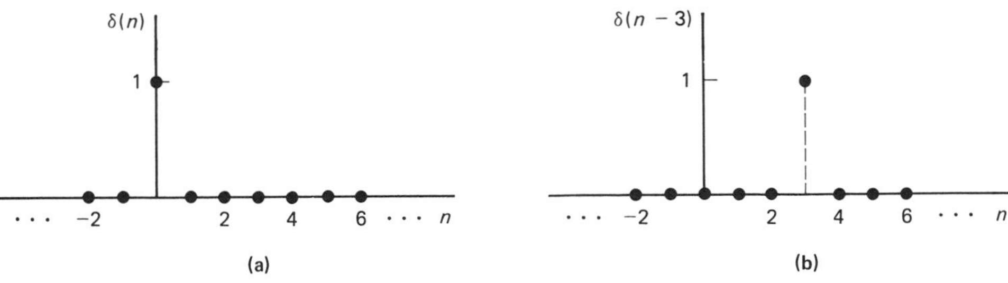

Figure 1.17 (a) The unit-sample sequence $\delta(n)$ and (b) the shifted unit-sample sequence $\delta(n-3)$.

The *unit-step sequence* $u(n)$ has the representation

$$u(n) = \begin{cases} 0 & n < 0 \\ 1 & n \geq 0 \end{cases} \tag{1.13}$$

Its graph is shown in Fig. 1.18(a). The sequence $u(n-k)$, given by

$$u(n-k) = \begin{cases} 0 & n < k \\ 1 & n \geq k \end{cases}$$

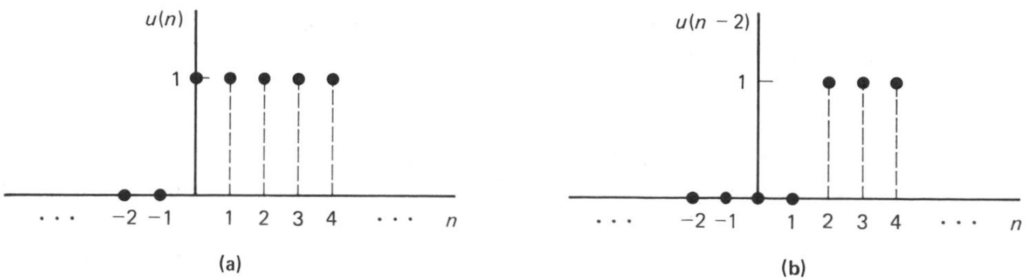

Figure 1.18 (a) The unit-step sequence $u(n)$ and (b) the shifted unit-step sequence $u(n-2)$.

where k is an integer, is simply the unit-step sequence shifted k units to the right or delayed by k samples. For example, the shifted or delayed unit-step sequence $u(n-2)$ is shown in Fig. 1.18(b).

Time-Domain Representations

The unit-sample sequence can be used to obtain a very useful representation of a discrete-time signal. For example, suppose we have the sequence defined by

$$x(n) = \begin{cases} -1 & n = -2, -1 \\ 2 & n = 1, 2 \\ 0 & \text{otherwise} \end{cases}$$

Noting that $\delta(n-k) = 0$ except when $n = k$, we see that a suitable expression for $x(n)$ is

$$x(n) = -\delta(n+2) - \delta(n+1) + 2\delta(n-1) + 2\delta(n-2)$$

Extending this method to a general sequence $x(n)$, we can obtain the two expressions:

$$\begin{aligned} x(n) &= \sum_{k=-\infty}^{\infty} x(k)\,\delta(n-k) \\ &= \sum_{k=-\infty}^{\infty} x(n-k)\,\delta(k) \end{aligned} \tag{1.14}$$

It should be noted that the second sum can be obtained from the first sum by a change of index. Since $\delta(n-k) = 0$ except when $n = k$, there is only one nonzero term in each sum. In each case, the only nonzero term remaining is $x(n)$.

Examples of Sequences

In particular, if $x(n) = u(n)$, then Eq. (1.14) becomes

$$\begin{aligned} u(n) &= \sum_{k=0}^{\infty} \delta(n-k) \\ &= \sum_{k=-\infty}^{n} \delta(k) \end{aligned}$$

We can also express the unit-sample sequence in terms of the unit-step sequence. The desired result is

$$\delta(n) = u(n) - u(n-1) \tag{1.15}$$

With the expressions of Eqs. (1.14) and (1.15), we can express any sequence in terms of the unit-step sequence. Substituting Eq. (1.15) into Eq. (1.14), we obtain

$$\begin{aligned} x(n) &= \sum_{k=-\infty}^{\infty} x(k)u(n-k) - \sum_{k=-\infty}^{\infty} x(k)u(n-k-1) \\ &= \sum_{k=-\infty}^{\infty} x(n-k)u(k) - \sum_{k=-\infty}^{\infty} x(n-k)u(k-1) \end{aligned} \tag{1.16}$$

Some Signal Classifications

A sequence $x(n)$ is said to be *causal* if $x(n) = 0$ for $n < 0$. The unit-step sequence $u(n)$, which is a causal sequence, is quite useful in writing expressions for other causal sequences. For example, $x(n) = a^n u(n)$ is a causal sequence since $x(n) = 0$ for $n < 0$. Another way to express this sequence is

$$x(n) = \begin{cases} a^n & n \geq 0 \\ 0 & n < 0 \end{cases}$$

and so the use of the step sequence provides a much simpler expression for $x(n)$.

The *unit-pulse sequence* $p(n, n_0)$, defined by

$$p(n, n_0) = u(n) - u(n - n_0) \tag{1.17}$$

where n_0 is a positive integer, has the values

$$p(n, n_0) = \begin{cases} 1 & 0 \leq n \leq n_0 - 1 \\ 0 & \text{otherwise} \end{cases}$$

In particular, we see from Eq. (1.17) that if $n_0 = 1$, then $p(n, 1) = \delta(n)$. The unit-pulse sequence is useful in forming new sequences. For example, the product $p(n, n_0)\, x(n)$, given by

$$p(n, n_0)\, x(n) = \begin{cases} x(n) & 0 \leq n \leq n_0 - 1 \\ 0 & \text{otherwise} \end{cases}$$

is a sequence that retains only certain terms of $x(n)$ while reducing the other terms to zero.

A sequence $x(n)$ is said to be *periodic* with period N if $x(n) = x(n + N)$ for all n. For example, the sequence defined by

$$x(n) = \begin{cases} x(0) \neq 0 & n = 3k,\ k = 0, \pm 1, \pm 2, \ldots \\ 0 & \text{otherwise} \end{cases}$$

is periodic with period $N = 3$. Its graph is shown in Fig. 1.19. A second example of a periodic sequence is

$$x(n) = A \cos \pi n/10 \tag{1.18}$$

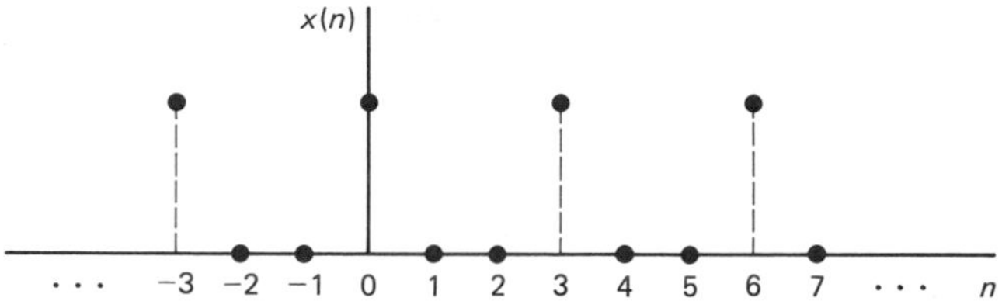

Figure 1.19 Example of a periodic sequence.

where $N = 20$. This is true since

$$\begin{aligned} x(n+N) &= A \cos [\pi(n+N)/10] \\ &= A \cos (\pi n/10 + 2\pi) \\ &= A \cos \pi n/10 \end{aligned}$$

The unit-pulse sequence is quite useful in obtaining an expression for a periodic sequence. If we define a new sequence $x_1(n)$ by

$$x_1(n) = \begin{cases} x(n) & 0 \le n \le N-1 \\ 0 & \text{otherwise} \end{cases}$$

then $x_1(n) = x(n)p(n, N)$ where N is the period. Noting that $x_1(n - N)$ is simply the sequence $x_1(n)$ shifted to the right by N units, we see that $x(n) = x_1(n - N)$ for $N \le n \le 2N - 1$. Continuing in this fashion, it follows that

$$\begin{aligned} x(n) &= \sum_{k=-\infty}^{\infty} x_1(n-kN) \\ &= \sum_{k=-\infty}^{\infty} x(n-kN)p(n-kN, N) \end{aligned} \tag{1.19}$$

Applying this expression to the example of Eq. (1.18), we obtain

$$x(n) = \sum_{k=-\infty}^{\infty} A \cos \frac{\pi(n-20k)}{10} p(n-20k, 20)$$

EXERCISES

1.3.1 From the two sequences $\{x(n)\} = \{(\frac{1}{2})^n\}$ and $\{y(n)\} = \{(-\frac{1}{2})^n\}$, determine the sequences:

(a) $4\{x(n)\}$
(b) $\{x(n)\} + \{y(n)\}$
(c) $\{x(n)\}\{y(n)\}$
(d) $\{y(n-4)\}$

1.3.2 Express the sequence defined by

$$x(n) = \begin{cases} -2 & n = -1, 0, 1 \\ 4 & n = -2, 2 \\ 0 & \text{otherwise} \end{cases}$$

as a weighted sum of unit-sample sequences.

1.3.3 Express the sequence of Ex. 1.3.2 as a weighted sum of unit-step sequences.

1.3.4 Determine the period of the periodic sequence

$$x(n) = 2 \sin (\pi n/5)$$

1.3.5 Prove or disprove that

$$x(n) = 5 \sin (2n + \pi/4)$$

is periodic. If the sequence is periodic, determine its period.

1.4 FREQUENCY-DOMAIN REPRESENTATION OF DISCRETE-TIME SIGNALS

Fourier Transform of a Sequence

Another equally important representation of a discrete-time signal $x(n)$ is its Fourier transform. To obtain such a representation, we shall start by finding the Fourier transform of a sampled analog signal $x_a^*(t)$, which was discussed in Sec. 1.2. Its Fourier transform is given by

$$\begin{aligned}\mathscr{F}\{x_a^*(t)\} &= \int_{-\infty}^{\infty} x_a^*(t)e^{-j\omega t}\,dt \\ &= \int_{-\infty}^{\infty} \sum_{n=-\infty}^{\infty} x_a(t)\,\delta_a(t - nT)e^{-j\omega t}\,dt \\ &= \sum_{n=-\infty}^{\infty} \int_{-\infty}^{\infty} x_a(t)e^{-j\omega t}\,\delta_a(t - nT)\,dt\end{aligned}$$

Using the sifting property of the impulse function, this expression can be reduced to

$$\mathscr{F}\{x_a^*(t)\} = \sum_{n=-\infty}^{\infty} x_a(nT)e^{-jn\omega T}$$

If we let $x(n) = x_a(nT)$, then we define its Fourier transform to be $\mathscr{F}\{x_a^*(t)\}$ when $T = 1$. In general, the Fourier transform of the discrete-time signal $x(n)$ is given by

$$X(e^{j\omega}) = \mathscr{F}\{x(n)\} = \sum_{n=-\infty}^{\infty} x(n)e^{-j\omega n} \tag{1.20}$$

This expression is very similar to the complex Fourier series that is probably familiar to most readers. The Fourier transform $X(e^{j\omega})$ is periodic with period 2π, as can be easily verified from Eq. (1.20).

Convergence of Complex Series

In Eq. (1.20), we are dealing with a series of complex numbers. Since this may be a new concept to the reader, we shall give a definition for convergence for the series

$\sum_{n=0}^{\infty} Z_n$, where Z_n is a complex number and in Eq. (1.20) is given by $Z_n = x(n)e^{-j\omega n}$. This series converges to the complex number S if for any $\epsilon > 0$, there exists a positive integer N_ϵ such that

$$|S - S_N| < \epsilon \text{ for } N > N_\epsilon \tag{1.21}$$

where S_N is the Nth partial sum defined by

$$S_N = \sum_{n=0}^{N-1} Z_n$$

We also express convergence of the series to S by writing

$$S = \lim_{N\to\infty} S_N = \sum_{n=0}^{\infty} Z_n \tag{1.22}$$

In many cases, convergence of complex series can be related to the convergence of real series. Frequently, we can determine if the complex series converges without actually knowing the limit S to which it converges.

Usually, when we encounter a doubly infinite series such as Eq. (1.20), we write it as a sum

$$\sum_{n=-\infty}^{\infty} z_n = \sum_{n=-\infty}^{-1} z_n + \sum_{n=0}^{\infty} z_n$$

and consider each sum separately. In particular, we can replace n by $-n$ in the first sum on the right-hand side and obtain

$$\sum_{n=-\infty}^{\infty} z_n = \sum_{n=1}^{\infty} z_{-n} + \sum_{n=0}^{\infty} z_n$$

Each of these series has the form that we have just considered.

In most of the cases that we will consider, $x(n)$ is a causal sequence and therefore the Fourier transform is given by

$$X(e^{j\omega}) = \sum_{n=0}^{\infty} x(n)e^{-j\omega n}$$

A Complex Geometric Series

A special series that we shall encounter quite frequently is the *geometric series* whose partial sum is

$$S_N = \sum_{n=0}^{N-1} z_0^n \tag{1.23}$$

Using the results of Eq. (1.5), which are valid here, we obtain

$$S_N = \begin{cases} N & z_0 = 1 \\ \dfrac{z_0^N - 1}{z_0 - 1} & z_0 \neq 1 \end{cases}$$

From the first relation, we see that $\lim_{N\to\infty} S_N = \infty$ and the series for $z_0 = 1$ diverges. From the second relation for $z_0 \neq 1$, we see that

$$|S_N| = \frac{|z_0^N - 1|}{|z_0 - 1|} \geq \frac{|z_0|^N - 1}{|z_0 - 1|}$$

and, consequently, $\lim_{N\to\infty} |S_N| = \infty$ if $|z_0| > 1$.

If $|z_0| < 1$, then $\lim_{N\to\infty} |z_0|^N \to 0$ and, as we shall see, the series converges. In this case, a logical guess for the sum of the series is $S = 1/(1 - z_0)$. To show that this is the case, we consider

$$|S - S_N| = \left| \frac{1}{1 - z_0} - \frac{z_0^N - 1}{z_0 - 1} \right|$$

$$= \frac{|z_0|^N}{|1 - z_0|}$$

It follows that $|S - S_N| < \epsilon$ for $N > N_\epsilon$, if we choose N_ϵ so that

$$|z_0|^{N_\epsilon} < \epsilon\, |1 - z_0|$$

An appropriate value of N_ϵ is one for which $N_\epsilon > \alpha$, where $|z_0|^\alpha = \epsilon |1 - z_0|$.

The last case to be considered for Eq. (1.23) is when $|z_0| = 1$. Then $z_0 = e^{j\theta}$, $0 < \theta < 2\pi$, and we must consider

$$S_N = \sum_{n=0}^{N-1} e^{jn\theta} = \frac{e^{jN\theta} - 1}{e^{j\theta} - 1}$$

Since $e^{jN\theta} = \cos N\theta + j \sin N\theta$, it should be clear that the sequence of partial sums S_N does not converge.

Summarizing the above results, we have shown that the geometric series with S_N given by Eq. (1.23) converges to

$$\sum_{n=0}^{\infty} z_0^n = \frac{1}{1 - z_0} \qquad |z_0| < 1 \tag{1.24}$$

for $|z_0| < 1$ and diverges otherwise.

Examples of Fourier Transforms

Some examples illustrating the computation of the Fourier transform are now given. If $x(n) = \delta(n)$, then

$$\mathcal{F}\{\delta(n)\} = X(e^{j\omega}) = \sum_{n=-\infty}^{\infty} \delta(n) e^{-j\omega n} = 1$$

For $x(n) = \alpha^n u(n)$, we have

$$\mathcal{F}\{\alpha^n u(n)\} = X(e^{j\omega}) = \sum_{n=-\infty}^{\infty} \alpha^n u(n) e^{-jn\omega}$$

$$= \sum_{n=0}^{\infty} (\alpha e^{-j\omega})^n$$

This is a special case of the complex geometric series of Eq. (1.24) with $z_0 = \alpha e^{-j\omega}$. Since $|z_0| = |\alpha e^{-j\omega}| = |\alpha| \cdot |e^{-j\omega}| = |\alpha|$, we find from Eq. (1.24) that

$$\mathcal{F}\{\alpha^n u(n)\} = \frac{1}{1 - \alpha e^{-j\omega}} \qquad |\alpha| < 1 \tag{1.25}$$

As another example of a Fourier transform, consider the sequence

$$x(n) = \begin{cases} 1 & 0 \le n \le 6 \\ 0 & \text{otherwise} \end{cases}$$

The Fourier transform of this sequence is given by

$$X(e^{j\omega}) = \sum_{n=0}^{6} e^{-jn\omega}$$

which by Eq. (1.5) can be written

$$X(e^{j\omega}) = \frac{1 - e^{-j7\omega}}{1 - e^{-j\omega}}$$

This expression for $X(e^{j\omega})$ can be simplified to

$$\begin{aligned} X(e^{j\omega}) &= \frac{e^{-j7\omega/2}}{e^{-j\omega/2}} \frac{e^{j7\omega/2} - e^{-j7\omega/2}}{e^{j\omega/2} - e^{-j\omega/2}} \\ &= e^{-j3\omega} \frac{\sin 7\omega/2}{\sin \omega/2} \end{aligned} \tag{1.26}$$

Inverse Fourier Transform

The sequence $x(n)$ is called the *inverse Fourier transform* of $X(e^{j\omega})$ and we write

$$x(n) = \mathcal{F}^{-1}\{X(e^{j\omega})\} \tag{1.27}$$

An expression for $x(n)$ in terms of $X(e^{j\omega})$ can be obtained formally by multiplying Eq. (1.20) by $e^{j\omega m}$ and integrating over any interval of length 2π. This results in

$$\int_{\omega_0}^{\omega_0+2\pi} X(e^{j\omega})e^{j\omega m}\, d\omega = \int_{\omega_0}^{\omega_0+2\pi} \sum_{n=-\infty}^{\infty} x(n)e^{-j\omega(n-m)}\, d\omega$$

Assuming that the order of integration and summation can be interchanged, this equation becomes

$$\sum_{n=-\infty}^{\infty} x(n) \int_{\omega_0}^{\omega_0+2\pi} e^{-j\omega(n-m)}\, d\omega = \int_{\omega_0}^{\omega_0+2\pi} X(e^{j\omega})e^{j\omega m}\, d\omega$$

Noting that

$$\int_{\omega_0}^{\omega_0+2\pi} e^{-j\omega(n-m)}\, d\omega = \begin{cases} 0 & n \ne m \\ 2\pi & n = m \end{cases} \tag{1.28}$$

we see that there is only one nonzero term in the summation, which occurs when $n = m$, and so

$$x(m) = \frac{1}{2\pi} \int_{\omega_0}^{\omega_0+2\pi} X(e^{j\omega})e^{j\omega m}\, d\omega \tag{1.29}$$

In many cases, it is convenient to take $\omega_0 = -\pi$.

The series of Eq. (1.20) converges uniformly to a continuous function of ω, and the procedure for obtaining $x(n)$ is valid if $x(n)$ is absolutely summable; that is, if

$$\sum_{n=-\infty}^{\infty} |x(n)| < \infty$$

Examples of Inverse Fourier Transforms

An important example where the Fourier transform is known and it is desired to recover the sequence $x(n)$ is that where

$$X(e^{j\omega}) = \begin{cases} 1 & |\omega| \le \omega_c \\ 0 & \omega_c < |\omega| < \pi \end{cases} \tag{1.30}$$

Using Eq. (1.29) with $\omega_0 = -\pi$, we obtain

$$x(n) = \frac{1}{2\pi}\int_{-\omega_c}^{\omega_c} e^{j\omega n}\, d\omega = \frac{1}{2\pi(jn)}[e^{j\omega_c n} - e^{-j\omega_c n}]$$

which simplifies to

$$x(n) = \frac{\sin \omega_c n}{n\pi}$$

In some cases, we may be able to obtain $x(n)$ by merely expanding $X(e^{j\omega})$ in a series of the form of Eq. (1.20) and comparing coefficients of $e^{-j\omega n}$. For example, if

$$X(e^{j\omega}) = \frac{1}{1 - \frac{1}{2}e^{-j\omega}}$$

we obtain, by using Eq. (1.24),

$$X(e^{j\omega}) = \sum_{n=0}^{\infty} (\tfrac{1}{2})^n e^{-j\omega n}$$

Comparison of this expression with Eq. (1.20) results in

$$x(n) = (\tfrac{1}{2})^n u(n)$$

As a final example of determining $x(n)$, consider

$$X(e^{j\omega}) = e^{-j\omega}[\tfrac{1}{2} + \tfrac{1}{2}\cos\omega]$$

Noting that

$$\cos\omega = \tfrac{1}{2}(e^{j\omega} + e^{-j\omega})$$

we can write

$$\begin{aligned} X(e^{j\omega}) &= e^{-j\omega}[\tfrac{1}{2} + \tfrac{1}{4}(e^{j\omega} + e^{-j\omega})] \\ &= \tfrac{1}{4} + \tfrac{1}{2}e^{-j\omega} + \tfrac{1}{4}e^{-j2\omega} \end{aligned}$$

Therefore it follows that $x(0) = x(2) = \frac{1}{4}$, $x(1) = \frac{1}{2}$, and $x(n) = 0$ otherwise.

Representations of $X(e^{j\omega})$

The Fourier transform $X(e^{j\omega})$ is a complex function of ω and can be expressed in the *rectangular* form:

$$X(e^{j\omega}) = X_R(e^{j\omega}) + jX_I(e^{j\omega}) \tag{1.31}$$

where $X_R(e^{j\omega}) = \text{Re}[X(e^{j\omega})]$ is the *real part*, and $X_I(e^{j\omega}) = \text{Im}[X(e^{j\omega})]$ is the *imaginary part* of $X(e^{j\omega})$, respectively. Also, we can use the *polar*-form representation:

$$X(e^{j\omega}) = |X(e^{j\omega})|e^{j\phi(\omega)} = |X(e^{j\omega})| \underline{/\phi(\omega)} \tag{1.32}$$

where $|X(e^{j\omega})|$ is the *magnitude*, and $\phi(\omega)$ is the *angle* or *phase* of $X(e^{j\omega})$. We also use the notation

$$\phi(\omega) = \arg X(e^{j\omega})$$

Using Euler's relation,

$$e^{j\phi(\omega)} = \cos\phi(\omega) + j\sin\phi(\omega) \tag{1.33}$$

we see that the rectangular and polar forms of $X(e^{j\omega})$ are related by

$$X_R(e^{j\omega}) = |X(e^{j\omega})| \cos\phi(\omega) \tag{1.34}$$

$$X_I(e^{j\omega}) = |X(e^{j\omega})| \sin\phi(\omega) \tag{1.35}$$

$$|X(e^{j\omega})|^2 = [X_R(e^{j\omega})]^2 + [X_I(e^{j\omega})]^2 \tag{1.36}$$

$$\tan\phi(\omega) = X_I(e^{j\omega})/X_R(e^{j\omega}) \tag{1.37}$$

Examples of the Polar and Rectangular Forms

For the sequence $\alpha^n u(n)$, α real, we can use Euler's relation and write Eq. (1.25) in the form

$$X(e^{j\omega}) = \frac{1}{1 - \alpha\cos\omega + j\alpha\sin\omega} = \frac{1 - \alpha\cos\omega - j\alpha\sin\omega}{1 + \alpha^2 - 2\alpha\cos\omega}$$

Therefore, we have

$$|X(e^{j\omega})| = \frac{1}{(1 + \alpha^2 - 2\alpha\cos\omega)^{1/2}}$$

$$\phi(\omega) = -\tan^{-1}\frac{\alpha\sin\omega}{1 - \alpha\cos\omega}$$

$$X_R(e^{j\omega}) = \frac{1 - \alpha\cos\omega}{1 + \alpha^2 - 2\alpha\cos\omega}$$

$$X_I(e^{j\omega}) = -\frac{\alpha\sin\omega}{1 + \alpha^2 - 2\alpha\cos\omega}$$

Clearly, $|X(e^{j\omega})|$ and $X_R(e^{j\omega})$ are even functions of ω and $\phi(\omega)$ and $X_I(e^{j\omega})$ are odd functions of ω.

For the example of Eq. (1.30), the magnitude is given by

$$|X(e^{j\omega})| = \begin{cases} 1 & |\omega| \leq \omega_c \\ 0 & \omega_c < |\omega| < \pi \end{cases}$$

and the phase is given by

$$\phi(\omega) = \arg X(e^{j\omega}) = 0$$

The magnitude $|X(e^{j\omega})|$ is illustrated graphically in Fig. 1.20, where it is seen to be periodic of period 2π. Also it is an even function of ω.

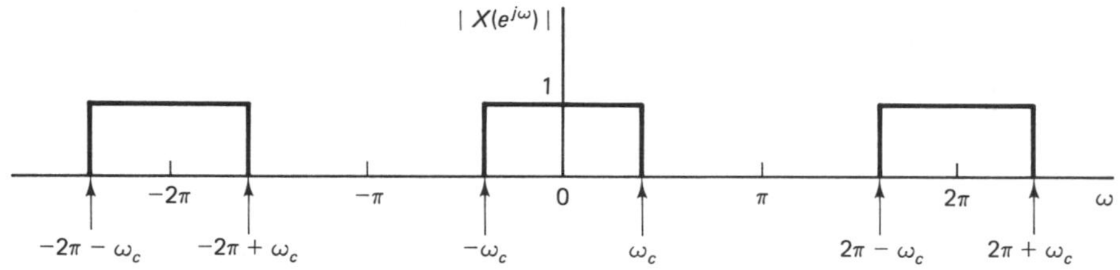

Figure 1.20 Graph of $|X(e^{j\omega})|$.

Even though we are dealing with discrete-time signals, the Fourier transform $X(e^{j\omega})$ of $x(n)$ is defined for all real ω. That is, it is a function of the continuous frequency ω. This is a feature shared by discrete-time and continuous-time signals.

Properties of $X(e^{j\omega})$

For the example of Eq. (1.25), we have shown that $X_R(e^{j\omega})$ and $|X(e^{j\omega})|$ are even functions of ω, whereas $X_I(e^{j\omega})$ and $\phi(\omega)$ are odd functions of ω. This is true for real sequences, as we now see. For the case where $x(n)$ is real, which will be of considerable interest to us, we can substitute Euler's relation into Eq. (1.20) to obtain

$$X_R(e^{j\omega}) = \sum_{n=-\infty}^{\infty} x(n) \cos \omega n \tag{1.38}$$

$$X_I(e^{j\omega}) = -\sum_{n=-\infty}^{\infty} x(n) \sin \omega n \tag{1.39}$$

Since $\cos(-\omega)n = \cos \omega n$ and $\sin(-\omega)n = -\sin \omega n$, we have

$$X_R(e^{-j\omega}) = \sum_{k=-\infty}^{\infty} x(n) \cos(-\omega)n = \sum_{k=-\infty}^{\infty} x(n) \cos \omega n = X_R(e^{j\omega})$$

$$X_I(e^{-j\omega}) = \sum_{k=-\infty}^{\infty} x(n) \sin(-\omega)n = -\sum_{k=-\infty}^{\infty} x(n) \sin \omega n = -X_I(e^{j\omega})$$

Therefore, it follows that $X_R(e^{j\omega})$ is an *even* function of ω and $X_I(e^{j\omega})$ is an *odd* function of ω. When these relations are substituted into Eqs. (1.36) and (1.37), we see that $|X(e^{j\omega})|$ is an *even* function of ω and $\phi(\omega)$ is an *odd* function of ω. Since $X(e^{j\omega})$ is periodic of period 2π in ω and since $|X(e^{j\omega})|$ and $\phi(\omega)$ are even and odd functions, respectively, we only need to know $X(e^{j\omega})$ over the interval $0 \leq \omega \leq \pi$ to completely determine its value for all ω.

In the case of $X(e^{j\omega})$ given in Eq. (1.26), we see that the magnitude, which is an even function of ω, is given by

$$\begin{aligned} |X(e^{j\omega})| &= \left|\frac{\sin 7\omega/2}{\sin \omega/2}\right| \\ &= \frac{\sin 7\omega/2}{\sin \omega/2} \qquad 0 \leq \omega < 2\pi/7,\ 4\pi/7 \leq \omega < 6\pi/7 \\ &= -\frac{\sin 7\omega/2}{\sin \omega/2} \qquad 2\pi/7 \leq \omega < 4\pi/7,\ 6\pi/7 \leq \omega \leq \pi \end{aligned}$$

The phase, which is an odd function of ω, is given by

$$\begin{aligned} \phi(\omega) &= -3\omega \qquad\qquad 0 \leq \omega < 2\pi/7,\ 4\pi/7 \leq \omega < 6\pi/7 \\ &= -3\omega + \pi \qquad 2\pi/7 \leq \omega < 4\pi/7,\ 6\pi/7 \leq \omega \leq \pi \end{aligned}$$

Graphs of these functions are shown in Fig. 1.21. It should be noted that the phase is *piecewise linear*.

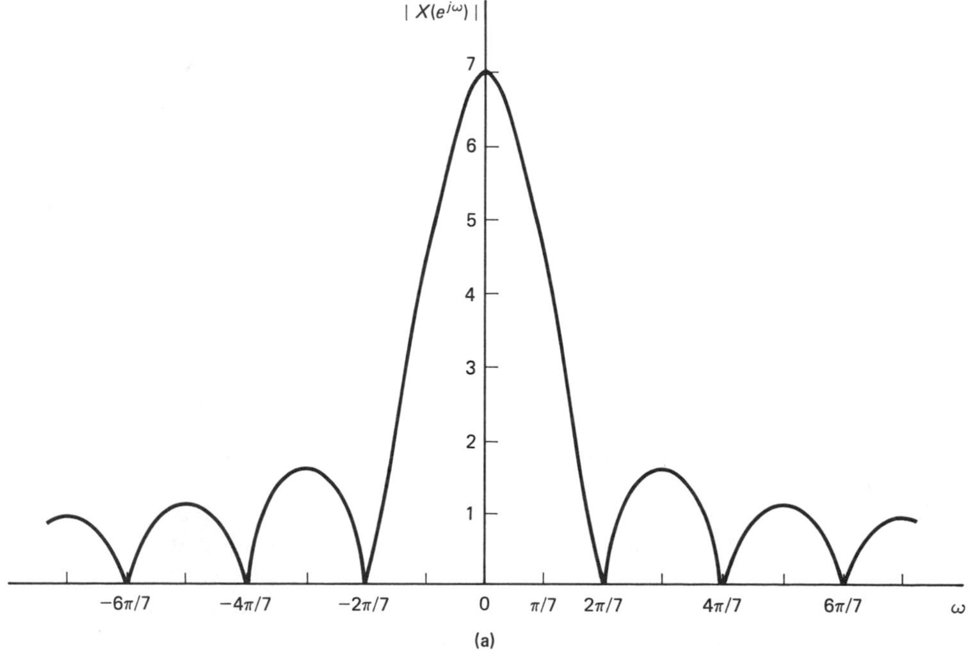

Figure 1.21 (a) Magnitude and (b) phase for $X(e^{j\omega})$ given in Eq. 1.26.

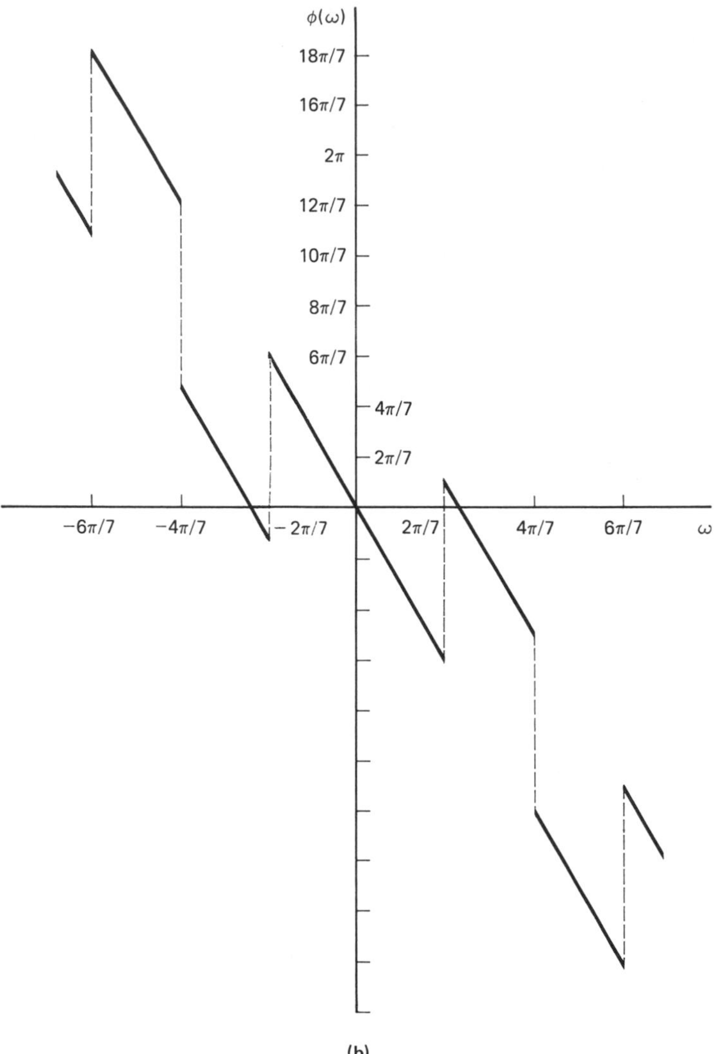

(b)

Figure 1.21 *(continued)*

Frequency Responses

The behavior of the Fourier transform $X(e^{j\omega})$ as a function of ω is of considerable interest in the analysis of system performance. The quantity $X(e^{j\omega})$ is generally called the *frequency* response, $|X(e^{j\omega})|$ is the *magnitude* or *amplitude* response, and $\phi(\omega)$ is the *phase* response. For example, the magnitude and phase responses for $X(e^{j\omega})$ of Eq. (1.26) are given in Fig. 1.21. From these graphs, the properties of $X(e^{j\omega})$ can be clearly seen.

EXERCISES

1.4.1 If $x(1) = x(3) = 2x(2) = 2$ and $x(n) = 0$ otherwise, determine:
(a) $X(e^{j\omega})$
(b) $|X(e^{j\omega})|$
(c) $\phi(\omega)$

Discuss the linearity of the phase $\phi(\omega)$.

1.4.2 Repeat Ex. 1.4.1 if $x(1) = -x(3) = 2$, and $x(2) = 0$.

1.4.3 Determine $x(n)$ if

$$X(e^{j\omega}) = \begin{cases} e^{-j3\omega/2} & |\omega| \leq 1 \\ 0 & 1 < |\omega| \leq \pi \end{cases}$$

1.4.4 Find $X(e^{j\omega})$ if $x(n) = \frac{1}{2}[(\frac{1}{2})^n + (-\frac{1}{4})^n]u(n)$.

1.5 DISCRETE-TIME SIGNALS OBTAINED BY SAMPLING

We have often mentioned that discrete-time signals are frequently obtained by sampling continuous-time signals. In Fig. 1.12, we gave an example of a system in which an analog signal is sampled, the resulting discrete-time signal is processed, and the processed discrete-time signal is then converted into a continuous-time signal. In this section, we consider the relation between the analog signal and the sampled discrete-time signal in the time and frequency domains.

Relation Between the Fourier Transforms of Analog and Digital Signals

Consider the analog signal $x_a(t)$ and its Fourier transform given by

$$X_a(j\Omega) = \int_{-\infty}^{\infty} x_a(t)e^{-j\Omega t}\,dt \tag{1.40}$$

$$x_a(t) = \frac{1}{2\pi}\int_{-\infty}^{\infty} X_a(j\Omega)e^{j\Omega t}\,d\Omega \tag{1.41}$$

We obtain a discrete-time signal $x(n) = x_a(nT)$ by sampling the analog signal once every T seconds, where T is the sampling period. The reciprocal of T is the sampling frequency or sampling rate. The discrete-time sequence is then related to the Fourier transform $X_a(j\Omega)$ by the relation

$$x(n) = x_a(nT) = \frac{1}{2\pi}\int_{-\infty}^{\infty} X_a(j\Omega)e^{j\Omega nT}\,d\Omega \tag{1.42}$$

Since we are interested in relating the Fourier transforms of $x(n)$ and $x_a(t)$, we need to compare Eq. (1.42) with the expression of Eq. (1.29), which, for $\omega_0 = -\pi$, is

$$x(n) = \frac{1}{2\pi}\int_{-\pi}^{\pi} X(e^{j\omega})e^{j\omega n}\,d\omega \tag{1.43}$$

Comparing these expressions, we note that the variables and limits of integration are different. This suggests writing Eq. (1.42) as a sum of integrals

$$x(n) = \frac{1}{2\pi}\sum_{k=-\infty}^{\infty}\int_{(2k-1)\pi/T}^{(2k+1)\pi/T} X_a(j\Omega')e^{j\Omega' nT}\,d\Omega'$$

This expression can be made more like Eq. (1.43) if we make the change of variable

$$\Omega' = \Omega + 2\pi k/T$$

which results in

$$x(n) = \frac{1}{2\pi}\sum_{k=-\infty}^{\infty}\int_{-\pi/T}^{\pi/T} X_a(j\Omega + j2\pi k/T)e^{j\Omega nT}\,d\Omega \tag{1.44}$$

since $\exp(j2\pi kn) = 1$. Now, making the change of variable $\omega = \Omega T$ in Eq. (1.44) gives

$$x(n) = \frac{1}{2\pi}\sum_{k=-\infty}^{\infty}\frac{1}{T}\int_{-\pi}^{\pi} X_a(j\omega/T + j2\pi k/T)e^{j\omega n}\,d\omega \tag{1.45}$$

Assuming that we can interchange the order of integration and summation, we obtain the desired relation

$$x(n) = \frac{1}{2\pi}\int_{-\pi}^{\pi}\frac{1}{T}\sum_{k=-\infty}^{\infty} X_a(j\omega/T + j2\pi k/T)e^{j\omega n}\,d\omega \tag{1.46}$$

Comparing the two expressions for $x(n)$ given in Eqs. (1.43) and (1.46), we see that in terms of the frequency ω,

$$X(e^{j\omega}) = \frac{1}{T}\sum_{k=-\infty}^{\infty} X_a(j\omega/T + j2\pi k/T) \tag{1.47}$$

and in terms of the analog frequency $\Omega = \omega/T$,

$$X(e^{j\Omega T}) = \frac{1}{T}\sum_{k=-\infty}^{\infty} X_a(j\Omega + j2\pi k/T) \tag{1.48}$$

The frequency response $X(e^{j\omega})$ is obtained by superposing the shifted components of $X_a(j\Omega)$, where the shifts are multiples of $2\pi/T$ both to the left and to the right. This is illustrated in both Figs. 1.22 and 1.23. These figures will be discussed later.

If we replace the Fourier transform $X(e^{j\omega})$ in Eq. (1.47) by its series expansion, we obtain

$$\sum_{n=-\infty}^{\infty} x(n)e^{-j\omega n} = \frac{1}{T}\sum_{k=-\infty}^{\infty} X_a(j\omega/T + j2\pi k/T)$$

Now substituting $x(n) = x_a(nT)$ and $\omega = \Omega T$ results in *Poisson's formula:*

$$\sum_{n=-\infty}^{\infty} x_a(nT)e^{-jn\Omega T} = \frac{1}{T}\sum_{k=-\infty}^{\infty} X_a(j\Omega + j2\pi k/T) \tag{1.49}$$

There are two cases involving $X_a(j\Omega)$ that we shall distinguish. The first case occurs when $x_a(t)$ is bandlimited with the highest frequency Ω_a present less than or equal to π/T. As an illustration, consider the function $X_a(j\Omega)$ shown in Fig. 1.22(a). For this signal, the functions $X_a(j\Omega + j2\pi k/T)$ do not overlap and $X(e^{j\Omega T})$ is shown in Fig. 1.22(b). The importance of this case is that since there is no overlap, the spectrum $X_a(j\Omega)$ can be recovered from $X(e^{j\Omega T})$ by using a low-pass filter that blocks all frequencies above π/T.

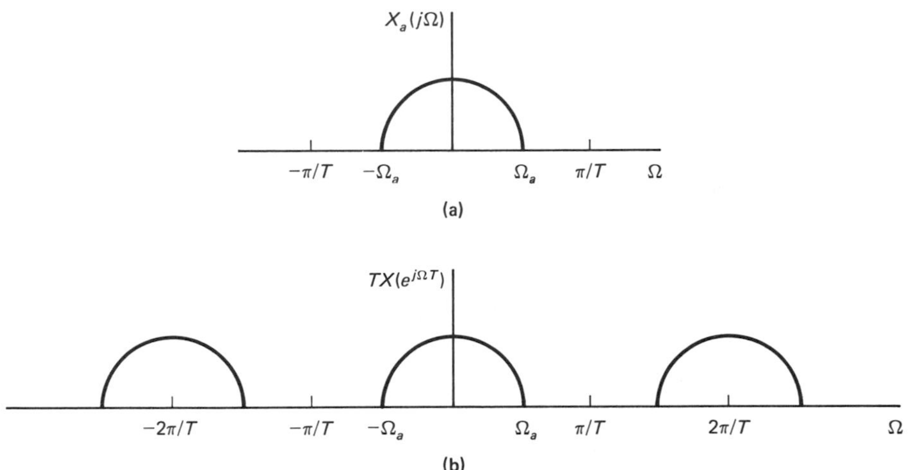

Figure 1.22 (a) Graph of $X_a(j\Omega)$ and (b) graph of $X(e^{j\Omega T})$, which corresponds to the sampled signal.

The second case of interest arises when $x_a(t)$ contains a frequency greater than π/T. In this case, the functions $X_a(j\Omega + j2\pi k/T)$ overlap and we cannot hope to recover $X_a(j\Omega)$ from $X(e^{j\Omega T})$. This case is illustrated in Fig. 1.23. The upper fre-

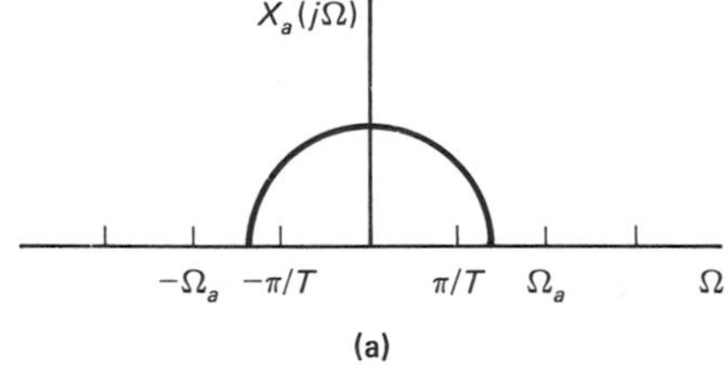

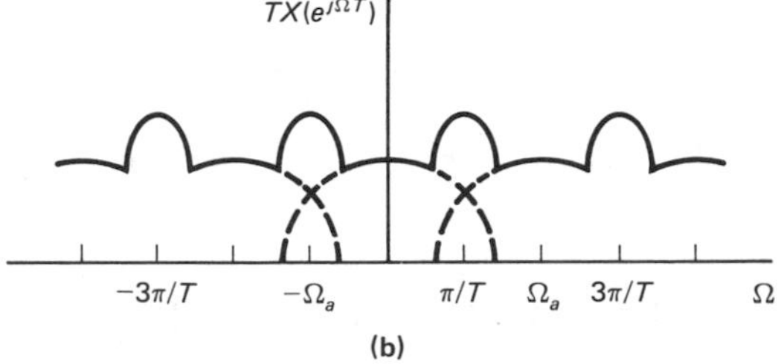

Figure 1.23 (a) Graph of $X_a(j\Omega)$ and (b) graph of corresponding $X(e^{j\Omega T})$ when there is aliasing.

quencies of $X_a(j\Omega)$ are reflected into the lower frequencies of $X(e^{j\Omega T})$, resulting in the phenomenon called *aliasing*.

Examples of Aliasing

As a simple illustration of aliasing that can occur as a result of uniform sampling, consider the sinusoid

$$f_a(t) = A \cos (2\pi/T)(k + \alpha)t$$

where k is an integer, and $0 < \alpha < 0.5$. If we sample this function at the frequency $1/T$ Hz, the resulting sampled sequence is

$$f_a(nT) = A \cos 2\pi (k + \alpha)n = A \cos 2\pi\alpha n$$

These samples are the same as those obtained by sampling the sinusoid

$$g_a(t) = A \cos (2\pi/T)\alpha t$$

Consequently, since $f_a(nT) = g_a(nT)$, it is impossible to determine which analog function produced the sample values. When k is a positive integer, the higher frequency $2\pi (k + \alpha)/T$ of $f_a(t)$ is aliased into the lower frequency $2\pi\alpha/T$ of $g_a(t)$ as a result of the sampling process. Another example of analog signals that have the same sample values arises when

$$f_a(t) = g_a(t) + \sin \pi t/T$$

A more general situation is considered in Ex. 1.5.3 [157].

Recovery of the Analog Signal

In the case of a bandlimited signal with highest frequency Ω_a, we can avoid aliasing by sampling at a frequency $\Omega_s = 2\pi/T$, so that

$$\Omega_s = 2\pi/T \geq 2\Omega_a$$

The frequency Ω_a is usually referred to as the *Nyquist frequency*. Consequently, aliasing is avoided if we sample at a frequency Ω_s that is at least twice the Nyquist frequency. When this sampling criterion is met, we have

$$X(e^{j\Omega T}) = \frac{1}{T}X_a(j\Omega) \qquad -\pi/T \leq \Omega \leq \pi/T$$

or, since $\omega = \Omega T$,

$$X(e^{j\omega}) = \frac{1}{T}X_a(j\omega/T) \qquad -\pi \leq \omega \leq \pi$$

Since $X_a(j\Omega) = 0$ outside the interval $-\pi/T \leq -\Omega_a \leq \Omega \leq \Omega_a \leqslant \pi/T$, we can write

$$x_a(t) = \frac{1}{2\pi}\int_{-\infty}^{\infty} X_a(j\Omega)e^{j\Omega t}\, d\Omega = \frac{1}{2\pi}\int_{-\pi/T}^{\pi/T} X_a(j\Omega)e^{j\Omega t}\, d\Omega \tag{1.50}$$

In view of the relation between $X_a(j\Omega)$ and $X(e^{j\Omega T})$, this expression becomes

$$x_a(t) = \frac{T}{2\pi}\int_{-\pi/T}^{\pi/T} X(e^{j\Omega T})e^{j\Omega t}\,d\Omega \tag{1.51}$$

Replacing $X(e^{j\Omega T})$ by its value as given by Eq. (1.20), where $\omega = \Omega T$ and $x(n) = x_a(nT)$, we obtain

$$\begin{aligned} x_a(t) &= \frac{T}{2\pi}\int_{-\pi/T}^{\pi/T}\left[\sum_{k=-\infty}^{\infty} x_a(kT)e^{-j\Omega Tk}\right]e^{j\Omega t}\,d\Omega \\ &= \sum_{k=-\infty}^{\infty} x_a(kT)\left[\frac{T}{2\pi}\int_{-\pi/T}^{\pi/T} e^{j\Omega(t-kT)}\,d\Omega\right] \end{aligned}$$

Carrying out the indicated integration results in

$$\begin{aligned} x_a(t) &= \sum_{k=-\infty}^{\infty} x_a(kT)\frac{T}{\pi}\frac{1}{t-kT}\left[\frac{e^{(j\pi/T)(t-kT)} - e^{-(j\pi/T)(t-kT)}}{2j}\right] \\ &= \sum_{k=-\infty}^{\infty} x_a(kT)\frac{\sin(\pi/T)(t-kT)}{(\pi/T)(t-kT)} \end{aligned} \tag{1.52}$$

In terms of the *sinc function* (pronounced "sink") defined by

$$\operatorname{sinc} u = \frac{\sin \pi u}{\pi u} \tag{1.53}$$

we can express the previous interpolation formula as

$$x_a(t) = \sum_{k=-\infty}^{\infty} x_a(kT)\operatorname{sinc}(1/T)(t-kT) \tag{1.54}$$

In practice, an analog signal is never really bandlimited and so aliasing always occurs. Whenever these effects tend to become too severe, the analog signal is usually passed through a low-pass guard filter to obtain a signal that is more nearly bandlimited.

The results given can be summarized as the *Sampling Theorem* [150], which can be stated as follows.

> If a signal $x_a(t)$ has a bandlimited frequency response $X_a(j\Omega)$, such that $X_a(j\Omega) = 0$ for $\Omega \geq \Omega_a$, then $x_a(t)$ can be uniquely reconstructed from equally spaced samples $x_a(nT)$, $-\infty < n < \infty$, if $1/T \geq 2(\Omega_a/2\pi) = \Omega_a/\pi$. That is, the sampling frequency $\Omega_s = 2\pi/T$ should be at least twice the Nyquist frequency Ω_a. The formula for the recovered signal $x_a(t)$ is given in Eq. (1.54).

Further Uses of Sinc *u*

The sinc function frequently arises when $x(n)$ is computed from a frequency response having a piecewise-constant magnitude response. For this case, the interval $0 \leq \omega \leq \pi$ can be subdivided into a finite number of subintervals over which

$|X(e^{j\omega})|$ is constant. The function $X(e^{j\omega})$ of Eq. (1.30) is such an example. The coefficients were given by

$$x(n) = \frac{\sin \omega_c n}{n\pi}$$

Comparing this expression with Eq. (1.53), we see that $\pi u = \omega_c n$ and, therefore,

$$x(n) = \frac{\omega_c}{\pi} \frac{\sin \omega_c n}{\omega_c n} = \frac{\omega_c}{\pi} \operatorname{sinc} \frac{\omega_c n}{\pi}$$

A graph of sinc u is shown in Fig. 1.24, where it can be observed that decaying ripples occur. The zeros of sinc u occur when u is any nonzero integer and its extrema occur approximately midway between consecutive zeros.

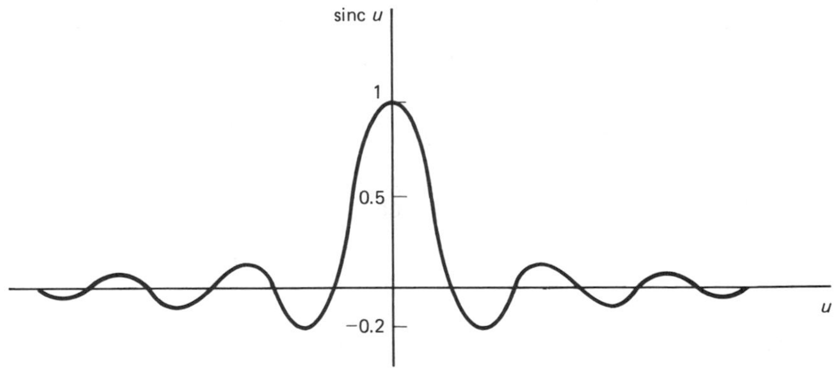

Figure 1.24 Graph of sinc u.

We shall see other applications of the sinc function in later chapters, especially when we consider filters.

EXERCISES

1.5.1 Show that if $X_a(j\Omega)$ is the Fourier transform of $x_a(t)$, then $2\pi x_a(-\Omega)$ is the Fourier transform of $X_a(jt)$.

1.5.2 Show that the Fourier transform of the analog function

$$p_a(t, b_1, b_2) = u_a(t - b_1) - u_a(t - b_2) \qquad b_2 > b_1$$

where the analog step function $u_a(t)$ is defined by

$$u_a(t) = \begin{cases} 1 & t \geq 0 \\ 0 & t < 0 \end{cases}$$

is given by

$$P_a(j\Omega, b_1, b_2) = (e^{-jb_1\Omega} - e^{-jb_2\Omega})/j\Omega$$

From this result, obtain $P_a(j\Omega, -b, b)$, and using the results of Ex. 1.5.1, show that the Fourier transform of $x_a(t) = 2b \operatorname{sinc}(bt/\pi)$ is

$X_a(j\Omega) = 2\pi p_a(\Omega, -b, b)$. Sketch the graph of $X(e^{j\Omega T})$ if (a) $\pi/T = 2b$ and (b) $\pi/T = 3b/4$.

1.5.3 If

$$f_a(t) = g_a(t) + \sum_k A_k \sin(k\pi t/T)$$

show that $f_a(nT) = g_a(nT)$, and, hence, $f_a(t)$ and $g_a(t)$ cannot be distinguished using their sample values.

1.5.4 Determine the zeros of sinc u. If u_k and u_{k+1} are consecutive zeros of sinc u, determine sinc $[(u_k + u_{k+1})/2]$.

1.6 THE DISCRETE FOURIER TRANSFORM

Fourier Representation of a Periodic Sequence

In addition to the representation of Eq. (1.14), which uses the impulse sequence, we also can obtain a Fourier series representation for a periodic sequence $x(n)$. This representation can be obtained by sampling a periodic analog signal $x_a(t)$ that has period $\omega_0 = 2\pi/NT$, where T is the sampling interval, and N is the period of $x(n)$. If $x_a(t)$ is represented by the Fourier series

$$x_a(t) = \sum_{k=-\infty}^{\infty} a_k e^{jk\omega_0 t}$$

then $x(n)$ is the sampled sequence

$$\begin{aligned} x(n) = x_a(nT) &= \sum_{k=-\infty}^{\infty} a_k e^{jk\omega_0 nT} \\ &= \sum_{k=-\infty}^{\infty} a_k e^{j2\pi kn/N} \end{aligned} \tag{1.55}$$

To see that this sequence is periodic, we note that

$$\begin{aligned} x(n+N) &= \sum_{k=-\infty}^{\infty} a_k e^{j2\pi k(n+N)/N} \\ &= \sum_{k=-\infty}^{\infty} a_k e^{j2\pi kn/N} \\ &= x(n) \end{aligned}$$

since $e^{j2\pi k} = 1$.

Further simplification of Eq. (1.55) can result by noting that there are only N distinct values of $e^{j2\pi nk/N}$, namely,

$$e^{j0}, e^{j2\pi/N}, e^{j4\pi/N}, \ldots, e^{j2\pi(N-1)/N}$$

Upon collecting the terms that have these exponentials as multipliers, we can write $x(n)$ as the finite sum

$$x(n) = \sum_{k=0}^{N-1} a'_k e^{j2\pi kn/N}$$

It has become standard practice to define

$$a'_k = \frac{1}{N}\tilde{X}(k)$$

and use $\tilde{x}(n)$ instead of $x(n)$ to emphasize that the sequence is periodic. Using this new notation, the Fourier series of Eq. (1.55) reduces to the finite series

$$\tilde{x}(n) = \frac{1}{N}\sum_{k=0}^{N-1} \tilde{X}(k)e^{j2\pi nk/N} \tag{1.56}$$

An Orthogonal Sequence

The sequence $\{e^{j2\pi nk/N}\}$ is an *orthogonal* sequence, satisfying the *orthogonality relation*

$$\sum_{n=0}^{N-1} e^{j2\pi nk/N}e^{-j2\pi nm/N} = N\,\delta(k - m) \qquad 0 \le k, m \le N - 1 \tag{1.57}$$

This sum is a special case of the sum of the geometric series, given by Eq. (1.5),

$$\sum_{n=0}^{N-1} a^n = \frac{a^N - 1}{a - 1} \qquad a \ne 1$$

In Eq. (1.57), we have $a = \exp\,[j2\pi(k - m)/N]$ and therefore it follows that $a^N = 1$. In case $k = m$, then $a = 1$ and the sum is N. Even though we have restricted m so that $0 \le m \le N - 1$, it is possible to obtain a similar result when m is any integer. In this case, the impulse function in Eq. (1.57) is replaced by $\delta(m - k - rN)$, where r is an integer.

Determination of $\tilde{X}(k)$

We can solve Eq. (1.56) for $\tilde{X}(k)$ by using the orthogonality relation of Eq. (1.57). Multiplying both sides of Eq. (1.56) by $\exp(-j2\pi nm/N)$ and summing, we obtain

$$\begin{aligned}
\sum_{n=0}^{N-1} \tilde{x}(n)e^{-j2\pi nm/N} &= \frac{1}{N}\sum_{n=0}^{N-1}\sum_{k=0}^{N-1} \tilde{X}(k)e^{j2\pi nk/N}e^{-j2\pi nm/N} \\
&= \frac{1}{N}\sum_{k=0}^{N-1} \tilde{X}(k)\sum_{n=0}^{N-1} e^{j2\pi nk/N}e^{-j2\pi nm/N} \\
&= \frac{1}{N}\sum_{k=0}^{N-1} \tilde{X}(k)N\,\delta(k - m) \\
&= \tilde{X}(m)
\end{aligned} \tag{1.58}$$

In obtaining this result, we have interchanged the order of summation of the two finite sums and have used the orthogonality relation given in Eq. (1.57). We also have assumed that $0 \leq m, k \leq N - 1$.

A Discrete Fourier-Transform Pair

The relations given in Eqs. (1.56) and (1.58) are called a *discrete Fourier-transform pair*. The *discrete Fourier transform* (DFT) is given by

$$\tilde{X}(k) = \text{DFT}\{\tilde{x}(n)\} = \sum_{n=0}^{N-1} \tilde{x}(n)e^{-j2\pi nk/N} \qquad k = 0, 1, \ldots, N - 1 \tag{1.59}$$

and the *inverse discrete Fourier transform* (IDFT) is defined by

$$\tilde{x}(n) = \text{IDFT}\{\tilde{X}(k)\} = \frac{1}{N}\sum_{k=0}^{N-1} \tilde{X}(k)e^{j2\pi nk/N} \qquad n = 0, 1, \ldots, N - 1 \tag{1.60}$$

The sequence $\tilde{X}(k)$ is periodic also with period N. However, in many applications we are concerned only with values of $\tilde{X}(k)$ and $\tilde{x}(n)$ for $0 \leq n, k \leq N - 1$.

Example of a DFT Calculation

To illustrate the calculation of a DFT, let us consider the sequence

$$\tilde{x}(n) = a^n \qquad 0 \leq n \leq N - 1$$

$$\tilde{x}(n + N) = \tilde{x}(n)$$

The DFT of this sequence is then

$$\begin{aligned}\tilde{X}(k) &= \sum_{n=0}^{N-1} a^n e^{-j2\pi nk/N} \\ &= \frac{1 - a^N e^{-j2\pi Nk/N}}{1 - ae^{-j2\pi k/N}} \\ &= \frac{1 - a^N}{1 - ae^{-j2\pi k/N}} \qquad k = 0, 1, \cdots, N - 1\end{aligned}$$

This result is obtained from Eq. (1.5), where the ratio is $ae^{-j2\pi k/N}$.

It may seem that the development of the DFT is rather artificial and consequently its applications are rather limited. This is not the case at all and we now give an example of a very important application.

Approximation of the Continuous-Time Fourier Transform

In the study of continuous-time systems, Fourier-transform theory is frequently used and we must evaluate the Fourier transform given by

$$F_a(j\Omega) = \int_{-\infty}^{\infty} f_a(t)e^{-j\Omega t}\, dt$$

In many cases, the function $f_a(t)$ is a causal function, so that $f_a(t) = 0$ for $t < 0$. In this case, the Fourier transform is

$$F_a(j\Omega) = \int_0^\infty f_a(t)e^{-j\Omega t}\,dt$$

When we use a digital computer in the analysis, we must approximate the integral by a finite series. One such approximation is (see Prob. 1.7)

$$F_a(j\Omega) = \Delta t \sum_{k=0}^{N-1} f_a(t_k)e^{-j\Omega t_k} \tag{1.61}$$

where $t_k = k\,\Delta t$, and there are N intervals of length Δt. Normally, we are interested in $F_a(j\Omega)$ at a set of discrete frequencies such as

$$F_a(j\Omega_n) = \Delta t \sum_{k=0}^{N-1} f_a(t_k)e^{-j\Omega_n t_k} \tag{1.62}$$

The inverse Fourier transform, given by

$$f_a(t) = \frac{1}{2\pi}\int_{-\infty}^{\infty} F_a(j\Omega)e^{j\Omega t}\,d\Omega$$

can be approximated using the same method as that used in Eq. (1.61), so that for N an even integer,

$$f_a(t) = \frac{\Delta\Omega}{2\pi}\sum_{n=-N/2}^{N/2-1} F_a(j\Omega_n)e^{j\Omega_n t}$$

At the sample time t_k, we have

$$f_a(t_k) = \frac{\Delta\Omega}{2\pi}\sum_{n=-N/2}^{N/2-1} F_a(j\Omega_n)e^{j\Omega_n t_k} \tag{1.63}$$

Let us now see how the quantities Δt, $\Delta\Omega$, and N can be selected and how they are related. If T_1 is the total time considered for the integral giving $F_a(j\Omega)$, then $T_1 = N\,\Delta t$ and $\Delta t = T_1/N$. Assuming that the highest frequency in $F_a(j\Omega)$ is Ω_a, then, from the sampling theorem discussed in Sec. 1.5, we should sample so that $(2\pi/T) \geq 2\Omega_a$, or, since $T = \Delta t$, so that $\Delta t \leq \pi/\Omega_a$. Choosing the largest value for Δt that satisfies this inequality, we see that the relation between the highest frequency Ω_a and the total time T_1 is

$$\Omega_a = \pi/\Delta t = \pi N/T_1 \tag{1.64}$$

Since the frequency band $-\Omega_a \leq \Omega \leq \Omega_a$ is divided into N intervals of length $\Delta\Omega$, we have $2\Omega_a = N\,\Delta\Omega$, and it follows that

$$\Delta\Omega = 2\Omega a/N = 2\pi N/NT_1 = 2\pi/T_1 \tag{1.65}$$

Once the values of T_1 and Ω_a are known, the value of N can be determined from Eq. (1.64). That is, N is a suitable even integer, such as a power of 2, for which

$$N \geq \Omega_a T_1/\pi \tag{1.66}$$

Now by using this value of N, the sample interval in the time domain is

$$\Delta t = T_1/N \tag{1.67}$$

and the sample interval in the frequency domain, from Eq. (1.65), is

$$\Delta\Omega = 2\Omega_a/N \tag{1.68}$$

In most cases, we take $t_k = k\,\Delta t$ and $\Omega_n = n\,\Delta\Omega$ in Eqs. (1.62) and (1.63).

We now show how Eqs. (1.62) and (1.63) can be related to the DFT and the IDFT. Noting from Eqs. (1.65) and (1.67) that

$$t_k\Omega_n = (k\,\Delta t)(n\,\Delta\Omega) = nk\frac{T_1}{N}\frac{2\pi}{T_1} = \frac{2\pi nk}{N} \tag{1.69}$$

we can use this value in Eqs. (1.62) and (1.63) to obtain

$$F_a(jn\,\Delta\Omega) = \Delta t \sum_{k=0}^{N-1} f_a(k\,\Delta t)e^{-j2\pi nk/N} \tag{1.70}$$

and

$$f_a(k\,\Delta t) = \frac{\Delta\Omega}{2\pi}\sum_{n=-N/2}^{N/2-1} F_a(jn\,\Delta\Omega)e^{j2\pi nk/N} \tag{1.71}$$

The last expression does not have the desired limits on the summation, which suggests that we write it as a sum of two terms:

$$f_a(k\,\Delta t) = \frac{\Delta\Omega}{2\pi}\left[\sum_{n=0}^{N/2-1} F_a(jn\,\Delta\Omega)e^{j2\pi nk/N} + \sum_{n=-N/2}^{-1} F_a(jn\,\Delta\Omega)e^{j2\pi nk/N}\right]$$

We note from Eq. (1.70) that $F_a(jn\,\Delta\Omega)$ is periodic in n with period N. Now substituting $n = m - N$ in the second sum of the preceding expression and using this periodicity property, we obtain

$$\begin{aligned} f_a(k\,\Delta t) &= \frac{\Delta\Omega}{2\pi}\left[\sum_{n=0}^{N/2-1} F_a(jn\,\Delta\Omega)e^{j2\pi nk/N} + \sum_{m=N/2}^{N-1} F_a(jm\,\Delta\Omega)e^{j2\pi mk/N}\right] \\ &= \frac{\Delta\Omega}{2\pi}\sum_{n=0}^{N-1} F_a(jn\,\Delta\Omega)e^{j2\pi nk/N} \end{aligned} \tag{1.72}$$

Comparing Eq. (1.70) with the DFT in Eq. (1.59), we see that $\tilde{x}(k)$ corresponds to $f_a(k\,\Delta t)\Delta t$. This suggests that we write Eq. (1.72) in the form

$$f_a(k\,\Delta t)\Delta t = \frac{\Delta\Omega\,\Delta t}{2\pi}\sum_{n=0}^{N-1} F_a(jn\,\Delta\Omega)e^{j2\pi nk/N}$$

Since, from Eq. (1.69) we have $\Delta\Omega\ \Delta t = 2\pi/N$, we can finally write

$$f_a(k\,\Delta t)\Delta t = \frac{1}{N}\sum_{n=0}^{N-1} F_a(jn\,\Delta\Omega)e^{j2\pi nk/N} \tag{1.73}$$

which has the same form as the IDFT of Eq. (1.60). If we now define the quantities

$$\begin{aligned} f_k &= f_a(k\,\Delta t)\Delta t \\ F_n &= F_a(jn\,\Delta\Omega) \end{aligned} \tag{1.74}$$

then Eqs. (1.70) and (1.73) reduce to the discrete Fourier-transform pair

$$F_n = \sum_{k=0}^{N-1} f_k e^{-j2\pi nk/N} \qquad n = 0, 1, \ldots, N-1 \tag{1.75}$$

$$f_k = \frac{1}{N}\sum_{n=0}^{N-1} F_n e^{j2\pi nk/N} \qquad k = 0, 1, \ldots, N-1 \tag{1.76}$$

To illustrate the use of the DFT in approximating the Fourier transform of a time-domain function, let us take $f_a(t) = e^{-t}u_a(t)$, where $u_a(t)$ is the analog unit-step function, defined in Ex. 1.5.2. If we choose $T_1 = 10$ and $\Omega_a = 100$, then $\Omega_a T_1/\pi = 1000/\pi = 318.3$. Taking $N = 320$, we obtain

$$\Delta\Omega = 200/320 = 0.625$$

$$\Delta t = 10/320 = 0.031$$

Using Eqs. (1.74) and (1.75), we obtain as the approximation to $F_a(jn\,\Delta\Omega)$

$$\begin{aligned} F_n &= \Delta t \sum_{k=0}^{319} e^{-k\Delta t} e^{-j\pi nk/160} \\ &= \frac{1}{32}\,\frac{1 - e^{-320\Delta t}}{1 - e^{-\Delta t}(\cos n\pi/160 - j\sin n\pi/160)} \\ &= \frac{1}{32}\,\frac{0.99995}{(1 - 0.96923\cos n\pi/160) + j0.96923\sin n\pi/160} \end{aligned}$$

When $n = 0$ and $n = 1$, we obtain

$$F_0 = 1.0155$$

$$F_1 = 0.85987\,\underline{/-0.55112\text{ rad}}$$

We know that $F_a(j\Omega) = 1/(j\Omega + 1)$. From this expression, we obtain

$$F_a(j0) = 1$$

and

$$\begin{aligned} F_a(j\,\Delta\Omega) &= \frac{1}{j\Delta\Omega + 1} = \frac{1}{j0.625 + 1} \\ &= 0.8480\,\underline{/-0.5586\text{ rad}} \end{aligned}$$

These values should be compared to the values F_0 and F_1.

There are numerous computer programs available for obtaining the DFT using fast Fourier transform algorithms, some of which are discussed in Chapter 8. In addition to the programs in [34], there are also three programs prepared by Brenner [14].

EXERCISES

1.6.1 Compute the DFT of the sequence $\{e^{-n}\}$ for $N = 5$.

1.6.2 Take $f_a(t) = e^{-2t}u_a(t)$ and $T_1 = 10$. Compute F_n for
(a) $\Omega_a = 20$
(b) $\Omega_a = 100$
In each case, compare F_0 and F_1 with $F_a(j0)$ and $F_a(j\,\Delta\Omega)$.

PROBLEMS

1.1 Verify the solution of Eq. (1.4).

1.2 Classify the following signals as discrete or continuous-time signals. Explain your answer.
(a) Temperature in a room.
(b) Closing price of a stock on the New York Stock Exchange. Plot the closing price of IBM for two weeks.
(c) Bill for each patron of a grocery store.
(d) Attendance at New York Yankees home games.
(e) Barometric pressure at the same location.

1.3 Determine the discrete-time signal obtained by sampling the continuous-time signal $f_a(t) = A \sin \omega t$ at times $t_k = kT$, $k = 0, 1, 2, \ldots$, with **(a)** $T = 0.1$ and **(b)** $T = 0.1/\pi$.

1.4 Solve the difference equation

$$y(k+2) = y(k+1) + y(k) \qquad k = 0, 1, 2, \ldots$$

by recursion with $y(0) = 0$ and $y(1) = 1$. The solution sequence is the Fibonacci numbers. Show that

$$y(k) = \frac{1}{5^{1/2}}\left[\left(\frac{1 + 5^{1/2}}{2}\right)^k - \left(\frac{1 - 5^{1/2}}{2}\right)^k\right]$$

is a solution.

1.5 Solve the difference equation

$$y(k) = 2\xi y(k-1) - y(k-2) \qquad k = 0, 1, 2, \ldots$$

with initial conditions **(a)** $y(-2) = -1$ and $y(-1) = 0$, and **(b)** $y(-2) = 2\xi^2 - 1$ and $y(-1) = \xi$. Hint: Let $\cos\theta = \xi$.

Answers

(a) $\dfrac{\sin (k+1)\theta}{\sin\theta}$ **(b)** $\cos k\theta$

1.6 Consider the problem of amortizing a debt of D dollars. Let $y(k)$ be the outstanding principal after the kth payment and $x(k)$ be the amount of the kth payment. Assuming an interest rate r per payment period, determine the difference equation that $y(k)$ satisfies. If $x(k) = R$, a constant, for $k = 1, 2, \ldots, N$, determine R so that the debt is repaid by N equal payments of amount R.

1.7 From the relation

$$dy/dt + y = x(t)$$

note that

$$\int_{t_1}^{t_2} \frac{dy}{dt}\, dt = \int_{t_1}^{t_2} (x - y)\, dt$$

Determine the difference equation for $y(kT)$ if $t_1 = (k - 1)T$ and $t_2 = kT$ for three integration schemes: **(a)** forward Euler, **(b)** backward Euler, and **(c)** trapezoidal, as shown in the figure. For each scheme take $T = 0.1$, $y(0) = 0$, and $x(t) = 1$ for $t \geq 0$ and compare the approximate solution with the actual solution.

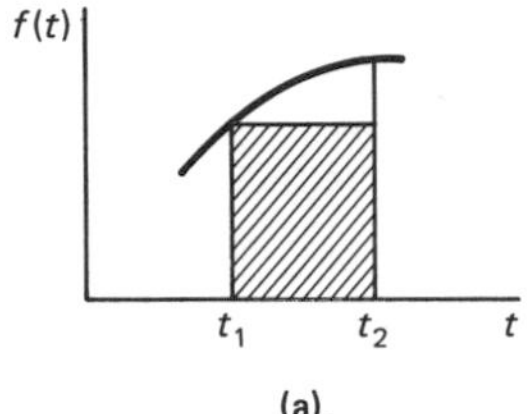

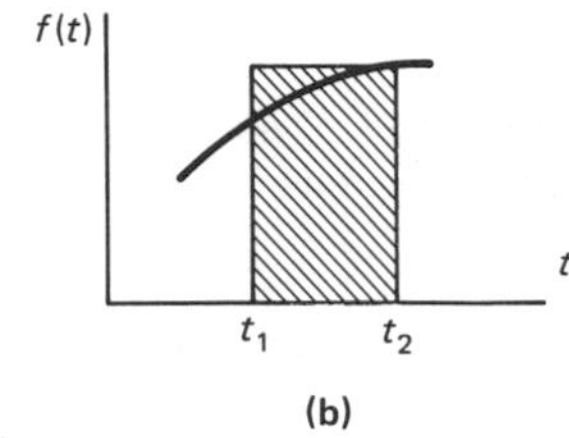

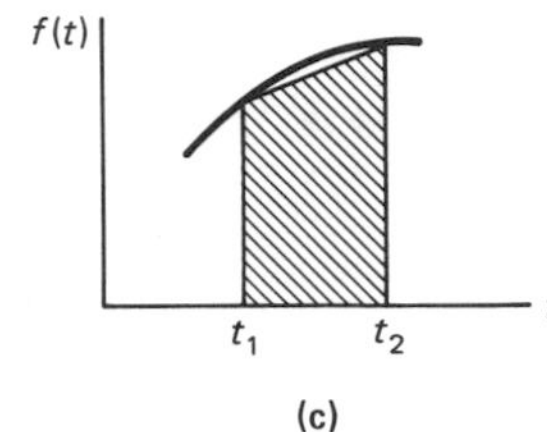

Problem 1.7

1.8 The network shown is made up of 1-ohm resistors. Find the equation for $i(k)$ in the kth loop. Determine the conditions for the zeroth loop and the nth loop.

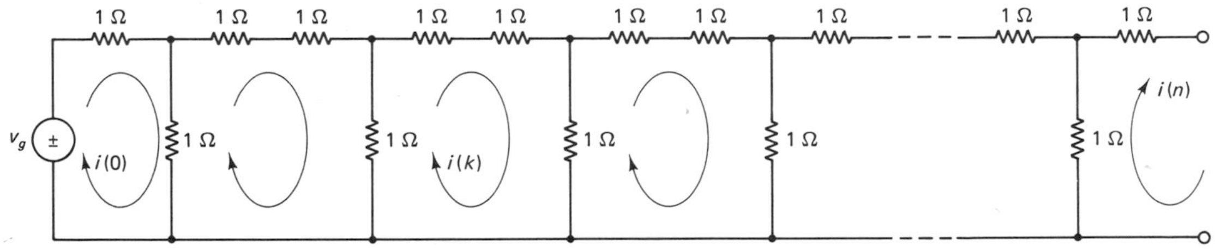

Problem 1.8

1.9 Determine the difference equation for $y(k)$, the determinant of the $k \times k$ tridiagonal matrix $\mathbf{A} = [a_{ij}]$ defined by

$$a_{ii} = a_{i,i+1} = a_{i,i-1} = 1$$
$$a_{ij} = 0 \text{ otherwise}$$

1.10 Repeat Prob. 1.9 if

$$a_{ii} = a,\ a_{i,i+1} = b,\ a_{i,i-1} = c$$
$$a_{ij} = 0 \text{ otherwise}$$

1.11 Express the sequence given by

$$x(0) = -x(2) = 2$$
$$x(n) = 0 \text{ otherwise}$$

as **(a)** a weighted sum of unit-sample sequences, and **(b)** a weighted sum of unit-step sequences.

1.12 Represent graphically the sequence given by

$$x(n) = \begin{cases} -3 & n = -2 \\ 1 & n = 0, 2 \\ -1 & n = -4, 4 \\ 0 & \text{otherwise} \end{cases}$$

Repeat Prob. 1.11 for this sequence.

1.13 Sketch the graph for the sequence

$$x(n) = 5 \sin \pi n/20$$

If the signal is periodic, determine its period.

1.14 Find the Fourier transform $X(e^{j\omega})$, $|X(e^{j\omega})|$, and $\phi(\omega)$, respectively, for the following sequences:

(a) the sequence of Prob. 1.11

(b) the sequence defined by

$$x(0) = -x(4) = 1$$
$$x(1) = -x(3) = 2$$
$$x(n) = 0 \text{ otherwise}$$

1.15 Repeat Prob. 1.14 for the sequences **(a)** $(\frac{1}{2})^n u(n)$, **(b)** $(\frac{1}{2})^n u(n-5)$, **(c)** $\delta(n)$, and **(d)** $\delta(n-5)$.

1.16 Repeat Prob. 1.15 for the sequence

$$x(0) = 2,\ x(1) = 4,\ x(2) = 2$$
$$x(n) = 0 \text{ otherwise}$$

1.17 Repeat Prob. 1.15 for the sequence

$$\begin{aligned} x(n) &= 1 \qquad 0 \le n \le N-1 \\ &= 0 \qquad \text{otherwise} \end{aligned}$$

Sketch $|X(e^{j\omega})|$ and $\phi(\omega)$.

1.18 Express the real sequence $x(n)$ as the sum of an even and an odd sequence:

$$x(n) = x_e(n) + x_o(n)$$

where $x_e(-n) = x_e(n)$ and $x_o(-n) = -x_o(n)$. Show that

$$x_e(n) = \tfrac{1}{2}[x(n) + x(-n)]$$
$$x_o(n) = \tfrac{1}{2}[x(n) - x(-n)]$$

1.19 Apply the results of Prob. 1.18 to the sequences of Probs. 1.17 and 1.11.

1.20 A complex sequence $x(n)$ is *conjugate symmetric* if

$$x(n) = x^*(-n)$$

and is *conjugate antisymmetric* if

$$x(n) = -x^*(-n)$$

where * represents the complex-conjugate operation. Show that a complex sequence is a sum of a conjugate symmetric sequence $x_e(n)$ and a conjugate antisymmetric sequence $x_o(n)$. Determine these sequences.

1.21 Show that

$$X(e^{j\omega}) = X_e(e^{j\omega}) + X_o(e^{j\omega})$$

where

$$X_e(e^{j\omega}) = \tfrac{1}{2}[X(e^{j\omega}) + X^*(e^{-j\omega})]$$

$$X_o(e^{j\omega}) = \tfrac{1}{2}[X(e^{j\omega}) - X^*(e^{-j\omega})]$$

$$X_e(e^{j\omega}) = X_e^*(e^{-j\omega})$$

$$X_o(e^{j\omega}) = -X_o^*(e^{-j\omega})$$

1.22 Apply Prob. 1.21 to the sequence of Prob. 1.17.

1.23 Determine the discrete Fourier transform of the sequences

(a) $\{1,0,1,0\}$

(b) $\{1,-1,1,-1\}$

1.24 Estimate the Fourier transform of the function $f_a(t) = e^{-2t}u_a(t)$ by using the discrete Fourier transform. Take $T_1 = 10$ s and $\Omega_a = 50$ rad/s.

2

ANALYSIS OF DISCRETE-TIME SYSTEMS

2.1 INTRODUCTION

In Chapter 1, we discussed discrete-time signals and described a discrete-time system as one that has input and output signals as well as signals within the system that are discrete-time signals. In this chapter, we give a formal definition of a discrete-time system and give several classifications. Emphasis is placed on the class of linear time-invariant systems that are described by a linear difference equation with constant coefficients. We consider classical techniques for solving these difference equations. In particular, we are concerned with finding the response to a unit-impulse sequence. This result is then used to find the response to a general input sequence by means of the convolution sum.

We define the frequency response of a system and use its properties to determine the output response to a sinusoidal input signal. As we will see, the frequency response is the Fourier transform of the output sequence due to a unit-impulse sequence. However, we will see how to determine the frequency response directly from the difference equation. One important result that we obtain in the chapter concerns conditions on a system that result in linear phase. This is a very desirable property and, while it cannot be obtained for analog systems composed of operational amplifiers and lumped resistors, capacitors, and inductors, it can be obtained for discrete-time systems.

2.2 SYSTEM DEFINITION AND EXAMPLES

Definition of a Discrete-Time System

A discrete-time system is a mapping, function, or algorithm that uniquely assigns an output sequence to each allowable input sequence. Representing the system by the mapping $\mathcal{T}$, we can write

$$\mathcal{T}\{x(n)\} = y(n) \tag{2.1}$$

where $x(n)$ is an input sequence, and $y(n)$ is the corresponding output sequence. A block-diagram representation of a discrete-time system is shown in Fig. 2.1. In the example of the savings account in Chapter 1, the system was described by the difference equation or recurrence relation, Eq. (1.3). Similarly, the voltages at the switching times of the switched-capacitor integrator shown in Fig. 1.16 satisfied the difference equation, Eq. (1.9). Even though a difference equation is involved, the switched capacitor is an analog device.

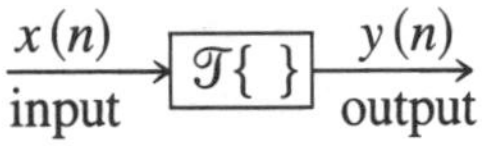

$x(n)$ input → $\mathcal{T}\{\ \}$ → $y(n)$ output

Figure 2.1 Block-diagram representation of a discrete-time system.

Example of a First-Order System

As another example, we consider the system described by the relation

$$y(n) - ay(n-1) = x(n) \tag{2.2}$$

We consider causal input sequences for which $x(n) = 0$ for $n < k$, $k \geq 0$. In order to obtain a unique output, we must also impose some restrictions on the output sequence. In particular, we assume that the output $y(n) = 0$ for $n < k$ also. In other words, the output is zero prior to the first nonzero input. Such a system is said to be causal, and it has the property that a causal input sequence produces a causal output sequence. This is discussed in more detail in Sec. 2.6. As we shall see later, this is a property that physically realizable systems must satisfy. A system will be assumed causal unless stated otherwise.

To illustrate how we might determine the output sequence for Eq. (2.2), we consider the case $x(n) = \delta(n)$. The output sequence can then be computed by recursion. Since $x(n) = 0$ for $n < 0$, we have $y(n) = 0$ for $n < 0$, resulting in

$$y(0) = x(0) + ay(-1) = 1$$

and

$$y(n) = ay(n-1) \qquad n \geq 1$$

From these two relations, it follows that $y(1) = ay(0) = a$, $y(2) = ay(1) = a^2$, and, in general,

$$y(n) = a^n u(n) \tag{2.3}$$

a result that can be easily established using mathematical induction. In Sec. 2.4, we show that the response of Eq. (2.2) to the input $x(n) = \delta(n - k)$ is

$$y(n) = a^{n-k}u(n - k) \tag{2.4}$$

where k is a positive integer with $y(n) = 0$ for $n < k$.

If we wish to obtain the *step response,* the response to the unit-step sequence $u(n)$, then we must solve the equation

$$y(n) = ay(n - 1) + u(n)$$

where $y(n) = 0$ for $n < 0$. Using recursion, we obtain

$$y(0) = ay(-1) + u(0) = 1$$

$$y(1) = ay(0) + u(1) = a + 1$$

$$y(2) = ay(1) + u(2) = a(a + 1) + 1 = a^2 + a + 1$$

$$y(3) = ay(2) + u(3) = a(a^2 + a + 1) + 1 = a^3 + a^2 + a + 1$$

and, in general, it appears that

$$y(n) = \left[\sum_{k=0}^{n} a^k\right]u(n)$$

A closed-form solution can be obtained using Eq. (1.5).

Central Difference Formula

Another example of a discrete-time system arises in approximating the derivative $x_a'(t)$ of a function $x_a(t)$ using the *central difference* formula

$$2Ty(n) = x(n + 1) - x(n - 1) \tag{2.5}$$

In this case, we have $x(n) = x_a(nT)$, and $y(n)$ is an approximation of the derivative of $x_a(t)$ evaluated at $t = nT$. If we take $x_a(t) = t^3 - t$, then the derivative at $t = nT$ is given by

$$x_a'(t)\big|_{t=nT} = x_a'(nT) = 3(nT)^2 - 1$$

and the approximation has the value

$$\begin{aligned} y(n) &= \frac{1}{2T}\{[(n + 1)^3T^3 - (n + 1)T] - [(n - 1)^3T^3 - (n - 1)T]\} \\ &= (3n^2 + 1)T^2 - 1 \end{aligned}$$

The error that results from using this approximation is given by

$$\begin{aligned} y(n) - x_a'(nT) &= [(3n^2 + 1)T^2 - 1] - (3n^2T^2 - 1) \\ &= T^2 \end{aligned}$$

If, for example, we take $T = 0.1$, then the values $x_a'(0) = -1$ and $x_a'(1) = 2$ are to be compared with $y(0) = -0.99$ and $y(10) = 2.01$. Clearly, the approximations improve as we take smaller values of T.

The expression for $y(n)$ gives the actual value of $x_a'(nT)$ if $x_a(t)$ is a polynomial of degree at most two. The reader is asked to verify this in Ex. 2.2.4.

EXERCISES

2.2.1 Find the response of the system of Eq. (1.3) with $x(k) = \delta(k)$ considered as the input. Assume $y(k) = 0$ for $k < 0$.

2.2.2 When data is "smoothed," the smoothed value $y(n)$ is the average of the $2m + 1$ values $x(n - m), \ldots, x(n - 1), x(n), x(n + 1), \ldots, x(n + m)$. Write the system equation for $y(n)$ if $m = 3$.

2.2.3 Let $x(n) = x_a(nT)$. If $y(n)$ is an approximation to the derivative of $x_a(t)$ at $t = nT$ and is the slope of the line joining the sampled values at $(n - 1)T$ and nT, write the equation describing the system.

2.2.4 Show that the central difference formula is exact if

$$x_a(t) = at^2 + bt + c$$

2.2.5 The solution of the differential equation

$$\frac{dy_a(t)}{dt} + \beta y_a(t) = x_a(t)$$

can be written

$$y_a(t) = e^{-\beta(t-t_0)}y_a(t_0) + \int_{t_0}^{t} e^{-\beta(t-\tau)}x_a(\tau)\,d\tau$$

If the input is a piecewise-constant function

$$x_a(t) = x_a(nT) = x(n) \qquad nT \le t < (n + 1)T$$

where $x(n)$ is a constant for each value of n, take $t = nT$ and $t_0 = (n - 1)T$ and find the equation describing the system whose response is $y(n) = y_a(nT)$ and whose input is $x(n)$.

2.3 LINEAR SYSTEMS

Definition of a Linear System

There are various ways that discrete-time systems can be classified. The first of these classifications that we consider is that of *linearity*. If $y_1(n)$ and $y_2(n)$ are the responses to the inputs $x_1(n)$ and $x_2(n)$, respectively, then a system is *linear* if and only if $a_1y_1(n) + a_2y_2(n)$ is the response to the input $a_1x_1(n) + a_2x_2(n)$ for arbitrary constants a_1 and a_2. In terms of the mapping $\mathcal{T}$, this condition can be expressed as

$$\begin{aligned}\mathcal{T}\{a_1x_1(n) + a_2x_2(n)\} &= a_1\,\mathcal{T}\{x_1(n)\} + a_2\,\mathcal{T}\{x_2(n)\}\\ &= a_1y_1(n) + a_2y_2(n)\end{aligned} \tag{2.6}$$

where

$$\mathcal{T}\{x_i(n)\} = y_i(n) \qquad i = 1, 2$$

The condition of Eq. (2.6) is often called the *principle of superposition*.

Example of a Linear System

The system described by Eq. (2.2) is an example of a linear system. In particular, the responses $y_1(n)$ and $y_2(n)$ satisfy the relations

$$y_1(n) - ay_1(n-1) = x_1(n)$$

$$y_2(n) - ay_2(n-1) = x_2(n)$$

Multiplying the first of these equations by a_1 and the second by a_2 and then adding the results, we obtain, after some rearrangement,

$$[a_1y_1(n) + a_2y_2(n)] - a[a_1y_1(n-1) + a_2y_2(n-1)] = a_1x_1(n) + a_2x_2(n)$$

From this relation, it is clear that the response due to the input $a_1x_1(n) + a_2x_2(n)$ is $a_1y_1(n) + a_2y_2(n)$ for arbitrary constants a_1 and a_2. Since $x_1(n) = x_2(n) = 0$ for $n < 0$, it follows that $a_1x_1(n) + a_2x_2(n) = 0$ for $n < 0$ also. Similarly, if $y_1(n) = y_2(n) = 0$ for $n < 0$, then $a_1y_1(n) + a_2y_2(n) = 0$ for $n < 0$ and is the response because of the uniqueness property.

Impulse and System Responses

The response to a unit-impulse or -sample sequence $\delta(n)$ is known as the *impulse response* of the system. It plays a very important role in the study of discrete-time systems just as its counterpart, the continuous-time impulse function, does in continuous-time systems. We will now see how the impulse response can be used to find the response of a linear system to a general input. Representing the input signal by Eq. (1.14), which gives $x(n)$ as a weighted sum of discrete-time impulses, we find that the system response is given by

$$y(n) = \mathcal{T}\{x(n)\} = \mathcal{T}\left\{\sum_{k=-\infty}^{\infty} x(k)\,\delta(n-k)\right\} \tag{2.7}$$

The linearity property of Eq. (2.6) can be extended to any finite sum of terms using mathematical induction (see Prob. 2.4). This result is given by

$$\mathcal{T}\left\{\sum_{k=-N}^{N} a_kx_k(n)\right\} = \sum_{k=-N}^{N} a_k\,\mathcal{T}\{x_k(n)\}$$

Assuming that the mapping $\mathcal{T}$ is such that this property is true for $N \to \infty$, we can write Eq. (2.7) in the form

$$\mathcal{T}\{x(n)\} = \sum_{k=-\infty}^{\infty} x(k)\mathcal{T}\{\delta(n-k)\} \tag{2.8}$$

This follows because k is a dummy summation index, whereas n is the variable indexing the input and output sequences.

The response to the shifted impulse sequence $\delta(n-k)$, denoted by $h(n, k)$, is defined by

$$h(n, k) = \mathcal{T}\{\delta(n-k)\} \tag{2.9}$$

It is interpreted as the nth sample of the response caused by the unit-impulse input whose kth sample is one and all others are zero. Using the notation of Eq. (2.9), we can write Eq. (2.8) as

$$y(n) = \mathcal{T}\{x(n)\} = \sum_{k=-\infty}^{\infty} x(k)h(n, k) \tag{2.10}$$

If we find the impulse response $h(n, k)$ for a system, then the response to the input $x(n)$ can be found from Eq. (2.10). However, it may not be an easy task to find $h(n, k)$ in closed form.

Examples of Impulse Responses

To illustrate the computation of $h(n, k)$, we consider two examples. For the first, consider the system described by

$$y(n) + y(n-1) = \delta(n-k) \qquad k \geq 0 \tag{2.11}$$

with $y(n) = 0$ for $n < k$. Using recursion as the method of solution, we find that

$$\begin{aligned} y(k) &= 1 - y(k-1) = 1 \\ y(k+1) &= -y(k) = -1 \\ y(k+2) &= -y(k+1) = 1 \\ &\vdots \\ y(k+m) &= (-1)^m \qquad m \geq 0 \end{aligned}$$

Writing this last expression in the form

$$y(k+m) = (-1)^m u(m)$$

we see that if $k + m = n$, then the response to $\delta(n-k)$ is

$$h(n, k) = (-1)^{n-k}u(n-k)$$

For a general input sequence $x(n)$, the response, obtained from Eq. (2.10), is

$$y(n) = \mathcal{T}\{x(n)\} = \sum_{k=-\infty}^{\infty} x(k)(-1)^{n-k}u(n-k)$$

In particular, if $x(n) = 2^{-n}u(n)$, the resulting response is

$$\begin{aligned} y(n) &= \sum_{k=-\infty}^{\infty} 2^{-k}u(k)(-1)^{n-k}u(n-k) \\ &= \left[\sum_{k=0}^{n} (-1)^n(-2)^{-k}\right]u(n) \\ &= (-1)^n \frac{(-\frac{1}{2})^{n+1} - 1}{(-\frac{1}{2}) - 1} u(n) \end{aligned}$$

As a second example, consider the system

$$y(n) - ny(n-1) = \delta(n-k) \qquad k \geq 0 \tag{2.12}$$

with $y(n) = 0$ for $n < k$. Again, using recursion, we obtain

$$\begin{aligned} y(k) &= 1 + ky(k-1) = 1 \\ y(k+1) &= (k+1)y(k) = k+1 \\ y(k+2) &= (k+2)y(k+1) = (k+2)(k+1) \\ y(k+3) &= (k+3)(k+2)(k+1) \\ &\vdots \\ y(k+m) &= (k+m)(k+m-1)\cdots(k+1)u(m) \end{aligned}$$

Substituting $k + m = n$ and recalling that the output is $h(n, k)$, the above result becomes

$$h(n, k) = \frac{n!}{k!}u(n-k) \qquad k \geq 0$$

For a causal input $x(n)$, the system response is given by

$$\begin{aligned} y(n) &= \sum_{k=-\infty}^{\infty} x(k)\frac{n!}{k!}u(n-k) \\ &= \left[\sum_{k=0}^{n} x(k)n!/k!\right]u(n) \end{aligned}$$

Comparing the responses for these two examples, we see that the second example, Eq. (2.12), is much more complicated than the first example, Eq. (2.11). The main reason for this increased complexity is the varying coefficient in Eq. (2.12). In the next section, we classify systems as time-invariant or time-varying. This is the distinguishing characteristic of the two examples just considered. We will see that Eq. (2.11) describes a time-invariant system, whereas Eq. (2.12) is time-varying.

Nonlinear Systems: A Square-Root Algorithm

A system that is not linear is said to be *nonlinear*. In other words, a nonlinear system does not satisfy Eq. (2.6). An example of a nonlinear discrete-time system is the difference equation

$$y(n) = \frac{1}{2}\left[y(n-1) + \frac{x(n)}{y(n-1)}\right] \tag{2.13}$$

which is related to an algorithm for computing the square root of a number [16].

The underlying algorithm is the result obtainable by using Newton's method to solve an equation. If we wish to determine $c^{1/2}$, $c > 0$, we do so by finding the positive zero of the function

$$f(y) = y^2 - c$$

Starting with the initial guess $y(n-1)$, the next approximation is obtained by replacing the original function by the tangent line described by

$$f(y) - f[y(n-1)] = f'[y(n-1)][y - y(n-1)]$$

where f' is the derivative. Now, setting $f(y) = 0$ and using $f'[y(n-1)] = 2y(n-1)$, we find that

$$y = y(n) = y(n-1) - \frac{f[y(n-1)]}{f'[y(n-1)]}$$

$$= y(n-1) - \frac{[y(n-1)]^2 - c}{2y(n-1)}$$

$$= \frac{1}{2}\left[y(n-1) + \frac{c}{y(n-1)}\right]$$

The difference equation, Eq. (2.13), is obtained by replacing c by $x(n)$. In other words, the square root of c is obtained as the asymptotic value of the response of the system to the constant input $x(n) = c$.

To see that this system does produce the square root of a number, let us take $x(n) = 2u(n)$ and $y(0) = 1$. Then it follows by recursion that

$$y(1) = \frac{1}{2}\left[y(0) + \frac{x(1)}{y(0)}\right] = \frac{1}{2}[1 + 2] = \frac{3}{2} = 1.5$$

$$y(2) = \frac{1}{2}\left[\frac{3}{2} + \frac{2}{3}(2)\right] = \frac{17}{12} = 1.4166667$$

$$y(3) = \frac{1}{2}\left[\frac{17}{12} + \frac{12}{17}(2)\right] = \frac{577}{408} = 1.4142157$$

$$y(4) = \frac{1}{2}\left[\frac{577}{408} + \frac{408}{577}(2)\right] = \frac{665{,}857}{470{,}832} = 1.4142136$$

This value for $y(4)$ is the value of $2^{1/2}$ given by a calculator.

To see that the system of Eq. (2.13) describing a square-root algorithm is nonlinear, suppose that $y_i(n)$ is the response due to the input $x_i(n)$, $i = 1, 2$. In other words, we have

$$2y_1(n) = y_1(n-1) + \frac{x_1(n)}{y_1(n-1)}$$

$$2y_2(n) = y_2(n-1) + \frac{x_2(n)}{y_2(n-1)}$$

Adding these two equations results in

$$2[y_1(n) + y_2(n)] - [y_1(n-1) + y_2(n-1)] = \frac{x_1(n)}{y_1(n-1)} + \frac{x_2(n)}{y_2(n-1)}$$

If the system is linear, the response to the input $x_1(n) + x_2(n)$ is $y_1(n) + y_2(n)$. However, the response to $x_1(n) + x_2(n)$ satisfies the relation

$$2y(n) - y(n-1) = \frac{x_1(n) + x_2(n)}{y(n-1)}$$

and it should be clear from the last equation that $y(n) \neq y_1(n) + y_2(n)$. Consequently, this system is nonlinear.

EXERCISES

2.3.1 Prove or disprove that the following systems are linear:
(a) $\mathcal{T}\{x(n)\} = e^{x(n)}$
(b) $\mathcal{T}\{x(n)\} = ax(n) + b$
(c) $\mathcal{T}\{x(n)\} = anx(n) + b$

2.3.2 Find the impulse response of the following systems, assuming $y(n) = 0$ for $n < 0$:
(a) $y(n) - ay(n-1) = x(n-1)$
(b) $y(n) - y(n-1) - y(n-2) = x(n)$
(c) $y(n) - y(n-2) = x(n) + x(n-1)$

2.3.3 Find the impulse response of the following systems, assuming $y(n) = 0$ for $n < 0$:
(a) $y(n) - ny(n-1) = x(n)$. Compare the result with the response of Eq. (2.12).
(b) $y(n) - 2\xi y(n-1) + 2(n-1)y(n-2) = x(n)$. The response $y(n)$ is a *Hermite* polynomial in the parameter ξ.

2.3.4 Find the impulse response of the system

$$y(n) - n^2 y(n-1) = x(n)$$

(a) by recursion and **(b)** by assuming a solution of the form $y(n) = (n!)^2 w(n)$. Assume that $y(n) = 0$ for $n < 0$.

2.3.5 Find a closed-form solution for the impulse response of

$$y(n) - (a+b)y(n-1) + aby(n-2) = x(n)$$

Assume $y(n) = 0$ for $n < 0$ and $a \neq b$.

2.3.6 Show that the system of Eq. (2.13) is not homogeneous.

2.4 TIME-INVARIANT SYSTEMS

Definition of a Time-Invariant System

There is a class of linear systems for which the impulse response is comparatively easy to find. These systems have the property

$$y(n-m) = \mathcal{T}\{x(n-m)\} \tag{2.14}$$

where m is a fixed integer, and $y(n) = \mathcal{T}\{x(n)\}$. A system satisfying Eq. (2.14) is said to be *time-invariant* or *shift-invariant*. In other words, a shift in the input sequence shifts the output sequence by the same amount. Otherwise, the system is *time-varying* or *shift-dependent*.

Let us now examine the impulse response of a time-invariant system. Setting $k = 0$ in Eq. (2.9) gives

$$h(n, 0) = \mathcal{T}\{\delta(n)\} \tag{2.15}$$

For simplicity, we denote

$$h(n) = h(n, 0)$$

so that

$$h(n) = \mathcal{T}\{\delta(n)\} \tag{2.16}$$

Now, applying the time-invariant property to Eq. (2.16), we obtain

$$h(n, m) = h(n - m) = \mathcal{T}\{\delta(n - m)\} \tag{2.17}$$

Summarizing, for a time-invariant system, the response to $\delta(n - m)$ can be found by first finding the response to $\delta(n)$ and then replacing n by $n - m$.

Response of a Linear Time-Invariant System

For a linear time-invariant system, the response is given by Eq. (2.10), with $h(n, k)$ given by Eq. (2.17). That is,

$$y(n) = \sum_{k=-\infty}^{\infty} x(k)h(n - k) \tag{2.18}$$

Convolution

The summation of Eq. (2.18) is commonly referred to as the *convolution sum,* or *convolution,* of the sequences $x(n)$ and $h(n)$, and is usually designated by

$$x(n) * h(n) = \sum_{k=-\infty}^{\infty} x(k)h(n - k) \tag{2.19}$$

By making the change of variable $m = n - k$, we readily obtain

$$\sum_{k=-\infty}^{\infty} x(k)h(n - k) = \sum_{m=-\infty}^{\infty} h(m)x(n - m) \tag{2.20}$$

which in the convolution notation is

$$x(n) * h(n) = h(n) * x(n)$$

Consequently, convolution is a *commutative* operation. Other properties of the convolution are left to the problems at the end of the chapter.

Example of a Time-Invariant System

As an example, let us show that the system of Eq. (2.2) is time-invariant. Using $z(n)$ as the response to $x(n - m)$, we have

$$z(n) - az(n - 1) = x(n - m) \tag{2.21}$$

where $x(n) = 0$ for $n < 0$, and $z(n)$ is zero until a nonzero input occurs. Since $x(n - m) = 0$ for $n < m$, it follows that $z(n) = 0$ for $n < m$ and

$$z(m) = x(0) = y(0)$$
$$z(m + 1) = x(1) + az(m) = x(1) + ay(0) = y(1)$$

where $y(n)$ is the response to Eq.(2.2). Following this procedure, we find

$$z(m + k) = y(k)$$

or

$$z(n) = y(n - m) \tag{2.22}$$

Since $y(n) = \mathcal{T}\{x(n)\}$ and $z(n) = \mathcal{T}\{x(n - m)\}$, it follows that the system is time-invariant.

For a general input $x(n)$, the response of Eq. (2.2) is then given by

$$\begin{aligned} y(n) &= \sum_{k=-\infty}^{\infty} x(k)h(n - k) \\ &= \sum_{k=-\infty}^{\infty} x(k)a^{n-k}u(n - k) \end{aligned}$$

where $h(n)$ is given by Eq. (2.3). Since $x(k) = 0$ for $k < 0$, we can start the summation at $k = 0$. Further, $u(n - k) = 0$ when $n - k < 0$ or $k > n$, so that the upper limit of the summation is $k = n$. The convolution sum then has the form

$$y(n) = \left[\sum_{k=0}^{n} x(k)a^{n-k}\right]u(n) \tag{2.23}$$

In most cases, the convolution sum cannot be expressed in closed form. One example where the sum can be obtained is the case where the system is that of Eq. (2.2) and the input is $x(n) = u(n)$. The output given by Eq. (2.23) becomes

$$\begin{aligned} y(n) &= \left[\sum_{k=0}^{n} u(k)a^{n-k}\right]u(n) \\ &= \left[\sum_{k=0}^{n} a^{-k}\right]a^{n}u(n) \end{aligned}$$

which involves the sum of a geometric sequence with ratio a^{-1}. Comparing this expression with the sum of the geometric progression of Eq. (1.5), we see that

$$\begin{aligned} y(n) &= a^{n}\frac{1 - a^{-n-1}}{1 - a^{-1}}u(n) \\ &= \frac{a^{n+1} - 1}{a - 1}u(n) \end{aligned} \tag{2.24}$$

provided $a \neq 1$. We note in passing that if $|a| > 1$, then the output sequence is an unbounded sequence that is almost always undesirable for a bounded input sequence. This concept is discussed more thoroughly in Sec. 2.8.

Example of a Time-Varying System

An example of a time-varying system is that of Eq. (2.12):

$$y(n) - ny(n-1) = x(n)$$

which we considered earlier in Sec. 2.3. If we denote the response to $x(n-k)$ by $y_1(n)$, then

$$y_1(n) - ny_1(n-1) = x(n-k)$$

For the moment, let us assume that $y_1(n) = y(n-k)$, where

$$y(n-k) - (n-k)y(n-k-1) = x(n-k)$$

Subtracting the last equation from the equation for $y_1(n)$ and noting that $y_1(n) = y(n-k)$, we obtain

$$ky(n-k-1) = ky_1(n-1) = 0$$

which is a contradiction. Therefore, $y_1(n) \neq y(n-k)$ and the system is time-varying or not shift-invariant. This property is a consequence of the varying coefficient $-n$ of $y(n-1)$.

The Square-Root Algorithm Revisited

The system of Eq. (2.13) describing a square-root algorithm can be shown to be shift-invariant even though it is nonlinear. (See Prob. 2.11.)

A Cascaded System

As a final example in this section, let us consider two systems that are interconnected, as shown in Fig. 2.2(a). We find it convenient to represent the system by its impulse response rather than by the operator $\mathcal{T}$, as we did in Fig. 2.1. The output of the first system, represented by its impulse response $h_1(n)$ in the block diagram, is the input to the second system, represented by its impulse response $h_2(n)$. When systems are interconnected in this manner, we say that they are *cascaded,* or *connected in cascade*. We determine the overall response $y(n)$ due to the input $x(n)$, which is also the input to the first system. If $w(n)$ is the output of the first system, then

$$w(n) = h_1(n) * x(n) \tag{2.25}$$

The output $y(n)$ is the response of the second system to the input $w(n)$ and is therefore given by

$$y(n) = h_2(n) * w(n)$$

Replacing $w(n)$ by its value from Eq. (2.25), we obtain for the system response

$$y(n) = h_2(n) * [h_1(n) * x(n)]$$

By using the associative property of the convolution given in Prob. 2.14(b),

$$h_2(n) * [h_1(n) * x(n)] = [h_2(n) * h_1(n)] * x(n)$$

it follows that

$$y(n) = [h_2(n) * h_1(n)] * x(n) \tag{2.26}$$

Since the response $y(n)$ is also given by

$$y(n) = h(n) * x(n)$$

where $h(n)$ is the impulse response for the overall system, we see by comparing these two expressions for $y(n)$ that

$$h(n) = h_2(n) * h_1(n) \tag{2.27}$$

which is the convolution of the impulse responses of each system. Since the convolution is commutative, we can also write

$$h(n) = h_1(n) * h_2(n) \tag{2.28}$$

which indicates that the same response is obtained if the ordering of the systems is reversed, as shown in Fig. 2.2(b).

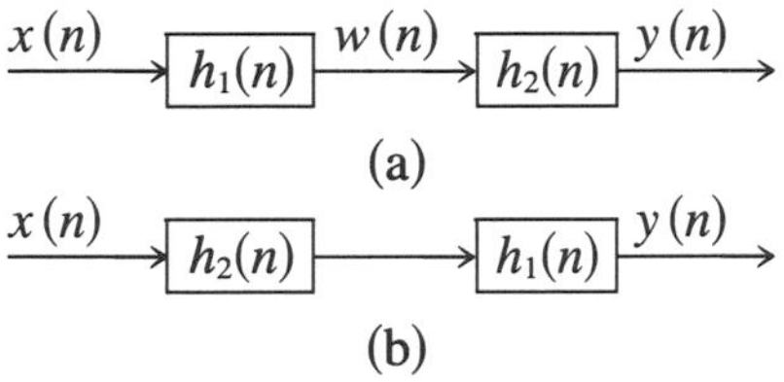

$x(n) \rightarrow$ [$h_1(n)$] $\xrightarrow{w(n)}$ [$h_2(n)$] $\xrightarrow{y(n)}$

(a)

$x(n) \rightarrow$ [$h_2(n)$] $\rightarrow$ [$h_1(n)$] $\xrightarrow{y(n)}$

(b)

Figure 2.2 (a) A cascaded system, and (b) the system with the subsystems reordered.

As an example of finding the impulse response for a cascaded system, let $h_1(n) = a^n u(n)$ and $h_2(n) = b^n u(n)$, where $a \neq b$. Then the impulse response is given by

$$\begin{aligned} h(n) = h_1(n) * h_2(n) &= \sum_{k=0}^{n} a^{n-k} u(n-k) b^k u(k) \\ &= a^n \left[\sum_{k=0}^{n} (b/a)^k \right] u(n) \end{aligned}$$

This involves the sum of a geometric series, and using Eq. (1.5) with ratio b/a, we obtain

$$h(n) = a^n \frac{(b/a)^{n+1} - 1}{b/a - 1} u(n) = \frac{b^{n+1} - a^{n+1}}{b - a} u(n)$$

For the most part, our concern is only with a special class of linear shift-invariant systems. These are described by a certain type of difference equation and are discussed in Sec. 2.6. As we will see, these systems can be analyzed rather easily.

EXERCISES

2.4.1 Prove or disprove that the systems of Ex. 2.3.1 are time-invariant.

2.4.2 Prove that the system of Ex. 2.3.2(a) is linear and time-invariant. Find $h(n, k)$ and use the convolution sum to find the response to a unit-step sequence.

2.4.3 Find $h(n, k)$, $k \geq 0$, for the system of Ex. 2.3.4.

2.4.4 Find $h(n)$ for the system

$$y(n) - ay(n-1) = x(n) + x(n-1)$$

by using **(a)** recursion and **(b)** superposition.

2.4.5 Find the impulse response of the cascaded system of Fig. 2.2(a) if $h_1(n) = (-\frac{1}{2})^n u(n)$ and $h_2(n) = (\frac{1}{2})^n u(n)$. Use the convolution sum to find the response to $x(n) = (\frac{1}{4})^n u(n)$.

2.4.6 Find the impulse response of the cascaded system of Fig. 2.2(a) if $h_1(n) = h_2(n) = a^n u(n)$.

2.4.7 Use the commutative property of the convolution to show that the system response is unchanged if the input and the system impulse response are interchanged.

2.5 FIR AND IIR SYSTEMS

Definitions

Linear time-invariant systems can be classified according to whether the impulse-response sequence is of finite duration or is of infinite duration, or, in other words, whether or not it has only a finite number of nonzero terms. If the impulse-response sequence is of finite duration, the system is called a *finite impulse-response* (FIR) system. An *infinite impulse-response* (IIR) system has an impulse response that is of infinite duration.

Example of an IIR System

The impulse response $h(n) = a^n u(n)$ given in Eq. (2.3) is of infinite duration since it is nonzero for $n \geq 0$. Consequently, the system of Eq. (2.2), which has Eq. (2.3) as its impulse response, is an IIR system.

Example of a FIR System

An example of a FIR system is a system for which

$$h(n) = \begin{cases} 2 & n = -1, 1 \\ 1 & n = 0 \\ 0 & \text{otherwise} \end{cases}$$

The response $y(n)$ to an input $x(n)$ has the form

$$\begin{aligned} y(n) &= \sum_{k=-\infty}^{\infty} x(n-k)h(k) \\ &= x(n+1)h(-1) + x(n)h(0) + x(n-1)h(1) \\ &= 2[x(n+1) + x(n-1)] + x(n) \end{aligned}$$

This relation can be thought of as a difference equation or recurrence relation expressing $y(n)$ in terms of values of the input only; it does not depend on other output values.

EXERCISES

2.5.1 Show that the system of Ex. 2.4.4 is an IIR system.

2.5.2 Find the difference equation or recurrence relation for the FIR system with impulse response

$$h(n) = \begin{cases} 1 & 0 \leq n \leq 3 \\ 0 & \text{otherwise} \end{cases}$$

Find the step response for the system.

2.6 SYSTEMS DESCRIBED BY A DIFFERENCE EQUATION

The Describing Difference Equation

We limit ourselves almost exclusively to the linear time-invariant systems whose behavior is described by the Nth *order linear difference equation with constant coefficients* given by

$$\sum_{k=0}^{N} a_k y(n-k) = \sum_{m=0}^{M} b_m x(n-m) \tag{2.29}$$

Equation (2.2) is of this form with $N = 1$. It is, therefore, a first-order difference equation with $a_0 = 1$, $a_1 = -a$, $b_0 = 1$, and $M = 0$. In order for Eq. (2.29) to define a unique input–output relation, we must specify N consecutive values for the output. Otherwise, we obtain a family of solutions. We must also beware that time-invariance and linearity are valid only for suitable initial conditions.

Causal Systems and Sequences Revisited

We are interested in systems that are physically realizable or causal. These are systems in which changes in the output do not precede changes in the input. The output of a causal system is dependent only on the present and past values of the input. For a linear time-invariant system, the causality condition is that $h(n) = 0$ for $n < 0$. (The proof of this is shown in Prob. 2.10.)

In most of the cases that we consider, the input sequence $x(n)$ is a causal sequence; that is, $x(n) = 0$ for $n < 0$. The output of a causal linear time-invariant system due to a causal input sequence is also a causal sequence, as can be seen from the convolution. Consequently, the response of a causal linear time-invariant system to a causal input is the special case of Eq. (2.18) given by

$$\begin{aligned} y(n) &= \left[\sum_{k=0}^{n} x(k)h(n-k)\right]u(n) \\ &= \left[\sum_{k=0}^{n} h(k)x(n-k)\right]u(n) \end{aligned} \tag{2.30}$$

It should be evident that the system of Eq. (2.2) with the imposed conditions is a causal system. Its step response given in Eq. (2.24) can be obtained from Eq. (2.30) with $h(n) = a^n u(n)$ and $x(n) = u(n)$.

When a causal system is described by Eq. (2.29), it will be linear and time-invariant. [Recall the system of Eq. (2.2).] In this case, we can write

$$y(n) = \frac{1}{a_0}\left[-\sum_{k=1}^{N} a_k y(n-k) + \sum_{m=0}^{M} b_m x(n-m)\right] \tag{2.31}$$

and determine the values of $y(n)$, $n \geq 0$, by recursion from the N prior values of the output and the present input $x(n)$ and the M prior input values. This is the technique used to solve Eq. (2.2). This ability to solve difference equations by recursion makes analysis of discrete-time systems generally easier than the analysis of continuous-time systems.

IIR and FIR Systems

It can be shown that the system described by Eq. (2.31) is an IIR system if some a_k, $1 \leq k \leq N$, is nonzero. If $a_k = 0$, $1 \leq k \leq N$, then the describing equation, Eq. (2.31), becomes

$$y(n) = \sum_{m=0}^{M} \frac{b_m}{a_0} x(n-m) \tag{2.32}$$

Comparing this expression with the convolution sum of Eqs. (2.18) and (2.20),

$$y(n) = \sum_{m=-\infty}^{\infty} h(m)x(n-m)$$

we see that

$$h(m) = \begin{cases} b_m/a_0 & m = 0, 1, \ldots, M \\ 0 & \text{otherwise} \end{cases} \tag{2.33}$$

Consequently, the system described by Eq. (2.31) with $a_k = 0$, $1 \leq k \leq N$, is a FIR system.

Examples

As an example, let us find the response for several inputs for the system described by the difference equation

$$y(n) = x(n) + 2x(n-1) + x(n-2) \tag{2.34}$$

The impulse response, determined from Eq. (2.33), is given by

$$h(0) = h(2) = 1 \qquad h(1) = 2 \qquad h(n) = 0 \qquad n \geq 3$$

The general expression for the response due to a causal input given in Eq. (2.30) then becomes

$$y(0) = h(0)x(0)$$

$$y(1) = h(0)x(1) + h(1)x(0)$$

$$y(n) = \sum_{k=0}^{2} h(k)x(n-k) \qquad n > 1$$

Of course, these relations can be obtained directly from Eq. (2.34). In particular, the *step response,* which is the response to a unit-step input sequence, is given by

$$y(0) = (1)(1) = 1$$

$$y(1) = (1)(1) + (2)(1) = 3$$

$$y(n) = (1)(1) + (2)(1) + (1)(1) = 4 \qquad n > 1$$

Next let $x(n) = e^{j\omega n}u(n)$. Then the response is

$$y(n) = \sum_{k=0}^{n} h(k)e^{j\omega(n-k)}u(n-k)$$

$$= e^{j\omega n}\sum_{k=0}^{n} e^{-j\omega k}h(k) \qquad n \geq 0$$

In particular, for the system of Eq. (2.34), we have

$$y(0) = 1,\ y(1) = e^{j\omega}(1 + 2e^{-j\omega})$$

$$y(n) = e^{jn\omega}(1 + 2e^{-j\omega} + e^{-2j\omega})$$

$$= 4e^{j(n-1)\omega}\cos^2 \omega/2 \qquad n \geq 2$$

As a final example in this section, consider the system

$$y(n) - y(n-1) = x(n) + x(n-1) \tag{2.35}$$

It can be shown that $h(n) = 2u(n) - \delta(n)$ and that the response to $x(n) = e^{j\omega n}u(n)$ is

$$\begin{aligned} y(n) &= \sum_{k=0}^{n} [2u(k) - \delta(k)]e^{j\omega(n-k)} \\ &= e^{j\omega n}\left[2e^{-j\omega n/2}\frac{\sin (n+1)\omega/2}{\sin \omega/2} - 1\right]u(n) \end{aligned} \tag{2.36}$$

We leave the determination of $h(n)$ and the response $y(n)$ to Prob. 2.16.

So far, the only technique that we have for solving difference equations is by recursion. In the next section, we examine *classical techniques* for solving equations of the form of Eq. (2.29). In Chapter 3, we introduce the z-transform and see that it is a tool that also can be used to solve these difference equations.

EXERCISES

2.6.1 An FIR system has an impulse response

$$h(n) = \begin{cases} 1 & 0 \le n \le 4 \\ 0 & \text{otherwise} \end{cases}$$

Find the response to the following.
(a) A general input $x(n)$
(b) The input $x(n) = e^{jn\omega}u(n)$, where ω is a parameter
(c) A unit-step input $x(n) = u(n)$
(d) The input $x(n) = 2^{-n}u(n)$

2.6.2 Consider the system

$$y(n) - 2y(n-1) = x(n)$$

Assume that $h(n) = 0$ for $n < 0$ and for $n > 3$ and show that a contradiction is obtained.

2.6.3 Find the impulse response of the causal system

$$y(n) - 2y(n-1) = x(n) + x(n-1)$$

Use the convolution sum to find the step response $[x(n) = u(n)]$.

2.7 SOLUTION OF DIFFERENCE EQUATIONS

The General Solution

In this section, we give a brief discussion of the *classical method* for solving linear difference equations with constant coefficients. (For a complete discussion of difference equations, see [82].) The equation to be studied is

$$y(n) + a_1 y(n-1) + \cdots + a_N y(n-N) = f(n) \tag{2.37}$$

We assume that $a_N \neq 0$ to ensure that the order of Eq. (2.37) is N, where the *order* is the difference in the highest and lowest arguments of the variable y that appear. The solution of Eq. (2.37) can be written as the sum

$$y(n) = y_h(n) + y_p(n) \tag{2.38}$$

where $y_h(n)$, the *natural response,* is a solution of the *homogeneous* equation

$$y_h(n) + a_1 y_h(n-1) + \cdots + a_N y_h(n-N) = 0 \tag{2.39}$$

and $y_p(n)$ the *forced response,* is a *particular* solution of the *nonhomogeneous* equation

$$y_p(n) + a_1 y_p(n-1) + \cdots + a_N y_p(n-N) = f(n) \tag{2.40}$$

It can be easily verified by direct substitution that Eq. (2.38) is a solution of Eq. (2.37).

The solution of Eq. (2.37) is obtained by finding $y_h(n)$ and $y_p(n)$ that satisfy Eqs. (2.39) and (2.40), respectively, and then adding them to obtain $y(n)$ given by Eq. (2.38). As we will see, $y_h(n)$ contains N arbitrary coefficients. The solution of Eq. (2.38) is referred to as the *general solution* of Eq. (2.37).

The Natural Response and the Characteristic Equation

To solve Eq. (2.39) for $y_h(n)$, we assume a solution of the form

$$y_h(n) = Aq^n \tag{2.41}$$

where $A \neq 0$. Substituting this expression into Eq. (2.39) results in

$$Aq^n(1 + a_1q^{-1} + \cdots + a_Nq^{-N}) = 0$$

We are interested in a nontrivial solution, $y_h(n) \neq 0$, and, consequently, q must be a root of the equation

$$1 + a_1q^{-1} + \cdots + a_Nq^{-N} = 0$$

or, equivalently,

$$q^N + a_1q^{N-1} + \cdots + a_{N-1}q + a_N = 0 \tag{2.42}$$

This latter equation is usually referred to as the *characteristic equation* corresponding to Eq. (2.39) or (2.37). By noting the relation between the coefficients of Eq. (2.42) and Eq. (2.39) or (2.37), the characteristic equation can be obtained immediately from the difference equation. That is, the term $y(n-k)$ is replaced by q^{N-k} to obtain Eq. (2.42).

Distinct Roots

For the case where Eq. (2.42) has N distinct roots, $q_1, q_2, \ldots, q_N$, it can be easily shown from the linearity of Eq. (2.39) that the most general solution is of the form

$$y_h(n) = A_1q_1^n + A_2q_2^n + \cdots + A_Nq_N^n \tag{2.43}$$

Fibonacci Numbers

As an example, let us consider the equation

$$y(n) = y(n-1) + y(n-2) \tag{2.44}$$

The solution of this equation is known as the *Fibonacci numbers* and the usual initial values are $y(0) = 0$ and $y(1) = 1$. The Fibonacci numbers, which can be obtained quite readily from Eq. (2.44) by recursion, are

$$0, 1, 1, 2, 3, 5, 8, 13, 21, \ldots$$

Any Fibonacci number is the sum of the two preceding numbers. If we are interested in $y(100)$, we must compute all the prior numbers in the sequence if we use recursion. However, using the technique of assuming a solution of the form of Eq. (2.41), we obtain the characteristic equation

$$q^2 - q - 1 = 0$$

whose roots are

$$q_1, q_2 = \frac{1 \pm 5^{1/2}}{2}$$

The general solution of Eq. (2.44) is then

$$y(n) = A_1\left(\frac{1 - 5^{1/2}}{2}\right)^n + A_2\left(\frac{1 + 5^{1/2}}{2}\right)^n \qquad n \geq 0$$

The initial conditions $y(0) = 0$ and $y(1) = 1$ require that

$$A_1 + A_2 = 0$$

$$A_1\left(\frac{1 - 5^{1/2}}{2}\right) + A_2\left(\frac{1 + 5^{1/2}}{2}\right) = 1$$

and, consequently,

$$A_1 = -A_2 = -1/5^{1/2}$$

The nth Fibonacci number is therefore given by

$$y(n) = \frac{1}{5^{1/2}}\left[\left(\frac{1 + 5^{1/2}}{2}\right)^n - \left(\frac{1 - 5^{1/2}}{2}\right)^n\right] \qquad n \geq 0 \tag{2.45}$$

a formula that can be used to determine any number in the sequence without having to compute any other number in the sequence.

Repeated Roots

If the characteristic equation has repeated roots, say q_1 is repeated m times, then the general solution of the homogeneous equation contains the term

$$q_1^n(A_{11} + A_{12}n + A_{13}n^2 + \cdots + A_{1m}n^{m-1}) \tag{2.46}$$

For each repeated root, there is a term of this form in $y_h(n)$. For example, the general solution of

$$y(n) - 4y(n-1) + 4y(n-2) = 0$$

is of the form

$$y_h(n) = (A_{11} + A_{12}n)2^n$$

Complex Roots

If the coefficients of Eq. (2.42) are real, then any complex roots that occur are complex conjugate pairs. We can still use the expressions of Eqs. (2.43) and (2.46) for the solution, but a real $y_h(n)$ requires some complex coefficients. A more convenient form can be obtained if we express the complex roots in polar form. For example, if

$$q_1, q_2 = a \pm jb$$

is a nonrepeated pair of complex roots, we can use the polar form

$$q_1, q_2 = re^{\pm j\theta} \tag{2.47}$$

to obtain

$$\begin{aligned} A_1 q_1^n + A_2 q_2^n &= A_1(re^{j\theta})^n + A_2(re^{-j\theta})^n \\ &= r^n[A_1 e^{jn\theta} + A_2 e^{-jn\theta}] \end{aligned}$$

Substituting Euler's formula into this expression results in

$$A_1 q_1^n + A_2 q_2^n = r^n[(A_1 + A_2) \cos n\theta + j(A_1 - A_2) \sin n\theta]$$

If we now make the substitutions

$$B_1 = A_1 + A_2$$

$$B_2 = j(A_1 - A_2)$$

then the portion of the solution due to the complex roots q_1 and q_2 is

$$r^n(B_1 \cos n\theta + B_2 \sin n\theta) \tag{2.48}$$

Since this expression contains only real functions, a real $y_h(n)$ produces real coefficients B_1 and B_2.

If the complex roots q_1 and q_2 having the polar form of Eq. (2.47) are repeated m times, then using Eq. (2.46) and Euler's formula results in

$$\begin{aligned} r^n[(A_1 + A_2 n + \cdots + A_m n^{m-1}) \cos n\theta \\ + (A_{m+1} + A_{m+2} n + \cdots + A_{2m} n^{m-1}) \sin n\theta] \end{aligned}$$

which is the portion of $y_h(n)$ due to q_1 and q_2.

As an example of complex roots, consider the difference equation

$$y(n) + 2y(n-1) + 2y(n-2) = 0$$

The characteristic equation is

$$q^2 + 2q + 2 = 0$$

with roots

$$\begin{aligned} q_1, q_2 &= -1 \pm j1 \\ &= 2^{1/2} \exp(\pm j3\pi/4) \end{aligned}$$

The solution is then given by

$$y(n) = (2^{n/2})[A_1 \cos(3n\pi/4) + A_2 \sin(3n\pi/4)]$$

The Impulse Response Obtained by the Classical Method

The technique given in this section can be used to obtain a closed-form expression for the impulse response. Since, in this case, the right-hand side of Eq. (2.29) is zero for $n > M$ when $x(n) = \delta(n)$, we can obtain $h(n)$ by solving the homogeneous equation and imposing the initial conditions to determine the arbitrary constants. For the system of Eq. (2.2), the impulse response satisfies the equation

$$y(n) - ay(n-1) = \delta(n)$$

For $n > 0$, this reduces to the homogeneous equation

$$y(n) - ay(n-1) = 0$$

which has the solution

$$y(n) = A_1 a^n \qquad n > 0$$

with A_1 arbitrary. If we choose A_1 so that $y(0) = 1$, then we have the impulse response that is valid for $n \geq 0$. Imposing this condition on $y(n)$, we obtain

$$y(0) = 1 = A_1$$

Therefore, the impulse response is

$$h(n) = a^n u(n)$$

which agrees with Eq. (2.3) obtained by recursion.

As another example, let us find the impulse response of the system

$$8y(n) - 6y(n-1) + y(n-2) = 8x(n) \tag{2.49}$$

The characteristic equation is

$$8q^2 - 6q + 1 = 0$$

with roots $q_1 = \frac{1}{2}$ and $q_2 = \frac{1}{4}$. Therefore, we have

$$h(n) = A_1(\tfrac{1}{2})^n + A_2(\tfrac{1}{4})^n \tag{2.50}$$

If $h(n) = 0$ for $n < 0$, then from the difference equation, we obtain $h(0) = 1$ and $h(1) = \frac{3}{4}$. Imposing these restrictions on Eq. (2.50) gives $A_1 = 2$ and $A_2 = -1$.

In the cases considered thus far, the solution, which was valid for values of n sufficiently large, contained enough arbitrary constants that could be determined so as to satisfy the conditions for $n \geq 0$. This may not always be the case, and in some cases, some terms involving impulses must be added. For example, consider finding the impulse response of the system given previously in Eq. (2.35):

$$y(n) - y(n-1) = x(n) + x(n-1)$$

For $x(n) = \delta(n)$, we have

$$y(n) - y(n-1) = 0 \qquad n \geq 2$$

with solution

$$y(n) = A$$

where A is a constant. Since we must have $y(0) = 1$, $y(1) = 2$, and $y(n) = 2$ for $n \geq 2$, we see that the solution is

$$y(n) = 2u(n) - \delta(n)$$

The impulse sequence $\delta(n)$ is chosen because any two consecutive terms must satisfy the difference equation.

Impulse Response of Eq. (2.29)

Now that we have looked at some simple examples, we are ready to consider the general system of Eq. (2.29). Setting $x(n) = \delta(n)$, we obtain

$$\sum_{k=0}^{N} a_k y(n-k) = \sum_{m=0}^{M} b_m \delta(n-m)$$

which for $n > M$ reduces to the homogeneous equation

$$\sum_{k=0}^{N} a_k y(n-k) = 0$$

If the roots of the characteristic equation are distinct, the solution of this equation has the form

$$y(n) = \sum_{k=1}^{N} A_k q_k^n$$

This expression can be used to obtain $y(n), y(n-1), \ldots, y(n-N)$ only when the argument is nonnegative. Consequently, it can be used only when $n \geq N$ and satisfies the equation for arbitrary values of $A_1, A_2, \ldots, A_N$. We thus have an expression satisfying the nonhomogeneous equation when $n \geq N$ and $n > M$. If $M < N$, then we can determine $A_1, A_2, \ldots, A_N$, so that $y(n)$ is a solution for $n = 0, 1, 2, \ldots, N-1$. If $M = N$, then we have $N + 1$ conditions (since $n > M$) and only N arbitrary constants. In this case, we take

$$y(n) = \sum_{k=1}^{N} A_k q_k^n + A_0 \delta(n)$$

and determine $A_1, A_2, \ldots, A_N$ to obtain the proper values for $y(1), y(2), \ldots, y(N)$. We then determine A_0 to obtain the correct value of $y(0)$. Since N consecutive values of $y(n)$ are needed for the difference equation and $\delta(n) = 0$ for $n \neq 0$, we add the term $A_0 \delta(n)$ rather than an impulse at some other value of n.

In Eqs. (2.2) and (2.49), we have $N > M$ and so there is no impulse function to be added to the solution of the homogeneous equation. However, in Eq. (2.35), $N = M$ and an impulse function is required.

As a final example of finding the impulse response, consider the system

$$y(n) - 3y(n-1) + 2y(n-2) = x(n) + 3x(n-1) + 2x(n-2)$$

In this case, $N = M = 2$ and the impulse response contains an impulse $\delta(n)$. For $x(n) = \delta(n)$, we have

$$y(n) - 3y(n-1) + 2y(n-2) = 0 \qquad n > 2$$

with characteristic equation

$$q^2 - 3q + 2 = (q-1)(q-2) = 0$$

The impulse response then has the form

$$y(n) = A_1 + A_2 2^n + A_0 \delta(n)$$

From the difference equation,

$$y(n) - 3y(n-1) + 2y(n-2) = \delta(n) + 3\delta(n-1) + 2\delta(n-2)$$

we have $y(0) = 1$, $y(1) = 6$, and $y(2) = 18$. Imposing these constraints on $y(n)$ results in

$$y(0) = 1 = A_1 + A_2 + A_0$$

$$y(1) = 6 = A_1 + 2A_2$$

$$y(2) = 18 = A_1 + 4A_2$$

The solution of this set of equations is $A_1 = -A_2 = -6$ and $A_0 = 1$. The impulse response is then given by

$$y(n) = h(n) = 6(2^n - 1)u(n) + \delta(n)$$

The Forced Response

We next consider solving the nonhomogeneous equation, Eq. (2.40). We consider only the case where $f(n)$ is a sum of terms of the form $P(n)q_s^n$, where $P(n)$ is a polynomial in n. The method to be used is known as the *method of undetermined coefficients* and it involves guessing a particular solution, generally of the same form as the terms in $f(n)$, which contains constants to be determined by requiring that $y_p(n)$ satisfy Eq. (2.40). The assumed solution must contain terms like those in $f(n)$ and also terms that can arise when $y_p(n)$ is substituted into Eq. (2.40). This procedure must be modified whenever q_s is a root of the characteristic equation.

If q_s is not a root of the characteristic equation, then for a term $P(n)q_s^n$ in $f(n)$, the forced response contains a term $Q(n)q_s^n$, where $Q(n)$ is a polynomial of the same degree as $P(n)$. If q_s is a root of the characteristic equation of multiplicity r, then the forced response contains a term $n^r Q(n)q_s^n$. The polynomial $Q(n)$ has arbitrary coefficients.

Examples of Forced Responses

Let us consider the system given by Eq. (2.2) and determine the step response. In this case, we have

$$y(n) - ay(n-1) = u(n) \qquad a \neq 1$$

and, in particular,

$$y(n) - ay(n-1) = 1 \qquad n \geq 0 \tag{2.51}$$

Since the right-hand side is a constant, we try

$$y_p(n) = A$$

where A is a constant to be determined, so that

$$A - a(A) = 1$$

For $a \neq 1$, this results in

$$A = \frac{1}{1 - a}$$

and

$$y_p(n) = \frac{1}{1 - a}$$

If we wish to find the complete response, we must also find $y_h(n)$. The characteristic equation is

$$q - a = 0$$

and, therefore,

$$y_h(n) = A_1 a^n$$

The complete solution is

$$y(n) = A_1 a^n + \frac{1}{1 - a} \qquad n \geq 0$$

The constant A_1 is determined from the value of $y(0)$, which is in turn obtained from

$$y(0) - ay(-1) = 1$$

Since $y(-1) = 0$, we have $y(0) = 1$. Using this condition, we obtain

$$y(0) = 1 = A_1 + \frac{1}{1 - a}$$

and, therefore,

$$A_1 = 1 - \frac{1}{1 - a} = \frac{-a}{1 - a}$$

The step response of Eq. (2.2) is, therefore,

$$y(n) = \left[-\frac{a}{1 - a} a^n + \frac{1}{1 - a} \right] u(n)$$

$$= \frac{1 - a^{n+1}}{1 - a} u(n)$$

which agrees with the result in Eq. (2.24).

As another example of determining a forced response, consider the system

$$y(n) - 2y(n - 1) = (-1)^n$$

The forced response has the form

$$y_p(n) = A(-1)^n$$

and satisfies the difference equation if

$$A(-1)^n - 2A(-1)^{n-1} = (-1)^n$$

Consequently, we have $A = \frac{1}{3}$ and $y_p(n) = \frac{1}{3}(-1)^n$.

As an example of finding a forced response when q_s is a root of the characteristic equation, consider the system

$$y(n) - ay(n-1) = a^n u(n)$$

Since q_s is a root of the characteristic equation of multiplicity one and $P(n) = 1$, we assume a forced response

$$y_p(n) = Ana^n$$

Substituting this expression into the difference equation, we obtain

$$Ana^n - aA(n-1)a^{n-1} = a^n$$

Therefore $A = 1$ and $y_p = na^n$.

As a final example of finding a forced response, consider the system

$$y(n) - y(n-1) = \cos 2n \tag{2.52}$$

In this case, $f(n) = \cos 2n$ and is of the form suitable for using undetermined coefficients since

$$2 \cos 2n = e^{j2n} + e^{-j2n}$$

If we followed the procedure outlined before, we would try

$$y_p(n) = B_1 e^{j2n} + B_2 e^{-j2n} \tag{2.53}$$

where B_1 and B_2 are generally complex. It is somewhat simpler if we try instead

$$y_p(n) = A \cos 2n + B \sin 2n \tag{2.54}$$

which can be obtained from Eq. (2.53) using Euler's formula. When Eq. (2.54) is substituted into Eq. (2.52), we obtain

$$[A(1 - \cos 2) + B \sin 2] \cos 2n + [-A \sin 2 + B(1 - \cos 2)] \sin 2n = \cos 2n$$

Since $\cos 2n$ and $\sin 2n$ are linearly independent, we must have

$$A(1 - \cos 2) + B \sin 2 = 1$$

$$-A \sin 2 + B(1 - \cos 2) = 0$$

The solution of these equations is

$$A = \tfrac{1}{2}$$

$$B = \frac{\sin 2}{2(1 - \cos 2)} = \tfrac{1}{2} \cot 1$$

and so the difference equation, Eq. (2.52), has the forced response

$$\begin{aligned} y_p(n) &= \tfrac{1}{2} \cos 2n + \tfrac{1}{2} (\cot 1) \sin 2n \\ &= \sin (2n + 1)/(2 \sin 1) \end{aligned}$$

Analysis of an RC Circuit

As a final example in this section, we analyze the simple RC parallel circuit shown in Fig. 2.3(a). Replacing the capacitor by the circuit of Fig. 1.2(a), we obtain the re-

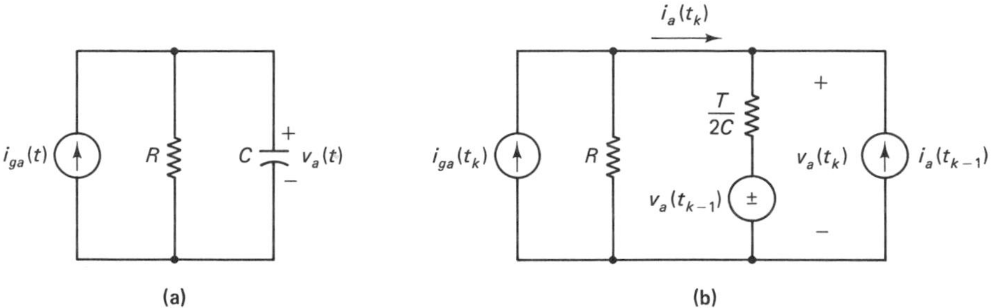

Figure 2.3 (a) An RC circuit and (b) a discretized model.

sistive circuit of Fig. 2.3(b). Applying Kirchhoff's current law at the upper node results in

$$\frac{1}{R}v_a(t_k) + \frac{2C}{T}[v_a(t_k) - v_a(t_{k-1})] = i_{ga}(t_k) + i_a(t_{k-1})$$

Setting $t_k = kT$, $v_a(t_k) = v(k)$, $i_{ga}(t_k) = i_g(k)$, and $i_a(t_k) = i(k)$, the equation becomes

$$v(k) = \left(\frac{2C}{T} + \frac{1}{R}\right)^{-1}\left[\frac{2C}{T}v(k-1) + i_g(k) + i(k-1)\right] \tag{2.55}$$

In a similar manner, we obtain the capacitor current

$$i(k) = i_g(k) - v(k)/R \tag{2.56}$$

We can use Eqs. (2.55) and (2.56) to analyze the circuit of Fig. 2.3(b) or we can replace k by $k - 1$ in Eq. (2.56) and obtain $i(k - 1)$. This expression can then be substituted into Eq. (2.55), resulting in a difference equation in $v(k)$. The result of these substitutions is

$$v(k) = \left(\frac{2C}{T} + \frac{1}{R}\right)^{-1}\left[\frac{2C}{T}v(k-1) + i_g(k) + i_g(k-1) - \frac{1}{R}v(k-1)\right]$$

which can be simplified to

$$v(k) = \left(\frac{2C}{T} + \frac{1}{R}\right)^{-1}\left[\left(\frac{2C}{T} - \frac{1}{R}\right)v(k-1) + i_g(k) + i_g(k-1)\right] \tag{2.57}$$

To complete the discussion of this example, we take $v_a(0) = 0$, $R = 10$ ohms (Ω), $C = 0.1$ F, $T = 0.1$ seconds (s), and $i_{ga}(t) = 1$ A for $t \geq 0$. Substituting these values into the difference equation results in

$$v(k) = (1/2.1)[1.9v(k-1) + 2] \text{ V}, \qquad k > 0$$

The solution of this equation has the form

$$v(k) = A(1.9/2.1)^k + B$$

where the forced response is $v_p(k) = B$. It is found that $B = 10$. Since $v(0) = v_a(0) = 0$, it follows that $A = -10$ and, therefore,

$$v(k) = 10[1 - (1.9/2.1)^k] \text{ V}, \qquad k \geq 0 \tag{2.58}$$

It can be shown that the voltage $v_a(t)$ is given by

$$v_a(t) = 10(1 - e^{-t/RC}) \text{ V}, \qquad t \geq 0 \tag{2.59}$$

To check the accuracy of our discretized model, we compare values of $v(k)$ with $v_a(0.1k)$ for $RC = 1$ given by

$$v_a(0.1k) = 10(1 - e^{-0.1k}) \text{ V}, \qquad k \geq 0$$

The error due to the approximation used to obtain the discrete-time system is

$$\begin{aligned} v_a(0.1k) - v(k) &= 10[e^{-0.1k} - (1.9/2.1)^k] \\ &= 10[(0.90484)^k - (0.90476)^k] \end{aligned}$$

In particular, we find that

$$\begin{aligned} v(1) &= 0.95238 \qquad & v_a(0.1) &= 0.95163 \\ v(2) &= 1.81406 \qquad & v_a(0.2) &= 1.81269 \\ v(3) &= 2.59367 \qquad & v_a(0.3) &= 2.59182 \\ v(4) &= 3.29904 \qquad & v_a(0.4) &= 3.29680 \end{aligned}$$

where it can be seen that agreement is obtained in the first two decimal places. The value of T that is chosen usually has a marked effect on the accuracy of the approximations $v_a(kT) = v(k)$.

EXERCISES

2.7.1 Find the impulse response of the following causal systems.
(a) $y(n) - y(n-1) = x(n)$
(b) $y(n) - y(n-1) = x(n) + x(n-1)$

2.7.2 Find the impulse response of the following causal systems.
(a) $y(n) - 2\cos\theta\, y(n-1) + y(n-2) = x(n)$
(b) $y(n) - 2\cos\theta\, y(n-1) + y(n-2) = x(n) - \cos\theta\, x(n-1)$

In each case, $\cos\theta$ is a constant parameter. Compare the results with those of Prob. 1.5.

2.7.3 Find the step response for the systems of Ex. 2.7.1.

2.7.4 Find the response for the causal system

$$y(n) - y(n-1) = x(n)$$

for **(a)** $x(n) = u(n)$, and **(b)** $x(n) = 2^{-n}u(n)$.

Answers

(a) $(n+1)u(n)$ **(b)** $(2 - 2^{-n})u(n)$

2.7.5 Find the forced response for the causal system

$$y(n) + 2y(n-1) + y(n-2) = 9(2^n) - 8n + 4$$

Answer

$$4(2^n) - 2n - 1$$

2.7.6 Find $v(k)$ and $v_a(kT)$ for **(a)** $T = 1$ s and **(b)** $T = 0.5$ s for the circuit of Fig. 2.3 and compare the values. Compare these results with those obtained in the example with T = 0.1 s.

2.8 STABILITY

Bounded Sequences

In this section, we are concerned with sequences, say $x(n)$, that have the property

$$|x(n)| \leq M_0$$

where M_0 is a constant. Such a sequence is said to be *bounded*. In the casc of digital sequences, bounded sequences are the only ones that can be stored in registers without introducing serious errors. Consequently, the input sequence to a physical system should be bounded and we will be concerned with determining when the system produces a bounded output sequence. It is important to be able to predict this happening from properties of $h(k)$ or from the difference equation describing the system.

Examples of Bounded and Unbounded Responses

In practical applications, the input sequences are bounded and it is desirable that bounded output sequences be produced. This is not always the case, as can be seen for the system

$$y(n) - y(n - 1) = x(n) \tag{2.60}$$

which was considered in Ex. 2.7.4. For the bounded input sequence $x(n) = 2^{-n}u(n)$, the output sequence is $y(n) = (2 - 2^{-n})u(n)$, a bounded sequence. However, the bounded sequence $x(n) = u(n)$ produces the unbounded sequence $y(n) = (n + 1)u(n)$. For this system, some bounded input sequences produce bounded output sequences, whereas others produce unbounded output sequences.

Definition of Stability

It is important to know from a description of the system when it will always produce a bounded output sequence corresponding to a bounded input sequence. This system property is referred to as *stability*. A system is *stable* whenever a bounded input sequence always produces a bounded output sequence.

Examples of Stability

As another example, consider the system

$$y(n) - \tfrac{1}{4}y(n - 1) = x(n) \tag{2.61}$$

Using the same inputs as applied to the system of Eq. (2.60), we find that the response to $x(n) = u(n)$ is $y(n) = [(4 - 4^{-n})/3]u(n)$ and the response to $x(n) = 2^{-n}u(n)$ is $y(n) = (2^{-n+1} - 4^{-n})u(n)$. In this case, both output sequences are bounded. This, of course, gives no guarantee that any other bounded input will produce a bounded output.

Clearly, we need criteria that can be easily applied to a representation of the system to determine if the system is stable. In this section, we develop a criterion applicable to the time-domain representation $h(k)$.

Criterion for Stability

We now show that a linear time-invariant system is stable if and only if

$$\sum_{k=-\infty}^{\infty} |h(k)| < \infty \tag{2.62}$$

where, of course, $h(k)$ is the impulse-response sequence. First, we assume that Eq. (2.62) is satisfied and show that the system is stable. Let $x(n)$ be any bounded input with $|x(n)| \leq M_0$ for some real positive number M_0. The output $y(n)$ is given by

$$y(n) = \sum_{k=-\infty}^{\infty} x(n-k)h(k) \tag{2.63}$$

It follows that

$$|y(n)| = \left| \sum_{k=-\infty}^{\infty} x(n-k)h(k) \right| \leq \sum_{k=-\infty}^{\infty} |x(n-k)||h(k)|$$

and because of the boundedness of $x(n)$,

$$|y(n)| \leq M_0 \sum_{k=-\infty}^{\infty} |h(k)|$$

Since $h(k)$ satisfies Eq. (2.62), the output sequence $y(n)$ is bounded and the system is stable.

We next show that if a system is stable, then its impulse response satisfies Eq. (2.62). The easiest proof is by contradiction. Let us assume that Eq. (2.62) is *not* satisfied for a stable system and show that this assumption leads to a contradiction. Consider the bounded input sequence defined by

$$\begin{aligned} x(n-k) &= 1 \qquad h(k) > 0 \\ &= -1 \quad\; h(k) < 0 \\ &= 0 \qquad h(k) = 0 \end{aligned}$$

Since $|x(k)|$ is either 0 or 1, the input sequence is bounded. The resulting output sequence is

$$y(n) = \sum_{k=-\infty}^{\infty} x(n-k)h(k) = \sum_{k=-\infty}^{\infty} |h(k)|$$

for our input. We assumed that Eq. (2.62) is not satisfied, and, consequently, $y(n)$ is not bounded. This contradicts the assumption that every bounded input produces a bounded output. Therefore, Eq. (2.62) must be satisfied. This concludes the proof that Eq. (2.62) is a necessary and sufficient condition for stability.

Determination of Stability for Previous Examples

The impulse response for the system of Eq. (2.60) is $h(n) = u(n)$, and, clearly,

$$\sum_{k=-\infty}^{\infty} |h(k)| = \sum_{k=0}^{\infty} u(k)$$

is a divergent series. Therefore, the system is unstable and there will be bounded input sequences that produce unbounded output sequences. One such sequence, of course, is $x(n) = u(n)$. The system of Eq. (2.61) has an impulse response $h(n) = 4^{-n}u(n)$. Consequently, for Eq. (2.61), we have

$$\sum_{k=-\infty}^{\infty} |h(k)| = \sum_{k=0}^{\infty} 4^{-k} = \frac{1}{1-\frac{1}{4}} = \frac{4}{3}$$

Therefore, Eq. (2.62) is satisfied for the system of Eq. (2.61) and it is stable. Every bounded input sequence will consequently produce a bounded output sequence. On the other hand, Eq. (2.62) is not satisfied for the system of Eq. (2.60) and so this system is unstable.

The system of Eq. (2.61) is a special case of the general first-order system of Eq. (2.2):

$$y(n) - ay(n-1) = x(n)$$

Substituting the impulse-response sequence $h(k) = a^k u(k)$ of Eq. (2.3) into the left-hand side of Eq. (2.62) results in

$$\sum_{k=0}^{\infty} |a|^k = \lim_{N\to\infty} \sum_{k=0}^{N} |a|^k = \lim_{n\to\infty} \frac{1-|a|^{N+1}}{1-|a|}$$

Since the series converges for $|a| < 1$, the system is then stable for $|a| < 1$ and unstable for $|a| \geq 1$.

A Stability Criterion Based on the Characteristic Roots

Since we are limiting the systems to those described by Eq. (2.29), we note that the impulse response satisfies the homogeneous equation

$$\sum_{k=0}^{N} a_k h(n-k) = 0 \qquad n > M$$

when at least one a_k, $k > 0$, is nonzero. If the corresponding characteristic roots are distinct, the impulse response is found, from Sec.2.7 for $N \geq M$, to have the form

$$h(n) = (A_1 q_1^n + A_2 q_2^n + \cdots + A_N q_N^n)u(n) + A_0\,\delta(n) \tag{2.64}$$

where, in general, $A_0, A_1, \ldots, A_N$ are determined so that Eq. (2.64) is valid for $n = 0, 1, 2, \ldots, N$. If $M < N$, then $A_0 = 0$ and only N initial values of $h(n)$ are required to determine $A_1, A_2 \ldots, A_N$. If there are repeated characteristic roots, then $h(n)$ contains terms of the form $n^r q_i^n$, where r is an integer varying from zero to one less than the multiplicity of the root q_i. Clearly, $\Sigma_{k=-\infty}^{\infty}|h(k)|$ converges if and only if the characteristic roots have magnitude less than one.

Stability of FIR Systems

A FIR system described by Eq. (2.29) when $a_k = 0$, $k = 1, 2, \ldots, N$, is always stable since there are only a finite number of nonzero samples $h(n)$ and the series of Eq. (2.62) always converges. This is one of the properties that makes FIR systems appealing.

EXERCISES

2.8.1 Given the causal system

$$y(n) - y(n-1) = x(n) + x(n-1)$$

Find the response to the following:
(a) $x(n) = u(n)$
(b) $x(n) = 2^{-n}u(n)$

What conclusion, if any, can be drawn concerning the stability of the system?

2.8.2 Find the impulse response $h(n)$ of the system in Ex. 2.8.1 and determine the stability of the system.

2.8.3 Determine the stability of the system

$$y(n) - \tfrac{5}{2}y(n-1) + y(n-2) = x(n) - x(n-1)$$

2.9 FREQUENCY RESPONSE

Sinusoidal Sequences

In Chapter 1, we discussed periodic signals and their Fourier representation. In this section, we determine the response of a system to the particular periodic signal

$$x(n) = e^{j\omega n} \tag{2.65}$$

where ω is a real parameter. As we will see, this analysis can provide an important insight into the system's behavior for arbitrary inputs. (The reader who has studied continuous-time systems is well aware of the importance of considering sinusoidal inputs.) We can obtain the theoretical signal of Eq. (2.65) by sampling the continuous-time signal

$$x_a(t) = e^{j\omega t} \tag{2.66}$$

with sampling interval $T = 1$. From a practical standpoint, we cannot generate the complex exponential of Eq. (2.66), but we can generate and sample the sinusoid

$$x_a(t) = \cos \omega t \tag{2.67}$$

resulting in the discrete-time signal

$$x(n) = \cos \omega n \tag{2.68}$$

Since we are dealing with linear time-invariant systems and since we have Euler's formula,

$$e^{j\omega n} = \cos \omega n + j \sin \omega n$$

we can obtain the same information about the system by using the sinusoid of Eq. (2.68) or the complex exponential of Eq. (2.65).

We are concerned primarily with systems described by the difference equation, Eq. (2.29), in which the coefficients a_k, $k = 0, 1, \ldots, N$, and b_m, $m = 0, 1, \ldots, M$, are real numbers. If $y_1(n)$ is a solution of Eq. (2.29) when $x(n) = e^{j\omega n}$, then since

$$\cos \omega n = \text{Re}(e^{j\omega n})$$

it follows from the linearity property that $y(n) = Re[y_1(n)]$ is the solution when $x(n) = \cos \omega n$. The reader is asked to verify this in Prob. 2.22. We use the exponential because the resulting analysis can be carried out easier.

Frequency Response

The response due to the exponential input of Eq. (2.65) is given by the convolution sum

$$y_1(n) = \sum_{k=-\infty}^{\infty} h(k)e^{j\omega(n-k)}$$

which can be rewritten

$$y_1(n) = e^{j\omega n} \sum_{k=-\infty}^{\infty} h(k)e^{-j\omega k} \tag{2.69}$$

Referring to Eq. (1.20), we note that

$$H(e^{j\omega}) = \sum_{k=-\infty}^{\infty} h(k)e^{-j\omega k} \tag{2.70}$$

is the Fourier transform of the impulse-response sequence and, as a consequence, the output sequence is given by

$$y_1(n) = H(e^{j\omega})e^{j\omega n} \tag{2.71}$$

The function $H(e^{j\omega})$ is called the *frequency response* of the system whose impulse response is $h(n)$.

Response to a Sinusoidal Input

To expand further on the role of the frequency response $H(e^{j\omega})$, let us consider its polar representation

$$H(e^{j\omega}) = |H(e^{j\omega})|\, e^{j\phi(\omega)}$$

which is discussed in Sec. 1.4. Using this expression for $H(e^{j\omega})$, we can write Eq. (2.71) in the form

$$y_1(n) = |H(e^{j\omega})|\, e^{j[\omega n + \phi(\omega)]}$$

It is now clear that $|H(e^{j\omega})|$ is a multiplication factor for the amplitude and $\phi(\omega)$ provides the difference in the phase of the input and output signals. If we had applied an input sequence

$$x(n) = A \cos \omega n$$

the resulting output sequence would have been

$$y(n) = \text{Re}[y_1(n)] = A|H(e^{j\omega})| \cos [\omega n + \phi(\omega)]$$

We do not propose to get into a more detailed discussion of convergence of the series of Eq. (2.70) than was given in Sec. 1.4. However, note that Eq. (2.70) converges for all values of ω for which the series

$$\sum_{k=-\infty}^{\infty} |h(k)e^{-j\omega k}|$$

converges. Since, from Euler's formula, it follows that

$$|e^{-j\omega k}|^2 = \cos^2 \omega k + \sin^2 \omega k = 1$$

we can write

$$\sum_{k=-\infty}^{\infty} |h(k)e^{-j\omega k}| = \sum_{k=-\infty}^{\infty} |h(k)|$$

Moreoever, it follows that

$$|H(e^{j\omega})| \le \sum_{k=-\infty}^{\infty} |h(k)e^{-j\omega k}| = \sum_{k=-\infty}^{\infty} |h(k)| \tag{2.72}$$

and, therefore, the series of Eq. (2.70) converges whenever the series $\Sigma_{k=-\infty}^{\infty} h(k)$ converges absolutely as mentioned in Sec. 1.4. It should be noted that this is also the condition for the system to be stable.

Example of a Frequency Response

As an example of the determination of the frequency response of a system, let us again consider the first-order system of Eq. (2.2). The impulse response for the system is given in Eq. (2.3) and, therefore, we have

$$H(e^{j\omega}) = \sum_{k=-\infty}^{\infty} a^k u(k)e^{-j\omega k} = \sum_{k=0}^{\infty} (ae^{-j\omega})^k$$

Assuming that $|a| < 1$, then the frequency response is

$$H(e^{j\omega}) = \frac{1}{1 - ae^{-j\omega}} = \frac{1}{1 - a \cos \omega + ja \sin \omega} \tag{2.73}$$

with magnitude

$$\begin{aligned} |H(e^{j\omega})| &= \frac{1}{[(1 - a \cos \omega)^2 + (a \sin \omega)^2]^{1/2}} \\ &= \frac{1}{[1 + a^2 - 2a \cos \omega]^{1/2}} \end{aligned} \tag{2.74}$$

and phase

$$\phi(\omega) = \arg H(e^{j\omega}) = -\tan^{-1}\frac{a \sin \omega}{1 - a \cos \omega}$$

Another expression can be obtained for the phase by first multiplying numerator and denominator by $e^{j\omega}$, resulting in

$$H(e^{j\omega}) = \frac{e^{j\omega}}{e^{j\omega} - a} \tag{2.75}$$

Now, from this expression, we obtain

$$\phi(\omega) = \omega - \tan^{-1}\frac{\sin \omega}{\cos \omega - a}$$

Geometric Interpretation

We can obtain a geometrical interpretation for $H(e^{j\omega})$ by noting that $e^{j\omega}$ is a point on a unit circle, as shown in Fig. 2.4. Then for $0 < a < 1$, we can identify $e^{j\omega} - a$ by the vector shown in the figure, and the magnitude and phase of $H(e^{j\omega})$ are given by

$$|H(e^{j\omega})| = \frac{|e^{j\omega}|}{|e^{j\omega} - a|} = \frac{1}{|e^{j\omega} - a|}$$

$$\phi(\omega) = \omega - \theta$$

where θ is identified in the figure.

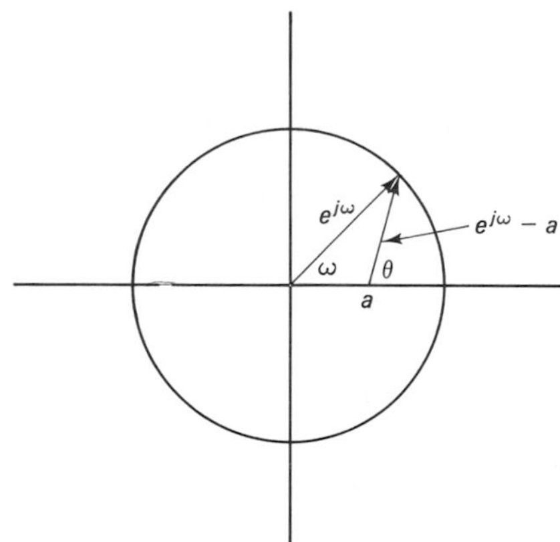

Figure 2.4 Geometric interpretation of Eq. (2.75).

Magnitude and Phase-Response Examples

Graphs of the magnitude and phase responses are shown in Fig. 2.5 for several values of a. Note from Fig. 2.4 that $\theta \geq \omega$ for $0 \leq \omega \leq \pi$ since θ is an exterior angle and ω is an interior angle of the same triangle. Using similar reasoning, $\theta \leq \omega$ for $\pi \leq \omega \leq 2\pi$. From Sec. 1.4, we only have to determine $|H(e^{j\omega})|$ and $\phi(\omega)$ for $0 \leq \omega \leq \pi$.

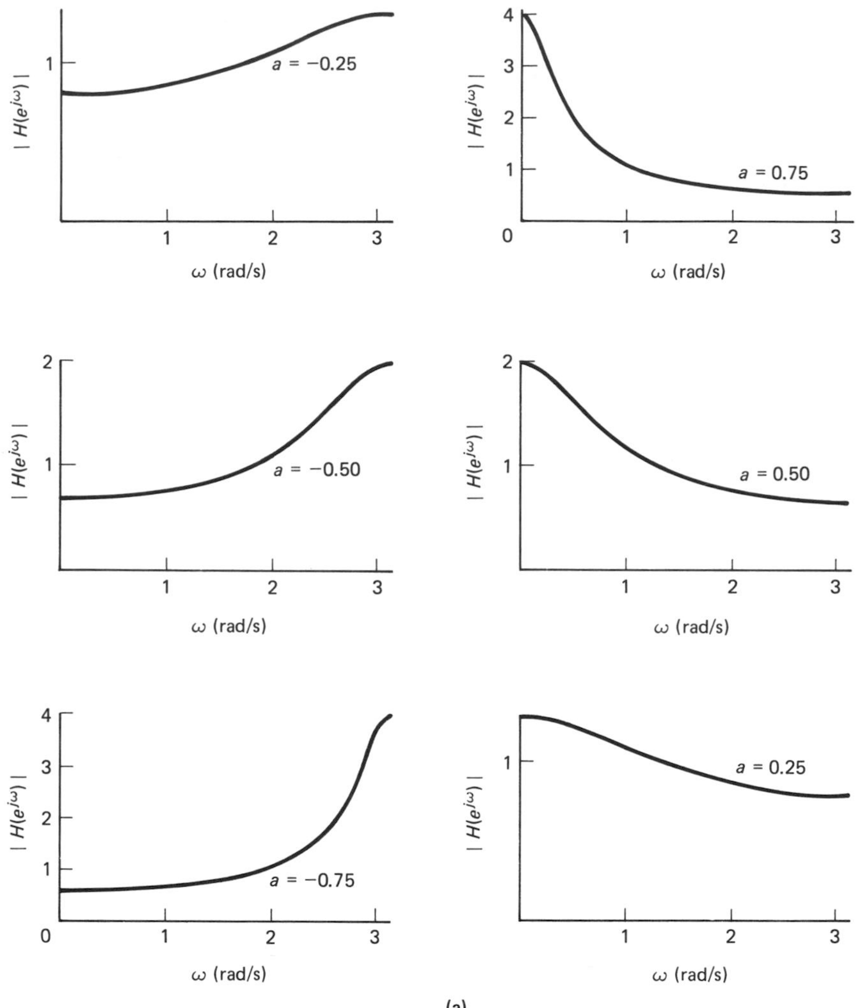

Figure 2.5 (a) Magnitude and (b) phase responses for the system of Eq. (2.73).

Frequency Response of a FIR System

As an example, we find the frequency response of the FIR system

$$y(n) = \tfrac{1}{2}x(n) + x(n-1) + \tfrac{1}{2}x(n-2)$$

We know from Sec. 2.6 that

$$h(0) = h(2) = \tfrac{1}{2} \qquad h(1) = 1 \qquad h(n) = 0 \text{ otherwise}$$

and, consequently,

$$\begin{aligned} H(e^{j\omega}) &= h(0) + h(1)e^{-j\omega} + h(2)e^{-2j\omega} \\ &= \tfrac{1}{2}(1 + e^{-2j\omega}) + e^{-j\omega} \end{aligned}$$

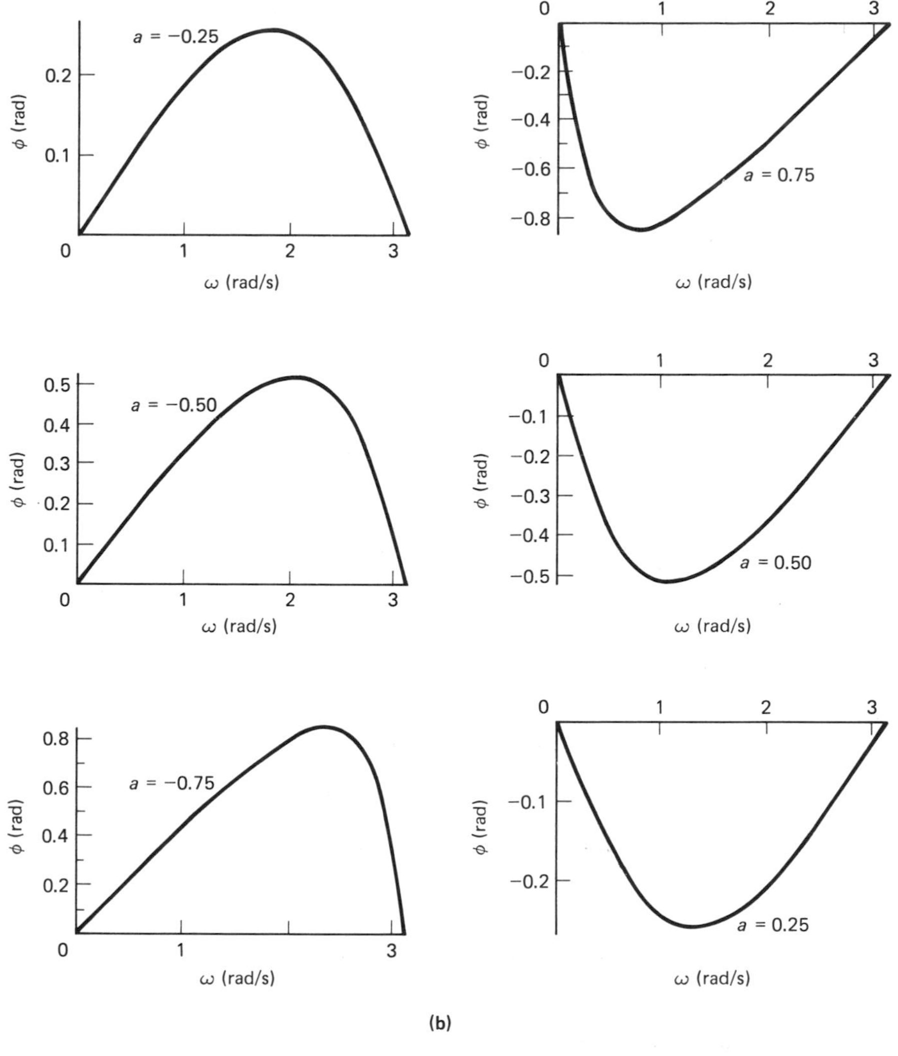

(b)

Figure 2.5 *(continued)*

$$= \tfrac{1}{2}e^{-j\omega}(e^{j\omega} + e^{-j\omega}) + e^{-j\omega}$$

$$= e^{-j\omega}(1 + \cos\omega)$$

From this expression, we have

$$|H(e^{j\omega})| = 1 + \cos\omega$$

$$\arg H(e^{j\omega}) = \phi(\omega) = -\omega$$

It should be noted that the phase response is *linear*. As we will see later, this is a result of the property

$$h(n) = h(N - 1 - n)$$

where, in the previous case, $N = 3$.

At first glance, it may appear that the units for the angles are not consistent. However, we should recall that whereas ω has units of rad/s, ωT is in radians, and, in our case, $T = 1$ s. Therefore, the angle ω is to be interpreted as (ω rad/s)(1 s) = ω rad.

Time Delay

Another important quantity used in evaluating system behavior is the *time delay* or *group delay* $\tau(\omega)$, defined by

$$\tau(\omega) = -\frac{d\phi(\omega)}{d\omega} \tag{2.76}$$

If ω is in rad/s, then $\tau(\omega)$ has the unit s^{-1}. For the previous example, $\phi(\omega) = -\omega$ and

$$\tau(\omega) = -\frac{d}{d\omega}(-\omega) = 1 \text{ s}^{-1}$$

If the phase response is linear, then the time delay is constant.

EXERCISES

2.9.1 Find the frequency response of the systems of Ex. 2.4.4 with $a = \frac{1}{2}$. Determine the magnitude and phase responses.

2.9.2 Find the frequency response for the FIR system of Ex. 2.6.1. Determine the magnitude and phase responses.

2.10 ANALYSIS OF A SECOND-ORDER SYSTEM

Impulse Response

Let us consider the second-order system

$$y(n) - y(n-1) + \tfrac{3}{16}y(n-2) = x(n) - \tfrac{1}{2}x(n-1) \tag{2.77}$$

The corresponding characteristic equation is

$$q^2 - q + \tfrac{3}{16} = 0 \tag{2.78}$$

with roots $q_1 = \frac{1}{4}$ and $q_2 = \frac{3}{4}$. Consequently, the impulse response has the form

$$h(n) = [A_1(\tfrac{1}{4})^n + A_2(\tfrac{3}{4})^n]u(n) \tag{2.79}$$

From the difference equation, we see that

$$h(0) = x(0) = 1$$

$$h(1) - h(0) = -\tfrac{1}{2}x(0)$$

Imposing these conditions on Eq. (2.79), we obtain the equations

$$A_1 + A_2 = 1$$

$$A_1(\tfrac{1}{4}) + A_2(\tfrac{3}{4}) = \tfrac{1}{2}$$

whose solution is

$$A_1 = A_2 = \tfrac{1}{2}$$

The impulse response obtained from Eq. (2.79) for these constants is

$$h(n) = [\tfrac{1}{2}(\tfrac{1}{4})^n + \tfrac{1}{2}(\tfrac{3}{4})^n]u(n)$$

Since the roots of the characteristic equation have magnitudes less than unity, we know that the impulse response is bounded.

Frequency Response

The frequency response is given by

$$\begin{aligned} H(e^{j\omega}) &= \sum_{n=0}^{\infty} [\tfrac{1}{2}(\tfrac{1}{4})^n + \tfrac{1}{2}(\tfrac{3}{4})^n]e^{-j\omega n} \\ &= \tfrac{1}{2}\sum_{n=0}^{\infty} [\tfrac{1}{4}e^{-j\omega}]^n + \tfrac{1}{2}\sum_{n=0}^{\infty} [\tfrac{3}{4}e^{-j\omega}]^n \\ &= \tfrac{1}{2}\frac{1}{1 - \tfrac{1}{4}e^{-j\omega}} + \tfrac{1}{2}\frac{1}{1 - \tfrac{3}{4}e^{-j\omega}} \end{aligned}$$

In order to exhibit the magnitude and phase responses more clearly, we express $H(e^{j\omega})$ in two forms:

$$H(e^{j\omega}) = \frac{e^{j\omega}(e^{j\omega} - \tfrac{1}{2})}{(e^{j\omega} - \tfrac{1}{4})(e^{j\omega} - \tfrac{3}{4})} \tag{2.80}$$

$$= \frac{e^{j\omega}(\cos\omega - \tfrac{1}{2} + j\sin\omega)}{[\cos\omega - \tfrac{1}{4} + j\sin\omega][\cos\omega - \tfrac{3}{4} + j\sin\omega]} \tag{2.81}$$

From Eq. (2.81), we obtain the analytic expressions

$$|H(e^{j\omega})| = \left[\frac{\tfrac{5}{4} - \cos\omega}{(\tfrac{17}{16} - \tfrac{1}{2}\cos\omega)(\tfrac{25}{16} - \tfrac{3}{2}\cos\omega)}\right]^{1/2}$$

$$\phi(\omega) = \omega + \tan^{-1}\frac{\sin\omega}{\cos\omega - \tfrac{1}{2}} - \tan^{-1}\frac{\sin\omega}{\cos\omega - \tfrac{1}{4}} - \tan^{-1}\frac{\sin\omega}{\cos\omega - \tfrac{3}{4}}$$

From Eq. (2.80), we can obtain the alternate expressions

$$|H(e^{j\omega})| = \frac{|e^{j\omega} - \tfrac{1}{2}|}{|e^{j\omega} - \tfrac{1}{4}|\cdot|e^{j\omega} - \tfrac{3}{4}|}$$

$$\phi(\omega) = \omega + \arg(e^{j\omega} - \tfrac{1}{2}) - \arg(e^{j\omega} - \tfrac{1}{4}) - \arg(e^{j\omega} - \tfrac{3}{4})$$

The quantities given in these expressions have the geometric interpretation shown in Fig. 2.6.

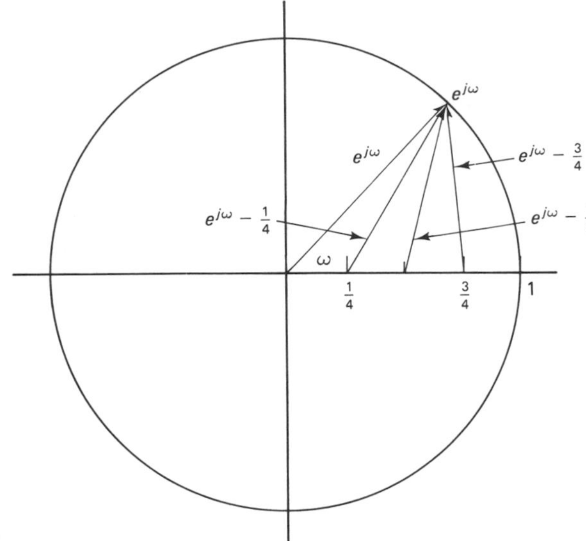

Figure 2.6 Geometric interpretation of $H(e^{j\omega})$.

EXERCISES

2.10.1 For the system

$$y(n) - \tfrac{3}{4}y(n-1) + \tfrac{1}{8}y(n-2) = x(n) - x(n-1)$$

find the impulse response and the frequency response. Give a geometric interpretation of $H(e^{j\omega})$ similar to that of Fig. 2.6.

2.10.2 For the system of Fig. 2.2(a), determine the frequency response $H(e^{j\omega})$ in terms of $H_1(e^{j\omega})$ and $H_2(e^{j\omega})$ if $h_1(n) = \alpha^n u(n)$ and $h_2(n) = \beta^n u(n)$ for the following cases:
(a) $\alpha = \frac{1}{2}$, $\beta = \frac{1}{4}$
(b) $\alpha = \beta = \frac{1}{2}$

2.11 LINEAR-PHASE FIR SYSTEMS

Desirability of Linear Phase

In applications such as speech processing and data transmission, it is desirable to design a filter that has linear phase. For example, linear phase prevents dispersion of pulses. To see the effect of linear phase, consider a discrete-time system whose system function is

$$H(e^{j\omega}) = |H(e^{j\omega})|e^{-j\omega\tau}$$

For a sinusoidal input, $x(n) = A\cos n\omega_0$, we have for the output

$$y(n) = \text{Re}[H(e^{j\omega_0})Ae^{jn\omega_0}]$$

$$= \text{Re}[|H(e^{j\omega_0})|e^{-j\tau\omega_0}Ae^{jn\omega_0}]$$

$$= A|H(e^{j\omega_0})|\cos(n - \tau)\omega_0$$

If ω_0 is in the passband of the system, then $|H(e^{j\omega_0})| = 1$ ideally, and the filter output is simply the input signal delayed by a certain amount. It is not possible to design a linear-phase analog filter from lumped parameter elements, but it is possible to design a linear-phase discrete-time filter.

Symmetry Condition

We now show that if the impulse response for a causal FIR system has the property that

$$h(n) = h(N - 1 - n) \tag{2.82}$$

where N is the length of the impulse-response sequence, then its phase is piecewise linear. Examples of impulse-response sequences for $N = 4$ and $N = 5$ that have the property of Eq. (2.82) are shown in Fig. 2.7. In both cases where $N = 4$ and $N = 5$ and in the general case, the impulse-response sequence satisfying Eq. (2.82) is symmetric about $(N - 1)/2$. For N odd, there is one sample, namely, $h[(N - 1)/2]$, that is not matched to any other sample.

For each of the cases where N is odd and N is even, we will obtain a simplified expression for the system function when $h(n)$ satisfies the symmetry property of Eq. (2.82). In each case, we will see that the phase is piecewise linear and that the magnitude is a series involving cosine functions.

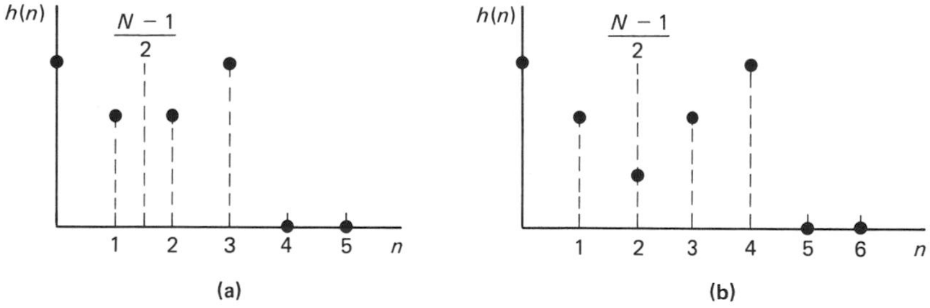

Figure 2.7 Impulse-response sequences satisfying Eq. (2.82) for (a) $N = 4$ and (b) $N = 5$.

Case Where *N* Is Odd

When N is odd, the system function becomes

$$H(e^{j\omega}) = \sum_{n=0}^{(N-3)/2} h(n)e^{-j\omega n} + h\left(\frac{N-1}{2}\right)e^{-j\omega(N-1)/2} + \sum_{n=(N+1)/2}^{N-1} h(n)e^{-j\omega n} \tag{2.83}$$

Using the symmetry property of Eq. (2.82) and changing the index $k = N - 1 - n$, we can write

$$\sum_{n=(N+1)/2}^{N-1} h(n)e^{-j\omega n} = \sum_{n=(N+1)/2}^{N-1} h(N-1-n)e^{-j\omega n}$$
$$= \sum_{k=0}^{(N-3)/2} h(k)e^{-j\omega(N-1-k)}$$

Substituting this sum, with k replaced by n, back into Eq. (2.83) results in the expression

$$H(e^{j\omega}) = h\left(\frac{N-1}{2}\right)e^{-j\omega(N-1)/2} + \sum_{n=0}^{(N-3)/2} h(n)(e^{-j\omega n} + e^{-j\omega(N-1-n)})$$
$$= e^{-j\omega(N-1)/2}\left[h\left(\frac{N-1}{2}\right) + \sum_{n=0}^{(N-3)/2} h(n)(e^{-j\omega[n-(N-1)/2]} + e^{j\omega[n-(N-1)/2]})\right]$$

By using the relation

$$e^{j\theta} + e^{-j\theta} = 2\cos\theta$$

the expression for the frequency response simplifies to

$$H(e^{j\omega}) = e^{-j\omega(N-1)/2}\left\{h\left(\frac{N-1}{2}\right) + 2\sum_{n=0}^{(N-3)/2} h(n)\cos\omega\left(n - \frac{N-1}{2}\right)\right\} \qquad N \text{ odd} \tag{2.84}$$

or, since $\cos(-\theta) = \cos\theta$, to the equivalent expression

$$H(e^{j\omega}) = e^{-j\omega(N-1)/2}\left\{h\left(\frac{N-1}{2}\right) + 2\sum_{n=0}^{(N-3)/2} h(n)\cos\omega\left(\frac{N-1}{2} - n\right)\right\} \tag{2.85}$$

Pseudomagnitude and Phase

Comparing Eq. (2.85) with the polar representation of $H(e^{j\omega})$, we see that the magnitude is

$$|H(e^{j\omega})| = |H_1(\omega)|$$

where $H_1(\omega)$ is a real quantity given by

$$H_1(\omega) = h\left(\frac{N-1}{2}\right) + 2\sum_{n=0}^{(N-3)/2} h(n)\cos\omega\left(\frac{N-1}{2} - n\right) \tag{2.86}$$

and the phase response is

$$\phi(\omega) = \begin{cases} -\left(\frac{N-1}{2}\right)\omega & H_1(\omega) > 0 \\ -\left(\frac{N-1}{2}\right)\omega + \pi & H_1(\omega) < 0 \end{cases} \tag{2.87}$$

Consequently, the phase is piecewise linear. We refer to $H_1(\omega)$ as the *pseudomagnitude* since it is real but not necessarily positive.

Examples Where *N* Is Odd

As an example of the frequency response of a linear-phase FIR system, consider

$$y(n) = ax(n) + bx(n-1) + cx(n-2) + bx(n-3) + ax(n-4) \tag{2.88}$$

For this system, $N = 5$, $h(0) = h(4) = a$, $h(1) = h(3) = b$, and $h(2) = c$. The corresponding frequency response, determined from Eqs. (2.84) and (2.86), has pseudomagnitude

$$\begin{aligned} H_1(\omega) &= 2h(0)\cos 2\omega + 2h(1)\cos \omega + h(2) \\ &= 2a\cos 2\omega + 2b\cos\omega + c \end{aligned}$$

For this simple case, we can determine the values of a, b, and c to give a somewhat desirable magnitude response. For example, we could specify conditions at $\omega = 0$, $\omega = \pi/2$, and $\omega = \pi$. The resulting constraints on $H_1(\omega)$ are

$$H_1(0) = 2a + 2b + c \tag{2.89}$$

$$H_1(\pi/2) = -2a + c \tag{2.90}$$

$$H_1(\pi) = 2a - 2b + c \tag{2.91}$$

If we subtract Eq. (2.91) from Eq. (2.89), we obtain

$$b = [H_1(0) - H_1(\pi)]/4 \tag{2.92}$$

Next, adding Eqs. (2.89) and (2.91) results in

$$4a + 2c = H_1(0) + H_1(\pi)$$

which when solved simultaneously with Eq. (2.90) results in

$$a = [H_1(0) + H_1(\pi) - 2H_1(\pi/2)]/8 \tag{2.93}$$

$$c = [H_1(0) + H_1(\pi) + 2H_1(\pi/2)]/4 \tag{2.94}$$

To complete this example, let us take $H_1(0) = \frac{5}{4}$, $H_1(\pi/2) = 0$, and $H_1(\pi) = -\frac{1}{4}$. Then from Eqs. (2.92)–(2.94), we obtain

$$\begin{aligned} a &= h(0) = (\tfrac{5}{4} - \tfrac{1}{4} - 0)/8 = \tfrac{1}{8} \\ b &= h(1) = [\tfrac{5}{4} - (-\tfrac{1}{4})]/4 = \tfrac{3}{8} \\ c &= h(2) = (\tfrac{5}{4} - \tfrac{1}{4} + 0)/4 = \tfrac{1}{4} \end{aligned}$$

For these values of a, b, and c, the pseudomagnitude is given by

$$H_1(\omega) = \tfrac{1}{4}(\cos 2\omega + 3\cos\omega + 1)$$

In order to determine the phase, we have to know where $H_1(\omega)$ changes sign. For this simple case, we can use the identity

$$\cos 2\omega = 2\cos^2\omega - 1 \tag{2.95}$$

to obtain

$$H_1(\omega) = \tfrac{1}{4}\cos\omega(2\cos\omega + 3)$$

Since $H_1(\omega) > 0$ for $0 \le \omega \le \pi/2$ and $H_1(\omega) < 0$ for $\pi/2 < \omega \le \pi$, for the phase response, we have

$$\phi(\omega) = \begin{cases} -2\omega & 0 \le \omega < \pi/2 \\ -2\omega + \pi & \pi/2 < \omega \le \pi \end{cases}$$

As another example related to the system of Eq. (2.88), suppose we start with $h(0) = h(4) = a = 1$, $h(1) = h(3) = b = \frac{5}{4}$, and $h(2) = c = -\frac{1}{4}$. For this impulse-response sequence, we have

$$H_1(\omega) = 2\cos 2\omega + \tfrac{5}{2}\cos\omega - \tfrac{1}{4}$$

and it follows that $H_1(0) = 4.25$, $H_1(\pi/2) = -2.25$, and $H_1(\pi) = -0.75$. If we want to determine where $H_1(\omega) = 0$, we use the identity of Eq. (2.95) to obtain the quadratic equation

$$H_1(\omega) = 2(2\cos^2\omega - 1) + \tfrac{5}{2}\cos\omega - \tfrac{1}{4} = 0$$

or equivalently

$$\cos^2\omega + \tfrac{5}{8}\cos\omega - \tfrac{9}{16} = 0$$

The solutions of this equation are $\cos\omega = -\frac{9}{8}, \frac{1}{2}$. Since $\cos\omega = -\frac{9}{8}$ does not give a real value of ω, we must have $\cos\omega = \frac{1}{2}$ and, therefore, $\omega = \pi/3$ rad is the only zero of $H_1(\omega)$ in the interval $0 \le \omega \le \pi$.

Case Where *N* Is Even

For the case where N is even, the symmetry condition of Eq. (2.82) allows us to write

$$\begin{aligned} H(e^{j\omega}) &= \sum_{n=0}^{N/2-1} h(n)e^{-j\omega n} + \sum_{n=N/2}^{N-1} h(n)e^{-j\omega n} \\ &= \sum_{n=0}^{N/2-1} h(n)e^{-j\omega n} + \sum_{n=N/2}^{N-1} h(N-1-n)e^{-j\omega n} \end{aligned}$$

If we let $k = N - 1 - n$ in the last summation, we obtain

$$\begin{aligned} H(e^{j\omega}) &= \sum_{n=0}^{N/2-1} h(n)e^{-j\omega n} + \sum_{k=0}^{N/2-1} h(k)e^{-j\omega(k-N+1)} \\ &= \sum_{n=0}^{N/2-1} h(n)[e^{-j\omega n} + e^{j\omega(n-N+1)}] \end{aligned}$$

which can be simplified to

$$H(e^{j\omega}) = e^{-j\omega(N-1)/2} H_1(\omega)$$

where

$$H_1(\omega) = 2\sum_{n=0}^{N/2-1} h(n)\cos\omega\left(\frac{N-1}{2} - n\right) \tag{2.96}$$

The procedure for obtaining this result is the same as was used to obtain Eq. (2.85) for N odd. Once again, we see that the symmetry condition of Eq. (2.82) has resulted in a piecewise linear phase.

Examples Where N Is Even

As an example of a linear-phase FIR system with N even, consider

$$y(n) = ax(n) + bx(n-1) + bx(n-2) + ax(n-3)$$

For this system, $N = 4$, $h(0) = h(3) = a$, and $h(1) = h(2) = b$. Using Eq. (2.96), we obtain

$$\begin{aligned} H_1(\omega) &= 2h(0)\cos\frac{3}{2}\omega + 2h(1)\cos\frac{\omega}{2} \\ &= 2\left(a\cos\frac{3}{2}\omega + b\cos\frac{\omega}{2}\right) \end{aligned}$$

and so the frequency response is

$$H(e^{j\omega}) = 2e^{-j3\omega/2}\left(a\cos\frac{3}{2}\omega + b\cos\frac{\omega}{2}\right)$$

If we use the trigonometric identity

$$2\cos A\cos B = \cos(A+B) + \cos(A-B)$$

with $A = \omega$ and $B = \omega/2$, then

$$\begin{aligned} \cos\frac{3}{2}\omega &= 2\cos\omega\cos\frac{\omega}{2} - \cos\frac{\omega}{2} \\ &= (2\cos\omega - 1)\cos\frac{\omega}{2} \end{aligned}$$

When this is substituted in the previous expression for $H_1(\omega)$, the result is

$$H_1(\omega) = 2(2a\cos\omega + b - a)\cos\omega/2$$

which is of a different form from that given in Eq. (2.96). The zeros of $H_1(\omega)$ occur when $\cos\omega/2 = 0$ and when $2\text{a}\cos\omega + b - a = 0$.

Summary

In this section, we have shown that if the symmetry condition $h(N-1-n) = h(n)$ is satisfied, then the phase response is piecewise linear. The frequency response is given by

$$H(e^{j\omega}) = e^{-j\omega(N-1)/2}H_1(\omega)$$

where the pseudomagnitude $H_1(\omega)$ is given by Eq. (2.85) for N odd and by Eq. (2.96) for N even. In the ideal case, where

$$|H_1(\omega)| = \begin{cases} 1 & |\omega| \le \omega_c \\ 0 & \omega_c < |\omega| \le \pi \end{cases}$$

an input $x(n) = A \cos \omega_0 n$ results in the response $y(n) = A \cos [n - (N-1)/2]\omega_0$ if $|\omega_0| < \omega_c$ and $y(n) = 0$ if $\omega_c < |\omega_0| \le \pi$. Consequently, the sinusoid with frequency ω_0 in the range $|\omega_0| \le \omega_c$ is simply delayed by the amount $[(N-1)/2]\omega_0$.

EXERCISES

2.11.1 If $h(0) = h(2) = a$, $h(1) = b$, and $N = 3$, determine a and b so that $H_1(0) = 0$ and $H_1(\pi) = 1$. For these values of a and b, find $H_1(\omega)$ and $\phi(\omega)$.

2.11.2 A linear-phase FIR filter has

$$H_1(\omega) = \cos \frac{\omega}{2} + \frac{1}{2} \cos \frac{3\omega}{2}$$

Determine the impulse response $h(n)$.

2.11.3 Show that if $h(N-1-n) = -h(n)$, then for N odd,

$$H(e^{j\omega}) = e^{-j\omega(N-1)/2} e^{j\pi/2} \sum_{n=1}^{(N-1)/2} c(n) \sin n\omega$$

where $c(n) = 2h[(N-1)/2 - n]$, $n = 1, 2, \ldots, (N-1)/2$ and $h[(N-1)/2] = 0$.

PROBLEMS

2.1 Find the impulse response for the following causal systems:

(a) $y(n) - \frac{1}{2}y(n-1) = x(n)$

(b) $y(n) - \frac{1}{2}y(n-1) = x(n) + x(n-1)$

(c) $y(n) - y(n-1) - y(n-2) = x(n-2)$

2.2 Find the impuse response for the following causal systems:

(a) $y(n) = x(n) - 2x(n-1) + x(n-2)$

(b) $y(n) = x(n) + 2x(n-1) - x(n-2)$

2.3 Prove or disprove that the following systems are linear.

(a) $\mathcal{T}\{x(n)\} = g(n)x(n)$

(b) $\mathcal{T}\{x(n)\} = \Sigma_{k=0}^{n} x(k)$

(c) $\mathcal{T}\{x(n)\} = x(n - n_0)$, n_0 a fixed integer

(d) $\mathcal{T}\{x(n)\} = a[x(n)]^2 + bx(n)$

(e) $\mathcal{T}\{x(n)\} = n[x(n)]^2$

(f) $\mathcal{T}\{x(n)\} = x_e(n)$, the even part of $x(n)$

(g) $\mathcal{T}\{x(n)\} = |x(n)|$

2.4 For a linear system, show that

$$\mathcal{T}\left\{ \sum_{k=-N}^{N} a_k x_k(n) \right\} = \sum_{k=-N}^{N} a_k \mathcal{T}\{x_k(n)\}$$

2.5 Use recursion to find the impuse response of the causal system

$$y(n) + y(n-1) = x(n) - 2x(n-1)$$

Then find the step response using **(a)** recursion and **(b)** the convolution sum.

2.6 Show that the central difference formula is exact if

$$x_a(t) = at^2 + bt + c$$

2.7 Using the algorithm in Eq. (2.13) to approximate $3^{1/2}$, show that y(4) $\approx$ 1.732058 (the value obtained from a calculator).

2.8 Find the impulse response of the causal system

$$y(n) - 2\cos\theta\, y(n-1) + y(n-2) = \sin\theta\, x(n-1)$$

where θ is a constant. Use mathematical induction and the identity

$$2\sin\alpha\cos\beta = \sin(\alpha+\beta) + \sin(\alpha-\beta)$$

Then use the convolution sum to find the step response.

2.9 Determine which of the systems of Prob. 2.3 are time-invariant.

2.10 The response of a causal system depends on the present and past values of the input sequence. Apply this property to the convolution sum for a linear time-invariant system and show that $h(n) = 0$ for $n < 0$.

2.11 Represent the system describing a square-root algorithm in Eq. (2.13) by

$$\mathcal{T}\{x(n)\} = y(n)$$

and let $y_1(n)$ be defined by

$$\mathcal{T}\{x(n-m)\} = y_1(n)$$

If $y(n) = 0$ for $n < 0$ with $y(0) = a$ and $y_1(n) = 0$ for $n < m$ with $y_1(m) = a$, show that the system is time-invariant.

2.12 Use recursion and find the impulse response of the causal system

$$y(n) - (2n-1)y(n-1) - \xi^2 y(n-2) = x(n) + \xi x(n-1)$$

where ξ is a parameter, and $y(n)$ is somtimes called a *Bessel* polynomial (see Sec. 5.10).

2.13 Find the response to the causal system

$$y(n) - n^2 y(n-1) = \delta(n-m) \qquad m \geq 0$$

Use **(a)** recursion and **(b)** $y(k+m) = [(k+m)!]^2 w(k+m)$.

2.14 Show that

(a) $x_1(n) * [x_2(n) + x_3(n)] = x_1(n) * x_2(n) + x_1(n) * x_3(n)$

(b) $x_1(n) * [x_2(n) * x_3(n)] = [x_1(n) * x_2(n)] * x_3(n)$

2.15 Find the impulse response of the causal system described by

$$y(n) + y(n-1) = x(n) - x(n-1)$$

Also find the step response using **(a)** recursion and **(b)** the convolution sum.

2.16 Find the impulse response of the system of Eq. (2.35) and verify the solution $y(n)$ given in Eq. (2.36).

2.17 A FIR system has the impulse-response sequence

$$h(n) = \begin{cases} (\frac{1}{2})^n & 0 \leq n \leq 3 \\ 0 & \text{otherwise} \end{cases}$$

Find the response to **(a)** a step input, **(b)** an input $x(n) = (-\frac{1}{2})^n u(n)$, and **(c)** $x(n) = \cos n\omega$, where ω is a parameter.

2.18 Repeat Prob. 2.17 if

$$h(n) = \begin{cases} \dfrac{\sin(n-1)}{\pi(n-1)} & n = 0, 2 \\ \dfrac{1}{\pi} & n = 1 \\ 0 & \text{otherwise} \end{cases}$$

2.19 Find the response for the following causal systems using the methods of Sec. 2.7.
(a) $y(n) + y(n-1) = \delta(n) - \delta(n-1)$
(b) $y(n) + y(n-1) = u(n) - u(n-1)$
(c) $y(n) + 3y(n-1) + 2y(n-2) = \delta(n)$
(d) $y(n) + 3y(n-1) + 2y(n-2) = u(n)$
(e) $y(n) + 3y(n-1) + 2y(n-2) = \delta(n) + 2\delta(n-1)$
(f) $y(n) + 3y(n-1) + 2y(n-2) = \delta(n) + 2\delta(n-1) - \delta(n-2)$
(g) $y(n) + y(n-1) = \cos 2n$. Hint: Try $y_p(n) = A\cos 2n + B\sin 2n$.
(h) $y(n) + 2y(n-1) = n + 1$
(i) $y(n) - (n-1) = u(n)$. Hint: Try $y_p(n) = An$.
(j) $y(n) + 4y(n-1) + 4y(n-2) = \delta(n) + 4\delta(n-1) + 3\delta(n-2)$
(k) $y(n) - 2y(n-1) + y(n-2) = n$

2.20 Find the impulse response of the system of Prob. 2.19(j) using superposition.

2.21 Find the response of the following causal systems:
(a) $y(n) + 2y(n-1) + 2y(n-2) = \delta(n)$
(b) $y(n) - 4y(n-1) + 16y(n-2) = u(n)$

2.22 Show that if $y_1(n) = \mathcal{T}\{e^{j\omega n}\}$, where $\mathcal{T}\{\ \}$ is given by Eq. (2.29), then

$$y(n) = \mathcal{T}\{\cos \omega n\} = \mathrm{Re}[y_1(n)]$$

Write $y_1(n) = \mathrm{Re}[y_1(n)] + j\,\mathrm{Im}[y_1(n)] = v_1(n) + jv_2(n)$ and note that $\mathrm{Re}[ay_1(n)] = a\,\mathrm{Re}[y_1(n)]$ for a real.

2.23 Find the frequency response of the systems given by
(a) $y(n) - \frac{1}{2}y(n-1) = x(n)$
(b) $y(n) + \frac{1}{2}y(n-1) = x(n) - x(n-1)$
(c) $y(n) = x(n) - x(n-1) + x(n-2)$
In each case, determine the magnitude, phase, and time delay.

2.24 Given

$$H(e^{j\omega}) = \frac{e^{j\omega} - a}{e^{j\omega} - b}$$

where a and b are real with $a \neq b$. Show that $|H(e^{j\omega})|^2$ is constant if $ab = 1$ and determine its value. Determine $\phi(\omega)$ in simplified form and then find $\tau(\omega)$.

2.25 Given

$$H(e^{j\omega}) = \frac{Q(e^{j\omega})}{P(e^{j\omega})} = \frac{\sum_{k=0}^{N} a_k e^{-jk\omega}}{\sum_{k=0}^{N} a_{N-k} e^{-jk\omega}} = \frac{\sum_{k=0}^{N} a_k e^{-jk\omega}}{\sum_{k=0}^{N} a_k e^{-j(N-k)\omega}}$$

where the coefficients a_k, $k = 0, 1, \ldots, N$, are real. Since

$$|H(e^{j\omega})|^2 = H(e^{j\omega})H(e^{-j\omega})$$

show that $|H(e^{j\omega})|^2 = 1$. Note that

$$\sum_{k=0}^{N} a_k e^{-jk\omega} = e^{-jN\omega} \sum_{k=0}^{N} a_{N-k} e^{jk\omega}$$

A frequency response having constant magnitude is called an *all-pass* function. If

$$P(e^{j\omega}) = 3 - e^{-j\omega}$$

determine $Q(e^{j\omega})$.

2.26 Show that the frequency response of the system

$$3y(n) - y(n-1) = -x(n) + 3x(n-1)$$

is an all-pass function. Find the time delay in the series form

$$\tau(\omega) = \sum_{k=0}^{\infty} a_k \omega^{2k}$$

2.27 Repeat Prob. 2.26 for the system

$$10y(n) - 5y(n-1) + y(n-2) = x(n) - 5x(n-1) + 10x(n-2)$$

Note in particular that $a_1 = 0$. This time delay is said to be *maximally flat* [77, 181, 183].

2.28 The frequency response of a linear-phase FIR filter is given by

$$H(e^{j\omega}) = e^{-j3\omega}(0.21145 + 0.386568 \cos \omega + 0.28856 \cos 2\omega + 0.21432 \cos 2\omega)$$

Determine the filter coefficients $h(n)$.

2.29 The frequency response of a linear-phase FIR filter is given by

$$H(e^{j\omega}) = e^{-j2\omega}(0.30256 + 0.5000 \cos \omega + 0.29744 \cos 2\omega)$$

Determine the filter coefficients $h(n)$. Sketch the graph of $e^{j2\omega}H(e^{j\omega})$.

2.30 If the frequency response of a linear-phase filter is

$$H(e^{j\omega}) = e^{-j3\omega/2} \cos \frac{\omega}{2}(0.78679 \cos \omega - 0.31320)$$

determine $h(n)$.

2.31 Determine the frequency response $H(e^{j\omega})$ for the causal system

$$y(n) - \tfrac{1}{4}y(n-1) - \tfrac{3}{8}y(n-1) = x(n) + x(n-1)$$

2.32 Two systems with impulse responses $h_1(n)$ and $h_2(n)$ are cascaded as shown in Fig. 2.2(a). Determine $h(n)$ using the convolution sum if $h_1(n) = \alpha^n u(n)$ and $h_2(n) = \beta^n u(n)$, $\alpha \neq \beta$. Then determine $Y(e^{j\omega})$ in terms of $X(e^{j\omega})$, $H_1(e^{j\omega})$, and $H_2(e^{j\omega})$ for the special case $\alpha = \frac{1}{2}$, $\beta = \frac{1}{4}$, and $x(n) = (-\frac{1}{2})^n u(n)$.

THE z-TRANSFORM

3.1 INTRODUCTION

In the study of discrete-time systems described by a linear difference equation with constant coefficients such as Eq. (2.29), the z-transform has a role similar to that of the Laplace transform in continuous-time systems. Applying the z-transform to an equation such as Eq. (2.29) transforms the difference equation into an algebraic equation in the z-transforms of $y(n)$ and $x(n)$. Knowing $x(n)$ and its z-transform allows us to solve the algebraic equation for the z-transform of $y(n)$ and ultimately find $y(n)$. In many cases, we do not know $x(n)$, but can obtain valuable information about the system's behavior from the ratio of the z-transforms of $y(n)$ and $x(n)$.

In discussing z-transforms, we could use the approach of Laplace and De Moivre in their probability studies or use the approach taken in the study of sampled-data control systems, which makes use of a train of impulses. The latter concept is illustrated in Prob. 3.42. We take the former approach and define the z-transform in a purely mathematical sense. What we call the z-transform is called a *generating function* in various areas of mathematics.

3.2 DEFINITION OF THE z-TRANSFORM

Definitions

Given the sequence $x(n)$, we define its z-transform by

$$Z\{x(n)\} = X(z) = \sum_{n=-\infty}^{\infty} x(n)z^{-n} \tag{3.1}$$

where z is taken to be a complex variable. This expression is sometimes referred to as the *two-sided* z-transform since the summation is over all the integers. If $x(n)$ is a causal sequence, $x(n) = 0$ for $n < 0$, then its z-transform is

$$Z\{x(n)\} = X(z) = \sum_{n=0}^{\infty} x(n)z^{-n} \tag{3.2}$$

This expression is sometimes referred to as the *one-sided* z-transform. For the most part, we are concerned with causal sequences, so that we will be using Eq. (3.2) most of the time. In Sec. 1.4, we discussed the convergence of the series of Eqs. (3.1) and (3.2) and obtained the sum of the series of Eq. (3.2) when it is a geometric series with $x(n)z^{-n} = z_0^n$. Clearly, the z-transform exists only for the values of z for which the series of Eq. (3.1) [or of Eq. (3.2)] converges. The series of Eq. (3.1) is known as a *Laurent* series and converges to a unique function of z within its region of convergence.

Convergence

The series of Eq. (3.2) is said to converge *absolutely* when the series of real numbers

$$\sum_{n=0}^{\infty} |x(n)z^{-n}| \tag{3.3}$$

converges. It is well known that a series that converges absolutely also converges. That is, convergence of Eq. (3.3) guarantees convergence of Eq. (3.2). Two tests that are available for testing the convergence of real series, and are therefore applicable to Eq. (3.3), are the ratio test and the root test, which follow.

> *Ratio Test*. If the sequence $\{|x_{n+1}/x_n|\}$ converges to $L < 1$, then the series $\Sigma_{n=0}^{\infty} x_n$ converges absolutely.
>
> *Root Test*. If the sequence $\{(|x_n|)^{1/n}\}$ converges to $L < 1$, then the series $\Sigma_{n=0}^{\infty} x_n$ converges absolutely.

In applying the ratio test to Eq.(3.2), we see that

$$\lim_{n\to\infty} |x_{n+1}/x_n| = \lim_{n\to\infty} \left|\frac{x(n+1)z^{-n-1}}{x(n)z^{-n}}\right|$$

$$= \left[\lim_{n\to\infty} \left|\frac{x(n+1)}{x(n)}\right|\right] |z|^{-1}$$

The convergence criterion requires that

$$|z|^{-1} \lim_{n\to\infty} \left|\frac{x(n+1)}{x(n)}\right| < 1$$

and from this, we obtain the region of convergence as

$$|z| > \lim_{n\to\infty} \left|\frac{x(n+1)}{x(n)}\right| = R$$

The series converges outside the circle with center at the origin and radius R. This is illustrated in Fig. 3.1.

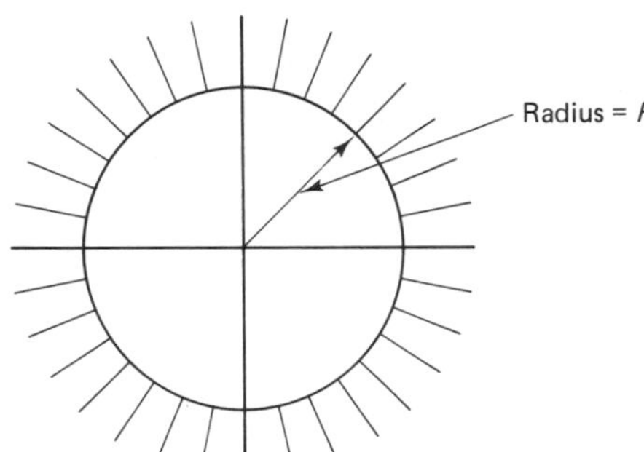

Figure 3.1 Region of convergence of Eq. (3.2).

Examples of z-Transforms

We now give several examples of z-transforms of useful sequences. First, consider the unit-impulse sequence $\delta(n)$. Its z-transform is given by

$$Z\{\delta(n)\} = \sum_{n=-\infty}^{\infty} \delta(n) z^{-n} = 1 \tag{3.4}$$

As was mentioned in Sec. 1.4, we not only must know when the series for $X(z)$ converges, but we normally need to know its sum as well. One series whose sum is known is the geometric series of Eq. (1.24), for which

$$\sum_{n=0}^{\infty} z_0^n = \frac{1}{1-z_0} \qquad |z_0| < 1 \tag{3.5}$$

We use this relation frequently in this chapter to find z-transforms.

One of the most important sequences that we consider is the geometric sequence $x(n) = a^n u(n)$. Its z-transform is given by

$$Z\{a^n u(n)\} = \sum_{n=0}^{\infty} a^n z^{-n} = \sum_{n=0}^{\infty} (az^{-1})^n$$

This is a geometric series, such as that considered in Eq. (3.5), with the ratio $z_0 = az^{-1}$. Consequently, the series converges for $|az^{-1}| < 1$ to the sum

$$Z\{a^n u(n)\} = \frac{1}{1-az^{-1}} \qquad |az^{-1}| < 1 \tag{3.6}$$

The unit-step sequence $u(n)$ is a special case of the geometric sequence with $a = 1$. Therefore, its z-transform is obtained from Eq. (3.6) with $a = 1$ and is given by

$$Z\{u(n)\} = \frac{1}{1 - z^{-1}} \qquad |z^{-1}| < 1 \tag{3.7}$$

Another special case of the geometric series arises when $a = e^{j\theta}$. Then, from Eq. (3.6),

$$Z\{e^{jn\theta}u(n)\} = \frac{1}{1 - e^{j\theta}z^{-1}} \qquad |z^{-1}| < 1 \tag{3.8}$$

and we can obtain the z-transforms

$$Z\{\cos n\theta\, u(n)\} = \frac{1 - z^{-1}\cos\theta}{1 - 2z^{-1}\cos\theta + z^{-2}} \tag{3.9}$$

$$Z\{\sin n\theta\, u(n)\} = \frac{z^{-1}\sin\theta}{1 - 2z^{-1}\cos\theta + z^{-2}} \tag{3.10}$$

by equating real and imaginary parts in Eq. (3.8), where we assume that z is real. (See Prob. 3.2.)

Inverse *z*-Transform

As we will see later, we need to find the sequence $x(n)$ that corresponds to a given z-transform $X(z)$. We call the sequence $x(n)$ the *inverse* z-transform of $X(z)$ and write

$$Z^{-1}\{X(z)\} = x(n) \tag{3.11}$$

Using some of the transforms that we have developed thus far, we can write

$$Z^{-1}\left\{\frac{1}{1 - z^{-1}}\right\} = u(n) \qquad |z| > 1 \tag{3.12}$$

$$Z^{-1}\{1\} = \delta(n) \tag{3.13}$$

$$Z^{-1}\left\{\frac{1}{1 - az^{-1}}\right\} = a^n u(n) \qquad |az^{-1}| < 1 \tag{3.14}$$

Note that we have included the values of z for which $X(z)$ is defined.

We will discuss methods for obtaining inverse z-transforms in a later section, but we note here that one method is to simply obtain the series expansion of $X(z)$ and identify $x(n)$ as the coefficient of z^{-n}. This method is used in what follows.

Example Involving a Noncausal Sequence

To see why specification of the range of values of z for which $X(z)$ is defined is important, let us consider

$$X(z) = \frac{1}{1 - z^{-1}}$$

If the region of definition is $|z| > 1$, then we obtain Eq. (3.12). However, if $X(z)$ is defined for $|z| < 1$, then we must obtain a series expansion that is valid for $|z| < 1$. The series must contain positive powers of z and we must write

$$\frac{1}{1 - z^{-1}} = -\frac{z}{1 - z} = -z \sum_{n=0}^{\infty} z^n = \sum_{n=1}^{\infty} (-1)z^n$$
$$= \sum_{n=-\infty}^{-1} (-1)z^{-n}$$

Consequently, $x(n) = 0$ for $n \geq 0$, and $x(n) = -1$ for $n < 0$, and we can write

$$Z^{-1}\left\{\frac{1}{1 - z^{-1}}\right\} = -u(-n - 1) \qquad |z| < 1$$

which is a noncausal sequence.

Convergence of the Two-Sided z-Transform

If it becomes necessary to consider a series of the form of Eq. (3.1), then we write

$$\sum_{n=-\infty}^{\infty} x(n)z^{-n} = \sum_{n=-\infty}^{-1} x(n)z^{-n} + \sum_{n=0}^{\infty} x(n)z^{-n}$$
$$= \sum_{n=1}^{\infty} x(-n)z^{n} + \sum_{n=0}^{\infty} x(n)z^{-n}$$

and determine the region of convergence for each series. The first series converges for $|z| < R_2$ and the second for $|z| > R_1$. Then Eq. (3.1) converges for $R_1 < |z| < R_2$ provided $R_1 < R_2$.

In general, causal sequences have z-transforms of the form

$$\sum_{n=0}^{\infty} x(n)z^{-n}$$

which converges for $|z| > R_1$, and noncausal sequences produce positive powers of z in $X(z)$, resulting in an annular region of convergence. In most cases that we will consider, $x(n)$ is a causal sequence. Consequently, we assume that the z-transforms are valid in a region $|z| > R$, unless it is otherwise stated.

EXERCISES

3.2.1 Find the z-transform of

(a) $u(n - 1)$

(b) $2^n \delta(n - 3)$

(c) $\cos(n\pi/2)\, u(n)$

3.2.2 Find the inverse z-transform of

(a) $\dfrac{1}{1 - 2z^{-1}}$

(b) $\dfrac{1 - \frac{1}{2}z^{-1}}{1 - z^{-1} + z^{-2}}$

(c) $\dfrac{(\frac{1}{2})^{1/2}z^{-1}}{1 - (2)^{1/2}z^{-1} + z^{-2}}$

3.2.3 If $X(z) = z^{-1}/(1 - 2z^{-1})$ for $|z| < 2$, use the appropriate series and find $x(n)$. Note that $x(n)$ is a noncausal sequence.

3.3 PROPERTIES OF THE z-TRANSFORM

In this section, we consider some of the more important properties of the z-transform. We then use these properties in conjunction with the transforms derived in Sec. 3.2 to obtain other useful transforms. These properties and some of the more useful transforms are summarized in Table 3.1.

TABLE 3.1 TABLE OF z-TRANSFORMS AND OPERATIONS

1. $x(n)$	$X(z)$
2. $\delta(n)$	1
3. $a^n u(n)$	$\dfrac{1}{1 - az^{-1}}$
4. $a^n \cos n\theta\, u(n)$	$\dfrac{1 - az^{-1}\cos\theta}{1 - 2az^{-1}\cos\theta + a^2z^{-2}}$
5. $a^n \sin n\theta\, u(n)$	$\dfrac{az^{-1}\sin\theta}{1 - 2az^{-1}\cos\theta + a^2z^{-2}}$
6. $na^n u(n)$	$\dfrac{az^{-1}}{(1 - az^{-1})^2}$
7. $n(n - 1)a^n u(n)$	$\dfrac{2a^2z^{-2}}{(1 - az^{-1})^3}$
8. $a_1x_1(n) + a_2x_2(n)$	$a_1X_1(z) + a_2X_2(z)$
9. $x(n - m)$	$z^{-m}X(z)$
10. $x_1(n) * x_2(n)$	$X_1(z)X_2(z)$
11. $a^n x(n)$	$X(a^{-1}z)$
12. $nx(n)$	$z^{-1}\dfrac{dX(z)}{dz^{-1}}$

Linearity

First, we show that taking the z-transform is a linear operation. This is proved by noting that

$$
\begin{aligned}
Z\{a_1x_1(n) + a_2x_2(n)\} &= \sum_{n=-\infty}^{\infty} \{a_1x_1(n) + a_2x_2(n)\}z^{-n} \\
&= a_1 \sum_{n=-\infty}^{\infty} x_1(n)z^{-n} + a_2 \sum_{n=-\infty}^{\infty} x_2(n)z^{-n} \qquad (3.15) \\
&= a_1 Z\{x_1(n)\} + a_2 Z\{x_2(n)\}
\end{aligned}
$$

We can also show that the inverse z-transform is linear. That is,

$$Z^{-1}\{a_1 X_1(z) + a_2X_2(z)\} = a_1 Z^{-1}\{X_1(z)\} + a_2 Z^{-1}\{X_2(z)\} \qquad (3.16)$$

From Eq. (3.15), we know that

$$Z\{a_1x_1(n) + a_2x_2(n)\} = a_1 X_1(z) + a_2 X_2(z)$$

and, therefore,

$$a_1x_1(n) + a_2x_2(n) = Z^{-1}\{a_1 X_1(z) + a_2 X_2(z)\}$$

Replacing $x_1(n)$ and $x_2(n)$ by $Z^{-1}\{X_1(z)\}$ and $Z^{-1}\{X_2(z)\}$, respectively, results in Eq. (3.16). This result is quite useful in determining inverse z-transforms, as illustrated by the following example.

Consider the transform

$$X(z) = \frac{1}{(1 - \frac{1}{2}z^{-1})(1 - \frac{1}{4}z^{-1})}$$

To find $x(n)$, we use the linearity property of Eq. (3.16) and known transforms from Sec. 3.2, particularly Eq. (3.6). To carry out the inversion, we first obtain the partial-fraction expansion of $X(z)$, which has the form

$$X(z) = \frac{A}{1 - \frac{1}{2}z^{-1}} + \frac{B}{1 - \frac{1}{4}z^{-1}}$$

The coefficient A can be determined by multiplying both sides of this equation by $(1 - \frac{1}{2}z^{-1})$ and letting $z^{-1} \to 2$. In the limiting process, the term containing B approaches zero and only A is left on the right-hand side. A similar procedure can be used to determine B. Consequently, we obtain

$$A = \lim_{z^{-1}\to 2} (1 - \tfrac{1}{2}z^{-1})\frac{1}{(1 - \frac{1}{2}z^{-1})(1 - \frac{1}{4}z^{-1})} = 2$$

$$B = \lim_{z^{-1}\to 4} (1 - \tfrac{1}{4}z^{-1})\frac{1}{(1 - \frac{1}{2}z^{-1})(1 - \frac{1}{4}z^{-1})} = -1$$

and it follows that

$$x(n) = 2Z^{-1}\left\{\frac{1}{1 - \frac{1}{2}z^{-1}}\right\} - Z^{-1}\left\{\frac{1}{1 - \frac{1}{4}z^{-1}}\right\}$$

$$= \left[2(\tfrac{1}{2})^n - (\tfrac{1}{4})^n\right]u(n)$$

Shift Property

The next property that we consider is the effect on the z-transform that shifting the sequence produces. We are concerned with a shift of a causal sequence to the right primarily, so we examine

$$Z\{x(n - m)\} = \sum_{n=0}^{\infty} x(n - m)z^{-n}$$

where it is assumed that $x(n)$ is a causal sequence and m is a positive integer. In this case, we can change the index of the summation so that when n is replaced by $n + m$, the result is

$$\begin{aligned} Z\{x(n - m)\} &= \sum_{n=0}^{\infty} x(n)z^{-(n+m)} = z^{-m}\sum_{n=0}^{\infty} x(n)z^{-n} \\ &= z^{-m}Z\{x(n)\} = z^{-m}X(z) \end{aligned} \tag{3.17}$$

This property is quite useful when we consider the solution of difference equations using z-transforms in the next section.

As an example of the shift property, consider

$$X_1(z) = \frac{z^{-1}}{1 - \frac{1}{2}z^{-1}} = z^{-1}X(z)$$

where

$$X(z) = \frac{1}{1 - \frac{1}{2}z^{-1}}$$

Therefore, using Eq. (3.17) with $m = 1$ and $x(n) = 2^{-n}u(n)$, we obtain

$$x_1(n) = x(n - 1) = 2^{-n+1}u(n - 1)$$

This example also illustrates the importance of including $u(n)$ in the inverse transform. Note that $x_1(0) = 0$, a result that follows since $X_1(z)$ contains no constant term. If we had left off $u(n)$ in $x(n)$, we would have obtained the incorrect result $x_1(n) = 2^{-n+1}$, with $x_1(0) \neq 0$.

Convolution

In order to find the z-transform of a convolution sum, we recall the *Cauchy product* from elementary calculus, which for our situation is written

$$\left[\sum_{n=0}^{\infty} f(n)z^{-n}\right]\left[\sum_{n=0}^{\infty} g(n)z^{-n}\right] = \sum_{n=0}^{\infty}\left[\sum_{k=0}^{n} f(k)g(n-k)\right]z^{-n}$$
$$= \sum_{n=0}^{\infty}\left[\sum_{k=0}^{n} f(n-k)g(k)\right]z^{-n} \tag{3.18}$$

Since for causal sequences $f(n)$ and $g(n)$, we have

$$Z\{f(n) * g(n)\} = \sum_{n=0}^{\infty}\left[\sum_{k=0}^{n} f(n-k)g(k)\right]z^{-n}$$

it follows by applying Eq. (3.18) to the right-hand side that

$$Z\{f(n) * g(n)\} = \left[\sum_{n=0}^{\infty} f(n)z^{-n}\right]\left[\sum_{n=0}^{\infty} g(n)z^{-n}\right]$$
$$= Z\{f(n)\}Z\{g(n)\} = F(z)G(z) \tag{3.19}$$

We see then that the z-transform of the convolution of two sequences is the product of the z-transforms of the sequences. It is generally of more importance to note that from Eq. (3.19), we obtain the inverse z-transform of a product of z-transforms:

$$Z^{-1}\{F(z)G(z)\} = \sum_{k=0}^{n} f(n-k)g(k) = f(n) * g(n) \tag{3.20}$$

To illustrate the use of Eq. (3.20), let us again consider the transform

$$X(z) = \frac{1}{(1-\frac{1}{2}z^{-1})(1-\frac{1}{4}z^{-1})}$$

Defining the quantities

$$F(z) = \frac{1}{1-\frac{1}{2}z^{-1}}$$

$$G(z) = \frac{1}{1-\frac{1}{4}z^{-1}}$$

we have $X(z) = F(z)G(z)$, with $f(n) = 2^{-n}u(n)$ and $g(n) = 4^{-n}u(n)$. Consequently, from Eq. (3.20), the inverse z-transform is

$$x(n) = \sum_{k=0}^{n} [2^{-(n-k)}u(n-k)][4^{-k}u(k)]$$
$$= \left(2^{-n}\sum_{k=0}^{n} 2^{-k}\right)u(n)$$

The summation can be obtained from Eq. (1.5), where $a = 2^{-1}$. Consequently, it follows that

$$x(n) = 2^{-n}\frac{(2^{-n-1}-1)}{2^{-1}-1}u(n) = [2(\tfrac{1}{2})^n - (\tfrac{1}{4})^n]u(n)$$

which agrees with the result obtained previously using partial fractions.

As a second example of the use of the convolution sum, consider

$$X(z) = \frac{1}{(1 - az^{-1})^2} \tag{3.21}$$

Taking

$$F(z) = G(z) = \frac{1}{1 - az^{-1}}$$

with corresponding inverse z-transforms

$$f(n) = g(n) = a^n u(n)$$

we see that

$$\begin{aligned} x(n) = f(n) * g(n) &= \left(\sum_{k=0}^{n} a^k a^{n-k}\right) u(n) \\ &= (n + 1)a^n u(n) \end{aligned} \tag{3.22}$$

Multiplication by an Exponential

A property that is quite useful is obtained when we form a new sequence by multiplying a given sequence by an exponential. The z-transform that results is

$$Z\{a^n x(n)\} = \sum_{n=0}^{\infty} a^n x(n) z^{-n} = \sum_{n=0}^{\infty} x(n)(a^{-1}z)^{-n}$$

The last expression is just the z-transform of $x(n)$, with z replaced by $a^{-1}z$ or z^{-1} replaced by az^{-1}. Consequently, we have the result

$$Z\{a^n x(n)\} = X(a^{-1}z) \tag{3.23}$$

and, in addition,

$$Z^{-1}\{X(a^{-1}z)\} = a^n x(n) \tag{3.24}$$

As examples, by applying Eq. (3.23) to Eqs. (3.9) and (3.10), we obtain the more general transforms

$$Z\{a^n \cos n\theta\, u(n)\} = \frac{1 - az^{-1} \cos \theta}{1 - 2az^{-1} \cos \theta + a^2 z^{-2}} \tag{3.25}$$

$$Z\{a^n \sin n\theta\, u(n)\} = \frac{az^{-1} \sin \theta}{1 - 2az^{-1} \cos \theta + a^2 z^{-2}} \tag{3.26}$$

(The reader is asked to verify these results in Prob. 3.5.)

Let us consider the example

$$X(z) = \frac{1 + 2z^{-1}}{1 + 2z^{-1} + 4z^{-2}} \tag{3.27}$$

Comparing this expression to Eqs. (3.25) and (3.26), we see that $a = 2$ and

$\cos\theta = -\frac{1}{2}$ makes the denominators match. By taking $\theta = 2\pi/3$ radians, it follows that we must determine constants A and B so that

$$\frac{1 + 2z^{-1}}{1 + 2z^{-1} + 4z^{-2}} = \frac{A(1 + z^{-1})}{1 + 2z^{-1} + 4z^{-2}} + \frac{B(3^{1/2}z^{-1})}{1 + 2z^{-1} + 4z^{-2}}$$

since $a \sin\theta = (2)(3^{1/2}/2) = 3^{1/2}$. We find that $A = 1$ and $B = 1/3^{1/2}$. From Eqs. (3.25) and (3.26) and the linearity property, we have

$$x(n) = 2^n\left(\cos\frac{2\pi n}{3} + \frac{1}{3^{1/2}}\sin\frac{2\pi n}{3}\right)u(n) \tag{3.28}$$

Differentiation

The final operation on the z-transform that we consider is that of differentiation. It is known that a convergent power series can be differentiated termwise within its region of convergence. Since $X(z)$ is really a function of z^{-1}, it is generally easier to differentiate with respect to z^{-1}. Differentiating Eq. (3.2) and multiplying the result by z^{-1} gives

$$\begin{aligned} z^{-1}\frac{dX}{dz^{-1}} &= z^{-1}\sum_{n=0}^{\infty} nx(n)(z^{-1})^{n-1} \\ &= \sum_{n=0}^{\infty} nx(n)z^{-n} \end{aligned}$$

When this result is compared with the definition of Eq. (3.2) of the z-transform, we see that

$$Z\{nx(n)\} = z^{-1}\frac{dX}{dz^{-1}} \tag{3.29}$$

As an example of the use of this formula, we see from Eq. (3.6) that

$$\begin{aligned} Z\{na^n u(n)\} &= z^{-1}\frac{d}{dz^{-1}}(1 - az^{-1})^{-1} \\ &= z^{-1}(-1)(-a)(1 - az^{-1})^{-2} \\ &= \frac{az^{-1}}{(1 - az^{-1})^2} \end{aligned} \tag{3.30}$$

As another example of the differentiation property, let us reconsider finding $x(n)$ for

$$X(z) = (1 - az^{-1})^{-2}$$

given in Eq. (3.21). The relation in Eq. (3.30) suggests that we write the numerator in the form

$$1 = (1 - az^{-1}) + az^{-1}$$

so that

$$\frac{1}{(1-az^{-1})^2} = \frac{1-az^{-1}}{(1-az^{-1})^2} + \frac{az^{-1}}{(1-az^{-1})^2}$$
$$= \frac{1}{1-az^{-1}} + \frac{az^{-1}}{(1-az^{-1})^2}$$

Consequently, the inverse z-transform is given by

$$x(n) = (a^n + na^n)u(n)$$
$$= (n+1)a^n u(n)$$

which agrees with Eq. (3.22).

Other useful relations can be obtained by repeated differentiation. For example, differentiating $X(z)$ twice with respect to z^{-1} gives

$$\frac{d^2X(z)}{d(z^{-1})^2} = \sum_{n=0}^{\infty} n(n-1)x(n)z^{-(n-2)}$$

Now multiplying this expression by z^{-2}, we obtain the property

$$z^{-2}\frac{d^2X(z)}{d(z^{-1})^2} = \sum_{n=0}^{\infty} n(n-1)x(n)z^{-n}$$
$$= Z\{n(n-1)x(n)\} \tag{3.31}$$

If we take $x(n) = u(n)$, then applying this property results in

$$Z\{n(n-1)u(n)\} = z^{-2}\frac{d^2}{d(z^{-1})^2}(1-z^{-1})^{-1}$$
$$= \frac{2z^{-2}}{(1-z^{-1})^3}$$

The reader will be asked to develop more relations by repeated differentiation in the problems at the end of the chapter.

EXERCISES

3.3.1 Use convolution to find $x(n)$ if $X(z)$ is given by

(a) $\dfrac{1}{(1-\frac{1}{2}z^{-1})(1+\frac{1}{4}z^{-1})}$

(b) $\dfrac{1-\frac{1}{2}z^{-1}}{(1+\frac{1}{2}z^{-1})(1-z^{-1}+z^{-2})}$

Hint: Use $\cos\theta = \frac{1}{2}(e^{j\theta} + e^{-j\theta})$ and Ex. 3.2.2(b).

3.3.2 Find $x(n)$ if $X(z)$ is

(a) $\dfrac{1-z^{-1}}{(1+z^{-1})(1+\frac{1}{2}z^{-1})}$

(b) $\dfrac{2-3z^{-1}+9z^{-2}}{(1+z^{-1})(1-2z^{-1}+4z^{-2})}$

(c) $\dfrac{z^{-2}}{1 - \frac{1}{2}z^{-1}}$

(d) $\dfrac{1 + \frac{1}{2}z^{-1}}{1 - \frac{1}{2}z^{-1}}$

Hint: Use partial fractions in parts (a) and (b).

3.3.3 Apply Eq. (3.29) to (3.30) and find $Z\{n^2u(n)\}$. Note that $a = 1$.

3.4 COMPUTATION OF THE INVERSE z-TRANSFORM

In Sec. 3.2, we defined the inverse z-transform and gave some simple examples. In this section, we consider some techniques for finding inverse z-transforms and illustrate them with more examples.

Long Division

Since the z-transform is defined by the series of Eq. (3.1) or its special case, Eq. (3.2), one way to obtain the inverse z-transform is to expand $X(z)$ in the proper power series and identify the coefficient of z^{-n}, which is $x(n)$. We can obtain the power series by simply performing the indicated long division. For example, again consider the z-transform given in Eq. (3.27),

$$X(z) = \frac{1 + 2z^{-1}}{1 + 2z^{-1} + 4z^{-2}}$$

and let us carry out the indicated long division as follows:

$$\begin{array}{r l}
 & 1 - 4z^{-2} + 8z^{-3} + \cdots \\
1 + 2z^{-1} + 4z^{-2} \,\big) & 1 + 2z^{-1} \\
 & \underline{1 + 2z^{-1} + 4z^{-2}} \\
 & \qquad\quad - 4z^{-2} \\
 & \qquad\quad \underline{- 4z^{-2} - 8z^{-3} - 16z^{-4}} \\
 & \qquad\qquad\qquad 8z^{-3} + 16z^{-4}
\end{array}$$

Therefore, $x(0) = 1$, $x(1) = 0$, $x(2) = -4$, and $x(3) = 8$. The reader should verify that this result agrees with Eq. (3.28) for $0 \le n \le 3$.

The Cauchy Product and a Recurrence Relation

When the z-transform is a rational function of z^{-1} of the form

$$X(z) = \frac{b_0 + b_1z^{-1} + \cdots + b_mz^{-m}}{a_0 + a_1z^{-1} + \cdots + a_mz^{-m}} = \sum_{n=0}^{\infty} x(n)z^{-n} \tag{3.32}$$

it is possible to obtain a recurrence relation for $x(n)$. We have indicated that the numerator and denominator have the same degree. If this is not the case, we merely have some coefficients that are zero. Clearing Eq. (3.32) of fractions, we have

$$\sum_{n=0}^{\infty} b_n z^{-n} = \left(\sum_{n=0}^{\infty} a_n z^{-n}\right)\left(\sum_{n=0}^{\infty} x(n) z^{-n}\right)$$

where $a_n = b_n = 0$ for $n > m$. Applying the Cauchy product to the right-hand side results in

$$\sum_{n=0}^{\infty} b_n z^{-n} = \sum_{n=0}^{\infty} \left[\sum_{k=0}^{n} x(k) a_{n-k}\right] z^{-n}$$

These series are equal when coefficients of z^{-n} are equal, so that

$$b_n = \sum_{k=0}^{n} x(k) a_{n-k}$$

When this equation is solved for $x(n)$, we obtain

$$\begin{aligned} x(n) &= \frac{1}{a_0}\left[b_n - \sum_{k=0}^{n-1} x(k) a_{n-k}\right] \qquad n > 0 \\ x(0) &= b_0/a_0 \end{aligned} \tag{3.33}$$

where we have assumed that $a_0 \neq 0$. For the example just considered, we have $b_0 = 1$, $b_1 = 2$, and $b_n = 0$ for $n \geq 2$; $a_0 = 1$, $a_1 = 2$, $a_2 = 4$, and $a_n = 0$ for $n \geq 3$. Substituting these values into Eq. (3.33), we obtain

$$x(0) = b_0/a_0 = 1$$

$$x(1) = b_1 - x(0)a_1 = 2 - (1)(2) = 0$$

$$x(2) = b_2 - x(0)a_2 - x(1)a_1 = 0 - (1)(4) - (0)(2) = -4$$

$$x(3) = b_3 - x(0)a_3 - x(1)a_2 - x(2)a_1 = -(-4)(2) = 8$$

etc.

As the reader has perhaps noted, the use of long division or the Cauchy product does not lead to an explicit expression for $x(n)$ and we cannot find any term $x(n)$ without finding all the preceding terms. Equation (3.33) can be evaluated quite readily using a digital computer, but frequently the error that results is large.

Partial-Fraction Expansions

A very useful technique for finding the inverse z-transform of $X(z)$ given in Eq. (3.32) is to obtain a partial-fraction expansion for $X(z)$ and then use Eqs. (3.13), (3.14), (3.25), or (3.26) if there are simple linear factors or nonrepeated quadratic factors in the denominator. For the case of repeated factors, we have to develop formulas such as Eq. (3.30). The reader should realize that in using a partial-fraction expansion, we are using the linearity property of the inverse z-transform. In Sec. 3.3, we used partial fractions to find

$$\begin{aligned} Z^{-1}\left\{\frac{1}{(1-\frac{1}{2}z^{-1})(1-\frac{1}{4}z^{-1})}\right\} &= Z^{-1}\left\{\frac{2}{1-\frac{1}{2}z^{-1}}\right\} + Z^{-1}\left\{\frac{-1}{1-\frac{1}{4}z^{-1}}\right\} \\ &= [2(\tfrac{1}{2})^n - (\tfrac{1}{4})^n]u(n) \end{aligned}$$

As another example of the use of partial fractions, consider

$$X(z) = \frac{3 + \frac{11}{2}z^{-1} + 7z^{-2}}{(1 - \frac{1}{2}z^{-1})(1 + 2z^{-1} + 4z^{-2})}$$

which has a partial-fraction expansion of the form

$$X(z) = \frac{A}{1 - \frac{1}{2}z^{-1}} + \frac{B + Cz^{-1}}{1 + 2z^{-1} + 4z^{-2}}$$

We can determine A from the expression

$$A = \lim_{z^{-1}\to 2} (1 - \tfrac{1}{2}z^{-1})X(z) = \left.\frac{3 + \frac{11}{2}z^{-1} + 7z^{-2}}{1 + 2z^{-1} + 4z^{-2}}\right|_{z^{-1}=2} = 2$$

By using this value of A, the remaining term in the partial fraction can be determined by subtraction. This results in

$$\frac{B + Cz^{-1}}{1 + 2z^{-1} + 4z^{-2}} = X(z) - \frac{2}{1 - \frac{1}{2}z^{-1}}$$

$$= \frac{1 + \frac{3}{2}z^{-1} - z^{-2}}{(1 - \frac{1}{2}z^{-1})(1 + 2z^{-1} + 4z^{-2})}$$

$$= \frac{1 + 2z^{-1}}{1 + 2z^{-1} + 4z^{-2}}$$

Now applying Eqs. (3.14) and (3.28) to $X(z)$, which has the form

$$X(z) = \frac{2}{1 - \frac{1}{2}z^{-1}} + \frac{1 + 2z^{-1}}{1 + 2z^{-1} + 4z^{-2}}$$

we obtain

$$x(n) = \left[2(\tfrac{1}{2})^n + 2^n\left(\cos\frac{2\pi n}{3} + \frac{1}{3^{1/2}}\sin\frac{2\pi n}{3}\right)\right]u(n)$$

As another example, consider finding the inverse z-transform of

$$X(z) = \frac{2 + 5z^{-1} + 12z^{-2}}{(1 + 2z^{-1} + 4z^{-2})(1 + 4z^{-1} + 8z^{-2})}$$

The two factors in the denominator have complex zeros and so the partial-fraction expansion has the form

$$X(z) = \frac{A_1 + A_2z^{-1}}{1 + 2z^{-1} + 4z^{-2}} + \frac{A_3 + A_4z^{-1}}{1 + 4z^{-1} + 8z^{-2}}$$

We can determine the constants A_1, A_2, A_3, and A_4 by combining these two fractions and making the numerators in the two expressions for $X(z)$ identical. This results in the equations

$$1: \quad A_1 + A_3 = 2$$
$$z^{-1}: \quad 4A_1 + A_2 + 2A_3 + A_4 = 5$$
$$z^{-2}: \quad 8A_1 + 4A_2 + 4A_3 + 2A_4 = 12$$
$$z^{-3}: \quad 8A_2 + 4A_4 = 0$$

We can solve for A_3 and A_4 in the first and fourth equations and substitute these values into the second and third equations, which then contain only A_1 and A_2. The solution is

$$A_1 = A_2 = A_3 = 1 \qquad A_4 = -2$$

To find $x(n)$, we must use Eqs. (3.25) and (3.26). Comparing the denominators of the two fractions with the denominator in Eq. (3.25), we obtain for $a = 2$ and $\theta = 2\pi/3$

$$1 + 2z^{-1} + 4z^{-2} = 1 - 2(2)z^{-1} \cos 2\pi/3 + (2)^2 z^{-2}$$

and for $a = 2(2^{1/2})$ and $\theta = 3\pi/4$,

$$1 + 4z^{-1} + 8z^{-2} = 1 - 2[2(2^{1/2})]z^{-1} \cos 3\pi/4 + [2(2^{1/2})]^2 z^{-2}$$

For convenience, we write $X(z) = X_1(z) + X_2(z)$, where

$$X_1(z) = \frac{1 - 2z^{-1} \cos 2\pi/3}{1 - 2(2)z^{-1} \cos 2\pi/3 + (2)^2 z^{-2}}$$

and

$$X_2(z) = \frac{1 - 2(2^{1/2})z^{-1} \cos 3\pi/4 + K[2(2^{1/2})]z^{-1} \sin 3\pi/4}{1 - 2[2(2^{1/2})]z^{-1} \cos 3\pi/4 + [2(2^{1/2})]^2 z^{-2}}$$

In the numerator of $X_2(z)$, the constant K must be determined so that

$$1 - 2z^{-1} = (1 - az^{-1} \cos \theta) + K(az^{-1} \sin \theta)$$

where, for our case, $a = 2(2^{1/2})$ and $\theta = 3\pi/4$. Consequently, we have $K = -2$.

Using the relations in Eqs. (3.25) and (3.26), we obtain

$$\begin{aligned} x(n) &= x_1(n) + x_2(n) \\ &= [2^n \cos 2n\pi/3 + [2(2^{1/2})]^n(\cos 3n\pi/4 - 2 \sin 3n\pi/4)]u(n) \end{aligned}$$

In this example, we also could have determined the coefficients A_1 and A_2 by multiplying both sides of the equation by $1 + 2z^{-1} + 4z^{-2}$ and taking the limit as $z \to -1 + j3^{1/2}$, which is a zero of $1 + 2z^{-1} + 4z^{-2}$. The constants A_1 and A_2 are determined then by equating the real and imaginary parts in the limiting expressions. The same procedure could be used to determine A_3 and A_4.

The inverse z-transform that we have just considered could have been obtained using linear factors with complex poles. To illustrate this technique, consider the partial-fraction expression

$$X(z) = \frac{2 + z^{-1}}{(1 - \frac{1}{2}z^{-1})(1 + \frac{1}{4}z^{-2})} \tag{3.34}$$

$$= \frac{A_1}{1 - \frac{1}{2}z^{-1}} + \frac{A_2}{1 - j\frac{1}{2}z^{-1}} + \frac{A_3}{1 + j\frac{1}{2}z^{-1}}$$

It can be shown that $A_3 = A_2^*$, so that

$$x(n) = [A_1(\tfrac{1}{2})^n + A_2(j\tfrac{1}{2})^n + A_2^*(-j\tfrac{1}{2})^n]u(n)$$

Since

$$[A_2(j\tfrac{1}{2})^n]^* = A_2^*(-j\tfrac{1}{2})^n$$

it follows that

$$A_2(j\tfrac{1}{2})^n + A_2^*(-j\tfrac{1}{2})^n = 2 \text{ Re } [A_2(j\tfrac{1}{2})^n]$$

By evaluating the coefficients by the limiting process used previously, it follows that

$$A_1 = \frac{2 + z^{-1}}{1 + \frac{1}{4}z^{-2}}\Bigg|_{z^{-1}=2} = 2$$

$$A_2 = \frac{2 + z^{-1}}{(1 - \frac{1}{2}z^{-1})(1 + j\frac{1}{2}z^{-1})}\Bigg|_{z^{-1}=-j2} = -j1$$

Noting that

$$\begin{aligned} 2 \text{ Re } [A_2(j\tfrac{1}{2})^n] &= 2 \text{ Re } [(\tfrac{1}{2})^n e^{jn\pi/2} e^{-j\pi/2}] \\ &= 2(\tfrac{1}{2})^n \cos (n - 1)\pi/2 \\ &= 2(\tfrac{1}{2})^n \sin n\pi/2 \end{aligned}$$

we finally obtain for the inverse z-transform

$$x(n) = 2^{-n+1}(1 + \sin n\pi/2)u(n)$$

As a final example in this section, consider the function $X(z)$ and its partial-fraction expansion given by

$$X(z) = \frac{4 - 8z^{-1} + 6z^{-2}}{(1 - 2z^{-1})^2(1 + z^{-1})}$$

$$= \frac{A}{1 - 2z^{-1}} + \frac{B}{(1 - 2z^{-1})^2} + \frac{C}{1 + z^{-1}}$$

We can determine B and C directly from

$$B = \lim_{z^{-1}\to 1/2} (1 - 2z^{-1})^2 X(z) = 1$$

$$C = \lim_{z^{-1}\to -1} (1 + z^{-1})X(z) = 2$$

The coefficient A can be determined by combining the fractions and equating the constant terms in the resulting numerator and the numerator of $X(z)$. This results in

$$A + B + C = 4$$

Consequently, we obtain

$$A = 4 - B - C = 4 - 1 - 2 = 1$$

and

$$X(z) = \frac{1}{1 - 2z^{-1}} + \frac{1}{(1 - 2z^{-1})^2} + \frac{2}{1 + z^{-1}}$$

In the second term, we see that in order to use Eq. (3.30), we should write the numerator as

$$1 = (1 - 2z^{-1}) + 2z^{-1}$$

Then

$$X(z) = \frac{2}{1 - 2z^{-1}} + \frac{2z^{-1}}{(1 - 2z^{-1})^2} + \frac{2}{1 + z^{-1}}$$

and it follows that

$$x(n) = [2(2^n) + n(2^n) + 2(-1)^n]u(n)$$

We also could have determined A in the partial fraction after determining B and C by setting $z^{-1} = 0$. This results in the relation

$$4 = A + B + C$$

and it is not necessary to combine the fractions.

In this example, we could have anticipated the modification of the numerator of the repeated linear factor by writing

$$X(z) = \frac{A_1}{1 - 2z^{-1}} + \frac{B_1 z^{-1}}{(1 - 2z^{-1})^2} + \frac{C_1}{1 + z^{-1}}$$

where $C = C_1 = 2$ as before. Then we find B_1 from

$$(1 - 2z^{-1})^2 X(z)\bigg|_{z^{-1}=1/2} = B_1 z^{-1}\bigg|_{z^{-1}=1/2}$$

and so it follows that

$$1 = \tfrac{1}{2}B_1$$

or $B_1 = 2$. Then when we set $z^{-1} = 0$, we obtain

$$X(z)\bigg|_{z^{-1}=0} = 4 = A_1 + 0 + C_1$$

resulting in $A_1 = 2$. The partial-fraction expansion is then given by

$$X(z) = \frac{2}{1 - 2z^{-1}} + \frac{2z^{-1}}{(1 - 2z^{-1})^2} + \frac{2}{1 + z^{-1}}$$

which agrees with our previous result.

A Formula from Complex-Variable Theory

The last formula that we will give is based on complex-variable theory. By using this theory, it can be shown that

$$x(n) = \frac{1}{2\pi j} \oint_C X(z) z^{n-1}\, dz$$

where C is a counterclockwise closed contour in the region of convergence of $X(z)$ and encircling the origin of the z plane. This integral is usually evaluated using residues. We will not rely on this relation to obtain inverse z-transforms, but will use series expansions and partial fractions. The reader who is interested in using this method should consult a book on complex variable theory, such as [23].

EXERCISES

3.4.1 Find $x(n)$, $0 \le n \le 3$, using both long division and Eq. (3.33) for $X(z)$ given by

(a) $$\frac{2 - 3z^{-1} + 9z^{-2}}{(1 + z^{-1})(1 - 2z^{-1} + 4z^{-2})}$$

(b) $$\frac{(2 + 3z^{-1})}{(1 + z^{-1})(1 + \frac{1}{2}z^{-1})(1 - \frac{1}{4}z^{-1})}$$

3.4.2 Find $x(n)$ for $X(z)$ given in Ex. 3.4.1 using partial fractions.

3.4.3 Find $x(n)$ for $X(z)$ given by

$$X(z) = \frac{1}{(1 - z^{-1})^2}$$

Hint: $1 = 1 - z^{-1} + z^{-1}$.

3.4.4 Find $x(n)$ for $X(z)$ in Ex. 3.4.3 using convolution.

3.4.5 Find $x(n)$ for $X(z)$ given in Eq. (3.34) using a partial-fraction expansion with real coefficients.

3.4.6 **(a)** Find $x(n)$ from Eq. (3.27) using linear factors with complex zeros.
(b) Repeat part (a) for

$$X(z) = \frac{2 + 5z^{-1} + 12z^{-2}}{(1 + 2z^{-1} + 4z^{-2})(1 + 4z^{-1} + 8z^{-2})}$$

3.5 SOLUTION OF DIFFERENCE EQUATIONS AND TRANSFER FUNCTIONS

Solution of Difference Equations

In this section, we are concerned with solving difference equations of the form given in Eq. (2.29), which we repeat here:

$$\sum_{k=0}^{N} a_k y(n-k) = \sum_{m=0}^{M} b_m x(n-m) \tag{3.35}$$

where $x(n)$ and $y(n)$ are causal sequences. Taking the z-transform of both sides, we obtain

$$Z\left\{\sum_{k=0}^{N} a_k y(n-k)\right\} = Z\left\{\sum_{m=0}^{M} b_m x(n-m)\right\}$$

which by the linearity property of Eq. (3.15) can be written

$$\sum_{k=0}^{N} a_k Z\{y(n-k)\} = \sum_{m=0}^{M} b_m Z\{x(n-m)\}$$

Using the shift property of Eq. (3.17), this last equation becomes

$$\sum_{k=0}^{N} a_k z^{-k} Y(z) = \sum_{m=0}^{M} b_m z^{-m} X(z)$$

Since $Y(z)$ and $X(z)$ can be factored out of the summations, we can solve the resulting equation for $Y(z)$. This result can be written

$$Y(z) = \frac{\sum_{m=0}^{M} b_m z^{-m}}{\sum_{k=0}^{N} a_k z^{-k}} X(z) \tag{3.36}$$

The sequence $y(n)$ can be obtained from this expression once $X(z)$ is known.

Example

As an example, let us find the step response of the system of Eq. (2.2) given by

$$y(n) - ay(n-1) = x(n) = u(n)$$

Taking the z-transform of both sides and solving for $Y(z)$, we have, as in Eq. (3.36),

$$Y(z) = \frac{1}{(1 - az^{-1})(1 - z^{-1})}$$

The partial-fraction expansion is

$$Y(z) = \frac{1}{1-a}\left(\frac{1}{1 - z^{-1}} - \frac{a}{1 - az^{-1}}\right)$$

which has the inverse z-transform

$$y(n) = \frac{1}{1-a}(1 - a^{n+1})u(n)$$

The System Function

Let us define the function $H(z)$ by

$$H(z) = \frac{\sum_{m=0}^{M} b_m z^{-m}}{\sum_{k=0}^{N} a_k z^{-k}} \tag{3.37}$$

so that Eq. (3.36) can be written

$$Y(z) = H(z)X(z) \tag{3.38}$$

The function $H(z)$ is known as the *transfer function* of the system or as the *system function*. If we take as input the unit-impulse sequence, $x(n) = \delta(n)$, then from Eq. (3.4), we have $X(z) = 1$ and $Y(z) = H(z)$. Therefore, $H(z)$ is the z-transform of the impulse response

$$H(z) = Z\{h(n)\} \tag{3.39}$$

Now that we have defined the system function, we find it convenient to represent a system by $H(z)$, as shown in Fig. 3.2. In what follows, we use either $h(n)$ or $H(z)$ to describe a linear time-invariant system, as shown in Figs. 2.2 and 3.2.

From the theory of infinite series, we know that for $a_k \neq 0$ for some k, $1 \leq k \leq N$, the sequence $h(n)$ determined by Eqs. (3.37) and (3.39) is of infinite duration, as claimed in Sec. 2.6. Hence, Eq. (3.35) is an IIR system if some a_k other than a_0 is nonzero.

$x(n) \longrightarrow$ [$h(n)$] $\longrightarrow y(n)$

(a)

$X(z) \longrightarrow$ [$H(z)$] $\longrightarrow Y(z)$

(b)

Figure 3.2 Block-diagram representations of a system using (a) the impulse response and (b) the system function.

Example of a Transfer Function

As an example, consider the system of Eq. (2.2) with the impulse response given in Eq. (2.3). The difference equation is

$$y(n) - ay(n-1) = x(n) = \delta(n)$$

and the z-transform $Y(z)$ is

$$Y(z) = \frac{1}{1 - az^{-1}} = H(z)$$

The impulse response is obtained by applying Eq. (3.14) and the result is $h(n) = a^n u(n)$, which agrees with Eq. (2.3) as expected.

Convolution and the Transfer Function

The result of Eq. (3.38) could also have been obtained by transforming the convolution sum

$$y(n) = \sum_{k=0}^{n} h(k)x(n-k)$$

which is obtained for a causal input and a causal system. When we utilize the formula of Eq. (3.19) for the z-transform of a convolution sum, we obtain Eq. (3.38).

Poles and Zeros of $H(z)$

It is often useful to express $H(z)$ given by Eq. (3.37) in the factored form

$$H(z) = K\frac{\prod_{i=1}^{M}(1 - z_i z^{-1})}{\prod_{j=1}^{N}(1 - p_j z^{-1})} \tag{3.40}$$

where we have assumed that a_0 and b_0 are nonzero. Since $H(z_i) = 0$, we say that $z = z_i$ is a *zero* of $H(z)$, and since $\lim_{z\to p_j}|H(z)| = \infty$, we say that $z = p_j$ is a *pole* of $H(z)$. The form of Eq. (3.40) displays the finite zeros and poles of $H(z)$. Depending on the values of M and N, we can have poles and zeros at infinity or zero and these can be determined by expressing $H(z)$ as a ratio of polynomials in z. For example, if $N \geq M$, we can write Eq. (3.37) in the form

$$H(z) = \frac{\sum_{m=0}^{M} b_m z^{N-m}}{\sum_{k=0}^{N} a_k z^{N-k}} = \frac{Q(z)}{P(z)}$$

Now if $N = M$, $Q(z)$ and $P(z)$ are both polynomials in z of degree N and there are N finite zeros, which are the roots of $Q(z) = 0$, and N finite poles, which are the roots of $P(z) = 0$. If $N > M$, then $Q(z)$ has a factor z^{N-M} and so there are $N - M$ zeros at $z = 0$.

To illustrate these concepts, consider the transfer functions

$$H_1(z) = \frac{b_0 + b_1 z^{-1} + b_2 z^{-2}}{1 + a_1 z^{-1} + a_2 z^{-2}} = \frac{b_0 z^2 + b_1 z + b_2}{z^2 + a_1 z + a_2}$$

$$H_2(z) = \frac{b_0 + b_1 z^{-1}}{1 + a_1 z^{-1} + a_2 z^{-2}} = \frac{z(b_0 z + b_1)}{z^2 + a_1 z + a_2}$$

$$H_3(z) = \frac{b_0}{1 + a_1 z^{-1} + a_2 z^{-2}} = \frac{b_0 z^2}{z^2 + a_1 z + a_2}$$

with b_0, b_1, and a_2 nonzero. In all cases, there are two finite poles that are nonzero. The function $H_1(z)$ has two finite zeros that are nonzero, $H_2(z)$ has the zeros $z = 0$ and $z = -b_1/b_0$, and $H_3(z)$ has two zeros at $z = 0$.

In the case of a FIR system, the denominator of Eq. (3.40) is one and $H(z)$ has a pole or poles at $z = 0$.

Fibonacci Numbers

As another example illustrating the use of the z-transform in solving difference equations, let us consider generating the Fibonacci numbers given in Prob. 1.4 and in Eq. (2.45) as the impulse response of the system

$$y(n) - y(n-1) - y(n-2) = x(n-1)$$

For this case, the system function $H(z)$ given by Eq. (3.37) is found to be

$$H(z) = \frac{z^{-1}}{1 - z^{-1} - z^{-2}} \tag{3.41}$$

The poles in terms of z are solutions of the equation

$$z^2 - z - 1 = 0 \tag{3.42}$$

and are

$$p_{1,2} = \tfrac{1}{2}(1 \pm 5^{1/2}) \tag{3.43}$$

The zeros are at $z = 0$ and $z = \infty$. The factored form of $H(z)$ corresponding to Eq. (3.40) is

$$H(z) = \frac{z^{-1}}{(1 - p_1 z^{-1})(1 - p_2 z^{-1})} \tag{3.44}$$

and has the partial-fraction expansion

$$\begin{aligned} H(z) &= \frac{(p_1 - p_2)^{-1}}{1 - p_1 z^{-1}} + \frac{(p_2 - p_1)^{-1}}{1 - p_2 z^{-1}} \\ &= \frac{1}{p_1 - p_2}\left[\sum_{n=0}^{\infty} p_1^n z^{-n} - \sum_{n=0}^{\infty} p_2^n z^{-n}\right] \end{aligned}$$

The impulse-response sequence is, therefore,

$$\begin{aligned} h(n) &= \frac{1}{p_1 - p_2}[p_1^n - p_2^n]u(n) \\ &= \frac{1}{5^{1/2}}\left[\left(\frac{1 + 5^{1/2}}{2}\right)^n - \left(\frac{1 - 5^{1/2}}{2}\right)^n\right]u(n) \end{aligned}$$

which agrees with the result of Prob. 1.4.

Cascaded Systems

As we will see later, it is of considerable interest to consider $H(z)$ as a product of factors. For example, let us consider

$$H(z) = H_1(z)H_2(z) \tag{3.45}$$

Now from the relation

$$Y(z) = H(z)X(z) = H_1(z)H_2(z)X(z) \tag{3.46}$$

we see that by introducing a new variable $W(z)$ defined by

$$W(z) = H_2(z)X(z) \tag{3.47}$$

we can write

$$Y(z) = H_1(z)W(z) \tag{3.48}$$

These relations describe the cascaded system shown in Fig. 3.3, where $x(n)$ is the input to the first subsystem represented by the system function $H_2(z)$. Its response $w(n)$ is then the input to the second subsystem described by $H_1(z)$ and having response $y(n)$. This is the same system that was considered in Sec. 2.4 and shown in Fig. 2.2. Since $H_1(z)H_2(z) = H_2(z)H_1(z)$, we can interchange the order of the blocks in the system.

If the systems described by $h_1(n) = a^n u(n)$ and $h_2(n) = b^n u(n)$, $a \neq b$, are cascaded, then the overall system function is given by

$$
\begin{aligned}
H(z) &= H_1(z)H_2(z) \\
&= \frac{1}{1 - az^{-1}} \frac{1}{1 - bz^{-1}} \\
&= \frac{1}{1 - (a + b)z^{-1} + abz^{-2}}
\end{aligned}
$$

We can obtain the impulse-response sequence $h(n)$ from $H(z)$ by noting that

$$H(z) = \frac{\dfrac{1}{1 - (b/a)}}{1 - az^{-1}} + \frac{\dfrac{1}{1 - (a/b)}}{1 - bz^{-1}}$$

which has the inverse z-transform

$$h(n) = (a - b)^{-1}(a^{n+1} - b^{n+1})u(n)$$

This is the same result that was obtained in Sec. 2.4 using the convolution

$$h(n) = h_1(n) * h_2(n)$$

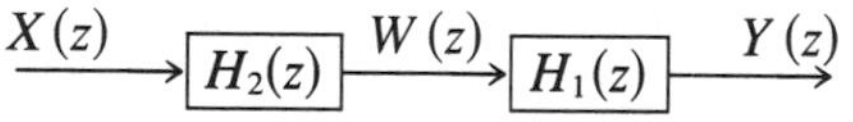

$X(z) \rightarrow$ [$H_2(z)$] $\xrightarrow{W(z)}$ [$H_1(z)$] $\xrightarrow{Y(z)}$

Figure 3.3 Representation of a cascaded system.

EXERCISES

3.5.1 Find $y(n)$ by transforming the difference equation if
(a) $y(n) + \frac{3}{4}y(n - 1) + \frac{1}{8}y(n - 2) = \delta(n - 1)$
(b) $y(n) - y(n - 1) = u(n) + u(n - 1)$

3.5.2 Find $H(z)$ and determine its poles and zeros if
(a) $y(n) + \frac{3}{4}y(n - 1) + \frac{1}{8}y(n - 2) = x(n) + x(n - 1)$
(b) $y(n) = x(n) + 3x(n - 1) + 2x(n - 2)$
(c) $y(n) + \frac{1}{2}y(n - 1) = x(n) + 2x(n - 1)$

3.5.3 Find $H(z)$ if the systems with $h_1(n) = (\frac{1}{2})^n u(n)$ and $h_2(n) = (-\frac{1}{2})^n u(n)$ are cascaded. Then determine $h(n) = Z^{-1}\{H(z)\}$. Compare the result with $h_1(n) * h_2(n)$.

3.6 STABILITY: FREQUENCY-DOMAIN CRITERION

Whenever $H(z)$ is used to describe a system, if we wish to study stability using the methods of Sec. 2.8, we first must find $h(n)$ and use Eq. (2.62). Rather than following this procedure, it is better to relate the condition of Eq. (2.62) to the frequency domain and obtain a criterion that can be applied directly to $H(z)$.

Stability Criterion

We now show that the system having transfer function $H(z)$ is stable if and only if $H(z)$ contains no poles on or outside the unit circle $|z| = 1$. First, we assume that the system is stable and show that the poles of $H(z)$ are all inside the unit circle $|z| = 1$. From the definition of $H(z)$, we obtain (for a causal system)

$$\begin{aligned} |H(z)| &= \left| \sum_{k=0}^{\infty} h(k) z^{-k} \right| \\ &\leq \sum_{k=0}^{\infty} |h(k)| |z|^{-k} \end{aligned}$$

which for $|z| \geq 1$ reduces to

$$|H(z)| \leq \sum_{k=0}^{\infty} |h(k)| \qquad |z| \geq 1 \tag{3.49}$$

Since the system is assumed to be stable, we see that

$$|H(z)| \leq \sum_{k=0}^{\infty} |h(k)| < \infty \qquad |z| \geq 1$$

Since $|H(z)|$ is finite for $|z| \geq 1$, there can exist no poles of $H(z)$ on or outside this unit circle.

Let us now assume that $H(z)$ has at least one pole z_1 in the region $|z| \geq 1$ and show that the system is unstable. As $z \to z_1$, we see that

$$\lim_{z \to z_1} |H(z)| = \infty$$

and it follows from Eq. (3.49) that the series $\Sigma_{k=0}^{\infty} |h(k)|$ diverges. As was shown in Sec. 2.8, divergence of this series, which is Eq. (2.62) for a causal system, indicates that the system is unstable.

From the criterion just developed, we see that system stability can be deduced from $H(z)$ by determining the locations of its poles. The poles of $H(z)$ in Eq. (3.37) are the roots of the equation

$$z^N \sum_{k=0}^{N} a_k z^{-k} = \sum_{k=0}^{N} a_k z^{N-k} = 0 \tag{3.50}$$

which is simply the characteristic equation of Eq. (2.29). Consequently, the criterion concerning the poles of $H(z)$ is equivalent to the criterion developed in Sec. 2.8 concerning the roots of the characteristic equation.

Example

As an example, consider the system of Eq. (3.41), which generates the Fibonacci numbers. The characteristic equation, obtained by equating the denominator of $H(z)$ to zero, is

$$P(z) = z^2 - z - 1 = 0$$

The poles of $H(z)$, which are the zeros of $P(z)$, are

$$z_1 = \tfrac{1}{2}(1 + 5^{1/2}) \qquad z_2 = \tfrac{1}{2}(1 - 5^{1/2})$$

Clearly, $|z_1| > 1$ and the system is unstable.

Stability Condition for Second-Order Systems

In Prob. 3.30, it is shown that necessary and sufficient conditions that a quadratic $P(z) = a_2 z^2 + a_1 z + a_0$ have no zeros on or outside the unit circle $|z| = 1$ are

$$P(1) > 0 \qquad P(-1) > 0 \qquad a_2 > |a_0| \tag{3.51}$$

For the case of the Fibonacci numbers, $P(1) = -1$, and, therefore, $P(z)$ has at least one zero on or outside the unit circle.

EXERCISES

3.6.1 Discuss the stability of the system described by

$$H(z) = \frac{1}{(1 - \frac{1}{2}z^{-1})(1 + \frac{1}{2}z^{-1})(1 - \frac{1}{4}z^{-1})}$$

3.6.2 Repeat Ex. 3.6.1 for the system

$$H(z) = \frac{1 + 2z^{-1}}{1 - z^{-1} + z^{-2}} \frac{1 - z^{-1}}{1 - \frac{1}{2}z^{-1} + \frac{1}{4}z^{-2}}$$

3.7 FREQUENCY RESPONSE FOR A STABLE SYSTEM

Obtaining the Frequency Response from $H(z)$

Let us now examine the relationship between the frequency response and the transfer function of a stable system. In this case, the series

$$H(z) = \sum_{k=0}^{\infty} h(k) z^{-k}$$

converges for $|z| \geq 1$ and in particular for $z = e^{j\omega}$. The frequency response is then given by

$$H(e^{j\omega}) = \sum_{k=0}^{\infty} h(k) e^{-j\omega k} = H(z)\Big|_{z=e^{j\omega}}$$

In other words, the frequency response can be obtained from the system function by setting $z = e^{j\omega}$. In particular, we can obtain from Eq. (3.37)

$$H(e^{j\omega}) = \frac{\sum_{m=0}^{M} b_m e^{-j\omega m}}{\sum_{k=0}^{N} a_k e^{-j\omega k}} \tag{3.52}$$

Also from Eq. (3.40) we can display the poles and zeros of $H(e^{j\omega})$ by writing

$$H(e^{j\omega}) = K \frac{e^{j(N-M)\omega} \prod_{i=1}^{M} (e^{j\omega} - z_i)}{\prod_{j=1}^{N} (e^{j\omega} - p_j)} \tag{3.53}$$

The magnitude and phase of $H(e^{j\omega})$ can be obtained geometrically as in Fig. 2.6, where again $e^{j\omega}$ is represented by the unit vector with phase angle ω rad = (ω rad/s)(1 s).

Example

We now give an example comparing the method presented here for finding the frequency response for a stable system from the transfer function to the direct method using the definition, Eq. (2.70), of the frequency response. Consider the system

$$8y(n) - 6y(n-1) + y(n-2) = 8x(n) \tag{3.54}$$

given previously in Eq. (2.49), which has the impulse response

$$h(n) = [2(2)^{-n} - 4^{-n}]u(n) \tag{3.55}$$

This system is stable and so, from Eqs. (3.52) and (3.54), we obtain the frequency response

$$H(e^{j\omega}) = \frac{8}{8 - 6e^{-j\omega} + e^{-j2\omega}} \tag{3.56}$$

Using the impulse response of Eq. (3.55) and the definition of the frequency response, we find that

$$\begin{aligned} H(e^{j\omega}) &= \sum_{n=0}^{\infty} 2(2)^{-n} e^{-j\omega n} + \sum_{n=0}^{\infty} (-1)(4)^{-n} e^{-j\omega n} \\ &= \frac{2}{1 - \frac{1}{2}e^{-j\omega}} + \frac{-1}{1 - \frac{1}{4}e^{-j\omega}} \\ &= \frac{8}{8 - 6e^{-j\omega} + e^{-j2\omega}} \end{aligned}$$

This result agrees with Eq. (3.56), which was obtained using Eq. (3.52) and the difference equation.

The methods that we have considered here are valid only for stable systems. For these cases, it is most convenient to obtain $H(e^{j\omega})$ in Eq. (3.52) directly from the difference equation or from $H(z)$. The linearity property of the Fourier transform and the result of Ex. 3.7.3 can be used to transform the difference equation.

EXERCISES

3.7.1 Determine $H(z)$ for the system

$$y(n) + \tfrac{1}{2}y(n-1) = x(n)$$

Discuss the system stability and, if possible, determine $H(e^{j\omega})$ from $H(z)$.

3.7.2 Determine $H(e^{j\omega})$ from $H(z)$ for the system

$$y(n) - \tfrac{1}{2}y(n-1) = x(n) - 2x(n-1)$$

Determine the magnitude and phase responses.

3.7.3 Show that if $\mathcal{F}\{x(n)\} = X(e^{j\omega})$, then $\mathcal{F}\{x(n-m)\} = e^{-j\omega m}X(e^{j\omega})$ for m a positive integer.

3.7.4 Find the poles of the system

$$y(n) - \tfrac{1}{4}y(n-1) + \tfrac{1}{4}y(n-2) - \tfrac{1}{16}y(n-3) = 2x(n) + 3x(n-1)$$

and determine if the system is stable. If possible, find $H(e^{j\omega})$ from the difference equation.

3.8 ZERO LOCATIONS OF LINEAR-PHASE FIR SYSTEMS

Locations of the Zeros

In Sec. 2.11, we saw that an FIR system has linear phase if the impulse response satisfies the condition

$$h(n) = h(N-1-n) \qquad n = 0, 1, \ldots, N-1 \tag{3.57}$$

In this section, we examine the location of the zeros of a linear-phase system function $H(z)$.

Suppose that $z_0 \neq 0$ is a finite zero of $H(z)$, that is,

$$H(z_0) = \sum_{n=0}^{N-1} h(n)z_0^{-n} = 0$$

We now show that z_0^{-1} is also a zero of $H(z)$. To see that this is true, we note that

$$H(z_0^{-1}) = \sum_{n=0}^{N-1} h(n)z_0^{n}$$

Using the symmetry property of Eq. (3.57), we can write

$$H(z_0^{-1}) = \sum_{n=0}^{N-1} h(N-1-n)z_0^{n}$$

If we now change the index so that $k = N-1-n$, then the above expression becomes

$$
\begin{aligned}
H(z_0^{-1}) &= \sum_{k=0}^{N-1} h(k) z_0^{N-1-k} \\
&= z_0^{N-1} \sum_{k=0}^{N-1} h(k) z_0^{-k} \\
&= z_0^{N-1} H(z_0)
\end{aligned}
$$

Since $z_0 \neq 0$ and $H(z_0) = 0$, we see that $H(z_0^{-1}) = 0$. Consequently, we can say that if z_0 is a zero of $H(z)$, then so is its reciprocal z_0^{-1}. We have shown that the zeros of a *symmetric* polynomial occur in reciprocal pairs z_0 and z_0^{-1}. This is a known result from the theory of equations.

In most applications, $h(n)$ is real and consequently complex zeros occur in complex conjugate pairs. If z_0 is a finite complex zero on the unit circle, then $z_0^* = z_0^{-1}$. If $|z_0| \neq 1$ for a complex z_0, it follows that z_0^*, z_0^{-1}, and $(z_0^*)^{-1}$ are also zeros of $H(z)$.

When $h(n)$ is real and satisfies the symmetry condition of Eq. (3.57), there are four distinct situations that arise for the finite zeros of $H(z)$. These are in the list that follows and are illustrated in Fig. 3.4.

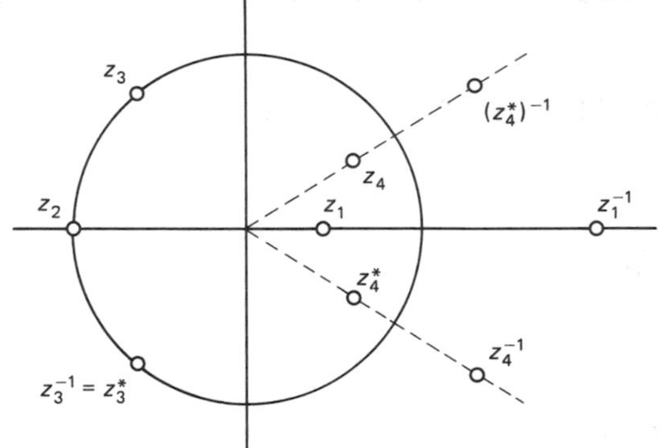

Figure 3.4 Locations of the zeros for the situations I–IV.

I. z_1 is a real zero with $|z_1| < 1$. Then z_1^{-1} is also a real zero and there are two zeros in this group.

II. $z_2 = -1$. Then $z_2^{-1} = z_2$ and this group contains only one zero. [Note that a simple zero at $z = 1$ corresponds to the factor $1 - z^{-1}$, which does not satisfy Eq. (3.57). An even number of zeros results in a power of $(1 - 2z^{-1} + z^{-2})$, which is symmetric.]

III. z_3 is a complex zero with $|z_3| = 1$. Then $z_3^{-1} = z_3^*$ and there are two zeros in this group.

IV. z_4 is a complex zero with $|z_4| \neq 1$. This group contains the four zeros z_4, z_4^{-1}, z_4^*, and $(z_4^*)^{-1}$.

Each factor formed from the zeros in these four groups exhibits the same coefficient symmetry as $H(z)$. In other words, each polynomial formed from the zeros of each group is a linear-phase polynomial. In particular, the second-degree polynomials are of the form

$$a_0 + a_1 z^{-1} + a_0 z^{-2}$$

and the fourth-degree polynomials have the form

$$a_0 + a_1 z^{-1} + a_2 z^{-2} + a_1 z^{-3} + a_0 z^{-4}$$

Example

As an example of a linear-phase FIR system, consider

$$z_1 = \tfrac{1}{2} \qquad z_2 = -1 \qquad z_3 = e^{j3\pi/4} \qquad z_4 = \tfrac{1}{2}e^{j\pi/4}$$

as defined in the groups I–IV. The corresponding system function is

$$\begin{aligned} H(z) &= (1 + z^{-1})[(1 - \tfrac{1}{2}z^{-1})(1 - 2z^{-1})][(1 - e^{j3\pi/4}z^{-1})(1 - e^{-j3\pi/4}z^{-1})] \\ &\quad \times [(1 - \tfrac{1}{2}e^{j\pi/4}z^{-1})(1 - \tfrac{1}{2}e^{-j\pi/4}z^{-1})(1 - 2e^{j\pi/4}z^{-1})(1 - 2e^{-j\pi/4}z^{-1})] \\ &= (1 + z^{-1})(1 - \tfrac{5}{2}z^{-1} + z^{-2})(1 + 2^{1/2}z^{-1} + z^{-2}) \\ &\quad \times \left(1 - \frac{5(2^{1/2})}{2}z^{-1} + \frac{25}{4}z^{-2} - \frac{5(2^{1/2})}{2}z^{-3} + z^{-4}\right) \end{aligned} \tag{3.58}$$

EXERCISES

3.8.1 Determine the linear-phase function of smallest degree with zeros at $z = -1$ and $(1/2^{1/2})(1 + j1)$.

3.8.2 Repeat Ex. 3.8.1 if there are zeros at $z = 1$ and $z = \frac{1}{2}$.

PROBLEMS

3.1 Find the z-transforms of the sequences

(a) $(\frac{1}{2})^n u(n)$
(b) $2^n u(-n)$
(c) $\delta(n - 3)$
(d) $u(n - 2)$
(e) $e^{jn\pi/2}u(n)$
(f) $(-1)^n u(n)$

3.2 Verify Eqs. (3.9) and (3.10).

3.3 Find the z-transforms of the sequences

(a) $f(n) = 0$ for $n < 0$
$= (\frac{1}{2})^n,\ 0 \le n \le 10$
$= (\frac{1}{4})^n,\ n > 10$

(b) $\sin(n\omega_0 + \theta)u(n)$

(c) $g(n) = (\frac{1}{2})^n f(n)$, where $f(n)$ is given in part (a).

3.4 Find $G(z)$ in terms of $F(z)$ if

$$g(n) = \left(\sum_{k=0}^{n} f(k)\right)u(n)$$

Answer: $G(z) = F(z)/(1 - z^{-1})$

3.5 Verify Eqs. (3.25) and (3.26).

3.6 Find the z-transforms of the sequences

(a) $[(\frac{1}{2})^n - (-\frac{1}{2})^n]u(n)$
(b) $[(\frac{1}{4})^n - \cos n\pi/4]\, u(n)$
(c) $2^n \sin (n\pi/4)\, u(n)$
(d) $2^n u(n-2)$
(e) $(\frac{1}{2})^n \sin (n\pi/2)\, u(n-1)$
(f) $\delta(n-1) - 2^n \cos n\pi\, u(n)$

3.7 If $x(n)$ is a causal sequence, find the z-transforms of the sequences

(a) $nu(n)$
Answer: $z^{-1}/(1-z^{-1})^2$.
(b) $n(n-1)x(n)$
(c) $n^2x(n)$
(d) $n(n-1)u(n)$
(e) $n^2a^nu(n)$
(f) $n^2\delta(n-1)$
(g) $n(n-1)(n-2)x(n)$
(h) $n^3u(n)$
(i) $u(n) - u(n-11)$
(j) $nu(n-1)$

3.8 Use the shift property and $Z\{nu(n)\}$ from Prob. 3.7(a) to find $Z\{n(n-1)u(n-1)\}$.

3.9 Find $Y(z)$ in terms of $X(z)$ if

$$y(n) = \beta x(n) + \alpha y(n-1)$$

and

$$y(n) = 0 \qquad n < 0$$

3.10 Find $Z\{x(n+k)\}$ if $x(n)$ is a causal sequence and k is a positive integer.
Answer: $z[x(z) - x(0)]$ for $k = 1$.

3.11 If $r(n)$ is the response to a unit-step sequence, find $R(z)$ in terms of $H(z)$.
Answer: $H(z)/(1-z^{-1})$.

3.12 Let $f(n) = u(n-1)$. Show that if $g(n) = n$, then $g(n) = \Sigma_{k=0}^{n} f(k)$. Find $G(z)$ by using Prob. 3.4.

3.13 Prove the initial-value theorem: for a causal sequence $f(n)$, show that

$$f(0) = \lim_{z\to\infty} F(z)$$

Note that $F(z) = f(0) + \Sigma_{n=1}^{\infty} f(n)z^{-n}$.

3.14 Prove the final-value theorem: for a causal sequence $f(n)$,

$$\lim_{n\to\infty} f(n) = f(\infty) = \lim_{z\to 1}(z-1)F(z)$$

Begin by showing that

$$Z\{f(n+1) - f(n)\} = (z-1)F(z) - zf(0)$$

and hence

$$\lim_{z\to 1}[(z-1)F(z) - zf(0)] = \lim_{N\to\infty}\sum_{n=0}^{N}[f(n+1) - f(n)]$$

The theorem is valid if $F(z)$ has no multiple poles on the unit circle and no poles outside the unit circle.

3.15 Find $f(0)$ and $f(\infty)$ using Probs. 3.13 and 3.14 for the sequence whose z-transform is

$$F(z) = \frac{z}{z-a} \qquad |a| \le 1$$

3.16 Find $Y(z)$ for the causal system with $h(n) = \alpha^n u(n)$ and input $x(n) = \beta^n u(n)$. Consider the following two cases:
(a) $\alpha \neq \beta$
(b) $\alpha = \beta$

3.17 Prove that

$$Z\left\{\sum_{k=-\infty}^{\infty} f(n-k)g(k)\right\} = F(z)G(z)$$

3.18 Find $H(z)$ for the following:

(a) The system of Prob. 3.9.

(b) $y(n) + 3y(n-1) + 2y(n-2) = x(n) - x(n-1)$

(c) $y(n) = x(n) - 2x(n-1) + x(n-2)$

3.19 Find $Y(z)$ for the following causal systems:

(a) $y(n) - \frac{3}{4}y(n-1) + \frac{1}{8}y(n-2) = u(n)$

(b) $y(n) - \frac{3}{4}y(n-1) + \frac{1}{8}y(n-2) = \delta(n-2)$

(c) $y(n) = u(n) + 2u(n-1) - 4u(n-2) + u(n-3)$

3.20 Let $f(n) = f(n+N)$, where N is a positive integer. Define

$$f_1(n) = \begin{cases} f(n) & 0 \le n \le N-1 \\ 0 & \text{otherwise} \end{cases}$$

so that

$$F_1(z) = \sum_{n=0}^{N-1} f(n)z^{-n}$$

Draw $f_1(n)$ if $N = 5$ and $f(n) = 2^{-n}u(n)$. Find $F(z)$ in terms of $F_1(z)$, assuming $f(n) = 0$ for $n < 0$.

3.21 Given $f(n)$ defined by

$$\begin{aligned} f(n) &= 0 && n < 0 \\ &= 1 && n = 0, 1, 2 \\ &= -1, && n = 3, 4, 5 \\ f(n+6) &= f(n) \end{aligned}$$

find $F_1(z)$ and $F(z)$.

3.22 Find $f(n)$, a causal sequence, if $F(z)$ is given by the following. *Hint:* Use long division to obtain a proper fraction in part (*k*).

(a) $\dfrac{6 + z^{-1}}{(1 + \frac{1}{4}z^{-1})(1 + \frac{1}{2}z^{-1})}$

(b) $\dfrac{14 - 14z^{-1} + 3z^{-2}}{(1 - \frac{1}{4}z^{-1})(1 - \frac{1}{2}z^{-1})(1 - z^{-1})}$

(c) $\dfrac{1 - 4z^{-1} - 11z^{-2}}{1 + 2z^{-1} - 5z^{-2} - 6z^{-3}}$

(d) $\dfrac{1 + 2z^{-1}}{1 - \frac{1}{2}z^{-1}}$

(e) $\dfrac{z^{-2} + z^{-3}}{1 + \frac{1}{2}z^{-1}}$

(f) $\dfrac{z^{-1}(1 + 5z^{-1})}{1 + \frac{1}{4}z^{-1} - \frac{1}{8}z^{-2}}$

(g) $\dfrac{z + 1}{z^{10}(z - \frac{1}{2})}$

(h) $\dfrac{z + 2}{z^2 - \frac{1}{4}z - \frac{1}{8}}$

(i) $3 + 2z^{-1} + 6z^{-4}$

(j) $\dfrac{1 + 2z^{-1}}{1 - 2z^{-1} + 4z^{-2}}$

(k) $\dfrac{z^2 + 5z + 3}{z^2 + \frac{1}{4}z - \frac{1}{8}}$

3.23 Repeat Prob. 3.22 for the following:

(a) $\dfrac{1 + z^{-1}}{1 - z^{-1} + z^{-2}}$

(b) $\dfrac{2z^{-1}}{1 - z^{-1} + z^{-2}}$

(c) $\dfrac{1 + z^{-1}}{(1 - z^{-1})(1 - z^{-1} + \frac{1}{4}z^{-2})}$

(d) $\dfrac{3 - 3z^{-1}}{(1 + z^{-1})(1 - z^{-1} + z^{-2})}$

(e) $\dfrac{1 - 3z^{-1} + 5z^{-2}}{(1 - z^{-1} + z^{-2})(1 - 2z^{-1} + 4z^{-2})}$

(f) $\dfrac{3 + 5z^{-2}}{(1 + z^{-1} + z^{-2})(1 - 2z^{-1} + 4z^{-2})}$

(g) $\dfrac{2z^{-1}}{(1 - \frac{1}{4}z^{-1})^2}$

(h) $\dfrac{1}{(1 - \frac{1}{2}z^{-1})^2}$

(i) $\dfrac{4z^{-3}}{(1 - z^{-1})^2}$

(j) $\dfrac{18}{(1 - 2z^{-1})^2(1 + z^{-1})}$

(k) $\dfrac{4 + z^{-1} + 6z^{-2} + 2z^{-3}}{(1 + z^{-1})^2(1 - z^{-1} + z^{-2})}$

(l) $\dfrac{5 - 2z^{-1} + z^{-2}}{(1 + z^{-1})^2(1 - z^{-1})^2}$

3.24 Find $y(n)$ in Prob. 3.9 if
(a) $x(n) = \delta(n)$
(b) $x(n) = u(n)$
(c) $x(n) = \cos n\pi/3\, u(n)$ and $\alpha = \frac{1}{2}$

3.25 Find the impulse response for the systems of Prob. 3.18(b) and (c) using the z-transform.

3.26 Find the impulse response for the following systems:
(a) $y(n) - \frac{3}{4}y(n-1) + \frac{1}{8}y(n-2) = x(n)$
(b) $y(n) = x(n) + 2x(n-1) - 4x(n-2) + x(n-3)$
(c) $y(n) - y(n-1) = x(n) + x(n-1)$
(d) $y(n) + 4y(n-1) + 4y(n-2) = x(n)$

3.27 Find $y(n)$ in Prob. 3.18(b) and (c) if
(a) $x(n) = u(n)$
(b) $x(n) = 2^{-n}u(n)$

3.28 Find $y(n)$ in Prob. 3.19(a) and (b).

3.29 System stability can be determined by testing the roots of the characteristic equation or, equivalently, the poles of the system function. Jury's test [84] provides a means for testing a polynomial to determine if all of its zeros lie within the unit circle without actually calculating them. For the polynomial

$$P(q) = a_n q^n + a_{n-1}q^{n-1} + \cdots + a_1 q + a_0$$

with $a_n > 0$, the following table is constructed.

AN ARRAY FOR JURY'S TEST

Row	q^0	q^1	q^2	$\cdots$	q^{n-k}	$\cdots$	q^{n-2}	q^{n-1}	q^n
1	a_0	a_1	a_2	$\cdots$	a_{n-k}	$\cdots$	a_{n-2}	a_{n-1}	a_n
2	a_n	a_{n-1}	a_{n-2}	$\cdots$	a_k	$\cdots$	a_2	a_1	a_0
3	b_0	b_1	b_2	$\cdots$	b_k	$\cdots$	b_{n-2}	b_{n-1}	
4	b_{n-1}	b_{n-2}	b_{n-3}	$\cdots$	$\cdots$	$\cdots$	b_1	b_0	
5	c_0	c_1	c_2	$\cdots$	c_k	$\cdots$	c_{n-2}		
6	c_{n-2}	c_{n-3}	c_{n-4}	$\cdots$	$\cdots$	$\cdots$	c_0		
$\vdots$					$\ddots$				
$2n-5$	s_0	s_1	s_2	s_3					
$2n-4$	s_3	s_2	s_1	s_0					
$2n-3$	r_0	r_1	r_2						

The coefficients appearing after the first two rows are given by

$$b_k = \begin{vmatrix} a_0 & a_{n-k} \\ a_n & a_k \end{vmatrix} \qquad c_k = \begin{vmatrix} b_0 & b_{n-1-k} \\ b_{n-1} & b_k \end{vmatrix}$$

$$d_k = \begin{vmatrix} c_0 & c_{n-2-k} \\ c_{n-2} & c_k \end{vmatrix} \qquad \cdots \qquad r_k = \begin{vmatrix} s_0 & s_{3-k} \\ s_3 & s_k \end{vmatrix}$$

The necessary and sufficient conditions for all zeros of $P(q)$ to be inside the unit circle $|q| = 1$ are

$$P(1) > 0 \qquad (-1)^n P(-1) > 0$$

$$a_n > |a_0|$$

$$|b_0| > |b_{n-1}|$$

$$|c_0| > |c_{n-2}|$$

$$\vdots$$

$$|r_0| > |r_2|$$

Use Jury's test to determine if the following polynomials have any zeros for which $|q| \geq 1$.

(a) $2q^4 + 3q^3 - 2q^2 + q + 1$
(b) $2q^4 + q^3 - 2q^2 + q + 1$
(c) $12q^3 + 8q^2 - 3q - 2$
(d) $2q^4 + 3q^3 + 2q^2 + q + 1$
(e) $2q^3 + 3q^2 - q + 1$
(f) $2q^3 - 3q^2 - q + 1$
(g) $2q^4 + q^3 + q^2 + 3q + 1$
(h) $2q^4 + q^3 + q^2 - 2q - 1$

3.30 Develop the constraints for the general second-degree polynomial

$$P(q) = a_2 q^2 + a_1 q + a_0$$

Apply this test to the following polynomials:

(a) $3q^2 + q + 2$
(b) $2q^2 + 3q + 1$

Answers:

$$P(1) = a_2 + a_1 + a_0 > 0$$

$$P(-1) = a_2 - a_1 + a_0 > 0$$

$$a_2 > |a_0|$$

(a) All zeros are inside the unit circle.
(b) At least one zero is on or outside the unit circle.

3.31 Another technique used for determining the location of zeros of a polynomial in z is to use a transformation. If we set

$$z = \frac{1 + s}{1 - s}$$

where $s = \sigma + j\Omega$, show that

$$|z| < 1 \text{ if and only if } \sigma < 0$$

$$|z| = 1 \text{ if and only if } \sigma = 0$$

$$|z| > 1 \text{ if and only if } \sigma > 0$$

If $P(z)$ is a polynomial of degree N in z, show that

$$P\left(\frac{1+s}{1-s}\right) = (1-s)^{-N}Q(s)$$

where $Q(s)$ is a polynomial in s of degree N. Obtain $Q(s)$ if

$$P(z) = 12z^3 + 8z^2 - 3z - 2$$

Hint: Note that

$$|z|^2 = \frac{(1+\sigma)^2 + \Omega^2}{(1-\sigma)^2 + \Omega^2}$$

3.32 In Prob. 3.31, we saw that the location of zeros of $P(z)$ could be determined from the location of the zeros of $Q(s)$, which was obtained from $P(z)$ by a transformation. *Routh's algorithm* [91] can be used to determine the location of the zeros of the polynomial

$$Q(s) = a_0 s^N + a_1 s^{N-1} + \cdots + a_{N-1}s + a_N$$

where $a_0 > 0$. The algorithm requires the construction of the *Routh array,* which is shown in the following table.

THE ROUTH ARRAY

s^n	a_0	a_2	a_4	$\cdots$
s^{n-1}	a_1	a_3	a_5	$\cdots$
s^{n-2}	b_1	b_2	b_3	$\cdots$
s^{n-3}	c_1	c_2	c_3	$\cdots$
s^{n-4}	d_1	d_2	d_3	$\cdots$
s^{n-5}	e_1	e_2	e_3	$\cdots$
$\vdots$	$\vdots$	$\vdots$	$\vdots$	$\ddots$
s^2	f_1	f_2		
s^1	g_1			
s^0	h_1			

The first two rows of the table contain the coefficients of $Q(s)$. Each row in the array corresponds to an even or an odd polynomial with the degree indicated by the power of s identifying the row. The entries in the table starting with the third row are calculated from the previous two rows using the relations

$$b_i = -\frac{\begin{vmatrix} a_0 & a_{2i} \\ a_1 & a_{2i+1} \end{vmatrix}}{a_1} \qquad c_i = -\frac{\begin{vmatrix} a_1 & a_{2i+1} \\ b_1 & b_{i+1} \end{vmatrix}}{b_1}$$

$$d_i = -\frac{\begin{vmatrix} b_1 & b_{i+1} \\ c_1 & c_{i+1} \end{vmatrix}}{c_1} \qquad e_i = -\frac{\begin{vmatrix} c_1 & c_{i+1} \\ d_1 & d_{i+1} \end{vmatrix}}{d_1} \qquad \cdots$$

for $i = 1, 2, 3, \ldots$. This pattern must be modified if a zero appears in the first column [91] and we consider this situation in Prob. 3.33. According to Routh's algorithm, the number of sign changes in the coefficients of the first column of the array is the number of zeros of $Q(s)$ in the right half of the s plane (RHP).

Show that the polynomial

$$Q(s) = s^4 + 6s^3 + 13s^2 + 12s + 4$$

has the Routh array shown below and therefore has no RHP zeros. It should be noted that a common positive factor can be divided out of a row without changing the sign of any elements in the first column. Divide out the factor 6 in the second row and complete the Routh array from this resulting row and the first row. Show that no sign changes occur in the first column of the resulting array.

s^4	1	13	4
s^3	6	12	
s^2	11	4	
s^1	$\frac{108}{11}$		
s^0	4		

3.33 Whenever a zero appears in the first column of a Routh array, the procedure given in Prob. 3.32 must be modified. Whenever an entire row of zeros appears, the polynomial corresponding to the previous row is a factor of $Q(s)$. This polynomial is then differentiated and its coefficients are used to replace the row of zeros. Construct the Routh array for the polynomial

$$Q(s) = s^4 + s^3 + 3s^2 + 2s + 2$$

Show that $s^2 + 2$ is a factor of $Q(s)$ and determine the nature of the zeros.

3.34 Use Routh's algorithm to determine the nature of the zeros of the following polynomials:

(a) $s^6 + 3s^5 + 2s^4 + 9s^3 + 5s^2 + 12s + 20$
(b) $s^5 + s^4 + 3s^3 + 3s^2 + 2s + 2$
(c) $2s^5 + 4s^4 + 5s^3 + 6s^2 + 5s + 2$
(d) $s^6 + 2s^5 + 4s^4 + 6s^3 + 5s^2 + 4s + 2$
(e) $s^4 + 4s^3 + 3s^2 + 8s + 2$
(f) $s^4 + 2s^3 + 11s^2 + 18s + 18$
(g) $s^6 + 3s^5 + 3s^4 + 4s^3 + 4s^2 + 4s + 2$

3.35 Whenever the Routh array for a polynomial $Q(s)$ has a first column zero in a row that is not all zeros, we can test either of the new polynomials (1) $(s + 1)Q(s)$ or (2) $s^N Q(1/s)$. Show that these polynomials have the same number of RHP zeros as does $Q(s)$. Use these polynomials to determine the nature of the zeros of the following polynomials:

(a) $s^4 + s^3 + 2s^2 + 2s + 5$
(b) $2s^4 + 2s^3 + 3s^2 + 3s + 2$

3.36 Use the transformation given in Prob. 3.31 and the Routh algorithm to determine the nature of the zeros with respect to the unit circle of the following polynomials:

(a) $2z^3 + 3z^2 - z + 1$
(b) $2z^4 + z^3 + 3z^2 + z + 1$
(c) $8z^3 + 4z^2 - 3z - 1$
(d) $4z^4 + 7z^3 + 2z^2 - 2$

3.37 Discuss the stability of the systems described by the following system functions.

(a) $\dfrac{1 + z^{-1} + z^{-2}}{1 - 1.5z^{-1} + z^{-2} - 0.5z^{-3} + 0.5z^{-4}}$

(b) $\dfrac{2 + z^{-1}}{1 + z^{-1} + 0.5z^{-2}}\,\dfrac{1 + z^{-1}}{1 + 2z^{-1} + 0.5z^{-2}}$

(c) $\dfrac{1}{1 - z^{-1} + 0.5z^{-2}}\,\dfrac{1 + 0.5z^{-1}}{2 + 2z^{-1} + z^{-2}}$

(d) $\dfrac{(1 + z^{-1})(1 + 0.5z^{-1})}{8 - 12z^{-1} - 10z^{-2} + 4z^{-3} + 2z^{-4}}$

(e) $\dfrac{1}{1 + 0.5z^{-1} - 0.25z^{-2} - 0.125z^{-3}}$

3.38 Determine $H(e^{j\omega})$ from $H(z)$ for the system

$$y(n) + \tfrac{1}{4}y(n-1) = x(n) - x(n-1)$$

Determine the magnitude and phase responses.

3.39 Determine $H(z)$ for the given systems. Discuss stability and, if possible, determine $H(e^{j\omega})$ from $H(z)$.

(a) $8y(n) - 6y(n-1) + y(n-2) = x(n) + x(n-1)$

(b) $y(n) + y(n-1) + 2y(n-2) = x(n)$

(c) $2y(n) + 3y(n-1) + y(n-2) + y(n-3) = x(n)$

3.40 If $H(z)$ has a zero at $z = 1$, $z = -1$, and $z = \frac{1}{2} + j\frac{1}{2}$, determine the lowest-degree $H(z)$ that has a linear phase.

3.41 Suppose that $H(z)$ has zeros at $\frac{3}{4}e^{-j\pi/2}$ and $2e^{-j\pi/4}$. Determine the lowest-degree $H(z)$ that has a linear phase.

3.42 A sampled signal $f^*(t)$ (* does not mean complex conjugation in this problem) is defined by

$$f^*(t) = f(t)s(t)$$

where $s(t)$ is a *train of impulses*,

$$s(t) = \sum_{n=-\infty}^{\infty} \delta(t - nT)$$

If $F^*(s)$ is the Laplace transform of $f^*(t)$, show that the z-transform is given by

$$F(z) = F^*(s)\big|_{z=e^{sT}}$$

3.43 Determine the relations between the parameters in the equivalent representations

$$H(e^{j\omega}) = \frac{1 + b_1e^{-j\omega} + b_2e^{-j2\omega}}{1 + a_1e^{-j\omega} + a_2e^{-j2\omega}} = \frac{(e^{j\omega} - r_0e^{j\phi_0})(e^{j\omega} - r_0e^{-j\phi_0})}{(e^{j\omega} - \gamma_0e^{j\theta_0})(e^{j\omega} - \gamma_0e^{-j\theta_0})}$$

Determine **(a)** $|H(e^{j\omega})|^2$, **(b)** $\phi(\omega)$, and **(c)** $\tau(\omega)$ in terms of r_0, ϕ_0, γ_0, and θ_0. These relations were used by Deczky [30] in a design procedure for IIR filters.

3.44 Apply the results of Prob. 3.43 to the system

$$y(n) + 2y(n-1) + \tfrac{1}{2}y(n-2) = \tfrac{1}{2}x(n) + 2x(n-1) + x(n-2)$$

4

REALIZATION OF DIGITAL SYSTEMS

4.1 INTRODUCTION

In Chapters 1 and 2, we discussed discrete-time signals and systems, emphasizing linear time-invariant systems described by a difference equation of the form

$$\sum_{k=0}^{N} a_k y(n-k) = \sum_{m=0}^{M} b_m x(n-m) \tag{4.1}$$

where the a_k and b_m are constants with $a_0 \neq 0$. In Chapter 3, we introduced the z-transform and saw that we could describe the system of Eq. (4.1) by its system function

$$H(z) = \frac{Y(z)}{X(z)} = \frac{\sum_{m=0}^{M} b_m z^{-m}}{\sum_{k=0}^{N} a_k z^{-k}} \tag{4.2}$$

It should be clear to the reader that we could start with Eq. (4.2), clear it of fractions to obtain

$$\sum_{k=0}^{N} a_k z^{-k} Y(z) = \sum_{m=0}^{M} b_m z^{-m} X(z)$$

and take the inverse z-transform to obtain Eq. (4.1). We can, therefore, start with either the describing relation of Eq. (4.1) or of Eq. (4.2) and obtain the other.

In practice, we are concerned with the digital system obtained from Eq. (4.1) by digitizing the coefficients and the input signal. This results in a digitized output signal. In this chapter, we consider some different computational algorithms for analyzing this digital system. Each computational algorithm can be implemented by using a digital computer or special-purpose digital hardware. In either case, the computational algorithm is specified by a set of basic computations. A digital-computer implementation requires a program that causes the computer to perform the required computations of the algorithm. An implementation using special-purpose digital hardware requires devices that perform the basic computations or operations. In order to implement the describing difference equations of the system, which have the form of Eq. (4.1), the basic operations needed are addition, delay, and multiplication by a constant. The particular computational algorithm chosen for the system implementation determines a structure or network consisting of an interconnection of devices that perform these basic operations. On the other hand, a chosen structure determines a computational algorithm.

Two structures can be equivalent if we assume that infinite-precision representations of the coefficients and variables are used. However, since finite-precision representations and arithmetic are used in the physical implementations, different structures generally give different results. In this chapter, we examine various structures for implementing the digital systems. In all cases, we use infinite-precision representations and arithmetic. The reader interested in the effects of finite-precision representations should consult [52, 74, 98, 99, 129, 184, 185].

Recursive and Nonrecursive Structures

The methods for realizing digital systems can be divided into two classes, *recursive* and *nonrecursive*. The functional relation between input and output sequences for a *recursive realization* has the form

$$y(n) = F[y(n-1), y(n-2), \ldots, x(n), x(n-1), \ldots]$$

For the linear time-invariant system of Eq. (4.1), the recursive realization has the form

$$y(n) = -\sum_{k=1}^{N} \frac{a_k}{a_0} y(n-k) + \sum_{m=0}^{M} \frac{b_m}{a_0} x(n-m) \tag{4.3}$$

The current output sample $y(n)$ is a function only of past outputs and present and past input samples. The input–output relation for a *nonrecursive realization* has the form

$$y(n) = F[x(n), x(n-1), \ldots]$$

For a linear time-invariant system, this relation would be

$$y(n) = \sum_{m=0}^{M} \frac{b_m}{a_0} x(n-m) \tag{4.4}$$

We note that Eq. (4.3) corresponds to an IIR system and Eq. (4.4) corresponds to a FIR system, as discussed in Chapter 2. It should be emphasized that classification based on the impulse-response duration and classification based on realization are two different concepts. For example, it is possible to realize a FIR system by a recursive realization. Such an example is considered in Prob. 4.34.

In this chapter, we discuss direct-form, cascade, and parallel realizations of IIR filters, and direct-form and cascade realizations of FIR filters. These are the most common types of realizations. In addition to these, we consider some of the more common ladder configurations. A number of other realizations have been given in the literature and some of these are given in the problems at the end of the chapter.

4.2 BLOCK DIAGRAMS AND SIGNAL-FLOW GRAPHS

Block-Diagram Representations

Once we have decided on the particular system structure, we will be interested in representing the system by a block diagram or a signal-flow graph. Since the basic operations that must be performed in analyzing the system of Eq. (4.1) are addition, delay, and multiplication by a constant, we require symbols for them. The block-diagram symbols for these operations are shown in Fig. 4.1. We show a symbol for subtraction in Fig. 4.1(d) that is frequently used in control systems. For the most part, we obtain the difference $x_1(n) - x_2(n)$ as the sum $x_1(n) + [-x_2(n)]$ and use the adder of Fig. 4.1(a). The delay operation represents the storing of the previous value in the sequence. We use z^{-1} in the delay representation of Fig. 4.1(b) since

$$Z\{x(n-1)\} = z^{-1}X(z)$$

The same symbols are used also to represent operations involving z-transforms. For

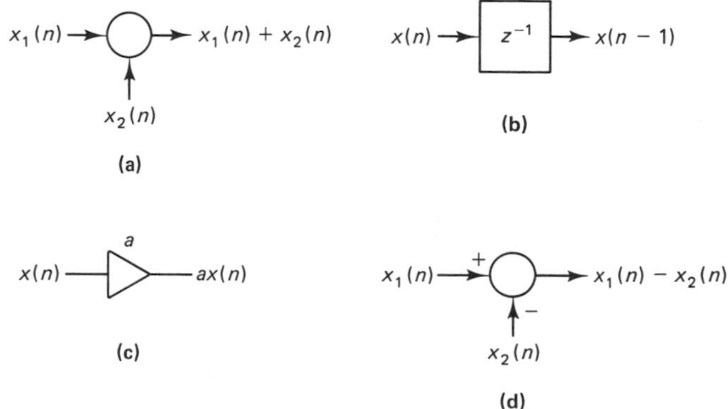

Figure 4.1 Symbols for (a) addition, (b) delay, (c) multiplication by a constant, and (d) subtraction.

example, if $X(z)$ is the input of the multiplier, then the output is $aX(z)$. The symbols used in signal-flow graphs are given later.

Example of a First-Order System

We now see how these basic elements or devices can be used to represent some simple systems by block diagrams. First, consider the system described by Eq. (2.2),

$$y(n) = x(n) + ay(n-1) \tag{4.5}$$

The term $y(n-1)$ is obtained as the output of a delay element with $y(n)$ as input, and $y(n)$ is the output of an adder with $x(n)$ and $ay(n-1)$ as the two inputs. The block diagram representing this system is shown in Fig. 4.2.

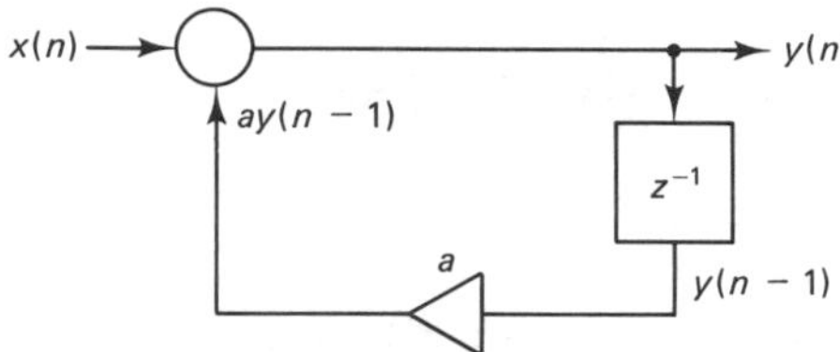

Figure 4.2 Block-diagram representation for the system of Eq. (4.5).

An important point concerns the output of the adder, which is given by Eq. (4.5) for this example. Since $x(n)$ and $y(n-1)$ are known *a priori* or have been calculated, we can compute $y(n)$. The computation could not be carried out, however, if an adder input depended on a term that has not yet been computed. The delay element in the loop* guarantees that this last situation is not encountered in this example. As we will see later, the inclusion of a delay element in each loop in a block diagram ensures that addition can be carried out in each case and makes possible the physical realization of the block-diagram representation of the system.

Example of a Second-Order System

As another example, consider the second-order system

$$y(n) + a_1 y(n-1) + a_2 y(n-2) = x(n) \tag{4.6}$$

The reader is asked in Prob. 4.1 to show that the block diagram given in Fig. 4.3 is a representation of this system. Note that there are two loops in the system and each loop contains a delay element. The outputs of the two adders are

$$[-a_1 y(n-1) - a_2 y(n-2)]$$

and

$$x(n) + [-a_1 y(n-1) - a_2 y(n-2)]$$

both of which can be computed at the nth sample.

*A loop is a closed path that is traced according to the signal flow that is in the direction from input to output for the devices shown in Fig. 4.1.

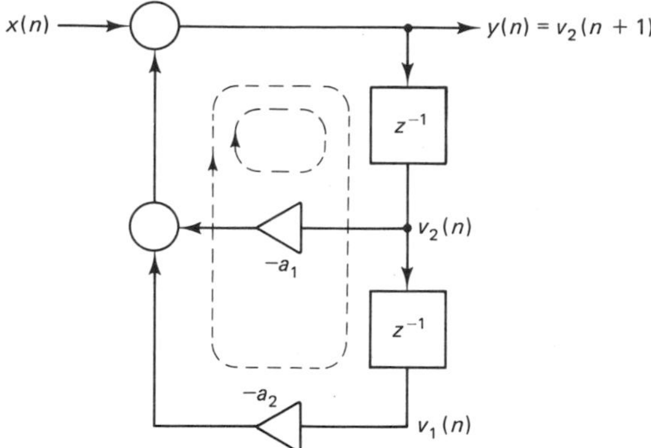

Figure 4.3 Block-diagram representation for the system of Eq. (4.6).

Example of a Cascaded System

Earlier in this chapter, we indicated that, in general, several realizations can be obtained for a system. Let us take $a_1 = 1$ and $a_2 = \frac{1}{4}$ in Eq. (4.6) and obtain a cascade realization as discussed in Sec. 3.5. For this system, we have

$$H(z) = H_1(z)H_2(z) = \frac{1}{(1 + \frac{1}{2}z^{-1})^2} \tag{4.7}$$

and so we can take

$$H_1(z) = H_2(z) = \frac{1}{1 + \frac{1}{2}z^{-1}}$$

Each of these subsystems is described by a difference equation such as Eq. (4.5) with $a = -\frac{1}{2}$. We can realize the system of Eq. (4.7) by cascading the two first-order subsystems, each having the block-diagram representation given in Fig. 4.2 with $a = -\frac{1}{2}$. The resulting representation is shown in Fig. 4.4.

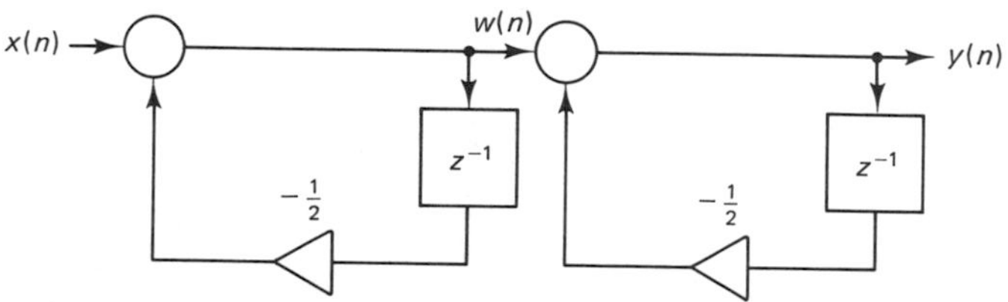

Figure 4.4 Cascade realization for the system of Eq. (4.7).

Example of a FIR System

The examples considered so far have been IIR systems. Let us next consider the FIR system described by

$$y(n) = b_0x(n) + b_1x(n-1) + b_2x(n-2) \tag{4.8}$$

One suitable representation of this system, shown in Fig. 4.5, is obtained where the computations are done according to the equation

$$y(n) = [b_0x(n) + b_1x(n - 1)] + b_2x(n - 2)$$

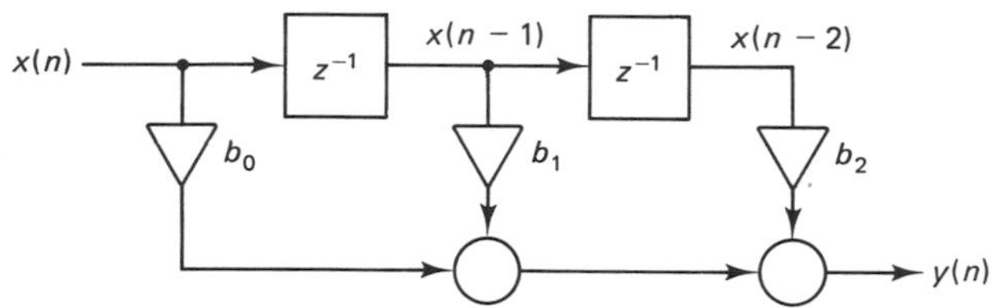

Figure 4.5 Block-diagram representation for the system of Eq. (4.8).

Signal-Flow Graphs

We can also represent the equations describing a discrete-time system by a *signal-flow graph,* which was introduced by Mason [101, 102]. For this purpose, let us assume that the equations have the form

$$x_i = \sum_{j=1}^{n} t_{ij}x_j \qquad i = 1, 2, \ldots, m \tag{4.9}$$

We obtain a graph whose *vertices* or *nodes* represent the variables and are labeled x_1, $x_2, \ldots, x_n$. For each vertex x_i, we construct an *edge* or a *branch* from vertex x_j to x_i in accordance with the ith equation of Eq. (4.9). These branches have *gains* $t_{ij} \neq 0$. We say that a signal flows from x_j to x_i with gain t_{ij}. The direction of signal flow is indicated by an arrow. For example, a single branch is shown in Fig. 4.6(a), with signal flow from x_2 to x_1, so that $x_1 = t_{12}x_2$. If $t_{ij} = 0$, then there is no branch from x_j to x_i. The equation

$$x_1 = t_{11}x_1 + t_{12}x_2 + t_{13}x_3 \tag{4.10}$$

is represented by the signal-flow graph of Fig. 4.6(b). We frequently leave the gain label off if it is unity.

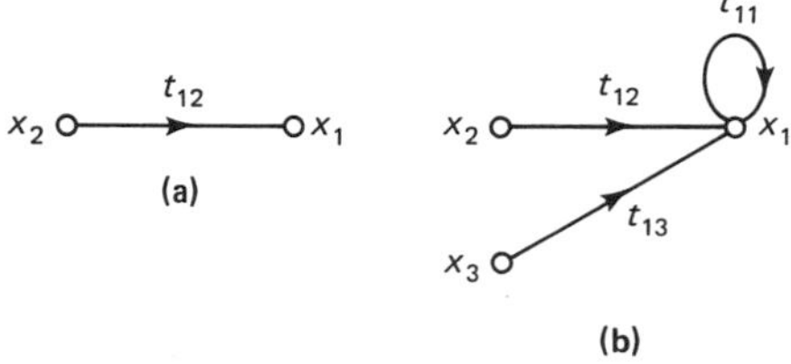

Figure 4.6 Signal-flow graphs (a) consisting of a single branch and (b) describing Eq. (4.10).

Each vertex x_i can be thought of as an adder that sums the incoming signals to obtain the value of x_i. Outgoing signals do not contribute to the sum. The resulting graph obtained in this way is Mason's signal-flow graph.

Normally, we construct a signal-flow graph from the equations relating the z-transforms of the variables. The basic operations of delay and multiplication by a constant are represented by the branch of Fig. 4.6(a), where $t_{ij} = z^{-1}$ for delay, and $t_{ij} = a$ for multiplication. We can also use difference equations to construct a signal-flow graph if we associate $X(z)$ and $z^{-1}X(z)$ with $x(n)$ and $x(n - 1)$, respectively.

Example of a First-Order System

As a first example of a signal-flow graph, let us consider the equation obtained by taking the z-transform of Eq. (4.5), which is

$$Y(z) = X(z) + az^{-1}Y(z) \tag{4.11}$$

The variables to be represented by nodes are $Y(z)$, $z^{-1}Y(z)$, and $X(z)$. The signal-flow graph representing Eq. (4.11) is shown in Fig. 4.7. Note that one node represents the sum $X(z) + az^{-1}Y(z)$. This added node makes the signal-flow graph resemble the block diagram of Fig. 4.2.

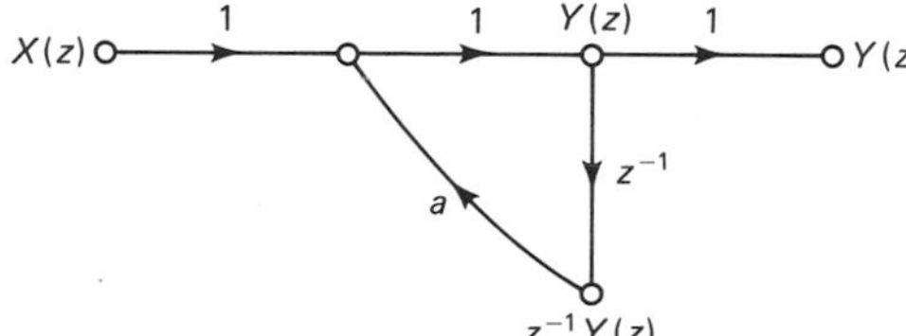

Figure 4.7 Signal-flow-graph representation for the system of Eq. (4.11).

Example of a Second-Order System

Another example of a signal-flow graph is shown in Fig. 4.8, which represents the z-transform of Eq. (4.6). In this case, we have nodes representing $X(z)$, $Y(z)$, $z^{-1}Y(z)$, $z^{-2}Y(z)$, and a node representing the sum $-a_1z^{-1}Y(z) - a_2z^{-2}Y(z) + X(z)$.

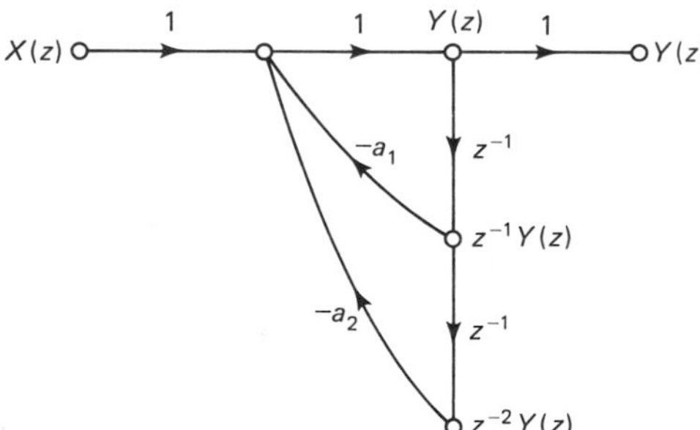

Figure 4.8 Signal-flow-graph representation for the system of Eq. (4.6).

Comparison of the signal-flow graphs of Figs. 4.7 and 4.8 with the corresponding block diagrams of Figs. 4.2 and 4.3 shows that there is very little difference. The main difference is that in the signal-flow graph, a node also represents an adder.

Mason's Formula

One advantage in using signal-flow graph representations is that a formula exists for determining the system outputs in terms of the inputs. Before giving this formula, we need to introduce some additional terminology. An input variable is represented by a *source node* (one that has only outgoing signals) and an output node is represented by a *sink node* (one that has only incoming signals). A *forward path* is any connected sequence of branches whose arrows are all in the same direction, with no node appearing more than once. The *gain* of a forward path is the product of the gains of the branches in the path. A *loop* is a closed forward path. Loops are said to be *touching* if they have at least one common node; otherwise, they are *nontouching*.

Mason's rule, or *formula,* for the graph gain between a source node and a sink node in a signal-flow graph is

$$G = \frac{Y}{X} = \frac{1}{\Delta} \sum_k G_k \Delta_k \tag{4.12}$$

where the *graph determinant* Δ, the *path factor* Δ_k for the kth forward path, and the forward-path gain G_k are defined as follows:

$\Delta = 1$ − (sum of all individual loop gains)
+ (sum of the products of loop gains of all possible nontouching loops taken two at a time)
− (sum of the products of loop gains of all possible nontouching loops taken three at a time)
+ . . . .

Δ_k = value of Δ for that part of the graph that remains after the kth forward path is removed. (Removal of a node also removes the branches connected to that node.)

G_k = path gain of the kth forward path from the source node to the sink node.

The summation is taken over all forward paths from the source node to the sink node.

Example of Mason's Formula

As an example of the use of Mason's formula, consider the signal-flow graph of Fig. 4.8. There is one forward path from $X(z)$ to $Y(z)$, shown in Fig. 4.9(a), and its gain

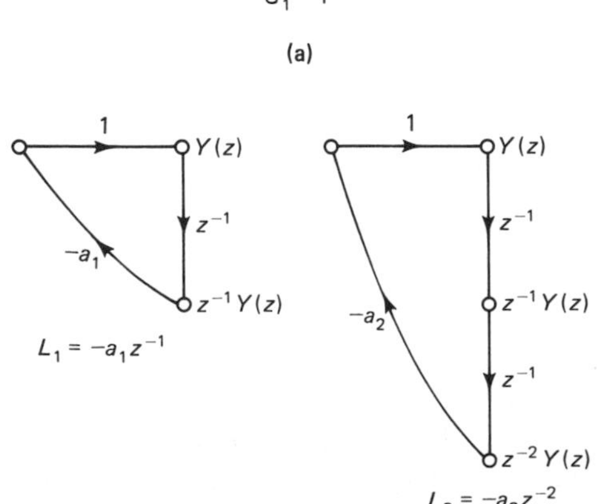

Figure 4.9 (a) Forward path and (b) loops of the signal-flow graph of Fig. 4.8.

is $G_1 = (1)(1)(1) = 1$. There are two loops, shown in Fig. 4.9(b), and their gains are $L_1 = (1)(z^{-1})(-a_1) = -a_1z^{-1}$ and $L_2 = (1)(z^{-1})(z^{-1})(-a_2) = -a_2z^{-2}$. Since these loops are touching, we have

$$\Delta = 1 - (L_1 + L_2) = 1 + a_1z^{-1} + a_2z^{-2}$$

When the forward path is removed from the signal-flow graph, the result is a graph that has two nodes and no branches. Therefore, $\Delta_1 = 1$ and

$$G = \frac{G_1\Delta_1}{\Delta} = \frac{1}{1 + a_1z^{-1} + a_2z^{-2}}$$

Consequently, the output has the representation

$$Y(z) = GX(z) = \frac{1}{1 + a_1z^{-1} + a_2z^{-2}}X(z)$$

In most cases, the block-diagram representation of a system is so similar to the corresponding signal-flow graph that the forward paths and loops can be determined directly from the block diagram. In these instances, we can use Mason's formula and determine the transfer function of the system directly from the block diagram. For example, in the block diagram of Fig. 4.3, there is one forward path from X to Y with $G_1 = 1$. When this path is removed, there are no loops left and so $\Delta_1 = 1$. There are two touching loops with gains $-a_1z^{-1}$ and $-a_2z^{-2}$.

General Second-Order System

An example of a signal-flow graph having more than one forward path is that of the general second-order system shown in Fig. 4.10(a). This system can be obtained

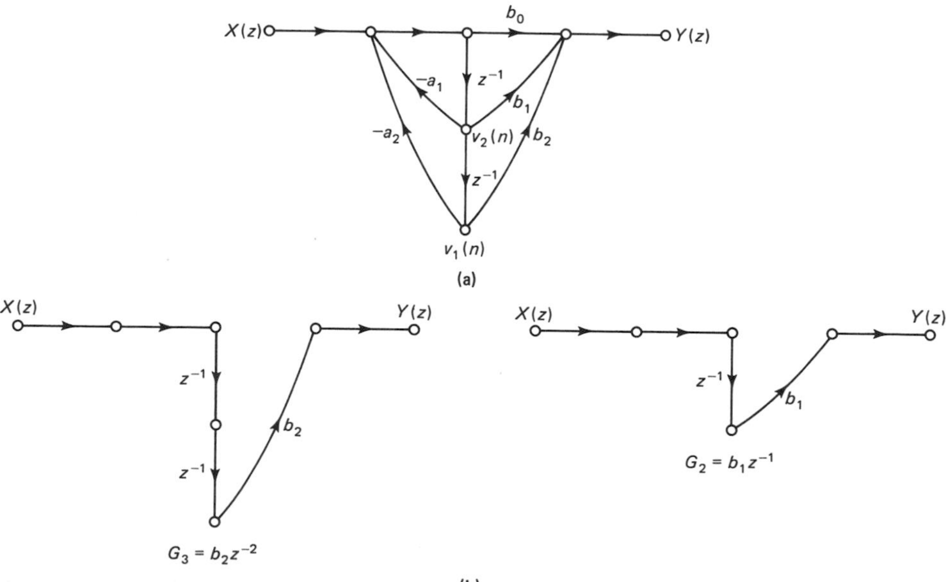

Figure 4.10 (a) Signal-flow graph of a general second-order system and (b) two of its forward paths.

from the graph of Fig. 4.8 by including the outputs of the delay elements in the system output. In addition to the forward path G_1, shown in Fig. 4.9(a), there are two additional forward paths, shown in Fig. 4.10(b), with gains $G_2 = (1)(z^{-1})(b_1) = b_1 z^{-1}$ and $G_3 = (1)(z^{-1})(z^{-1})(b_2) = b_2 z^{-2}$. The two signal-flow graphs have the same loops L_1 and L_2 shown in Fig. 4.9(b). When each of the forward paths is removed, the remaining subgraph has no branches and so $\Delta_1 = \Delta_2 = \Delta_3 = 1$. Therefore, the system function, given by Mason's formula, is

$$\begin{aligned} \frac{Y(z)}{X(z)} &= \frac{G_1\Delta_1 + G_2\Delta_2 + G_3\Delta_3}{1 - (L_1 + L_2)} \\ &= \frac{b_0 + b_1 z^{-1} + b_2 z^{-2}}{1 + a_1 z^{-1} + a_2 z^{-2}} \end{aligned} \tag{4.13}$$

Example of a More Complex System

As a final example in this section, let us determine the system function for the system whose block diagram is shown in Fig. 4.11(a). A signal-flow graph for this system using the indicated variables is shown in Fig. 4.11(b). The graph that remains after removal of the forward path from X to Y is shown in Fig. 4.11(c), and the loops of the signal-flow graph are shown in Fig. 4.11(d) with L_1 and L_3 nontouching. The loops of Fig. 4.11(c) are L_2 and L_3, which are touching. Consequently, the transfer function is given by

$$\begin{aligned} H(z) &= \frac{G_1\Delta_1}{\Delta} \\ &= \frac{(1)(1 - L_2 - L_3)}{1 - (L_1 + L_2 + L_3 + L_4) + L_1 L_3} \\ &= \frac{1 + 2z^{-1} + 2z^{-2}}{1 + (2z^{-1} - z^{-1} + z^{-2} + 2z^{-2}) - 2z^{-2}} \\ &= \frac{1 + 2z^{-1} + 2z^{-2}}{1 + z^{-1} + z^{-2}} \end{aligned}$$

This structure is taken from Hwang [71], where several types of structures are developed.

The reader should observe from Fig. 4.11 that the forward path and loops of the signal-flow graph can be obtained directly from the block diagram. These are designated as L_1', L_2', L_3', and L_4', where L_1' and L_3' are nontouching.

Summary

In this section, we have considered block-diagram and signal-flow-graph representations of a linear time-invariant discrete-time system. We have given Mason's formula, which provides a convenient means for obtaining the transfer function for a

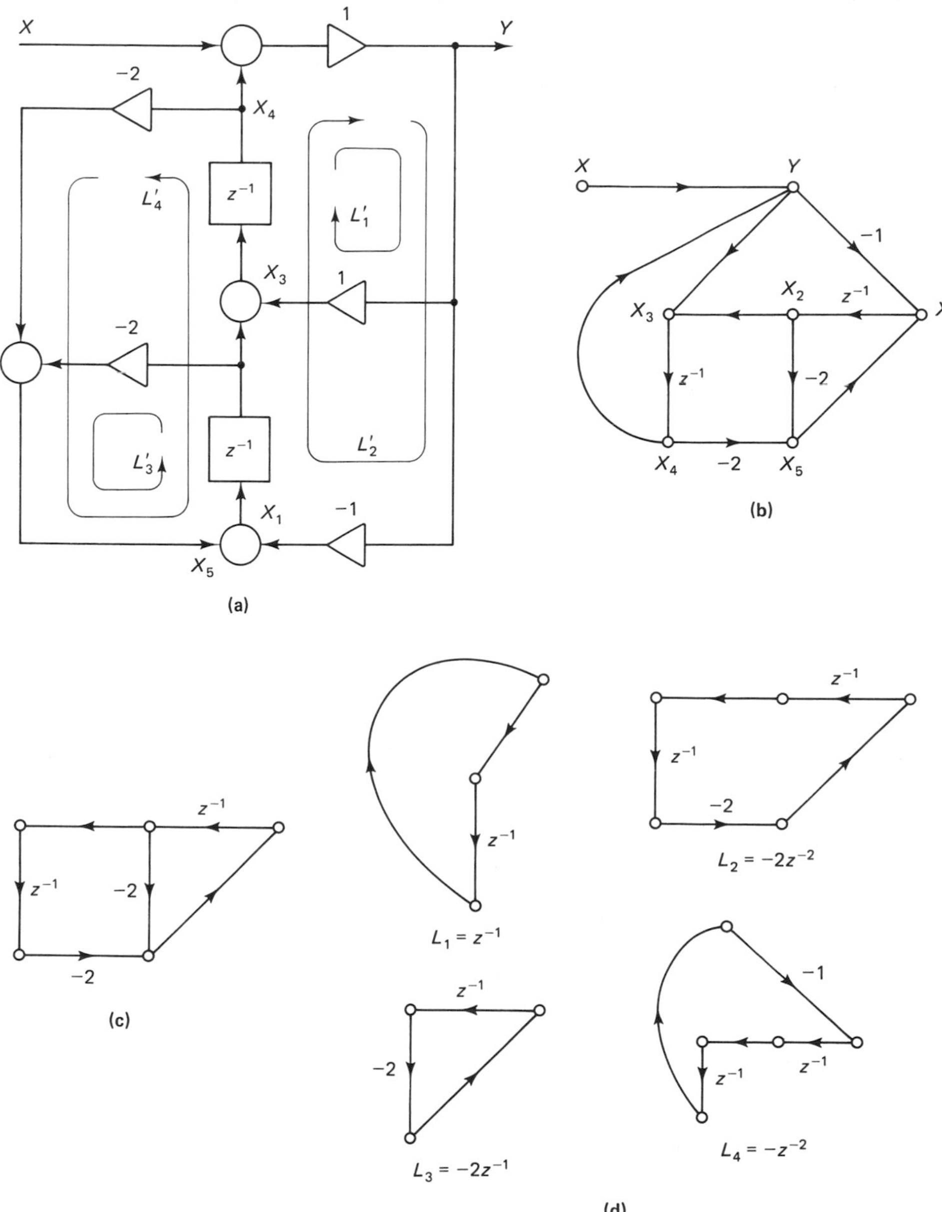

Figure 4.11 (a) Block diagram and (b) signal-flow graph of a system. (c) Subgraph and (d) loops of the signal-flow graph. (S.Y. Hwang, "Realization of Canonical Digital Networks".)

system represented by a signal-flow graph. In many cases, we can apply Mason's formula directly to a block-diagram representation of the system. As we will see later, we can also use block diagrams and signal-flow graphs to obtain a representation of the system that involves first-order difference equations.

EXERCISES

4.2.1 Draw block diagrams representing the following systems:

(a) $y(n) - \frac{1}{2}y(n-1) = x(n)$

(b) $y(n) + \frac{1}{2}y(n-1) = x(n) + \frac{1}{4}x(n-1)$

(c) $y(n) = x(n) + \frac{1}{2}x(n-1)$

4.2.2 Draw a signal-flow graph for the system shown and use Mason's formula to find

(a) $Y(z)/X(z)$

(b) $Y_1(z)/X(z)$

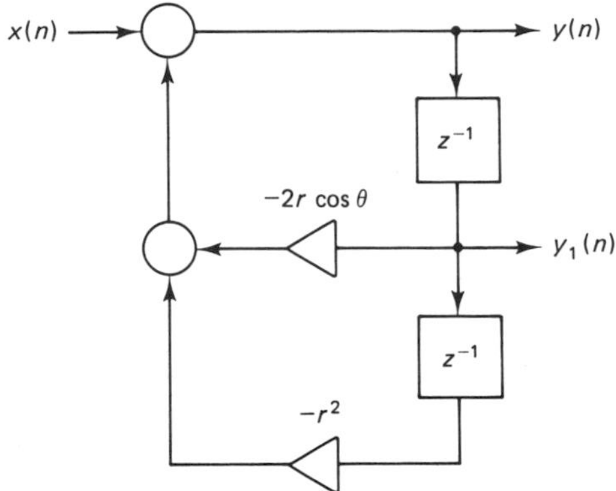

4.2.3 Draw a signal-flow graph for the system shown and find $Y(z)/X(z)$.

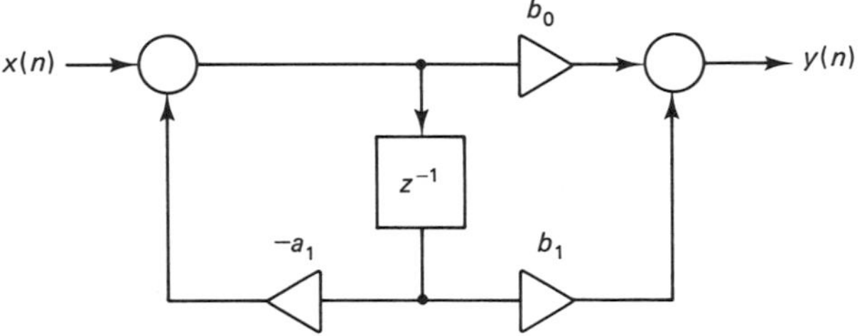

Answer:

$$\frac{b_0 + b_1 z^{-1}}{1 + a_1 z^{-1}}$$

4.2.4 Suppose that

$$\frac{Y(z)}{X(z)} = H(z) = H_1(z) + H_2(z)$$

Show that a realization is given by the system shown in the figure. Obtain a realization of this form using the partial-fraction expansion (see Sec. 3.4) for

$$H(z) = \frac{2 + 5z^{-1} + 12z^{-2}}{(1 + \frac{1}{2}z^{-1} + \frac{1}{4}z^{-2})(1 + \frac{1}{4}z^{-1} + \frac{1}{8}z^{-2})}$$

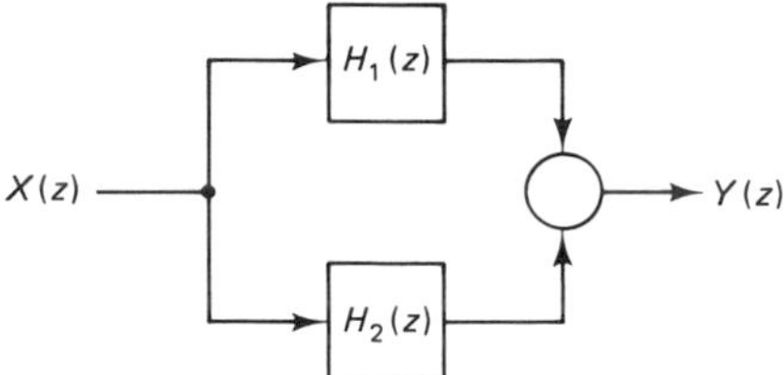

4.2.5 The transposition theorem for signal-flow graphs states that if all of the arrows are reversed with the input and output interchanged, the system transfer function remains unchanged. Use Mason's rule to prove this theorem, and apply the theorem to the signal-flow graph of Fig. 4.8.

4.3 MATRIX REPRESENTATIONS AND COMPUTABILITY

Matrix Representations

Let us assume for now that we have obtained a block-diagram representation of a discrete-time system. As mentioned previously, this actually provides us with an algorithm for computing the output sequence. In carrying out these computations, we quite naturally compute some intermediate quantities. The computational algorithm can be represented by a system of equations that arise from an identification of quantities in the block diagram. To further illustrate this point, let us identify the quantities $w_1(n)$, $w_2(n)$, and $w_3(n)$ for the system in Fig. 4.2 as shown in Fig. 4.12(a). The equations that define these quantities are

$$\begin{aligned} w_1(n) &= w_3(n) + x(n) \\ w_2(n) &= w_1(n-1) \\ w_3(n) &= aw_2(n) \\ y(n) &= w_1(n) \end{aligned} \tag{4.14}$$

which have the matrix representation

$$\begin{bmatrix} w_1(n) \\ w_2(n) \\ w_3(n) \end{bmatrix} = \begin{bmatrix} 0 & 0 & 1 \\ 0 & 0 & 0 \\ 0 & a & 0 \end{bmatrix} \begin{bmatrix} w_1(n) \\ w_2(n) \\ w_3(n) \end{bmatrix} + \begin{bmatrix} 0 & 0 & 0 \\ 1 & 0 & 0 \\ 0 & 0 & 0 \end{bmatrix} \begin{bmatrix} w_1(n-1) \\ w_2(n-1) \\ w_3(n-1) \end{bmatrix} + \begin{bmatrix} 1 \\ 0 \\ 0 \end{bmatrix} x(n)$$

$$y(n) = \begin{bmatrix} 1 & 0 & 0 \end{bmatrix} \begin{bmatrix} w_1(n) \\ w_2(n) \\ w_3(n) \end{bmatrix} \tag{4.15}$$

It is clear from Eqs. (4.14) that we cannot compute the variables $w_1(n)$, $w_2(n)$, and $w_3(n)$ in this order since $w_1(n)$ cannot be calculated until we know $w_3(n)$. In fact, the variables can be calculated in the order of their subscripts if and only if the right-hand side of the equation giving $w_1(n)$ is independent of $w_1(n)$, $w_2(n)$, and $w_3(n)$; the

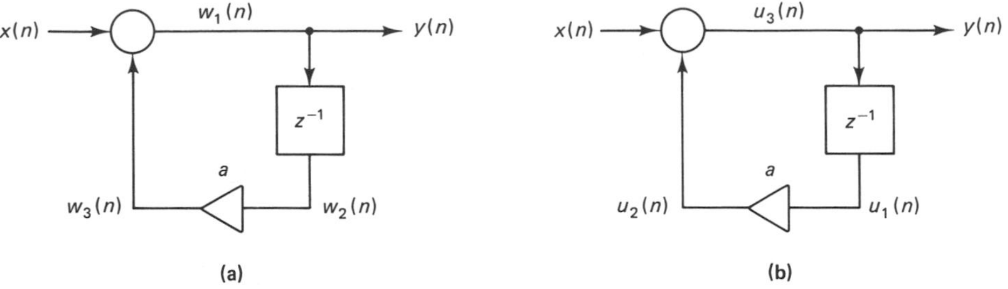

Figure 4.12 Two possible assignments of variables.

right-hand side of the equation giving $w_2(n)$ is independent of $w_2(n)$ and $w_3(n)$; and the right-hand side of the equation giving $w_3(n)$ is independent of $w_3(n)$. This is equivalent to saying that the matrix multiplying $[w_1(n) \quad w_2(n) \quad w_3(n)]^T$ (T means transpose) in Eq. (4.15) is lower triangular with zero diagonal elements. Clearly, these conditions are not satisfied for the choice of variables $w_1(n)$, $w_2(n)$, and $w_3(n)$.

In this example, it is possible to assign subscripted variables to the quantities to be computed in Fig. 4.2 so that they can be computed in the sequence determined by their subscripts. One such assignment is that shown in Fig. 4.12(b), where we use $u_1(n)$, $u_2(n)$, and $u_3(n)$. The resulting equations are

$$\begin{aligned} u_1(n) &= u_3(n-1) \\ u_2(n) &= au_1(n) \\ u_3(n) &= u_2(n) + x(n) \\ y(n) &= u_3(n) \end{aligned}$$

and

$$\begin{bmatrix} u_1(n) \\ u_2(n) \\ u_3(n) \end{bmatrix} = \begin{bmatrix} 0 & 0 & 0 \\ a & 0 & 0 \\ 0 & 1 & 0 \end{bmatrix} \begin{bmatrix} u_1(n) \\ u_2(n) \\ u_3(n) \end{bmatrix} + \begin{bmatrix} 0 & 0 & 1 \\ 0 & 0 & 0 \\ 0 & 0 & 0 \end{bmatrix} \begin{bmatrix} u_1(n-1) \\ u_2(n-1) \\ u_3(n-1) \end{bmatrix} + \begin{bmatrix} 0 \\ 0 \\ 1 \end{bmatrix} x(n)$$
$$y(n) = [0 \quad 0 \quad 1] \begin{bmatrix} u_1(n) \\ u_2(n) \\ u_3(n) \end{bmatrix} \tag{4.16}$$

Evidently, the conditions stated concerning the manner in which the variables appear on the right-hand side of the matrix equation are satisfied.

Computability

It should be noted that each of the variables $u_1(n)$, $u_2(n)$, and $u_3(n)$ is determined directly from a single equation. Whenever the system variables can be chosen so that this situation occurs, then the computations are greatly simplified. Whenever we can order the system variables and compute them in this sequence, we say that the system is *computable*. The ordering can be accomplished by subscripting the variables. The sequence subscripts indicate the order in which the variables are to be computed. For example, since $u_1(n)$, $u_2(n)$, and $u_3(n)$ can be computed in this order for the system of Fig. 4.12, we see that the system is computable.

In some cases, it is impossible to order the variables so that they can be calculated in sequence. A block diagram of this type is said to be *noncomputable*. An example of such a system is shown in Fig. 4.13. The reader is asked to verify this in Ex. 4.3.3.

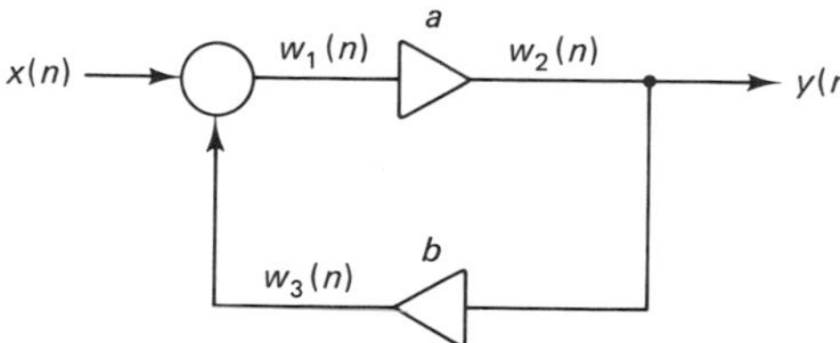

Figure 4.13 Example of a noncomputable block diagram.

Even though the system is noncomputable, it does not mean that the set of equations describing the system cannot be solved. It means that the variables cannot be obtained directly and that some method, such as Gaussian elimination, must be used to solve the equations simultaneously.

It can be shown that a necessary and sufficient condition for the computability of a block diagram is that every loop has at least one delay element. This was shown for signal-flow graphs [28], but the proof can be modified to apply to block diagrams. The block diagram shown in Fig. 4.13 has a loop with no delays, and, consequently, is noncomputable.

A computable block diagram represents a system that can be programmed on a digital computer or can be built using digital hardware. We refer to such a system as *physically realizable*. To determine when a system is not realizable, we check to see if there are loops with no delays in its block-diagram representation.

Matrix Description of a System

The system of Eqs. (4.16), as mentioned earlier, represents a computable block diagram. For a general computable block diagram (constructed using adders, delays, and multipliers), the describing matrix equation has the form

$$\mathbf{w}(n) = \mathbf{A}_c\mathbf{w}(n) + \mathbf{A}_d\mathbf{w}(n-1) + \mathbf{B}_c\mathbf{x}(n) \tag{4.17}$$

where $\mathbf{A}_c$ is a lower triangular matrix whose diagonal elements are zero. To complete the block-diagram description, we obtain the relation

$$\mathbf{y}(n) = \mathbf{C}\mathbf{w}(n) \tag{4.18}$$

We could take the z-transform of Eqs. (4.17) and (4.18) and determine the transforms of the output variables in terms of the transforms of the input variables.

In addition to the description given in Eqs. (4.17) and (4.18), there are other equally important descriptions that are expressed in a somewhat similar matrix form. In particular, we consider the *state-variable* description in Sec. 4.4. It is frequently used to describe discrete-time control systems. Some of the results that we obtain using state variables can be applied also to Eqs. (4.17) and (4.18).

For additional information on matrix representations and computability, the reader should see [126].

EXERCISES

4.3.1 For the system with variables assigned as shown, write the describing equations and determine if the variables can be calculated in sequence.

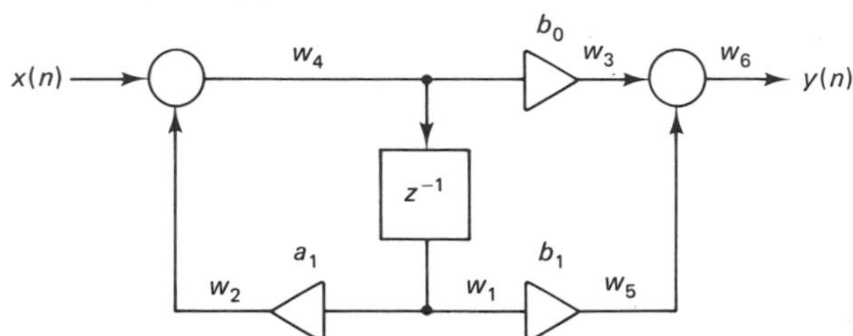

4.3.2 Since the only loop in the system of Ex. 4.3.1 has a delay element, the system is computable. Determine a variable assignment for which the variables can be computed in sequence.

4.3.3 Show that the system given in Fig. 4.13 is noncomputable. First, consider the set $w_1(n)$, $w_2(n)$, and $w_3(n)$. Note that if $u_i(n)$ and $u_j(n)$ are the input and output, respectively, of a multiplier, then we must have $i < j$ in order to satisfy the computability criterion.

4.3.4 If a system is computable, any set of variables can be transformed by permutations into a set that can be computed in sequence. Find a permutation matrix that transforms $w_1(n)$, $w_2(n)$, and $w_3(n)$ into $u_1(n)$, $u_2(n)$, $u_3(n)$ for the system of Fig. 4.12. In other words, express $u_1(n)$, $u_2(n)$, and $u_3(n)$ in terms of $w_1(n)$, $w_2(n)$, and $w_3(n)$.

4.4 STATE VARIABLES

Definition of State Variables

In many areas of applications where discrete-time systems arise, it is convenient to describe the system by a set of auxiliary variables known as *state variables*. A *state* of a system is a minimum set of numbers that must be specified at time $t = t_0$ so that the behavior of the system can be uniquely predicted for any time $t > t_0$ for any allowable input that is known for $t \geq t_0$. These numbers are called *state variables*. The vector having the state variables as components is known as the *state vector*. State variables can be useful in obtaining various representations of systems and are used frequently in computations concerning system behavior. For a more complete description of state variables, see [80, 123].

State-Variable Description

It is common practice when discussing discrete-time systems to choose the state-variable description

$$\mathbf{v}(n+1) = \mathbf{A}\mathbf{v}(n) + \mathbf{B}\mathbf{x}(n)$$
$$\mathbf{y}(n) = \mathbf{C}\mathbf{v}(n) + \mathbf{D}\mathbf{x}(n) \tag{4.19}$$

where $\mathbf{v}(n)$ is the state vector, $\mathbf{x}(n)$ is the input vector, $\mathbf{y}(n)$ is the output vector, and $\mathbf{A}$, $\mathbf{B}$, $\mathbf{C}$, and $\mathbf{D}$ are matrices whose orders are determined by the orders of $\mathbf{v}(n)$, $\mathbf{x}(n)$, and $\mathbf{y}(n)$. It should be noted that the future value of the state vector $\mathbf{v}(n+1)$ is determined from the present values of the state vector and the input vector.

Example of a State-Variable Representation

A set of state variables can be associated with a given system in a number of ways. Once such a set is chosen, we can describe the discrete-time system by the first-order difference equations of Eq. (4.19). As an example, let us consider the difference equation of Eq. (4.6):

$$y(n) + a_1 y(n-1) + a_2 y(n-2) = x(n) \tag{4.20}$$

If we define the new variables

$$v_2(n) = y(n-1)$$
$$v_1(n) = y(n-2) = v_2(n-1) \tag{4.21}$$

then the second-order difference equation can be described by the two first-order difference equations

$$v_1(n+1) = v_2(n)$$
$$v_2(n+1) = -a_1 v_2(n) - a_2 v_1(n) + x(n) \tag{4.22}$$

where the last equation is obtained by substituting Eq. (4.21) into Eq. (4.20).

State-Variable Selection from the Block Diagram

The system of Eq. (4.6) or Eq. (4.20) is represented by the block diagram of Fig. 4.3. We have identified the state variables $v_1(n)$ and $v_2(n)$ in this figure, where it can be seen that they are the outputs of the delay elements. This is a choice that is frequently used in defining state variables whenever a block diagram or a signal-flow graph is available. The inputs to the delay elements will then be the state variables evaluated at $n + 1$, and so the state equations can be obtained by writing the constraints satisfied at the inputs of the delay elements.

Finally, the output equation giving $y(n)$ in terms of the state variables and the input $x(n)$ for the system of Eq. (4.6) or Eq. (4.20) can be obtained directly from Eq. (4.20) or from the block diagram. From Eq. (4.20), we obtain

$$y(n) = -a_1 y(n-1) - a_2 y(n-2) + x(n)$$
$$= -a_1 v_2(n) - a_2 v_1(n) + x(n)$$

From the block diagram of Fig. 4.3, we see that

$$y(n) = v_2(n+1)$$
$$= -a_1 v_2(n) - a_2 v_1(n) + x(n) \tag{4.23}$$

The output equation together with Eq. (4.22) results in the state-variable description

$$\mathbf{v}(n+1) = \begin{bmatrix} 0 & 1 \\ -a_2 & -a_1 \end{bmatrix} \mathbf{v}(n) + \begin{bmatrix} 0 \\ 1 \end{bmatrix} x(n)$$
$$y(n) = [-a_2 \quad -a_1]\mathbf{v}(n) + x(n) \tag{4.24}$$

Comparing these equations with Eq. (4.19), we see that

$$\mathbf{A} = \begin{bmatrix} 0 & 1 \\ -a_2 & -a_1 \end{bmatrix} \qquad \mathbf{B} = \begin{bmatrix} 0 \\ 1 \end{bmatrix} \qquad \mathbf{C} = [-a_2 \quad -a_1] \qquad \mathbf{D} = [1] \tag{4.25}$$

Solution of the State Equations

Since Eq. (4.19) is a system of first-order difference equations, we can readily compute the values of the state vector by recursion. For example, we can obtain

$$\begin{aligned}
\mathbf{v}(1) &= \mathbf{A}\mathbf{v}(0) + \mathbf{B}\mathbf{x}(0) \\
\mathbf{v}(2) &= \mathbf{A}\mathbf{v}(1) + \mathbf{B}\mathbf{x}(1) \\
&= \mathbf{A}^2\mathbf{v}(0) + \mathbf{A}\mathbf{B}\mathbf{x}(0) + \mathbf{B}\mathbf{x}(1) \\
\mathbf{v}(3) &= \mathbf{A}\mathbf{v}(2) + \mathbf{B}\mathbf{x}(2) \\
&= \mathbf{A}^3\mathbf{v}(0) + \mathbf{A}^2\mathbf{B}\mathbf{x}(0) + \mathbf{A}\mathbf{B}\mathbf{x}(1) + \mathbf{B}\mathbf{x}(2)
\end{aligned}$$

From these equations, we can surmise that the general expression for $\mathbf{v}(n)$ is

$$\mathbf{v}(n) = \mathbf{A}^n\mathbf{v}(0) + \sum_{k=0}^{n-1} \mathbf{A}^{n-k-1}\mathbf{B}\mathbf{x}(k) \tag{4.26}$$

This equation provides us with a closed-form solution of the state equations, but it requires the calculation of powers of the matrix **A** that can be rather complicated.

z-Transform Solution

We can also use *z*-transforms to analyze the state equations. This can be done more systematically if we define the *z*-transform of a matrix. If $\mathbf{A}(n) = [a_{ij}(n)]$, the matrix whose *ij*th element is $a_{ij}(n)$, then the *z*-transform of **A** is defined to be

$$Z\{\mathbf{A}(n)\} = [Z\{a_{ij}(n)\}] \tag{4.27}$$

In other words, the *z*-transform of a matrix is obtained by transforming each element.

Using this definition and the *z*-transform given in Prob. 3.10 with $k = 1$, we can take the *z*-transform of Eq. (4.19) and obtain

$$z\mathbf{V}(z) - z\mathbf{v}(0) = \mathbf{A}\mathbf{V}(z) + \mathbf{B}\mathbf{X}(z)$$
$$\mathbf{Y}(z) = \mathbf{C}\mathbf{V}(z) + \mathbf{D}\mathbf{X}(z) \tag{4.28}$$

where $\mathbf{V}(z) = Z\{\mathbf{v}(n)\}$ and so forth, and **A, B, C,** and **D** are constant matrices. The reader is asked to develop these relations in Ex. 4.4.6. From the first of Eq. (4.28), we obtain

$$(\mathbf{I} - z^{-1}\mathbf{A})\mathbf{V}(z) = \mathbf{v}(0) + z^{-1}\mathbf{B}\mathbf{X}(z)$$

where $\mathbf{I}$ is the identity matrix of the proper order. We can solve this equation for $\mathbf{V}(z)$, obtaining

$$\mathbf{V}(z) = (\mathbf{I} - z^{-1}\mathbf{A})^{-1}[\mathbf{v}(0) + z^{-1}\mathbf{B}\mathbf{X}(z)]$$

where $(\mathbf{I} - z^{-1}\mathbf{A})^{-1}$ is the inverse of the matrix $(\mathbf{I} - z^{-1}\mathbf{A})$. Substituting this expression into the second of Eq. (4.28), we have

$$\mathbf{Y}(z) = \mathbf{C}(\mathbf{I} - z^{-1}\mathbf{A})^{-1}\mathbf{v}(0) + [\mathbf{C}(\mathbf{I} - z^{-1}\mathbf{A})^{-1}z^{-1}\mathbf{B} + \mathbf{D}]\mathbf{X}(z) \qquad (4.29)$$

If the initial state variable $\mathbf{v}(0)$ and the input $\mathbf{x}(n)$ are known, then $\mathbf{Y}(z)$ can be determined from this equation. Once $\mathbf{Y}(z)$ is known, then $\mathbf{y}(n)$ can be found by taking the inverse z-transform of each element of $\mathbf{Y}(z)$.

System-Transfer Matrix

If we take $\mathbf{v}(0) = \mathbf{0}$, we can define the *system-transfer matrix* $\mathbf{H}(z)$ by

$$\mathbf{Y}(z) = \mathbf{H}(z)\mathbf{X}(z) \qquad (4.30)$$

Comparing this equation with Eq. (4.29) with $\mathbf{v}(0) = \mathbf{0}$, we see that

$$\mathbf{H}(z) = \mathbf{C}(\mathbf{I} - z^{-1}\mathbf{A})^{-1}z^{-1}\mathbf{B} + \mathbf{D} \qquad (4.31)$$

To illustrate finding the system-transfer matrix, we will find $\mathbf{H}(z)$ for the system of Eq. (4.20) for the case where $a_1 = 3$ and $a_2 = 2$. Then

$$\mathbf{I} - z^{-1}\mathbf{A} = \begin{bmatrix} 1 & -z^{-1} \\ 2z^{-1} & 1 + 3z^{-1} \end{bmatrix}$$

and it follows that

$$(\mathbf{I} - z^{-1}\mathbf{A})^{-1} = \frac{1}{1 + 3z^{-1} + 2z^{-2}}\begin{bmatrix} 1 + 3z^{-1} & z^{-1} \\ -2z^{-1} & 1 \end{bmatrix}$$

Substituting these values of $\mathbf{A}$, $\mathbf{B}$, $\mathbf{C}$, and $\mathbf{D}$ into Eq. (4.31) results in the transfer function

$$\begin{aligned} \mathbf{H}(z) &= [-2 \quad -3]\begin{bmatrix} 1 + 3z^{-1} & z^{-1} \\ -2z^{-1} & 1 \end{bmatrix} z^{-1}\begin{bmatrix} 0 \\ 1 \end{bmatrix}(1 + 3z^{-1} + 2z^{-2})^{-1} + 1 \\ &= \frac{1}{1 + 3z^{-1} + 2z^{-2}} \end{aligned}$$

This result, of course, is the same obtained by taking the z-transform of Eq. (4.20). Since there is one input variable and one output variable, $\mathbf{H}(z)$ is a scalar or a 1×1 matrix.

Even though we can describe a system by many different sets of state variables, which give rise to different matrices $\mathbf{A}$, $\mathbf{B}$, $\mathbf{C}$, and $\mathbf{D}$, there is a unique system matrix $\mathbf{H}(z)$ that is independent of the choice of state variables.

Systems with Delayed Inputs

If past inputs such as $x(n-1)$, for example, appear in the difference equation, the procedure for obtaining a set of state variables is usually more complicated. As an example, consider the system

$$y(n) + a_1 y(n-1) + a_2 y(n-2) = b_0 x(n) + b_1 x(n-1) + b_2 x(n-2) \tag{4.32}$$

We now discuss two procedures for obtaining a state equation for this system.

State Variables Obtained from Block Diagrams

The first method evolves from a block-diagram or signal-flow graph representation of the system. As we have seen, one such representation is that shown in Fig. 4.10(a). If we define the state variables $v_1(n)$ and $v_2(n)$ as the outputs of the delay elements as shown, then we obtain the state equations by writing the constraints at the inputs of these delay elements. The resulting equations describing the relations between the state variables are

$$v_1(n+1) = v_2(n)$$

$$v_2(n+1) = x(n) - a_1 v_2(n) - a_2 v_1(n)$$

and the output equation is

$$\begin{aligned} y(n) &= b_0 v_2(n+1) + b_1 v_2(n) + b_2 v_1(n) \\ &= b_0[x(n) - a_1 v_2(n) - a_2 v_1(n)] + b_1 v_2(n) + b_2 v_1(n) \\ &= (b_2 - b_0 a_2) v_1(n) + (b_1 - b_0 a_1) v_2(n) + b_0 x(n) \end{aligned}$$

These equations can be written in the matrix form

$$\mathbf{v}(n+1) = \begin{bmatrix} 0 & 1 \\ -a_2 & -a_1 \end{bmatrix} \mathbf{v}(n) + \begin{bmatrix} 0 \\ 1 \end{bmatrix} x(n)$$

$$y(n) = [b_2 - b_0 a_2 \quad b_1 - b_0 a_1] \mathbf{v}(n) + b_0 x(n)$$

State Variables Obtained from the Difference Equation

There are numerous procedures that can be used to obtain state-variable representations of systems described by a difference equation. We give one such method here and consider some other methods in the exercises at the end of the section and the problems at the end of the chapter.

We develop the procedure using the second-order system of Eq. (4.32). We begin by transforming the equation and writing it as

$$Y(z) = b_0 X(z) + z^{-1}\{[b_1 X(z) - a_1 Y(z)] + z^{-1}[b_2 X(z) - a_2 Y(z)]\} \tag{4.33}$$

The state variables are defined by starting with the innermost term that has z^{-1} as a factor. This term is designated as $V_1(z)$, so that

$$V_1(z) = z^{-1}[b_2 X(z) - a_2 Y(z)] \tag{4.34}$$

Substituting this expression into Eq. (4.33) results in the relation

$$Y(z) = b_0 X(z) + z^{-1}[b_1 X(z) - a_1 Y(z) + V_1(z)] \tag{4.35}$$

Repeating the above procedure for this equation, we define $V_2(z)$ by

$$V_2(z) = z^{-1}[b_1 X(z) - a_1 Y(z) + V_1(z)] \tag{4.36}$$

so that Eq. (4.35) becomes

$$Y(z) = b_0 X(z) + V_2(z) \tag{4.37}$$

The state equations are now obtained by substituting Eq. (4.37) into Eqs. (4.34) and (4.36), resulting in

$$\begin{aligned} zV_1(z) &= b_2 X(z) - a_2[b_0 X(z) + V_2(z)] \\ &= -a_2 V_2(z) + [b_2 - b_0 a_2]X(z) \\ zV_2(z) &= b_1 X(z) - a_1[b_0 X(z) + V_2(z)] + V_1(z) \\ &= V_1(z) - a_1 V_2(z) + [b_1 - b_0 a_1]X(z) \end{aligned}$$

In order to obtain the state equations, we must find the inverse z-transforms of the previous equations. We can start with the relation

$$Z\{x(n+1)\} = zX(z) - zx(0)$$

which was obtained in Prob. 3.10 for $k = 1$. If we are working with the transfer functions, we assume that $x(0) = 0$ and, therefore, we have

$$x(n+1) = Z^{-1}\{zX(z)\}$$

We can now use this relation to find the inverse z-transform of $zV_1(z)$ and $zV_2(z)$. From these results and Eq. (4.37), we obtain the state equations

$$\begin{aligned} \mathbf{v}(n+1) &= \begin{bmatrix} 0 & -a_2 \\ 1 & -a_1 \end{bmatrix}\mathbf{v}(n) + \begin{bmatrix} b_2 - b_0 a_2 \\ b_1 - b_0 a_1 \end{bmatrix}x(n) \\ y(n) &= [0 \quad 1]\mathbf{v}(n) + b_0 x(n) \end{aligned} \tag{4.38}$$

The reader is asked to consider a third-order system in Ex. 4.4.4 and the general Nth-order system in Prob. 4.14.

Block Diagrams Obtained from the State Equations

We can use the state equations to construct a block-diagram or a signal-flow-graph representation for the system. In the case of Eq. (4.38), the state variables $v_1(n)$ and $v_2(n)$ are taken as the outputs of the delays. The state equations then define constraints at the inputs of the delays. The resulting block diagram is shown in Fig. 4.14.

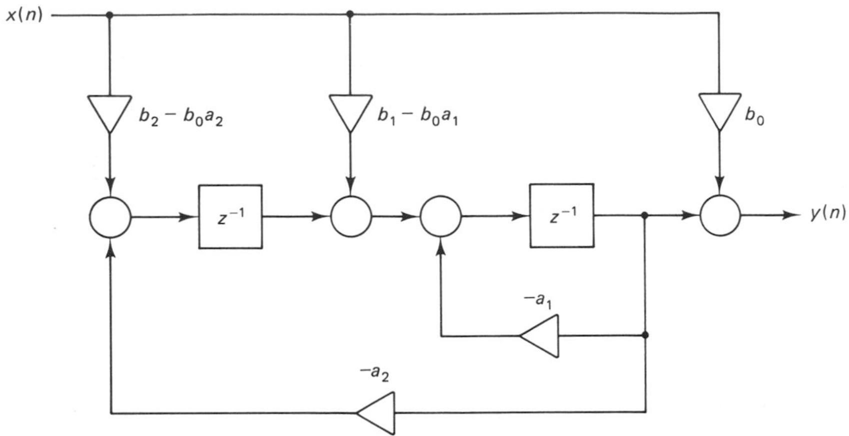

Figure 4.14 Block-diagram representation for the system of Eq. (4.38).

EXERCISES

4.4.1 Define $v(n) = w_1(n)$ in the system of Ex. 4.3.1 and obtain the state equations.

4.4.2 Obtain $H(z)$ for the system

$$\mathbf{v}(n+1) = \begin{bmatrix} 0 & 1 \\ 1 & 0 \end{bmatrix}\mathbf{v}(n) + \begin{bmatrix} 0 \\ 1 \end{bmatrix}x(n)$$

$$y(n) = [0 \quad 1]\mathbf{v}(n) + x(n)$$

4.4.3 Draw a block diagram representing the system of Ex. 4.4.2.

4.4.4 Obtain a state-variable representation of the form of Eq. (4.38) for the third-order system

$$\begin{aligned} y(n) + a_1 y(n-1) + a_2 y(n-2) + a_3 y(n-3) \\ = b_0 x(n) + b_1 x(n-1) + b_2 x(n-2) + b_3 x(n-3) \end{aligned}$$

4.4.5 Write the system function $H(z)$ for the system of Eq. (4.32) as

$$H(z) = \frac{Y(z)}{X(z)} = \frac{(b_0 + b_1 z^{-1} + b_2 z^{-2})W(z)}{(1 + a_1 z^{-1} + a_2 z^{-2})W(z)}$$

where $W(z)$ is determined as in Sec. 3.5, so that the numerators and denominators of the two fractions are equal. Define the state variables to be $V_2(z) = z^{-1}W(z)$ and $V_1(z) = z^{-2}W(z)$ and obtain the state equations.

4.4.6 Show that the z-transform of the ith equation of Eq. (4.19) is

$$zV_i(z) - zv_i(0) = \sum_{j=1}^{r} a_{ij} V_j(z) + \sum_{j=1}^{m} b_{ij} X_j(z)$$

and use this result to obtain Eq. (4.28).

4.5 DIRECT-FORM REALIZATION OF AN IIR SYSTEM

Development of the Direct-Form Realization

In the previous sections of this chapter, we have discussed structures for realizing discrete-time systems in general terms rather than considering special structures. We now begin to focus our attention on some of the more common structures that are presently used. This section is devoted to a consideration of a structure that can be obtained directly from the difference equation or system function describing the system. The structure is called the *direct-form* realization and it can be obtained without any calculations.

The structure can be obtained directly from the difference equation, Eq. (2.29),

$$\sum_{k=0}^{N} a_k y(n-k) = \sum_{m=0}^{M} b_m x(n-m) \tag{4.39}$$

with $a_0 = 1$. The reader is asked to carry out this procedure in Ex. 4.5.3. We proceed from the system function $H(z)$ as we did in Sec. 3.5 and Ex. 4.4.5. Therefore, we write

$$\frac{Y(z)}{X(z)} = H(z) = \frac{\sum_{m=0}^{M} b_m z^{-m}}{1 + \sum_{k=1}^{N} a_k z^{-k}} \frac{W(z)}{W(z)} \tag{4.40}$$

where $W(z)$ is chosen so that

$$\begin{aligned} Y(z) &= \left(\sum_{m=0}^{M} b_m z^{-m}\right) W(z) \\ X(z) &= \left(1 + \sum_{k=1}^{N} a_k z^{-k}\right) W(z) \end{aligned} \tag{4.41}$$

We can realize this system with input $X(z)$ and output $W(z)$ quite readily from the equivalent expression

$$W(z) = X(z) - \sum_{k=1}^{N} a_k[z^{-k}W(z)] \tag{4.42}$$

In realizing this system, we consider the variables $z^{-k}W(z)$, $k = 1, \ldots, N$, as the outputs of delay elements. Once these variables are obtained, we can then obtain $Y(z)$ from the expression given in Eq. (4.41).

Example of a Second-Order System

Before obtaining the direct-form realization for the general case, we first consider the second-order system

$$\begin{aligned} y(n) + a_1 y(n-1) + a_2 y(n-2) = \\ b_0 x(n) + b_1 x(n-1) + b_2 x(n-2) \end{aligned} \tag{4.43}$$

For this system, Eqs. (4.41) and (4.42) become

$$Y(z) = b_0 W(z) + b_1 z^{-1} W(z) + b_2 z^{-2} W(z) \tag{4.44}$$

and

$$W(z) = X(z) - a_1 z^{-1} W(z) - a_2 z^{-2} W(z) \tag{4.45}$$

The realization of Eq. (4.45) is shown in Fig. 4.15(a) and the realization of both Eqs. (4.44) and (4.45) is shown in Fig. 4.15(b). Of course, the realization of Fig. 4.15(b) is a realization of Eq. (4.43).

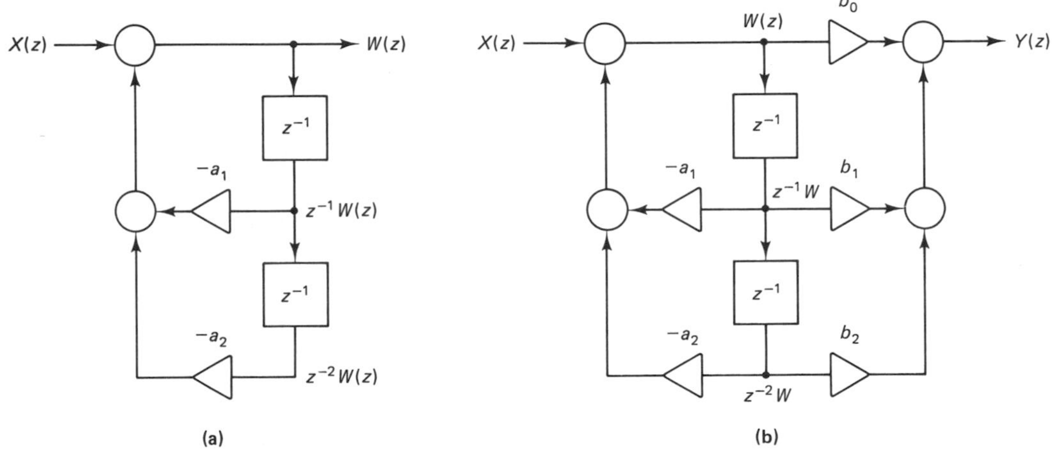

Figure 4.15 (a) A realization of Eq. (4.45) and (b) a realization of Eq. (4.43).

Direct-Form Realization for the General Case

In the case of Eq. (4.39) for $N = M$, we can proceed exactly as we have done for the special case where $N = M = 2$ and obtain the direct-form realization shown in Fig. 4.16. If $M < N$, then the corresponding multipliers are removed along with the unneeded adders. In this realization, there are N delays, which is the same number as the order of the difference equation, Eq. (4.39). Such a realization is said to be *canonical*. We also refer to this realization as the *direct-form* realization.

Parameter Quantization

As can be seen from Fig. 4.16, the direct-form realization is very easy to obtain. The coefficients that appear in the realization are the same as those in the difference equation. Therefore, no additional computations are required in going from the difference equation to the block diagram. Unfortunately, this convenient property is usually overshadowed by the poor accuracy that results when the coefficients and signals are discretized and represented by finite-length binary numbers. This last property, the computational inaccuracy, most frequently rules out the use of the direct-form realization in practice for large N. In Prob. 4.36, we consider this inac-

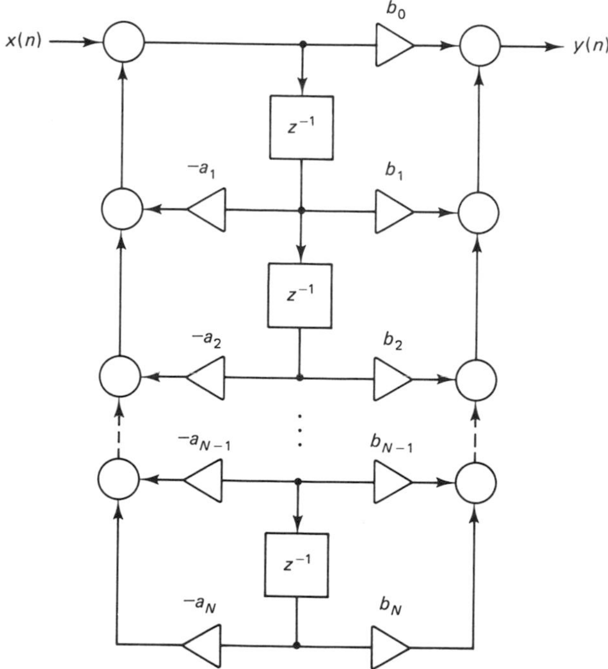

Figure 4.16 Direct-form realization of Eq. (4.39).

curacy problem, which is generally referred to as the *parameter quantization effect,* since it is caused by quantizing or discretizing the coefficients using representations with a finite number of binary positions. As we will see, the difficulty arises when poles (or zeros) are close together and becomes a more serious problem when the order of the system becomes large. The problem does not generally arise for second-order systems, which are almost always realized using the direct-form realization.

The system function of the second-order system of Eq. (4.43) is

$$H(z) = \frac{b_0 + b_1 z^{-1} + b_2 z^{-2}}{1 + a_1 z^{-1} + a_2 z^{-2}}$$

and the poles and zeros are solutions of the equations

$$z^2 + a_1 z + a_2 = 0$$

$$b_0 z^2 + b_1 z + b_2 = 0$$

respectively. In practice, the poles, as well as the zeros, are not located near each other.

More Examples

As an example, consider the system of Eq. (3.41) described by the system function

$$H(z) = \frac{z^{-1}}{1 - z^{-1} + \frac{1}{2} z^{-2}} \tag{4.46}$$

For this system, we have $N = 2$, $M = 1$, $b_0 = 0$, $b_1 = 1$, $a_1 = -1$ and $a_2 = \frac{1}{2}$. The direct-form realization is shown in Fig. 4.17.

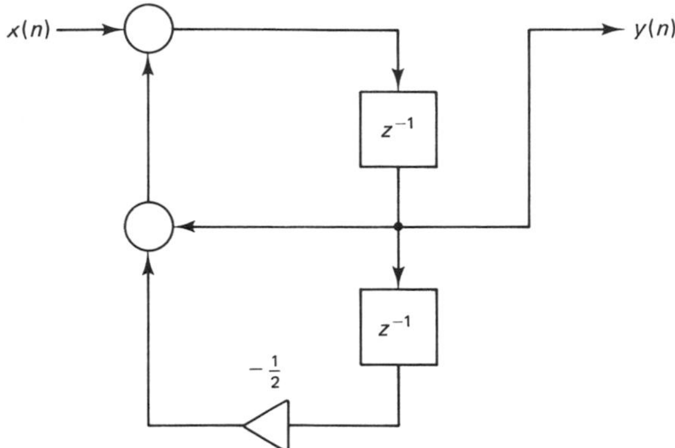

Figure 4.17 Direct-form realization of Eq. (4.46).

A more complicated example is the system with

$$H(z) = \frac{1 + \frac{1}{4}z^{-1} + \frac{1}{2}z^{-2} + \frac{1}{4}z^{-3}}{1 + z^{-1} + \frac{1}{4}z^{-2} - \frac{1}{4}z^{-3} + \frac{1}{2}z^{-4}} \tag{4.47}$$

For this system, we have $N = 4$, $M = 3$, $b_0 = 1$, $b_1 = \frac{1}{4}$, $b_2 = \frac{1}{2}$, $b_3 = \frac{1}{4}$, $a_1 = 1$, $a_2 = \frac{1}{4}$, $a_3 = -\frac{1}{4}$, and $a_4 = \frac{1}{2}$. The direct-form realization is given in Fig. 4.18.

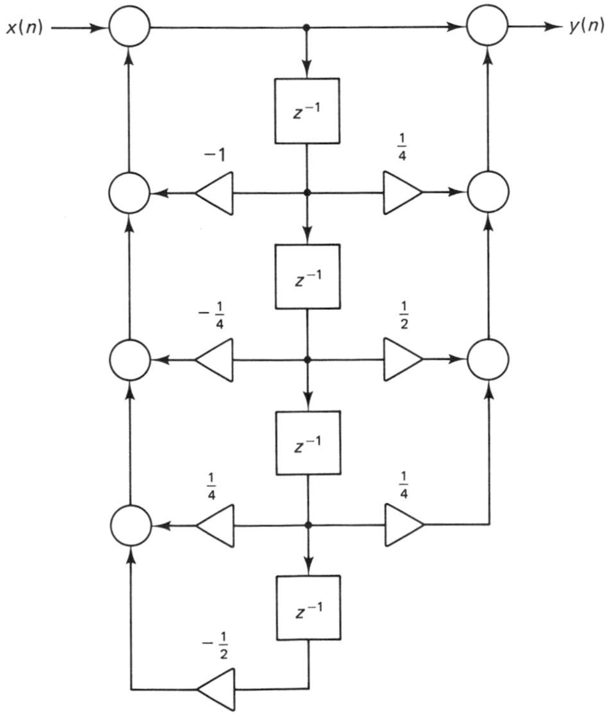

Figure 4.18 Direct-form realization of Eq. (4.47).

EXERCISES

4.5.1 Obtain a direct-form realization for

$$H(z) = \frac{1 + \frac{1}{4}z^{-1} + \frac{1}{2}z^{-2}}{1 + z^{-1} + \frac{1}{2}z^{-2} + \frac{1}{4}z^{-3} + \frac{1}{2}z^{-4}}$$

4.5.2 Give a direct-form realization for

$$H(z) = \frac{(1 + \frac{1}{2}z^{-1})(1 + \frac{1}{4}z^{-1})}{(1 - \frac{1}{2}z^{-1})(1 - \frac{1}{4}z^{-1})(1 - \frac{1}{8}z^{-1})}$$

4.5.3 Solve Eq. (4.39) for $y(n)$ and obtain a block-diagram representation for this system by delaying both $y(n)$ and $x(n)$ appropriately. Show that the representation is a cascading of two systems that can be interchanged so as to eliminate all except N delays.

4.6 CASCADE REALIZATION OF IIR SYSTEMS

Development of the Cascade Realization

In Secs. 2.4 and 3.5, we considered the cascading of two systems and gave block-diagram representations of such systems in Figs. 2.2 and 3.3. In these representations, the output of the first system is the input to the second system. In this section, we consider the general problem of realizing systems obtained by cascading two or more subsystems.

The *cascade* realization of an IIR system is obtained by first decomposing the system function $H(z)$ given in Eq. (4.2) into the product of several simpler transfer functions, as given by

$$\begin{aligned} H(z) &= AH_1(z)H_2(z) \cdots H_K(z) \\ &= A \prod_{k=1}^{K} H_k(z) \end{aligned} \tag{4.48}$$

where A is a constant, and it is assumed that $M \leq N$. The individual transfer functions are generally chosen to be either first-order sections with

$$H_k(z) = \frac{b_{k0} + b_{k1}z^{-1}}{1 + a_{k1}z^{-1}} \tag{4.49}$$

or second-order sections with

$$H_k(z) = \frac{b_{k0} + b_{k1}z^{-1} + b_{k2}z^{-2}}{1 + a_{k1}z^{-1} + a_{k2}z^{-2}} \tag{4.50}$$

where the coefficients are assumed to be real. In the case where $b_0 \neq 0$, then all the coefficients represented by b_{k0} in Eqs. (4.49) and (4.50) are generally taken to be unity, and in Eq. (4.48), we set $A = b_0$. In the case where $b_0 = 0$, we can either take some of the b_{k0} coefficients to be zero or we can introduce a certain number of delays at the input or the output of the system. We are assuming that $M \leq N$ in Eq. (4.2). If $M = N$, then all the sections are represented by Eq. (4.49) or Eq. (4.50) with $b_{k1} \neq 0$ in Eq. (4.49) and $b_{k2} \neq 0$ in Eq. (4.50). If $M < N$, then some of the coefficients are zero in the numerators of Eq. (4.49) and/or Eq. (4.50). For example, in Eq. (4.49), we could have $b_{k1} = 0$ for some sections, and in Eq. (4.50), we could have $b_{k2} = 0$ and also $b_{k1} = 0$ for some sections.

The second-order sections of Eq. (4.50) are used whenever possible. Of course, when N is odd, one first-order section is needed. The second-order denominators are formed by pairing two real poles or by pairing complex conjugate poles so that real coefficients result.

The sections represented by Eqs. (4.49) and (4.50) can be realized using the direct-form realization discussed in Sec. 4.5. The parameter-quantization effects mentioned there are generally not severe for second-order sections. In any event, we have no better choice when complex poles are present. The direct-form realization for Eq. (4.50) was given in Fig. 4.15(b) and is repeated here in Fig. 4.19. It should be noted that if $b_{k0} = 1$ for all sections, then we eliminate a number of multipliers equal to the number of sections. This is preferred since using fewer multipliers is more economical and saves computing time.

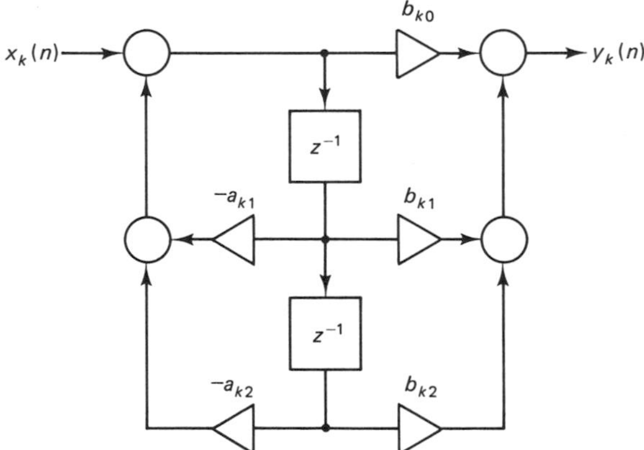

Figure 4.19 Direct-form realization of Eq. (4.50).

The general first-order section given in Eq. (4.49) has the realization given in Ex. 4.2.3 and again shown in Fig. 4.20. Again, if $b_{k0} = 1$, we have one less multiplication to perform per section. As mentioned earlier, a first-order section is used only if the denominator has odd degree and is associated with a real pole.

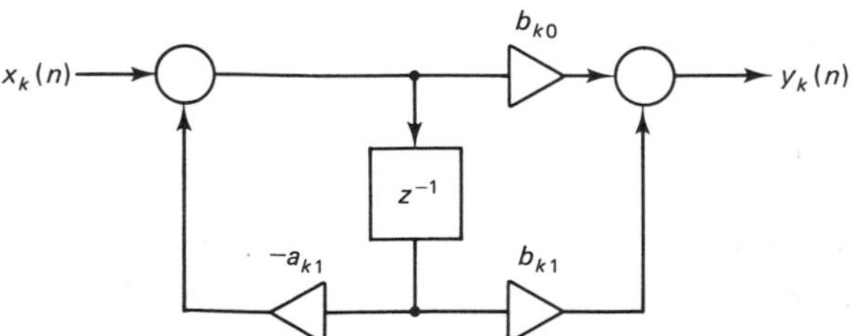

Figure 4.20 Realization of Eq. (4.49).

Examples of Cascaded Realizations

As an example of a fourth-order system, consider the system function

$$H(z) = \frac{1 + \frac{1}{2}z^{-1}}{(1 - z^{-1} + \frac{1}{4}z^{-2})(1 - z^{-1} + \frac{1}{2}z^{-2})} \tag{4.51}$$

As has been pointed out, as long as we use infinite-precision arithmetic, it makes no difference how the poles and zeros are paired nor does the order in which the sections are cascaded matter. However, in practice, we must use finite-precision arithmetic and then the pairings and orderings yield different results [136]. We choose the following pairing

$$H(z) = \frac{1 + \frac{1}{2}z^{-1}}{1 - z^{-1} + \frac{1}{4}z^{-2}} \frac{1}{1 - z^{-1} + \frac{1}{2}z^{-2}}$$

and order the sections so as to obtain the realization shown in Fig. 4.21.

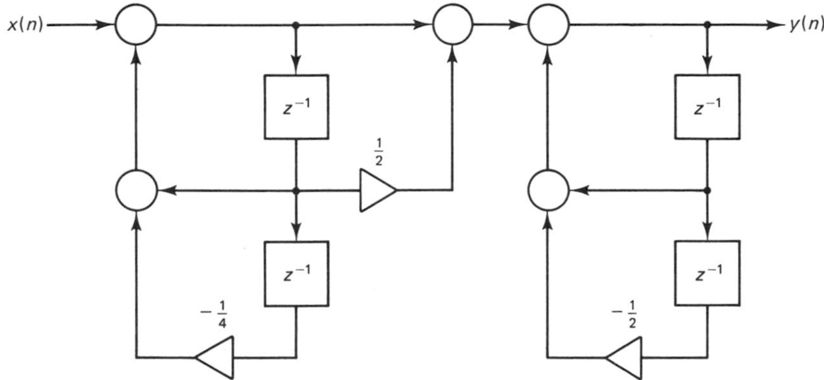

Figure 4.21 Cascade realization of Eq. (4.51).

As another example, we obtain a realization for the system function

$$H(z) = \frac{1 + \frac{1}{2}z^{-1}}{1 - \frac{1}{2}z^{-1}} \frac{1 - z^{-1} + \frac{1}{4}z^{-2}}{1 + z^{-1} + \frac{1}{4}z^{-2}} \frac{1 + z^{-1}}{1 + z^{-1} + \frac{1}{2}z^{-2}} \tag{4.52}$$

One possible realization in which the sections are ordered as in Eq. (4.52) is shown in Fig. 4.22.

As we will see in Chapter 5, where we discuss the bilinear-transformation technique for obtaining digital filters from analog filters, we generally start with an analog function in factored form and obtain a digital-system function that has the form of Eq. (4.48). The cascade realization is the most natural structure for digital filters obtained in this manner.

If we have a system function in the form of Eq. (4.2), then we must factor both numerator and denominator. In most cases, a digital computer or a programmable calculator would be needed to accomplish this.

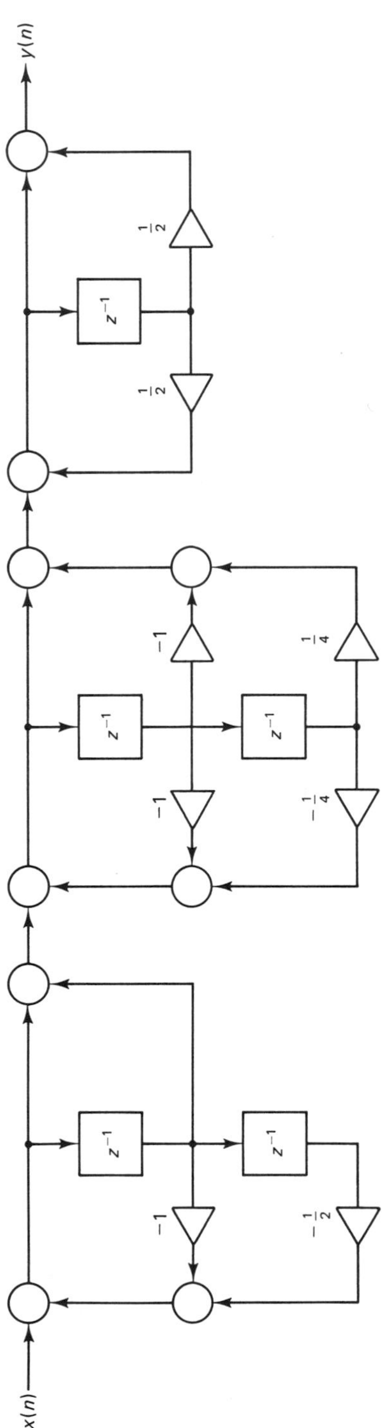

Figure 4.22 Cascade realization of Eq. (4.52).

EXERCISES

4.6.1 Using first-order sections, obtain a cascade realization for

$$H(z) = \frac{(1 + \frac{1}{2}z^{-1})(1 + \frac{1}{4}z^{-1})}{(1 - \frac{1}{2}z^{-1})(1 - \frac{1}{4}z^{-1})(1 - \frac{1}{8}z^{-1})}$$

4.6.2 Obtain a cascade realization using second-order sections where possible for

$$H(z) = \frac{1 + \frac{1}{4}z^{-1} + \frac{1}{2}z^{-2}}{(1 + \frac{1}{2}z^{-1})(1 + z^{-1} + \frac{1}{2}z^{-2})(1 + \frac{1}{2}z^{-1} + \frac{1}{4}z^{-2})}$$

4.7 PARALLEL-FORM REALIZATION OF IIR SYSTEMS

Development of the Parallel-Form Realization

The last of the common IIR system structures is the *parallel-form* realization. The structure given in Ex. 4.2.4 is an example of this form. In this, the input signal is processed separately by different subsystems. The system output is then a weighted sum of the outputs of the subsystems, which together constitute the overall system. In this section, we consider the general situation for IIR systems.

The parallel-form realization evolves from the partial-fraction expansion of $H(z)$, which generally has the form

$$\begin{aligned} H(z) = \sum_{k=0}^{M-N} c_k z^{-k} + \frac{b_{00}}{1 + a_{01}z^{-1}} \\ + \sum_{k=1}^{N_1} \frac{b_{k0} + b_{k1}z^{-1}}{1 + a_{k1}z^{-1} + a_{k2}z^{-2}} \end{aligned} \qquad (4.53)$$

where $N_1 = [N/2]$, the largest integer less than or equal to $N/2$. If $M < N$, then the term $\sum_{k=0}^{M-N} c_k z^{-k}$ does not appear, and if N is even, the term $b_{00}/(1 + a_{01}z^{-1})$ does not appear in Eq. (4.53). Since $Y(z) = H(z)X(z)$, the output $y(n)$ is a sum of terms resulting from Eq. (4.53). To see how the realization is obtained, let us consider $M \leq N$. If we define

$$H_k(z) = \frac{b_{k0} + b_{k1}z^{-1}}{1 + a_{k1}z^{-1} + a_{k2}z^{-2}} \qquad k = 0, 1, \ldots, N_1$$

with $b_{01} = a_{02} = 0$, then we can write

$$Y(z) = c_0 X(z) + \sum_{k=0}^{N_1} H_k(z)X(z) \qquad (4.54)$$

which is realized by the structure shown in Fig. 4.23. It should be noted that $x(n)$ is the input to each subsystem and that the output $y(n)$ is the sum of the subsystem outputs. Identifying $Y_k(z) = H_k(z)X(z)$, we obtain for the output

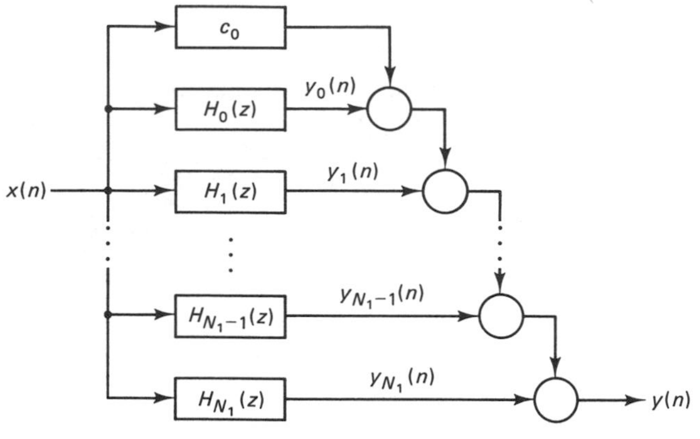

Figure 4.23 Realization of Eq. (4.54).

$$y(n) = c_0 x(n) + \sum_{k=0}^{N_1} y_k(n)$$

Example

As an example, again consider the system function $H(z)$ given in Eq. (4.51), whose partial-fraction expansion is

$$H(z) = \frac{5 - \frac{3}{2}z^{-1}}{1 - z^{-1} + \frac{1}{4}z^{-2}} + \frac{-4 + 3z^{-1}}{1 - z^{-1} + \frac{1}{2}z^{-2}}$$

Each of these terms is realized by a second-order section of the form shown in Fig. 4.19. The parallel-form realization is then shown in Fig. 4.24. If we write $H(z) = H_1(z) + H_2(z)$ and define $Y_i(z) = H_i(z)X(z)$, then the response is given by $y(n) = y_1(n) + y_2(n)$. The outputs $y_1(n)$ and $y_2(n)$ are identified in the figure.

In Chapter 5, we consider a method known as *impulse invariance* to obtain a digital filter from an analog filter. This method requires that the analog-system function be expanded into a partial fraction. Each term in the partial-fraction expansion is converted into a digital-system function. Consequently, this method gives $H(z)$ as a partial-fraction expansion and the parallel realization is the most natural realization to be used.

A Sampled Analog System

As a final example in this section, we obtain a discrete-time system by sampling the analog system shown in Fig. 4.25. Choosing capacitor voltages v_{a1} and v_{a2} as state variables, we can obtain the Kirchhoff's voltage equations

$$\begin{aligned} \frac{dv_{a1}}{dt} + \lambda_1 v_{a1} &= \lambda_1 v_{ga} \\ \frac{dv_{a2}}{dt} + \lambda_2 v_{a2} &= \lambda_2 v_{ga} \end{aligned} \tag{4.55}$$

where $\lambda_i = (R_i C_i)^{-1}$. Noting that $e^{\lambda_i t}$ is an integrating factor for $i = 1, 2$, we can obtain

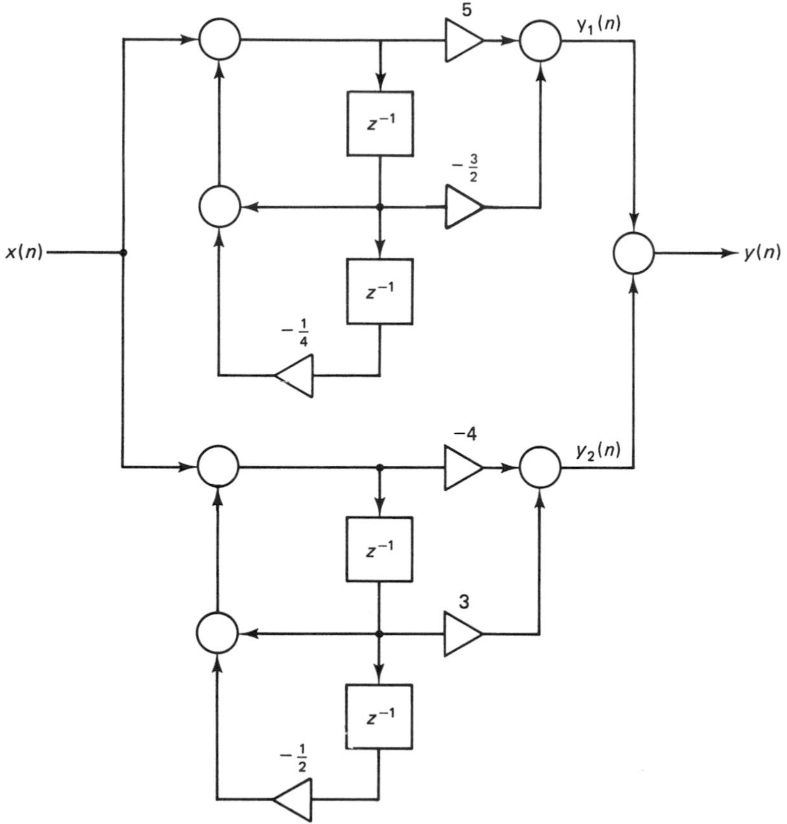

Figure 4.24 Parallel-form realization of the system of Eq. (4.51).

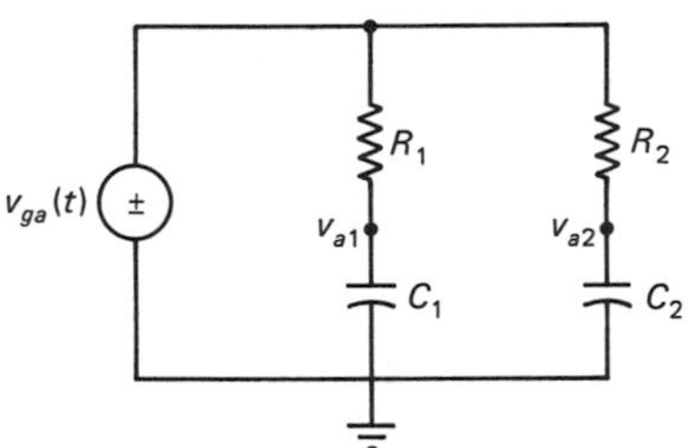

Figure 4.25 An RC circuit.

$$\int_{(n-1)T}^{nT} \frac{d}{dt}(v_{ai}e^{\lambda_i t})\, dt = \lambda_i \int_{(n-1)T}^{nT} v_{ga}e^{\lambda_i t}\, dt \tag{4.56}$$

If we assume that v_{ga} is the output of a sample-and-hold circuit,

$$v_{ga}(t) = v_{ga}(nT) \qquad nT \leq t < (n+1)T$$

then the discretized equation becomes

$$v_i(n)e^{\lambda_i nT} - v_i(n-1)e^{\lambda_i (n-1)T} = v_g(n-1)(e^{\lambda_i nT} - e^{\lambda_i (n-1)T})$$

where $v_{ai}(nT) = v_i(n)$, and $v_{ga}(nT) = v_g(n)$. Now solving for $v_i(n)$, we obtain

$$v_i(n) = e^{-\lambda_i T} v_i(n-1) + v_g(n-1)(1 - e^{-\lambda_i T}) \tag{4.57}$$

To complete the example, let us take

$$y_a(t) = \tfrac{1}{2} v_{a1}(t) + \tfrac{1}{4} v_{a2}(t)$$

so that the sampled output is

$$y(n) = \tfrac{1}{2} v_1(n) + \tfrac{1}{4} v_2(n) \tag{4.58}$$

A block-diagram representation of Eqs. (4.57) and (4.58) is shown in Fig. 4.26; this is a parallel-form realization. We have shown $v_1(n)$ and $v_2(n)$ in the figure, but we would normally replace the two multipliers that are in cascade by single multipliers.

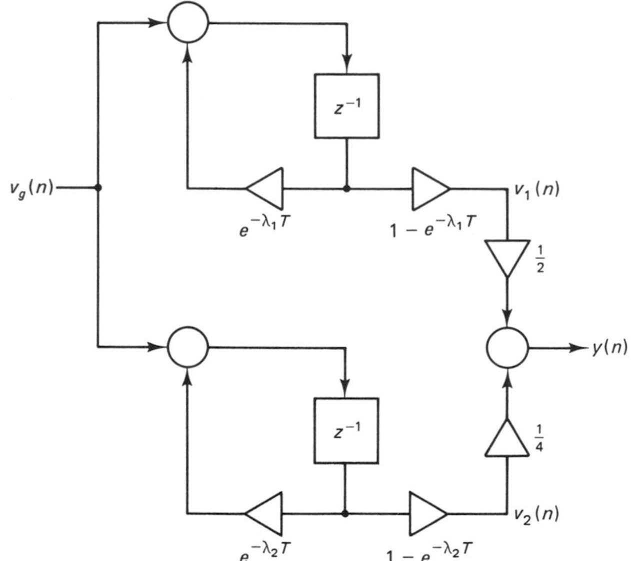

Figure 4.26 A realization of Eqs. (4.57) and (4.58).

EXERCISES

4.7.1 Using first-order sections, obtain a parallel-form realization of

$$H(z) = \frac{(1 + \frac{1}{2}z^{-1})(1 + \frac{1}{4}z^{-1})}{(1 - \frac{1}{2}z^{-1})(1 - \frac{1}{4}z^{-1})(1 - \frac{1}{8}z^{-1})}$$

4.7.2 Obtain a parallel-form realization of

$$H(z) = \frac{2 + z^{-1} + \frac{1}{4}z^{-2}}{(1 + \frac{1}{2}z^{-1})(1 + z^{-1} + \frac{1}{2}z^{-2})}$$

4.7.3 Approximate the integral on the right-hand side of Eq. (4.56) using the trapezoidal rule. Draw a block diagram using the resulting equations and Eq. (4.58).

4.8 LADDER STRUCTURES

It is known that analog filters realized using *ladder structures* have desirable coefficient-sensitivity properties [116, 181, 183]. By this, we mean that small changes in the filter parameters have little effect on its performance. It is reasonable to assume that digital filters can also be realized using ladder structures that are similar to those obtained for analog filters. Indeed, this is the case and there have been numerous methods for developing digital ladder structures in this manner [27, 39, 113, 114, 117]. In particular, we utilize continued-fraction expansions to obtain ladder realizations starting with the system function. Only a few of the various types of ladder structures are discussed in detail in this section. Other structures are considered in the problems at the end of the chapter. To obtain a more complete discussion of ladder structures for digital filters, the reader should also scc [113, 116].

In developing the digital ladder structures, we start with the system function $H(z)$, which we write

$$H(z) = \frac{a_N z^{-N} + a_{N-1} z^{-N+1} + \cdots + a_1 z^{-1} + a_0}{b_N z^{-N} + b_{N-1} z^{-N+1} + \cdots + b_1 z^{-1} + b_0} \tag{4.59}$$

The reader should note that this representation differs from the representation of Eq. (4.2). The reason for using Eq. (4.59) to represent $H(z)$ is to allow use of the Routh array considered in Probs. 3.32, 4.21, and 4.28 to more readily obtain the desired continued-fraction expansions of $H(z)$, which are used in the ladder realizations.

Continued-Fraction Expansion of $H(z)$

The first ladder structure that we consider is obtained from the continued-fraction expansion

$$H(z) = \alpha_0 + \cfrac{1}{\beta_1 z^{-1} + \cfrac{1}{\alpha_1 + \cfrac{1}{\beta_2 z^{-1} + \cdots + \cfrac{1}{\alpha_N}}}} \tag{4.60}$$

where the coefficients can be determined from the elements of the first column of the Routh array (see Prob. 3.32 with $s = z^{-1}$) shown in Table 4.1.

The first row of the array contains the numerator coefficients and the second row contains the denominator coefficients. The remaining rows, though labeled differently, are obtained as outlined in Prob. 3.32. In particular, the continued-fraction coefficients are given by

$$\alpha_0 = a_N/b_N, \ \beta_1 = b_N/c_{N-1}, \ \alpha_1 = c_{N-1}/d_{N-1}, \ \ldots \tag{4.61}$$

The conditions for the existence of this realization are given by Mitra and Sherwood [116].

TABLE 4.1

z^{-N}	a_N	a_{N-1}	a_{N-2}	$\cdots$	a_1	a_0
z^{-N}	b_N	b_{N-1}	b_{N-2}	$\cdots$	b_1	b_0
z^{-N+1}	c_{N-1}	c_{N-2}	c_{N-3}	$\cdots$	c_0	
z^{-N+1}	d_{N-1}	d_{N-2}	d_{N-3}	$\cdots$	d_0	
z^{-N+2}	e_{N-2}	e_{N-3}	e_{N-4}	$\cdots$		
z^{-N+2}	f_{N-2}	f_{N-3}	f_{N-4}	$\cdots$		
$\vdots$	$\vdots$			$\ddots$		
1	g_0					
1	h_0					

Example of a Continued Fraction

As an example, consider the system function

$$H(z) = \frac{2z^{-2} + 3z^{-1} + 1}{z^{-2} + z^{-1} + 1} \tag{4.62}$$

This is an unstable system but is used to demonstrate the realization procedure. In order to obtain the continued fraction of the form of Eq. (4.60), we first construct the Routh array of Table 4.1 for this system function. From Eq. (4.62), we see that $N = 2$, $a_2 = 2$, $a_1 = 3$, $a_0 = 1$, and $b_2 = b_1 = b_0 = 1$. The resulting Routh array is

z^{-2}	2	3	1
z^{-2}	1	1	1
z^{-1}	1	-1	
z^{-1}	2	1	
1	$-\frac{3}{2}$		
1	1		

From this table, we see that $c_1 = 1$, $d_1 = 2$, $e_0 = -\frac{3}{2}$, and $f_0 = 1$, and so the continued-fraction coefficients as determined from Eq. (4.61) are

$$\alpha_0 = a_2/b_2 = 2/1 = 2$$

$$\beta_1 = b_2/c_1 = 1/1 = 1$$

$$\alpha_1 = c_1/d_1 = 1/2$$

$$\beta_2 = d_1/e_0 = 2/(-3/2) = -4/3$$

$$\alpha_2 = e_0/f_0 = (-3/2)/1 = -3/2$$

Consequently, the continued-fraction expansion is

$$H(z) = 2 + \cfrac{1}{z^{-1} + \cfrac{1}{\frac{1}{2} + \cfrac{1}{-\frac{4}{3}z^{-1} + \cfrac{1}{-\frac{3}{2}}}}}$$

Realization of a Ladder Structure

Now that we see how to obtain the continued-fraction expansion, let us obtain a realization of $H(z)$. We first write

$$\begin{aligned} Y(z) &= H(z)X(z) \\ &= \left[\alpha_0 + \frac{1}{\beta_1 z^{-1} + 1/T_1(z)}\right]X(z) \\ &= \alpha_0 X(z) + H_1(z)X(z) \end{aligned} \tag{4.63}$$

where

$$H_1(z) = \frac{1}{\beta_1 z^{-1} + 1/T_1(z)} \tag{4.64}$$

and

$$T_1(z) = \alpha_1 + \cfrac{1}{\beta_2 z^{-1} + \alpha_2 + \cfrac{1}{\beta_3 z^{-1} + \cdots + \cfrac{1}{\alpha_N}}} \tag{4.65}$$

As a first step in the realization of Eq. (4.60), we realize Eq. (4.63) in the parallel form shown in Fig. 4.27(a), where $Y_1(z)$ is the output of the subsystem having $H_1(z)$ as system function. The describing relation for this subsystem is

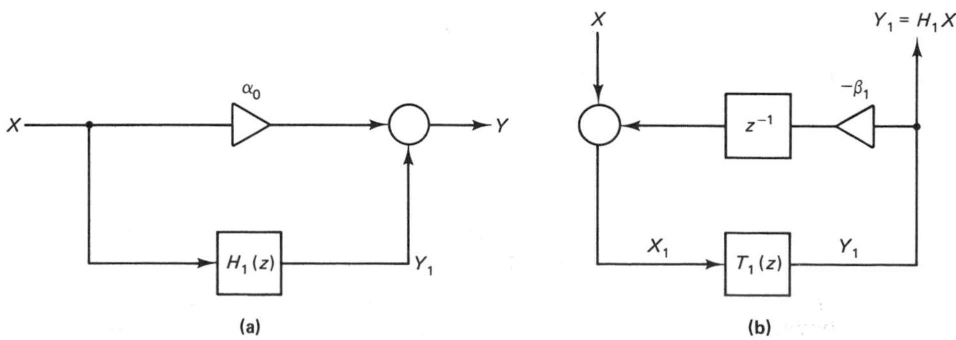

Figure 4.27 (a) A parallel-form realization of $H(z)$ of Eq. (4.63) and (b) a realization of $H_1(z)$ of Eq. (4.64).

$$\begin{aligned} Y_1(z) &= H_1(z)X(z) \\ &= \frac{1}{\beta_1 z^{-1} + [1/T_1(z)]} X(z) \end{aligned} \tag{4.66}$$

or in the equivalent form,

$$[\beta_1 z^{-1} + 1/T_1(z)]Y_1(z) = X(z)$$

For realization purposes, this equation is written

$$\begin{aligned} Y_1(z) &= T_1(z)[X(z) - \beta_1 z^{-1} Y_1(z)] \\ &= T_1(z)X_1(z) \end{aligned} \tag{4.67}$$

where

$$X_1(z) = X(z) - \beta_1 z^{-1} Y_1(z) \tag{4.68}$$

We can now obtain a realization of Eq. (4.66) using Eqs. (4.67) and (4.68), as shown in Fig. 4.27(b). The structures shown in Fig. 4.27 are then combined to form the structure of Fig. 4.28, which realizes $H(z)$ in terms of $T_1(z)$.

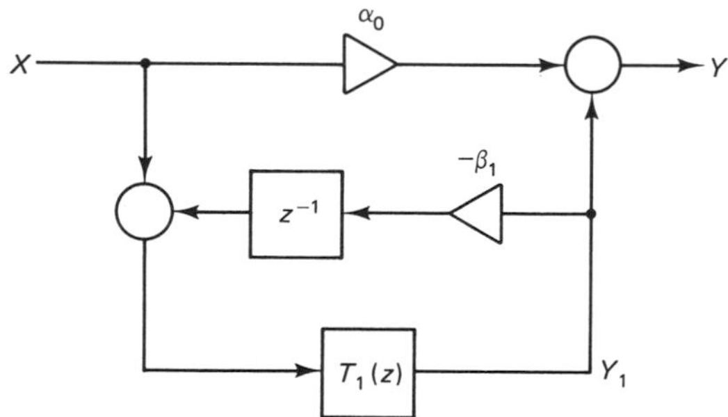

Figure 4.28 Structure for $H(z) = \alpha_0 + 1/[\beta_1 z^{-1} + 1/T_1(z)]$.

The next step in the realization of $H(z)$ is to obtain a realization of $T_1(z)$ that can be used in Fig. 4.28. If we write the expression for $T_1(z)$ in Eq. (4.65) as

$$T_1(z) = \frac{Y_1(z)}{X_1(z)} = \alpha_1 + \frac{1}{\beta_2 z^{-1} + 1/T_2(z)} \tag{4.69}$$

and compare this with the expression for $H(z)$ given in Eq. (4.63),

$$H(z) = \alpha_0 + \frac{1}{\beta_1 z^{-1} + 1/T_1(z)}$$

we see that $T_1(z)$ can be obtained from the structure for $H(z)$ shown in Fig. 4.28 if we replace α_0 by α_1, β_1 by β_2, and $T_1(z)$ by $T_2(z)$. The resulting structure for $T_1(z)$ is shown in Fig. 4.29. We can then replace the subsystem with system function $T_1(z)$ in Fig. 4.28 by the structure given in Fig. 4.29 and thus obtain the realization for $H(z)$ shown in Fig. 4.30. This procedure can be continued until we obtain the final structure for $H(z)$ shown in Fig. 4.31.

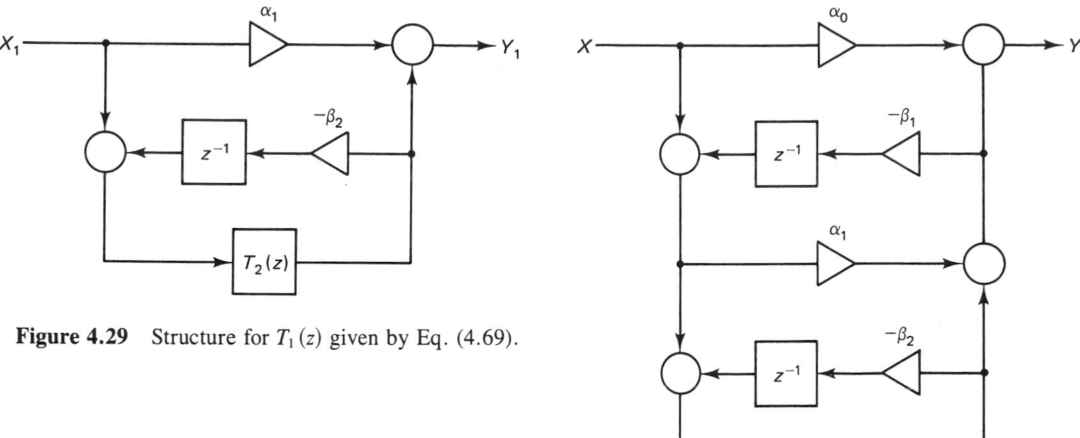

Figure 4.29 Structure for $T_1(z)$ given by Eq. (4.69).

Figure 4.30 Structure for Eqs. (4.63) and (4.69).

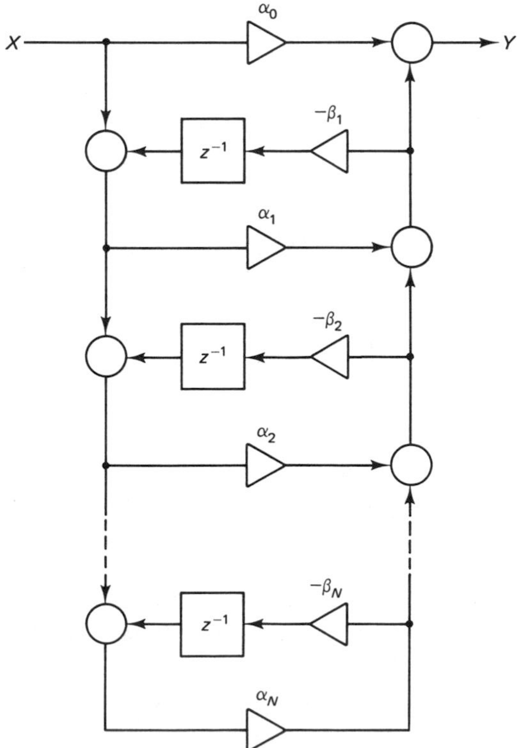

Figure 4.31 Continued-fraction realization of $H(z)$ given by Eq. (4.60).

Example of a Ladder Realization

For the system function given in Eq. (4.62), we have shown that $\alpha_0 = 2$, $\beta_1 = 1$, $\alpha_1 = \frac{1}{2}$, $\beta_2 = -\frac{4}{3}$, and $\alpha_2 = -\frac{3}{2}$. The ladder realization for this system that corresponds to Fig.4.31 is shown in Fig. 4.32.

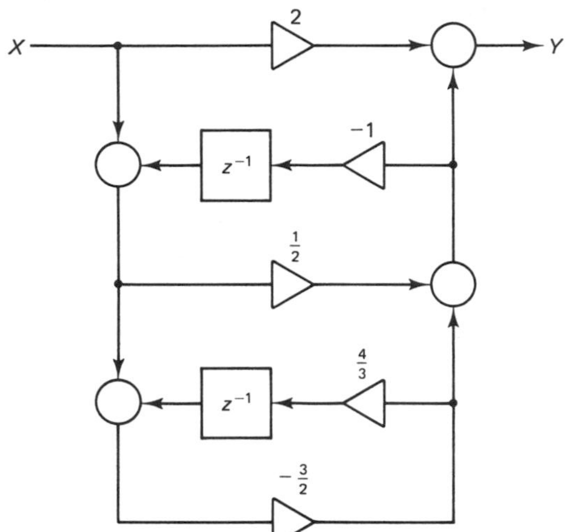

Figure 4.32 Continued-fraction realization of Eq. (4.62).

Two-Port Filter Realizations

The last technique that we consider in this section utilizes the *two-port filter* or *digital two-pair* shown in Fig. 4.33(a) [117]. The variables X_1 and X_2 are inputs and Y_1 and Y_2 are outputs. This network can be described by its *chain matrix*

$$\begin{bmatrix} X_1 \\ Y_1 \end{bmatrix} = \begin{bmatrix} A & B \\ C & D \end{bmatrix} \begin{bmatrix} Y_2 \\ X_2 \end{bmatrix} \tag{4.70}$$

or by its *transfer matrix*

$$\begin{bmatrix} Y_1 \\ Y_2 \end{bmatrix} = \begin{bmatrix} t_{11} & t_{12} \\ t_{21} & t_{22} \end{bmatrix} \begin{bmatrix} X_1 \\ X_2 \end{bmatrix} \tag{4.71}$$

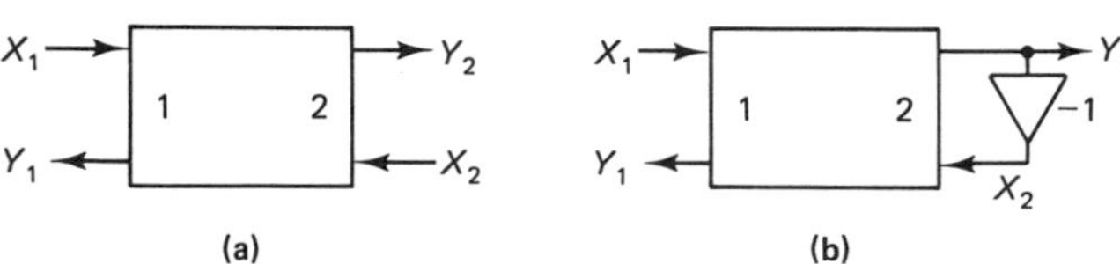

Figure 4.33 (a) A digital two-pair and (b) a constrained two-pair with $Y_2 = -X_2$.

In order to determine the relations between the two sets of parameters, we solve Eq. (4.70) for Y_1 and Y_2 in terms of X_1 and X_2 and compare the results with Eq. (4.71). These relations are (see Ex. 4.8.2)

$$\begin{aligned} Y_1 &= (C/A)X_1 + [D - (BC/A)]X_2 \\ Y_2 &= (1/A)X_1 - (B/A)X_2 \end{aligned} \tag{4.72}$$

Comparing these relations with Eq. (4.71), we find that

$$\begin{aligned} t_{11} &= C/A \qquad & t_{12} &= (AD - BC)/A \\ t_{21} &= 1/A \qquad & t_{22} &= -B/A \end{aligned} \tag{4.73}$$

To make this two-pair useful in system realizations, we must constrain the network. We choose to make

$$Y_2 = -X_2 \tag{4.74}$$

which results in the constrained network shown in Fig. 4.33(b). Taking Y_2 as output and X_1 as input, we can use Eq. (4.74) and the second equation of Eqs. (4.72) to obtain the system function

$$\begin{aligned} H(z) &= \frac{Y_2}{X_1} = \frac{1}{A - B} \\ &= \frac{1/A}{1 - (B/A)} \\ &= \frac{t_{21}}{1 + t_{22}} \end{aligned} \tag{4.75}$$

Parameter t_{22} can be computed from Eq. (4.71) and is given by

$$t_{22} = \left.\frac{Y_2}{X_2}\right|_{X_1=0}$$

This expression suggests that we set $X_1 = 0$ in Fig. 4.33(a) and synthesize Y_2/X_2, looking in at the output port Y_2–X_2.

Two-Pair Ladder Configuration

The two-pair ladder configuration shown in Fig. 4.34 can be used to realize a transfer function of the form

$$H(z) = \frac{Y_2}{X_1} = \frac{a_0}{a_N z^{-N} + a_{N-1} z^{-N+1} + \cdots + a_1 z^{-1} + a_0} \tag{4.76}$$

The number of elements T_i forming the ladder rungs is equal to N, the order of the filter. If N is odd, then T_0 is an element in the structure; it does not appear if N is even. The synthesis procedure depends on whether N is even or odd.

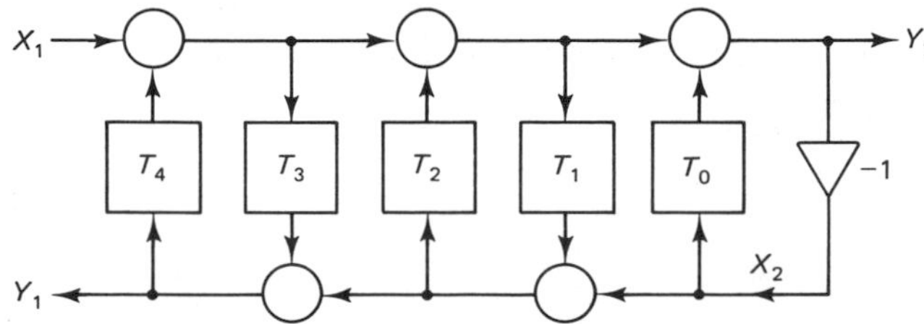

Figure 4.34 A ladder structure that can be used to realize Eq. (4.76) for $N = 5$.

Synthesis for *N* Odd

If N is odd, we write Eq. (4.76) as

$$H(z) = \frac{a_0}{P(z) + Q(z)} = \frac{a_0/P(z)}{1 + [Q(z)/P(z)]} \tag{4.77}$$

where $P(z)$ is an even polynomial, and $Q(z)$ is an odd polynomial in z^{-1}. Comparing this expression with Eq. (4.75), we see that $t_{22} = Q(z)/P(z)$ can be expanded in a continued fraction having the form of Eq. (4.60). The values of the T_i are obtained by making the continued fraction obtained from Eq. (4.77) identical to the continued fraction obtained for the unconstrained structure of Fig. 4.34.

Realization of a Third-Order System

To see how the synthesis procedure evolves, let us consider the third-order system described by

$$H(z) = \frac{a_0}{a_3 z^{-3} + a_2 z^{-2} + a_1 z^{-1} + a_0} \tag{4.78}$$

For this function, we have

$$\begin{aligned} t_{22} = \frac{Q(z)}{P(z)} &= \frac{a_3 z^{-3} + a_1 z^{-1}}{a_2 z^{-2} + a_0} \\ &= \alpha_0 z^{-1} + \frac{1}{\alpha_1 z^{-1} + 1/\alpha_2 z^{-1}} \end{aligned} \tag{4.79}$$

which can be determined from the Routh array, as shown in Prob. 4.28. For our particular case, the Routh array is

z^{-3}	$a_3 \quad a_1$
z^{-2}	$a_2 \quad a_0$
z^{-1}	$b_1 = (a_1 a_2 - a_3 a_0)/a_2$
1	$c_1 = a_0$

Consequently, the coefficients in the continued fraction are

$$\alpha_0 = a_3/a_2 \qquad \alpha_1 = a_2/b_1 \qquad \alpha_2 = b_1/c_1$$

We realize $H(z)$ given by Eq. (4.78) using the ladder structure of Fig. 4.34 with the three rung elements T_0, T_1, and T_2. With $X_1 = 0$ we find that

$$\begin{aligned} t_{22} = \frac{Y_2}{X_2} &= \frac{T_0(1 - T_1 T_2) + T_2}{1 - T_1 T_2} \\ &= T_0 + \frac{1}{-T_1 + 1/T_2} \end{aligned} \tag{4.80}$$

When we compare Eqs. (4.79) and (4.80), we see that

$$T_0 = \alpha_0 z^{-1} \qquad -T_1 = \alpha_1 z^{-1} \qquad T_2 = \alpha_2 z^{-1}$$

Consequently, the desired ladder structure is that shown in Fig. 4.35.

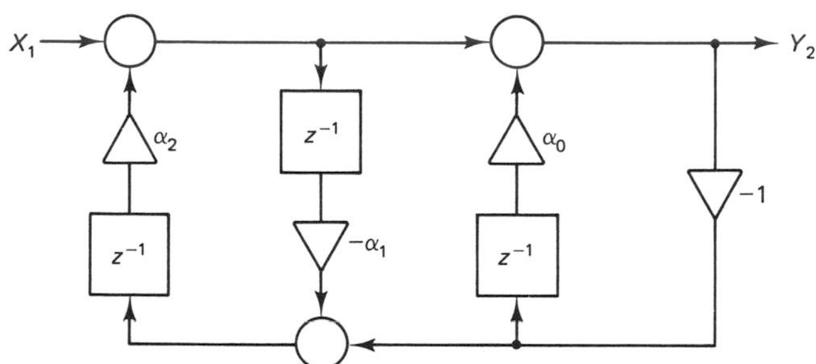

Figure 4.35 Ladder structure for the third-order system of Eq. (4.78).

Example of a Third-Order System

For the system described by

$$H(z) = \frac{2}{z^{-3} + 4z^{-2} + z^{-1} + 2} \tag{4.81}$$

we find that $b_1 = (a_1 a_2 - a_3 a_0)/a_2 = \frac{1}{2}$ and so

$$\alpha_0 = a_3/a_2 = \tfrac{1}{4} \qquad \alpha_1 = a_2/b_1 = 4/\tfrac{1}{2} = 8 \qquad \alpha_2 = b_1/c_1 = \tfrac{1}{2}/2 = \tfrac{1}{4}$$

The ladder structure for this $H(z)$ is shown in Fig. 4.36.

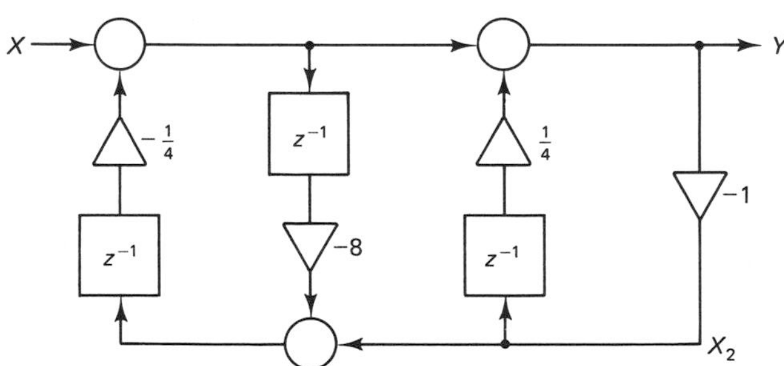

Figure 4.36 Ladder structure for the third-order system of Eq. (4.81).

The Case When *N* Is Even

When N is even, element T_0 is omitted from the ladder structure of Fig. 4.34. The realization is obtained by writing

$$t_{22} = \frac{Q(z)}{P(z)} = \frac{1}{[P(z)/Q(z)]}$$

Since $P(z)$ has higher degree than $Q(z)$, we can obtain a continued-fraction expansion for $P(z)/Q(z)$ having the form of Eq. (4.60). We consider the case of a fourth-order system in Prob. 4.30.

Other Cases

The cases for which $H(z)$ has finite zeros that are not zero will not be considered. The interested reader should consult [117].

EXERCISES

4.8.1 Given the system function

$$H(z) = \frac{2 + 8z^{-1} + 6z^{-2}}{1 + 8z^{-1} + 12z^{-2}}$$

obtain a realization of the form of Fig. 4.31.

4.8.2 Obtain Eq. (4.72) from Eqs. (4.70). *Hint:* Solve the first equation of Eqs. (4.70) for Y_2.

4.8.3 Obtain a realization of the form of Fig. 4.34 for

$$H(z) = \frac{1}{z^{-3} + 2z^{-2} + 2z^{-1} + 1}$$

4.9 BASIC REALIZATIONS OF FIR SYSTEMS

The realizations of the general FIR system are similar to those of IIR systems. The direct-form realization and the cascade realization are obtained in the same way as in Secs. 4.5 and 4.6. Parallel realizations are not used since they require more elements.

Direct-Form Realization

Assuming that the system function is

$$H(z) = \sum_{k=0}^{N-1} h(k)z^{-k} \tag{4.82}$$

the system response is

$$y(n) = \sum_{k=0}^{N-1} h(k)x(n - k) \tag{4.83}$$

This representation can be obtained from Eq. (4.39) by setting $b_k = h(k)$, $a_k = 0$, $k = 1, 2, \ldots, N$, and $M = N$. Its direct-form realization is shown in Fig. 4.37 and is simply a special case of the realization of Fig. 4.16.

Cascade Realization

The cascade realization is obtained from the factored form of $H(z)$, which we write

$$H(z) = \prod_{k=1}^{[N/2]} (b_{k0} + b_{k1}z^{-1} + b_{k2}z^{-2}) \tag{4.84}$$

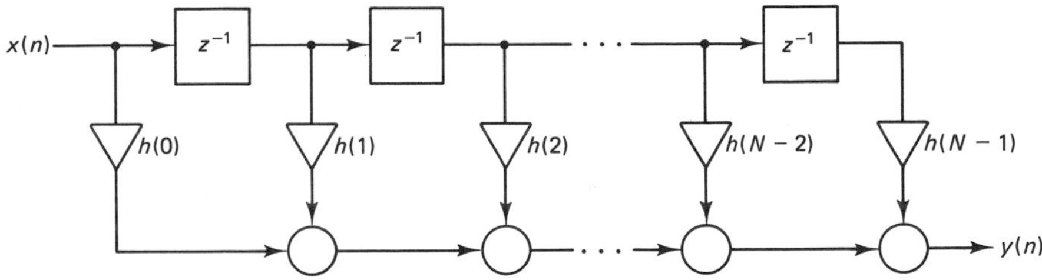

Figure 4.37 Direct-form realization of Eq. (4.82).

where $[N/2]$ is the largest integer less than or equal to $N/2$. If N is even ($N - 1$ is odd), one of the coefficients, b_{k2}, is zero. The realization is shown in Fig. 4.38.

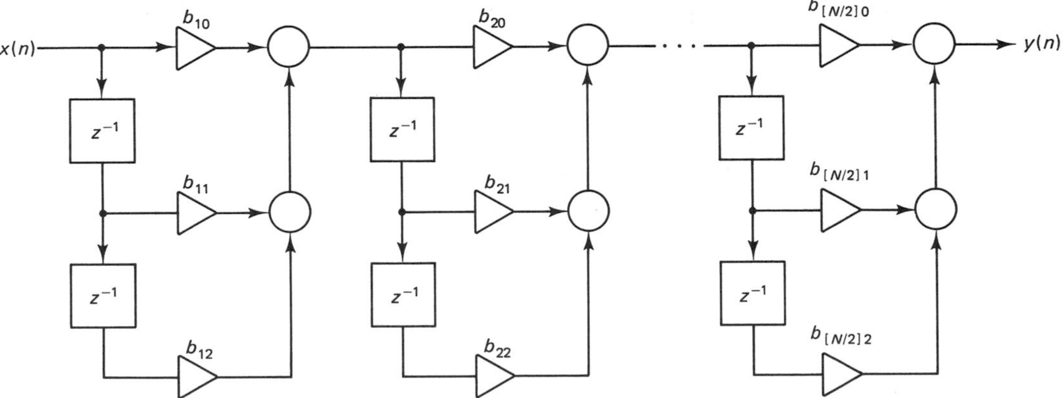

Figure 4.38 Cascade realization of Eq. (4.84).

Consider the system function

$$H(z) = (1 - z^{-1} - z^{-2})(1 - z^{-1} + \tfrac{1}{2}z^{-2}) \tag{4.85}$$

$$= 1 - 2z^{-1} + \tfrac{1}{2}z^{-2} + \tfrac{1}{2}z^{-3} - \tfrac{1}{2}z^{-4} \tag{4.86}$$

The cascade realization obtained from Eq. (4.85) is shown in Fig. 4.39 and the direct-form realization obtained from Eq. (4.86) is shown in Fig. 4.40.

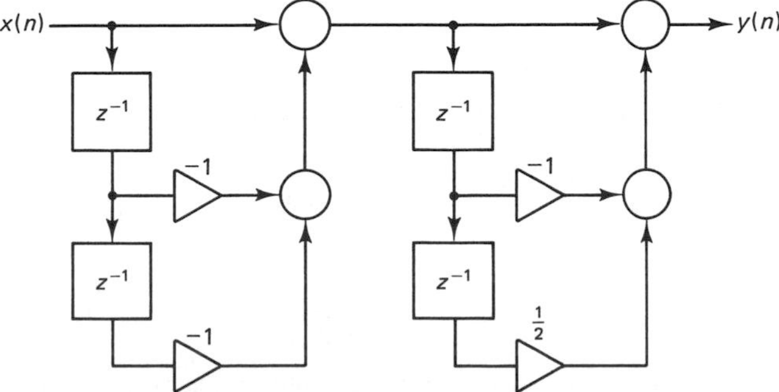

Figure 4.39 Cascade realization of Eq. (4.85).

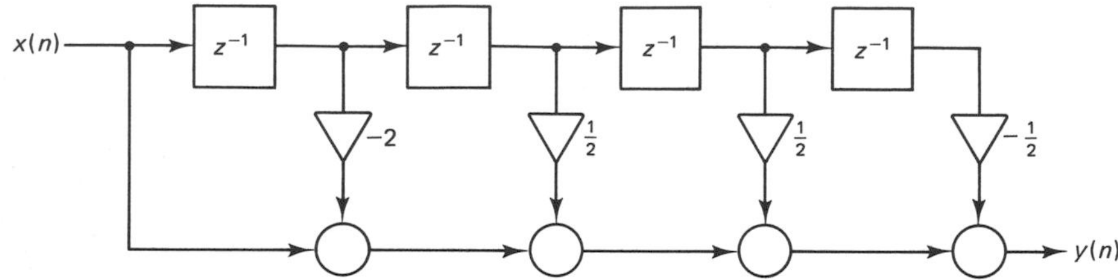

Figure 4.40 Direct-form realization of Eq. (4.86).

Linear Phase Realizations

We can realize linear-phase FIR filters using the general direct and cascade forms shown in Figs. 4.37 and 4.38, respectively. However, since the coefficients are equal in pairs,

$$h(n) = h(N - 1 - n) \tag{4.87}$$

it is possible to use this symmetry to reduce the number of multipliers required in the realization. Using this condition, we can write

$$\begin{aligned} H(z) &= \sum_{n=0}^{N-1} h(n)z^{-n} \\ &= \sum_{n=0}^{(N/2)-1} h(n)[z^{-n} + z^{-(N-1-n)}] \end{aligned} \tag{4.88}$$

for N even, and

$$H(z) = h\left(\frac{N-1}{2}\right)z^{-(N-1)/2} + \sum_{n=0}^{(N-3)/2} h(n)[z^{-n} + z^{-(N-1-n)}] \tag{4.89}$$

for N odd.

The realizations corresponding to Eqs. (4.88) and (4.89) are shown in Fig. 4.41 for N even and Fig. 4.42 for N odd, respectively. For N even, we see that $N/2$

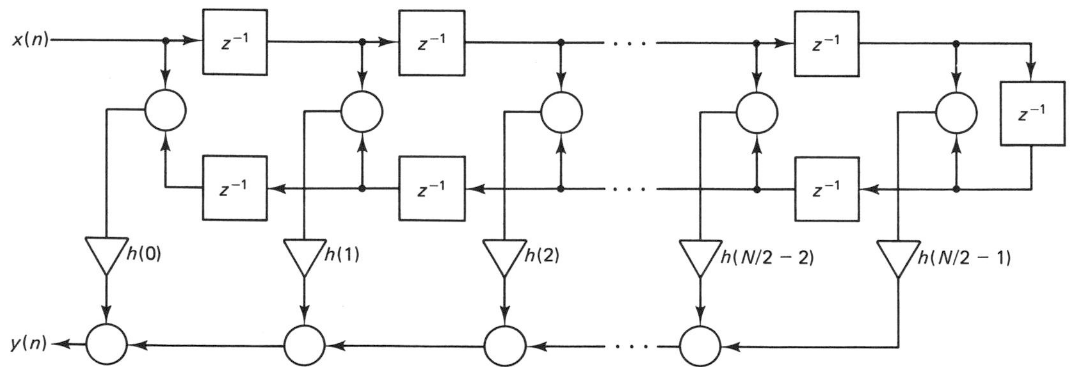

Figure 4.41 Direct-form realization of a linear-phase FIR system for N even.

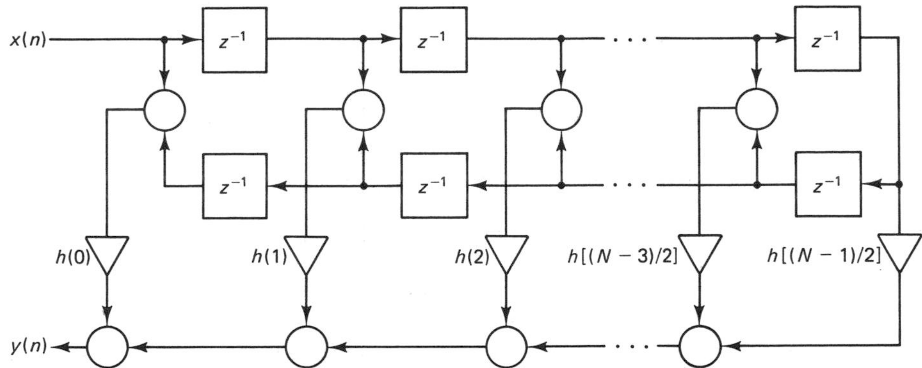

Figure 4.42 Direct-form realization of a linear-phase FIR system for N odd.

multipliers are required, and for N odd, $(N + 1)/2$. The general direct-form realization shown in Fig. 4.37 shows that N multipliers are required, and the cascade realization of Fig. 4.38 for N even requires $\frac{3}{2}N$. This last number can be reduced to $N + 1$ if we factor out the product $\Pi_{k=1}^{[N/2]} b_{k0}$. This is equivalent to adding one multiplier for this gain factor and eliminating the $N/2$ multipliers for the b_{k0} if N is even.

As an example, consider the system function

$$H(z) = \tfrac{1}{2} + \tfrac{1}{4}z^{-1} + \tfrac{1}{4}z^{-2} + \tfrac{1}{2}z^{-3} \tag{4.90}$$

Here $N = 4$ and the realization is shown in Fig. 4.43.

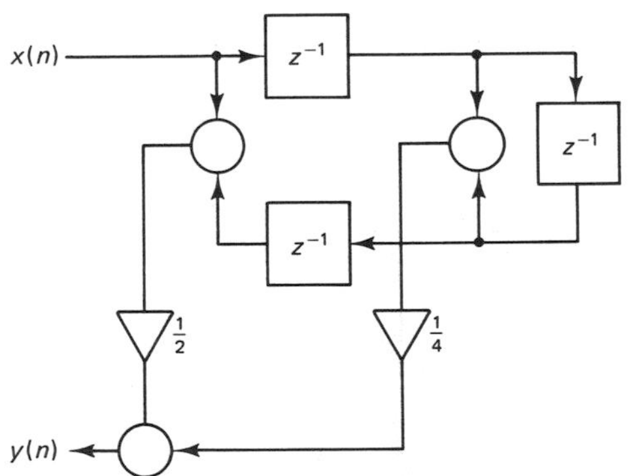

Figure 4.43 Direct-form realization of Eq. (4.90).

Cascade realizations can be obtained by factoring $H(z)$ as a product of factors that have the linear-phase symmetry property, which was discussed in Sec. 3.8. The factors involved can be first-, second-, or fourth- degree polynomials in z^{-1}. Each filter section is then realized as in Fig. 4.41 or Fig. 4.42, and the resulting sections are then cascaded. For example, the system function of Eq. (4.90) can be written

$$H(z) = (1 + z^{-1})(\tfrac{1}{2} - \tfrac{1}{4}z^{-1} + \tfrac{1}{2}z^{-2})$$

The corresponding cascade realization is shown in Fig. 4.44.

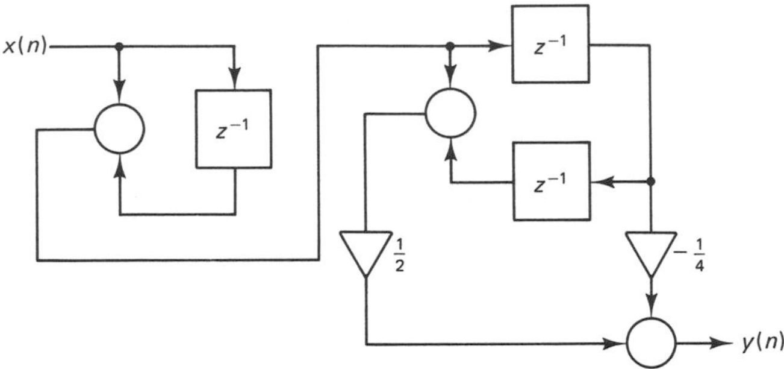

Figure 4.44 Cascade realization of Eq. (4.90).

EXERCISES

4.9.1 Realize the system function

$$H(z) = 1 + \tfrac{3}{4}z^{-1} + \tfrac{17}{8}z^{-2} + \tfrac{3}{4}z^{-3} + z^{-4}$$

by using the following:
(a) the general direct form of Fig. 4.37
(b) the linear-phase form of Fig. 4.42

4.9.2 Note that $H(z)$ in Ex. 4.9.1 has the factorization

$$H(z) = (1 + \tfrac{1}{2}z^{-1} + z^{-2})(1 + \tfrac{1}{4}z^{-1} + z^{-2})$$

Obtain a cascade realization using the minimum number of multipliers.

4.9.3 Realize the system function

$$H(z) = 1 + \tfrac{1}{2}z^{-1} + \tfrac{1}{4}z^{-2} + \tfrac{1}{4}z^{-3} + z^{-4}$$

using the direct form of Fig 4.37. Use a parallel realization for the representation

$$H(z) = (1 + \tfrac{1}{2}z^{-1} + \tfrac{1}{4}z^{-2} + \tfrac{1}{2}z^{-3} + z^{-4}) - \tfrac{1}{4}z^{-3}$$

Compare the number of adders, multipliers, and delays used in each realization.

PROBLEMS

4.1 Verify that the block diagram of Fig. 4.3 represents Eq. (4.6).

4.2 Draw a block-diagram representation for each of the following systems:
(a) $y(n) - \frac{1}{2}y(n-1) = x(n) + \frac{1}{2}x(n-1)$
(b) $y(n) = x(n) + \frac{1}{2}x(n-1) + x(n-2)$
(c) $y(n) + \frac{3}{4}y(n-1) + \frac{1}{8}y(n-2) = 2x(n) - \frac{1}{2}x(n-1)$
(d) $y(n) = ay(n-1) - ax(n) + x(n-1)$

4.3 Draw a block diagram that realizes the system of Prob. 4.2(d) with *only one* multiplier having multiplication factor other than -1.

4.4 Draw block diagrams for the following systems having input $x(n)$ and output $y(n)$:

(a) $w(n) = x(n) + \frac{1}{2}x(n-1)$
$y(n) + \frac{1}{4}y(n-1) = w(n)$

(b) $w(n) = x(n) + \frac{3}{4}x(n-1) + \frac{1}{8}x(n-2)$
$y(n) - \frac{3}{16}y(n-1) + \frac{1}{32}y(n-2) = w(n)$

(c) $w(n) + aw(n-1) = b_0x(n) + b_1x(n-1)$
$y(n) + \alpha y(n-1) = \beta_0 w(n) + \beta_1 w(n-1)$

(d) $w(n) + \frac{3}{8}w(n-1) + \frac{1}{32}w(n-2) = \frac{1}{2}x(n) + \frac{1}{4}x(n-1) + \frac{1}{8}x(n-2)$
$y(n) - \frac{1}{4}y(n-1) = \frac{1}{2}w(n) - \frac{2}{5}w(n-1) + \frac{3}{5}w(n-2)$

(e) $y_1(n) + a_1 y_1(n-1) = x(n) + b_1x(n-1)$
$y_2(n) + a_2 y_2(n-1) = x(n) + b_2x(n-1)$
$y(n) = y_1(n) + y_2(n)$

(f) $y_1(n) + \frac{1}{4}y_1(n-1) + \frac{1}{2}y_1(n-2) = 2x(n) + \frac{1}{2}x(n-1)$
$y_2(n) + \frac{1}{4}y_2(n-1) = x(n) - \frac{1}{2}x(n-1)$
$y_3(n) - \frac{1}{4}y_3(n-1) + \frac{1}{2}y_3(n-2) = \frac{1}{4}x(n)$
$y(n) = y_1(n) + y_2(n) + \frac{3}{2}y_3(n)$

(g) $w(n) = x(n) + a_1x(n-1)$,
$y(n) = b_0w(n) + b_1w(n-1)$

4.5 Determine the system function $H(z) = Y(z)/X(z)$ for each of the systems shown in the figure.

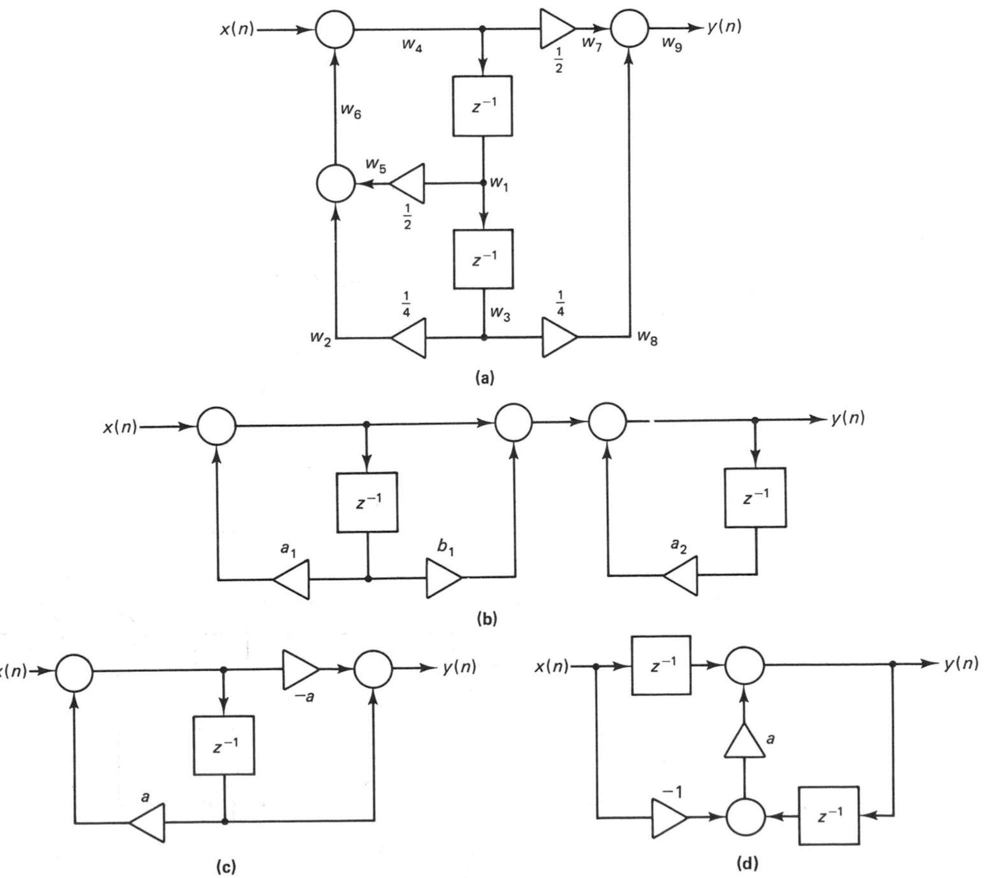

Problem 4.5

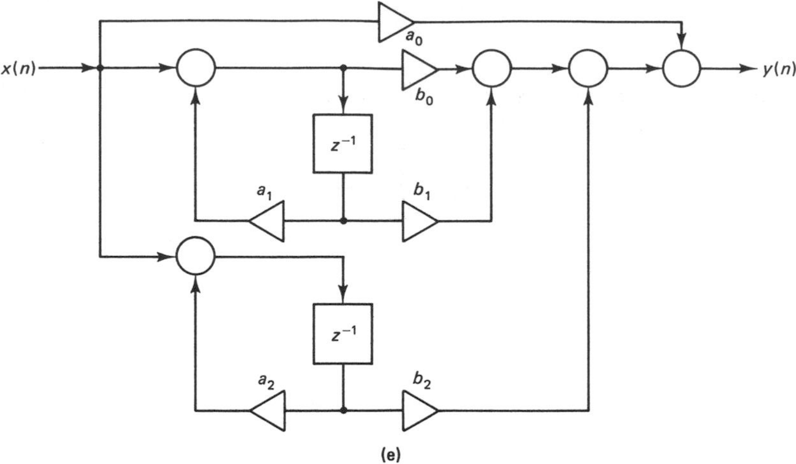

(e)

Problem 4.5 (cont.)

4.6 Draw a signal-flow graph for each of the systems shown in the figure and find $H(z) = Y(z)/X(z)$ using Mason's formula.

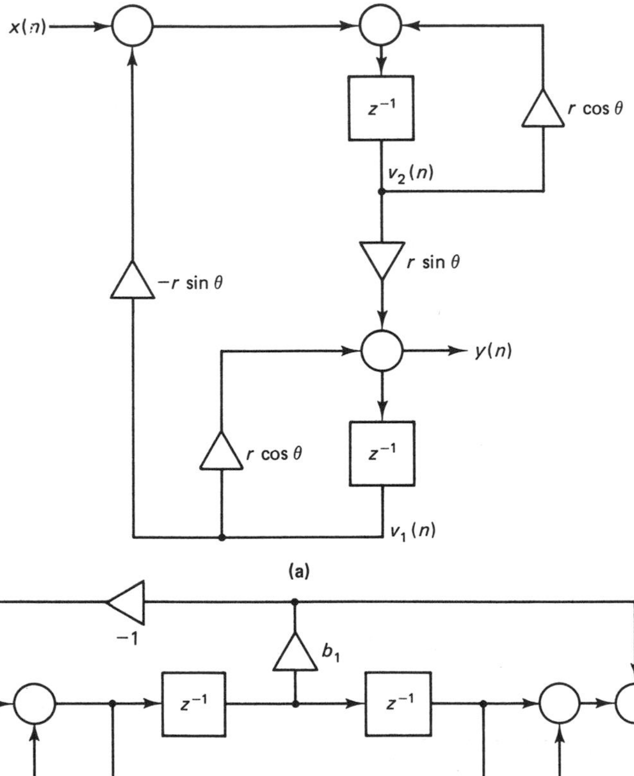

(a)

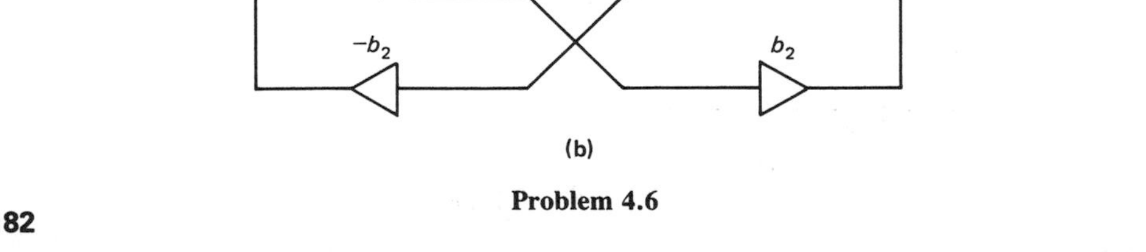

(b)

Problem 4.6

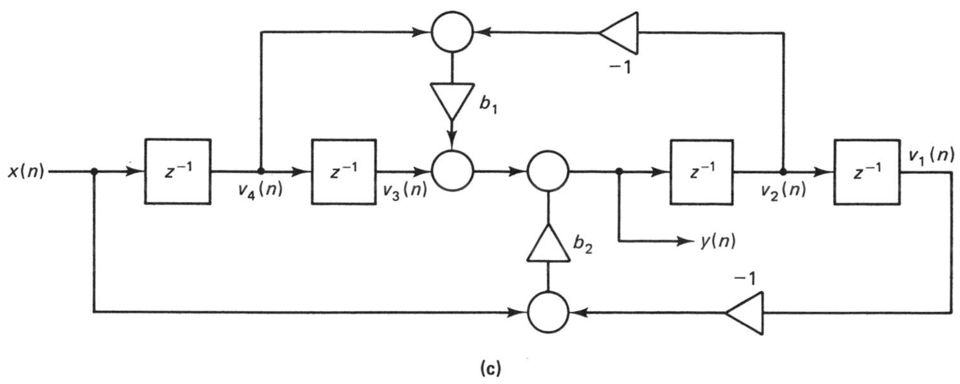

(c)

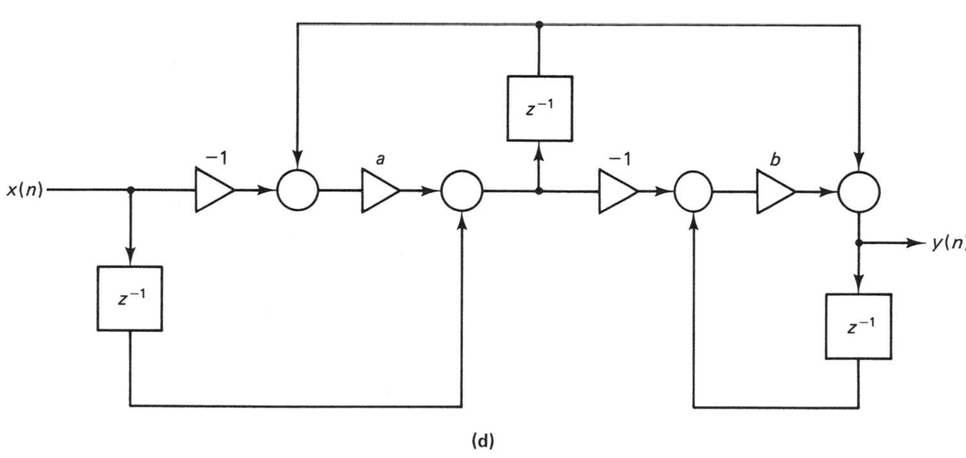

(d)

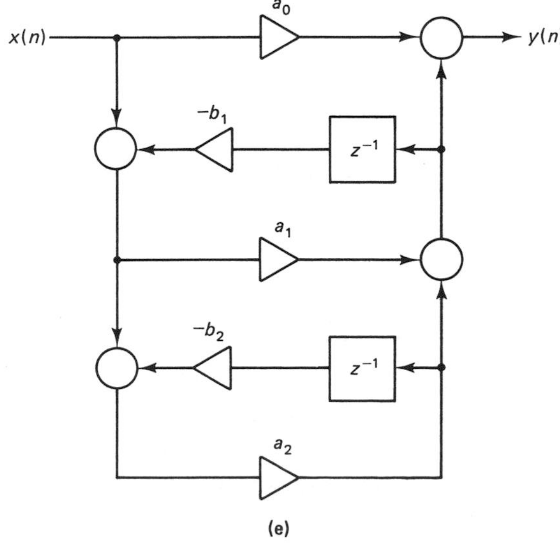

(e)

Problem 4.6 (cont.)

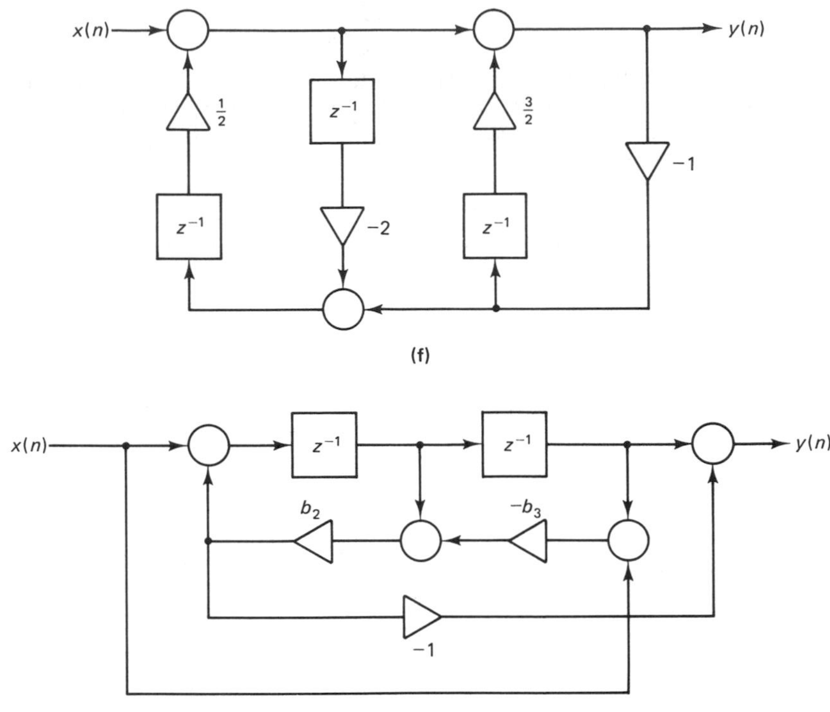

Problem 4.6 (cont.)

4.7 Interchange the variables w_3 and w_4 in Ex. 4.3.1 and determine if the new set of assigned variables can be computed in sequence.

4.8 Determine if the assigned variables for system (a) of Prob. 4.5 can be computed in sequence.

4.9 Assign variables to system (d) of Prob. 4.5 that can be computed in sequence. Each loop has a delay element so that such an assignment is possible. Write the matrix equations, Eqs. (4.17) and (4.18), for this variable assignment.

4.10 For system (a) of Prob. 4.5, define the state variables $v_1(n) = w_3(n)$ and $v_2(n) = w_1(n)$. Write the state equations and determine $H(z)$ from them.

4.11 For system (a) in Prob. 4.6, write the state equations using $v_1(n)$ and $v_2(n)$ as shown. Obtain $H(z)$ using Eq. (4.31).

4.12 For system (c) in Prob. 4.6, write the state equations using v_1, v_2, v_3, and v_4 as shown as the state variables.

4.13 The Fibonacci numbers given in Prob. 1.4 satisfy the difference equation

$$y(n) - y(n-1) - y(n-2) = x(n-1)$$

where $x(n) = \delta(n)$, $y(0) = 0$, and $y(1) = 1$. Obtain a state-variable representation of the form given in Eq. (4.38). Determine $y(n)$ by recursion using the state equations.

4.14 Obtain the state equations of the form of Eq. (4.38) for the general Nth-order difference equation.

Answers:

$$A = \begin{bmatrix} 0 & 0 & 0 & \cdots & 0 & -a_N \\ 1 & 0 & 0 & \cdots & 0 & -a_{N-1} \\ 0 & 1 & 0 & \cdots & 0 & -a_{N-2} \\ & & & \vdots & & \\ 0 & 0 & 0 & \cdots & 1 & -a_1 \end{bmatrix}$$

$$B = \begin{bmatrix} b_N - b_0 a_N \\ b_{N-1} - b_0 a_{N-1} \\ \cdots\cdots\cdots \\ b_1 - b_0 a_1 \end{bmatrix}$$

$$C = [0 \ \ 0 \ \ \cdots \ \ 0 \ \ 1] \qquad D = [b_0]$$

4.15 For the general second-order system, write

$$H(z) = \frac{Y(z)}{X(z)} = \frac{(b_0 + b_1 z^{-1} + b_2 z^{-2})W(z)}{(1 + a_1 z^{-1} + a_2 z^{-2})W(z)}$$

where $W(z)$ is determined to make the numerators and the denominators equal. Define the state variables as $V_2(z) = z^{-1}W(z)$ and $V_1(z) = z^{-2}W(z)$, and obtain the state equations.

4.16 Obtain a direct-form realization for the following systems:

(a) $y(n) + \frac{3}{4}y(n-1) + \frac{1}{8}y(n-2) = x(n) + x(n-1)$

(b) $H(z) = \dfrac{1 + \frac{3}{2}z^{-1} + \frac{1}{2}z^{-2}}{(1 + \frac{1}{2}z^{-1})(1 + \frac{1}{2}z^{-1} + \frac{1}{4}z^{-2})}$

(c) $H(z) = \dfrac{\frac{1}{2} + \frac{1}{2}z^{-1} + \frac{1}{4}z^{-2} + z^{-3}}{1 + \frac{1}{4}z^{-1} + \frac{1}{2}z^{-2} + \frac{1}{2}z^{-3}}$

4.17 Obtain a cascade realization for the following systems:

(a) $H(z) = \dfrac{1 + \frac{1}{4}z^{-1}}{(1 + \frac{1}{2}z^{-1})(1 + \frac{1}{2}z^{-1} + \frac{1}{4}z^{-2})}$

(b) $H(z) = \dfrac{(1 + \frac{3}{2}z^{-1} + \frac{1}{2}z^{-2})(1 - \frac{3}{2}z^{-1} + z^{-2})}{(1 + z^{-1} + \frac{1}{4}z^{-2})(1 + \frac{1}{4}z^{-1} + \frac{1}{2}z^{-2})}$

(c) $H(z) = \dfrac{(1 - \frac{1}{2}z^{-1})(1 - \frac{1}{2}z^{-1} + \frac{1}{4}z^{-2})}{(1 + \frac{1}{4}z^{-1})(1 + z^{-1} + \frac{1}{2}z^{-2})(1 - \frac{1}{4}z^{-1} + \frac{1}{2}z^{-2})}$

(d) $H(z) = \dfrac{(1 + z^{-1})^3}{(1 - \frac{1}{4}z^{-1})(1 - z^{-1} + \frac{1}{2}z^{-2})}$

4.18 Obtain parallel realizations for the systems of **(a)** Prob. 4.17(a) and **(b)** Prob. 4.17(d).

4.19 Obtain a parallel realization for the following systems:

(a) $H(z) = \dfrac{2 + z^{-1} + \frac{5}{4}z^{-2} + \frac{1}{4}z^{-3}}{(1 + \frac{1}{2}z^{-1} + \frac{1}{4}z^{-2})(1 - \frac{1}{2}z^{-1} + \frac{1}{2}z^{-2})}$

(b) $H(z) = \dfrac{(1 + z^{-1})(1 + 2z^{-1})}{(1 + \frac{1}{2}z^{-1})(1 - \frac{1}{4}z^{-1})(1 + \frac{1}{8}z^{-1})}$

4.20 Obtain a cascade realization of

$$H(z) = \frac{2 + z^{-1} + z^{-2}}{(1 + \frac{1}{2}z^{-1})(1 - \frac{1}{4}z^{-1})(1 + \frac{1}{8}z^{-1})}$$

4.21 It is desired to obtain a continued fraction of the form

$$\frac{a_n s^n + a_{n-1}s^{n-1} + \cdots + a_0}{b_n s^n + b_{n-1}s^{n-1} + \cdots + b_0} = \alpha_1 + \cfrac{1}{\alpha_2 s + \cfrac{1}{\alpha_3 + \cfrac{1}{\alpha_4 s + \ddots}}}$$

show that

$$\frac{a_n s^n + a_{n-1}s^{n-1} + \cdots + a_0}{b_n s^n + b_{n-1}s^{n-1} + \cdots + b_0} = \frac{a_n}{b_n} + \frac{c_{n-1}s^{n-1} + c_{n-2}s^{n-2} + \cdots +}{b_n s^n + b_{n-1}s^{n-1} + \cdots +}$$

4.22 Obtain a realization of

$$H(z) = \frac{z^{-2} + 2z^{-1} + 2}{z^{-2} + z^{-1} + 3}$$

having the form shown in Fig. 4.31.

4.23 If we write

$$H(z) = \frac{a_N z^N + a_{N-1}z^{N-1} + \cdots + a_1 z + a_0}{b_N z^N + b_{N-1}z^{N-1} + \cdots + b_1 z + b_0}$$

we can obtain the continued-fraction expansion

$$H(z) = \alpha_0 + \cfrac{1}{\beta_1 z + \cfrac{1}{\alpha_1 + \cfrac{1}{\beta_2 z + \ddots + \cfrac{1}{\beta_N z + 1/\alpha_N}}}}$$

where the coefficients α_i and β_i are determined from the Routh array of Table 4.1, where the rows are identified by the positive powers $z^N, z^{N-1}, \ldots$. Obtain the continued fraction above for

$$H(z) = \frac{z^2 + 3z + 2}{z^2 + z + 1}$$

4.24 Show that the transfer functions

$$H_1(z) = \frac{1}{\beta z + T(z)}$$

$$H_2(z) = \frac{1}{\alpha + T(z)}$$

are realized by the structures shown in the figure.

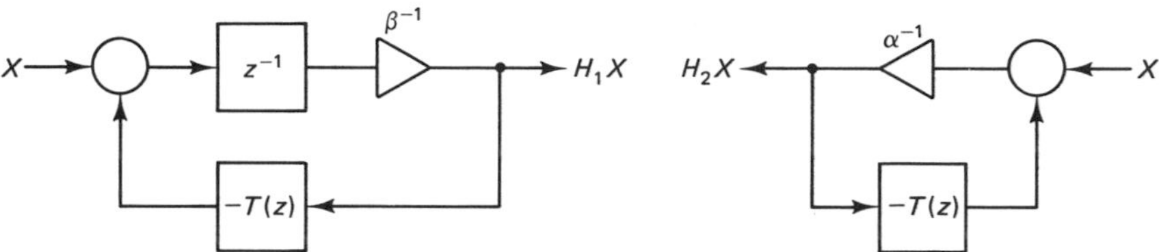

Problem 4.24

4.25 Use the realizations of Prob. 4.24 to obtain the realization shown in the figure for the continued-fraction representation of $H(z)$ given in Prob. 4.23.

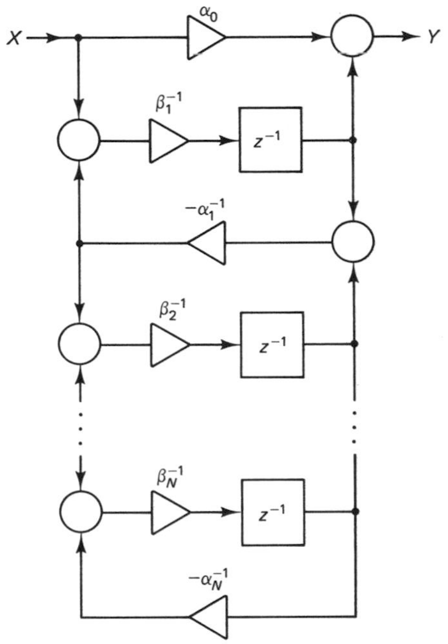

Problem 4.25

4.26 Obtain a realization of the type shown in Prob. 4.25 for

$$H(z) = \frac{z^2 + 3z + 2}{z^2 + z + 1}$$

considered in Prob. 4.23.

4.27 Another continued fraction of $H(z)$ has the form [114]

$$H(z) = A_0 + \cfrac{1}{B_1 z + A_1 + \cfrac{1}{B_2 z + A_2 + \ddots + \cfrac{1}{B_N z + A_N}}}$$

Show that the basic building block is that shown in part (a) of the figure and has the transfer function

$$G(z) = \frac{1}{Bz + A + T(z)}$$

If $N = 2$, show that the realization is that shown in part (b) of the figure. Show that this realization cannot be implemented.

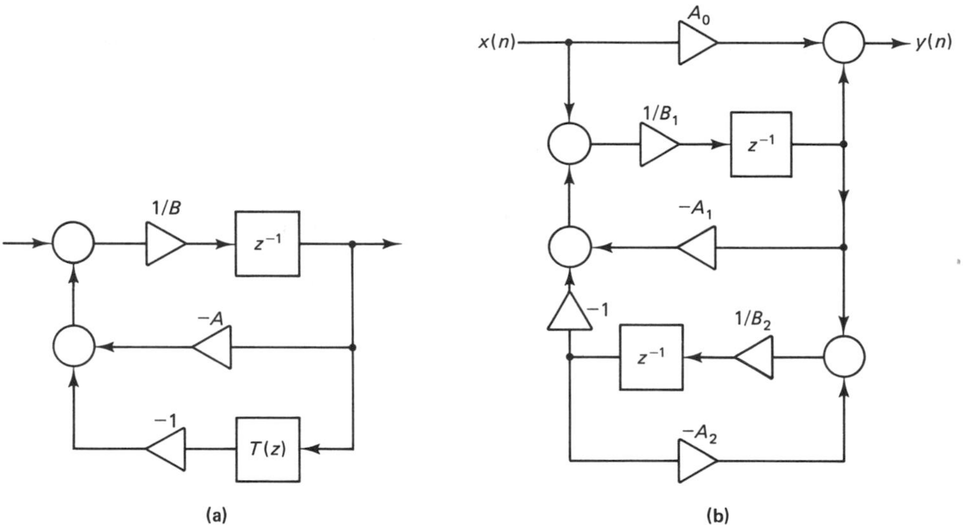

Problem 4.27

4.28 The long division required in a continued-fraction expansion can be carried out as follows:

$$\frac{a_0 s^n + a_2 s^{n-2} + \cdots +}{a_1 s^{n-1} + a_3 s^{n-3} + \cdots +} = \frac{a_0}{a_1} s + \frac{b_1 s^{n-2} + b_2 s^{n-4} + \cdots +}{a_1 s^{n-1} + a_3 s^{n-3} + \cdots +}$$

$$\frac{a_1 s^{n-1} + a_3 s^{n-3} + \cdots +}{b_1 s^{n-2} + b_2 s^{n-4} + \cdots +} = \frac{a_1}{b_1} + \frac{c_1 s^{n-3} + c_2 s^{n-5} + \cdots +}{b_1 s^{n-2} + b_2 s^{n-4} + \cdots +}$$

etc., where b_i, c_i, d_i, e_i, . . . are determined from the Routh array having a_0, a_2, . . . as first row and a_1, a_3, . . . as second row. If the continued fraction has the form

$$\frac{a_0 s^n + a_2 s^{n-2} + \cdots +}{a_1 s^{n-1} + a_3 s^{n-3} + \cdots +} = \alpha_1 s + \cfrac{1}{\alpha_2 s + \cfrac{1}{\alpha_3 s + \ddots + \cfrac{1}{\alpha_k s}}}$$

verify that $\alpha_1 = a_0/a_1$, $\alpha_2 = a_1/b_1$, $\alpha_3 = b_1/c_1$, $\alpha_4 = c_1/d_1$, ...

4.29 Obtain a ladder realization of the form shown in Fig. 4.34 for

$$H(z) = \frac{4}{2z^{-3} + 2z^{-2} + z^{-1} + 4}$$

4.30 For the system function

$$H(z) = \frac{a_0}{a_4 z^{-4} + a_3 z^{-3} + a_2 z^{-2} + a_1 z^{-1} + a_0}$$

obtain the continued fraction

$$\frac{P(z)}{Q(z)} = \alpha_0 z^{-1} + \cfrac{1}{\alpha_1 z^{-1} + \cfrac{1}{\alpha_2 z^{-1} + 1/\alpha_3 z^{-1}}}$$

For the ladder structure of Fig. 4.34 with T_0 missing, show that

$$t_{22} = \left.\frac{Y_2}{X_2}\right|_{X_1=0} = \cfrac{1}{-T_1 + \cfrac{1}{T_2 + \cfrac{1}{-T_3 + 1/T_4}}}$$

Finally, by equating

$$t_{22} = \frac{1}{P(z)/Q(z)} = \left.\frac{Y_2}{X_2}\right|_{X_1=0}$$

show that

$$T_1 = -\alpha_0 z^{-1} \qquad T_2 = \alpha_1 z^{-1} \qquad T_3 = -\alpha_2 z^{-1} \qquad T_4 = \alpha_3 z^{-1}$$

and that the realization is the structure shown in the figure.

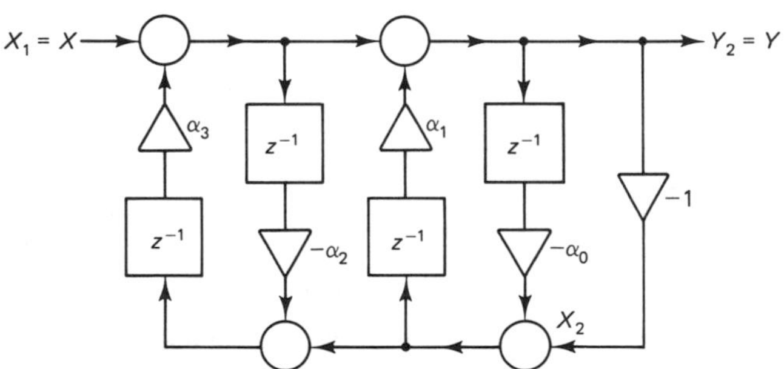

Problem 4.30

4.31 Using the procedure outlined in Prob. 4.30, obtain a realization of

$$H(z) = \frac{4}{z^{-4} + 2z^{-3} + 2z^{-2} - z^{-1} + 4}$$

4.32 Consider the system

$$Y(z) = \frac{b_0 z^3 + b_1 z^2 + b_2 z + b_3}{z^3 + a_1 z^2 + a_2 z + a_3} X(z)$$

Carry out one step in the division and obtain

$$Y = g_0 X + W_3$$

$$W_3 = \frac{\alpha_1 z^2 + \alpha_2 z + \alpha_3}{z^3 + a_1 z^2 + a_2 z + a_3} X$$

By writing the relation for W_3 in the form

$$z(z^2 + a_1 z + a_2)W_3 + a_3 W_3 = (\alpha_1 z^2 + \alpha_2 z + \alpha_3)X$$

show that

$$zW_3 + \frac{a_3 W_3}{z^2 + a_1 z + a_2} = g_1 X + \frac{\beta_2 z + \beta_3}{z^2 + a_1 z + a_2} X$$

Now define $W_2 = zW_3 - g_1 X$ and continue the previous process to obtain $W_1 = zW_2 - g_2 X$. Use the relations

$$W_3 = z^{-1}(g_1 X + W_2)$$

$$W_2 = z^{-1}(g_2 X + W_1)$$

$$W_1 = z^{-1}(g_3 X - a_1 W_1 - a_2 W_2 - a_3 W_3)$$

to obtain the realization [71] shown in figure. Apply this technique to

$$H(z) = \frac{2 + 8z^{-1} + 6z^{-2}}{1 + 8z^{-1} + 12z^{-2}}$$

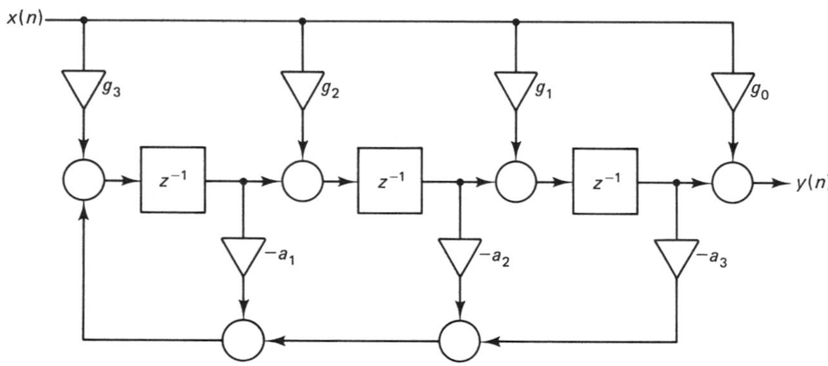

Problem 4.32

4.33 Realize the following system functions using a minimum number of multipliers:
(a) $H(z) = 1 + \frac{1}{2}z^{-1} + \frac{3}{4}z^{-2} + \frac{1}{2}z^{-3} + z^{-4}$
(b) $H(z) = 1 + \frac{1}{2}z^{-1} + \frac{1}{2}z^{-2} + z^{-3}$
(c) $H(z) = (1 + \frac{1}{2}z^{-1} + z^{-2})(1 + \frac{1}{4}z^{-1} + z^{-2})$
(d) $H(z) = 1 + \frac{1}{2}z^{-1} + \frac{1}{4}z^{-2} + z^{-3}$

4.34 The system function of a FIR filter is given by

$$H(z) = \sum_{n=0}^{N-1} h(n)z^{-n}$$

Suppose that (see the IDFT in Sec. 1.6)

$$h(n) = \frac{1}{N}\sum_{k=0}^{N-1} \tilde{H}(k)W_N^{-nk}$$

where $W_N = \exp(-j2\pi/N)$, and $\tilde{H}(N - k) = \tilde{H}^*(k)$. Show that

$$H(z) = \frac{1 - z^{-N}}{N}\sum_{k=0}^{N-1} \frac{\tilde{H}(k)}{1 - z^{-1}W_N^{-k}}$$

Obtain realizations using these two expressions for $H(z)$ if $N = 5$. Note that the first form is nonrecursive and the second is recursive. (See Sec. 6.6 on frequency sampling.)

4.35 A particular form of the DPLL given in Fig. 1.7 has the form shown in the figure. The input $n(k)/(2P)^{1/2}$ is noise. Obtain $\Phi(z)$ in terms of $\Theta(z)$ and $N(z)$. Then assume that $g[\phi(k)] = \phi(k)$ and $n(k) = 0$, and find $\Phi(z)/\Theta(z)$. If $D(z) = G_1 + G_2/(1 - z^{-1})$, show that

$$\frac{\Phi(z)}{\Theta(z)} = \frac{(z-1)^2}{(z-\alpha)^2 + \beta^2}$$

Using this system function, find the steady-state or forced response to the input $\theta(k) = \Delta\theta \sin k\omega_m T$. This corresponds to phase modulation.

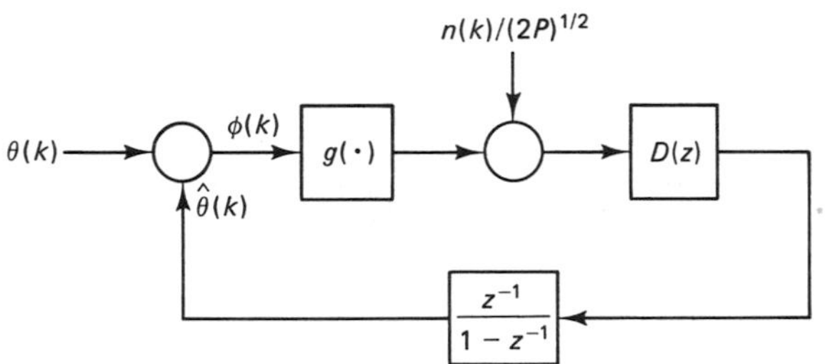

Problem 4.35

4.36 The system function $H(z)$ can be written

$$H(z) = \frac{\sum_{k=0}^{M} b_k' z^{-k}}{1 + \sum_{k=1}^{N} a_k z^{-k}} = \frac{A \sum_{k=0}^{M} b_k z^{-k}}{1 + \sum_{k=1}^{N} a_k z^{-k}} \qquad b_0 = 1$$

or, if it is desired to display the finite poles and zeros,

$$H(z) = \frac{A \prod_{k=1}^{M} (1 - z_k z^{-1})}{\prod_{k=1}^{N} (1 - p_k z^{-1})} = \frac{AQ(z)}{P(z)}$$

If Δp_i is the change in the ith pole, then

$$\Delta p_i = \sum_{j=1}^{N} \frac{\partial p_i}{\partial a_j} \Delta a_j$$

Show that [86, 126]

(a) $$\left.\frac{\partial P(z)}{\partial p_m}\right|_{z=p_i} = \begin{cases} 0 & i \neq m \\ -p_m^{-N} \prod_{\substack{k=1 \\ k\neq m}}^{N} (p_m - p_k) \end{cases}$$

(b) $$\left.\frac{\partial P(z)}{\partial a_k}\right|_{z=p_m} = -p_m^{-k}$$

(c) $$\left.\frac{\partial P(z)}{\partial a_k}\right|_{z=p_m} = \sum_{j=1}^{N} \left(\left.\frac{\partial P(z)}{\partial p_j}\right|_{z=p_m}\right)\left(\left.\frac{\partial p_j}{\partial a_k}\right|_{z=p_m}\right)$$

(d) $\dfrac{\partial p_m}{\partial a_k} = \dfrac{p_m^{N-k}}{\prod\limits_{\substack{k=1\\k\neq m}}^{N} (p_m - p_k)}$

From this, it can be concluded that when poles are close together, small changes in the coefficients a_k, $k = 1, 2, \ldots, N$, can cause large changes in the actual pole positions.

4.37 Draw a signal-flow graph for the lattice structure [54] shown in the figure and find the system function if

$$k_0 = -0.87559 \qquad k_1 = 0.83546 \qquad k_2 = -0.4583$$

$$\hat{v}_0 = 0.19031 \qquad \hat{v}_1 = 0.60635 \qquad \hat{v}_2 = 0.05269 \qquad \hat{v}_3 = 0.0154$$

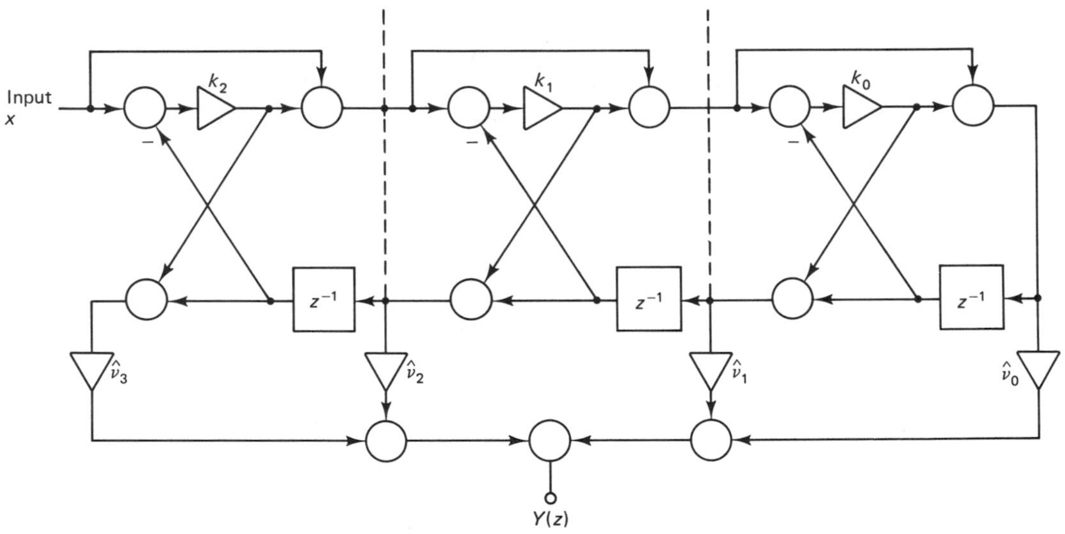

Problem 4.37

5

DESIGN OF INFINITE IMPULSE-RESPONSE DIGITAL FILTERS

5.1 INTRODUCTION

Digital filters considered in this chapter are linear time-invariant systems. Designing such digital filters involves the determination of a transfer function in z^{-1} that meets certain performance specifications. Once the transfer function is obtained, the next step in the design procedure is to realize the system using finite-precision arithmetic.

We first restrict ourselves to the problem of obtaining the transfer function corresponding to a causal system that provides an adequate approximation to the system specifications. In this chapter, we consider only infinite impulse-response (IIR) systems. The next chapter is devoted to finite impulse-response (FIR) systems.

In many applications of digital filtering, the specifications are given in the frequency domain. The wealth of analog-filter-design data and techniques that have evolved over the years are also based on frequency-domain specifications. For this reason, many design techniques for IIR digital filters utilizing analog-filter data have been developed, whereby transformations from the analog variable s to the digital variable z have been derived. The traditional approach is to convert the digital-filter specifications to corresponding specifications on the analog filter. The analog filter is then designed to meet these specifications and then the digital filter is obtained by applying the desired transformation to the analog-filter function. The pioneering work in this area was done by Kaiser [85,86].

There have been a number of different transformations developed whereby a digital filter is obtained from an analog filter. In this chapter, we discuss two of these techniques—impulse-invariant and bilinear transformations. Other transformations are considered in the problems.

5.2 INTRODUCTION TO FILTERS

A filter is generally a frequency-selective device. Signals having certain frequencies are passed, whereas signals having other frequencies are blocked or attenuated. Whether or not a signal is passed or blocked is determined by the system function

$$H(e^{j\omega}) = |H(e^{j\omega})|\underline{/\phi(\omega)}$$

The frequencies of signals that are passed through the filter are in ranges, or bands, called *passbands,* and those that are blocked are in bands called *stopbands*. For frequencies in the passband, the magnitude or amplitude $|H(e^{j\omega})|$ is relatively large and ideally is a constant. A stopband is characterized by magnitude $|H(e^{j\omega})|$, which is relatively small and ideally is zero. The magnitude response of an ideal *low-pass filter* having a single passband and a single stopband is illustrated in Fig. 5.1(a). Frequencies in the passband, $0 < \omega < \omega_c$, are passed, whereas the higher frequencies in the stopband, $\omega > \omega_c$, are blocked. The frequency ω_c between the two bands is the *cutoff frequency*.

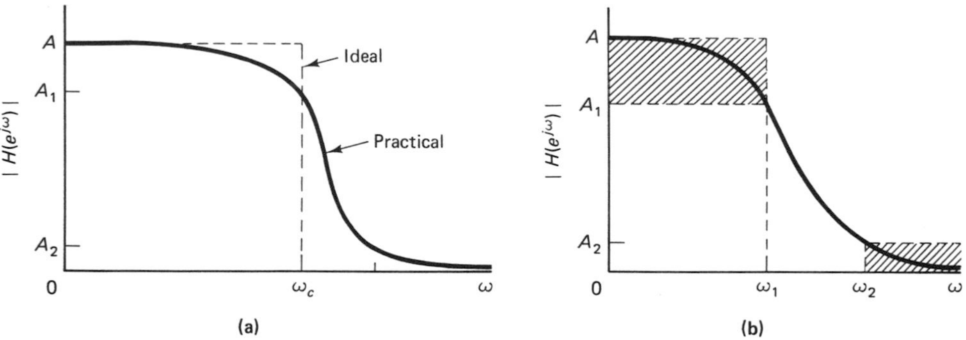

Figure 5.1 (a) Ideal and practical low-pass filter magnitudes. (b) A typical low-pass filter magnitude specification and a typical Butterworth response.

In practice, it is impossible to obtain the ideal response because of the sharp corners. A central problem in filter design is to obtain a practical response that is a suitable approximation to the ideal response. One such practical response is represented by the solid line in Fig. 5.1(a). In the practical case, the passband and stopband are not clearly demarcated and must be formally defined. The passband is the band of frequencies, $0 < \omega < \omega_1$, where $A_1 \leq |H(e^{j\omega})| \leq A$. The stopband is the band, $\omega > \omega_2$, where $0 \leq |H(e^{j\omega})| \leq A_2$. The frequency band, $\omega_1 < \omega < \omega_2$, between the passband and the stopband is the *transition band*. In this interval, the response continually decreases. Generally, A_1 is never less than $A/2^{1/2}$. (In some cases, we use ω_p and ω_s in place of ω_1 and ω_2.) The magnitude constraints and the passband, transition band, and stopband for a low-pass filter are shown in Fig.

5.1(b). A typical response is shown that satisfies these constraints. The cutoff frequency ω_c is usually taken to be ω_1, the passband frequency, or ω_{3dB}, the frequency at which $|H(e^{j\omega})| = A/2^{1/2}$.

For a given set of specifications, such as those shown in Fig. 5.1(b), there are numerous types of filters. The most popular types are the Butterworth, Chebyshev, inverse Chebyshev, and elliptic filters. Typical magnitude responses of these filters are shown in Fig. 5.1(b) for a Butterworth filter and in Fig. 5.2 for Chebyshev, inverse Chebyshev, and elliptic filters. The cutoff frequency for Butterworth and in-

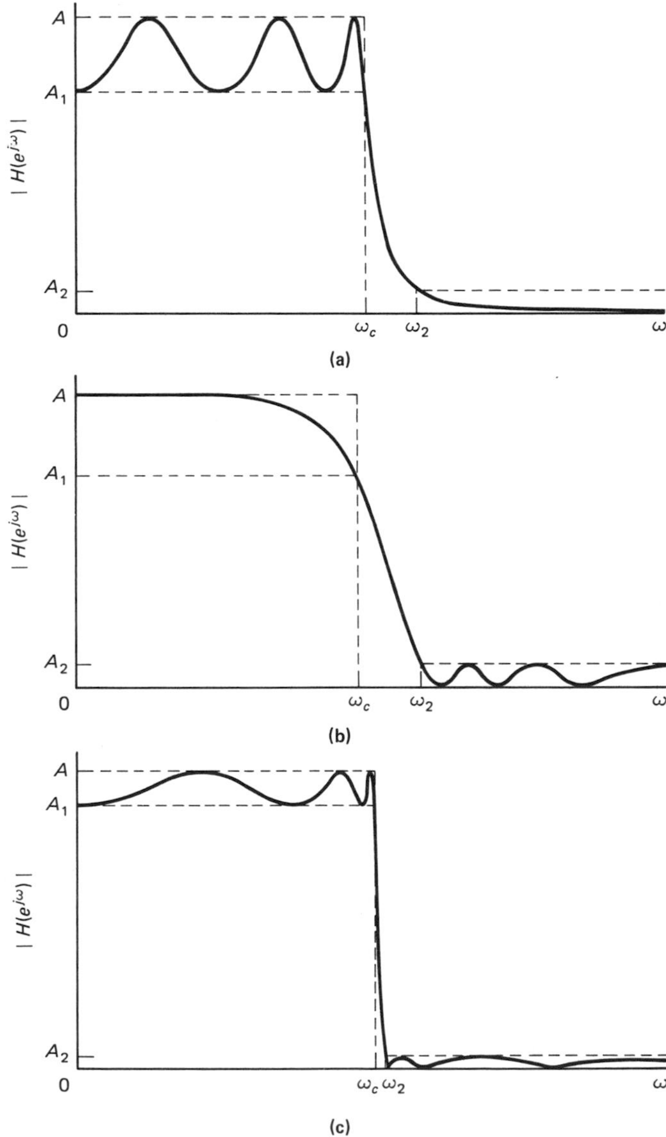

Figure 5.2 (a) A sixth-order Chebyshev response. (b) A sixth-order inverse Chebyshev response. (c) A sixth-order elliptic response.

verse Chebyshev filters is $\omega_c = \omega_{3dB}$, and for Chebyshev and elliptic filters, $\omega_c = \omega_1$. Each of these filters is discussed later in the chapter.

The other common types of frequency-selective filters are *high-pass* (which pass high frequencies and block low frequencies), *bandpass* (which pass a band of frequencies and block others), and *band-reject* (which block a band of frequencies and pass others). Ideal magnitude responses of these types of filters are shown in Fig. 5.3. These filters are obtained from low-pass filters using frequency transformations as discussed in Sec. 5.9.

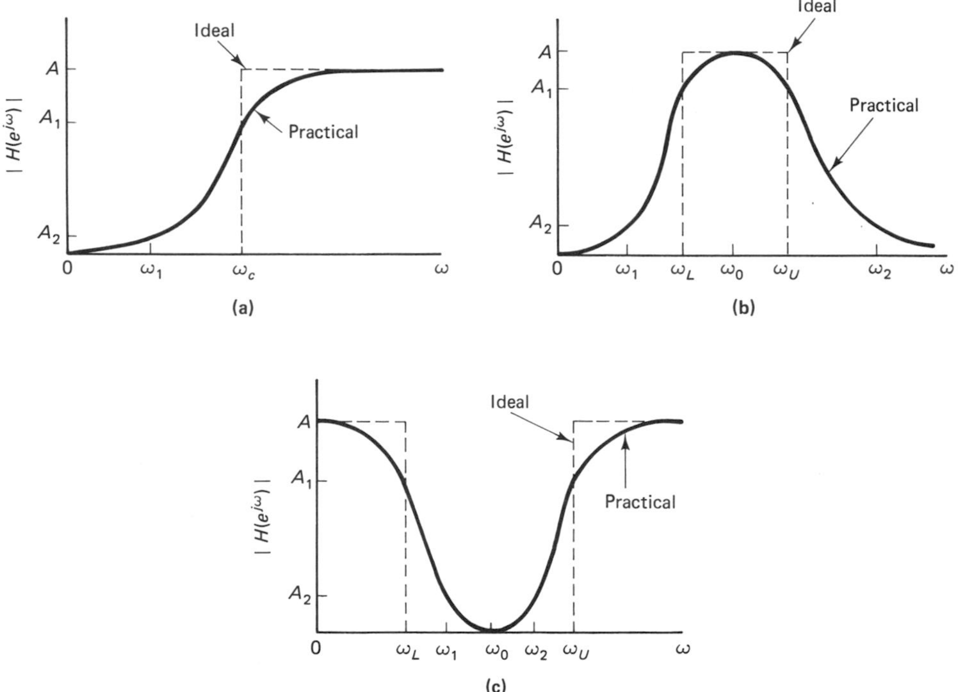

Figure 5.3 Ideal and practical (a) high-pass, (b) bandpass, and (c) band-reject responses.

In addition to frequency-selective filters, we can have filters in which the phase response $\phi(\omega)$ is the important characteristic. The output signal is an *undistorted* version of the input signal if it is the input signal *amplified* and/or *delayed* in time. This is the case if the magnitude response is constant and the phase response is *linear,* that is,

$$\phi(\omega) = -\tau_0 \omega$$

where τ_0 is a constant. As the phase response becomes more nonlinear, the output signal becomes more distorted. The *time delay* or *group delay* of a filter is defined by

$$\tau(\omega) = -\frac{d}{d\omega}\phi(\omega)$$

For a linear phase, as just given, the time delay is $\tau(\omega) = \tau_0$, a constant. Therefore, linear phase is accompanied by constant time delay. If $y(t)$ is the output and $x(t)$ is the input for a linear-phase system, then

$$y(t) = Ax(t - \tau_0)$$

where A is a constant, and τ_0 is the constant time delay. Linear phase is important in pulse transmission, where dispersion of pulses is to be avoided, and is useful in speech-processing applications, where precise time alignment is essential.

It is possible to obtain linear phase for a FIR filter, as we saw in Sec. 2.11. We cannot obtain linear phase for an IIR filter, but we can obtain a filter with a maximally flat time delay. This means that for an Nth-order filter, the first $2N - 1$ derivatives of $\tau(\omega)$ are zero at $\omega = 0$. In other words, we obtain a close approximation to a constant time delay for small values of ω. We give an example of a second-order filter with maximally flat time delay in Sec. 5.10.

Whenever the phase distortion is too severe, it is possible to design a *group delay equalizer,* which is cascaded with the original filter. This equalizer is an *all-pass* filter, which is characterized by constant magnitude.

5.3 IMPULSE-INVARIANT TRANSFORMATION

Development of the Transformation

The first method that we consider for obtaining a digital filter from an analog filter is called the *impulse-invariant transformation*. The impulse response of the digital filter is obtained by uniformly sampling the impulse response of the analog filter. This requires that

$$h(n) = h_a(nT) \tag{5.1}$$

where $h_a(t)$ is the analog-filter impulse response, and T is the sampling period.

To see how the transformation is effected, let us consider the distinct pole case with

$$H_a(s) = \sum_{i=1}^{N} \frac{A_i}{s - p_i} \tag{5.2}$$

If the poles are not distinct, the discussion is slightly altered. (This case is left to Prob. 5.7.) The impulse response $h_a(t)$ corresponding to Eq. (5.2) has the form

$$h_a(t) = \sum_{i=1}^{N} A_i e^{p_i t} u_a(t) \tag{5.3}$$

where $u_a(t)$ is the continuous-time step function. The corresponding digital impulse response obtained by applying Eq. (5.1) is

$$h(n) = h_a(nT) = \sum_{i=1}^{N} A_i e^{np_i T} u_a(nT) \tag{5.4}$$

The system function $H(z)$ of the digital filter is the z-transform of this sequence and is defined by

$$H(z) = Z\{h(n)\} = \sum_{n=0}^{\infty} h(n)z^{-n}$$

For our case, $h(n)$ is given by Eq. (5.4), and so the system function is

$$H(z) = Z\left\{\sum_{i=1}^{N} A_i e^{p_i n T}\right\}$$

$$= \sum_{i=1}^{N} A_i Z\{e^{p_i n T}\}$$

By using the z-transform of Eq. (3.6), this expression becomes

$$H(z) = \sum_{i=1}^{N} \frac{A_i}{1 - e^{p_i T} z^{-1}} \tag{5.5}$$

Comparing the expressions in Eqs. (5.2) and (5.5), we see that the impulse-invariant transformation is accomplished by the mapping

$$\frac{1}{s - p_i} \longrightarrow \frac{1}{1 - e^{p_i T} z^{-1}} \tag{5.6}$$

Relation Between Analog- and Digital-Filter Poles

It should be noted that the analog pole at $s = p_i$ is transformed by Eq. (5.6) into a digital pole at $z = e^{p_i T}$. Consequently, the poles of the analog filter are related to the corresponding poles of the digital filter by the relation

$$z = e^{sT} \tag{5.7}$$

If $s = \sigma + j\Omega$, then

$$z = e^{\sigma T} e^{j\Omega T}$$

and

$$|z| = e^{\sigma T} < 1 \qquad \sigma < 0$$
$$= 1 \qquad \sigma = 0$$
$$> 1 \qquad \sigma > 0$$

Therefore, analog-filter poles in the left-half plane (LHP) map into digital filter poles inside the circle $|z| = 1$, and right-half plane (RHP) analog poles map into digital-filter poles outside the circle $|z| = 1$. Clearly then, a stable digital filter is obtained from a stable analog filter using the impulse-invariant transformation. Also, analog poles that are on the imaginary axis map into digital-filter poles on the unit circle $|z| = 1$. It is also important to note that the zeros of the resulting digital filter are determined by the poles and the coefficients A_i that characterize the partial-fraction expansion of $H_a(s)$. They are not mapped in the same way that the poles are mapped, but are determined by Eq. (5.5).

Examples

As a simple example, consider the analog-filter transfer function

$$H_a(s) = \frac{1}{s + 1} \tag{5.8}$$

Using the impulse-invariant transformation of Eq. (5.6), the digital-filter transfer function that results is

$$H(z) = \frac{1}{1 - e^{-T}z^{-1}} = \frac{Y(z)}{X(z)} \tag{5.9}$$

The difference equation describing the digital filter is given by

$$y(n) = e^{-T}y(n - 1) + x(n)$$

To see how the zeros of a digital filter are obtained using the impulse-invariant transformation, consider the partial fraction

$$H_a(s) = \frac{(s + 2)}{(s + 1)(s + 3)} = \frac{1}{2}\left(\frac{1}{s + 1} + \frac{1}{s + 3}\right)$$

The corresponding digital-filter function is then

$$\begin{aligned} H(z) &= \frac{1}{2}\left(\frac{1}{1 - e^{-T}z^{-1}} + \frac{1}{1 - e^{-3T}z^{-1}}\right) \\ &= \frac{1 - e^{-2T}(\cosh T)z^{-1}}{(1 - e^{-T}z^{-1})(1 - e^{-3T}z^{-1})} \end{aligned}$$

It should be noted that the zero at $z = e^{-2T} \cosh T$ is not obtained by transforming the zero at $s = -2$ into a zero at $z = e^{-2T}$.

Comparison of Step Responses

Since the impulse responses $h(n)$ and $h_a(t)$ essentially characterize the digital and analog filters, respectively, it is natural to see how other properties of the two filters are related. For example, what is the relation between the step responses for the digital and analog filters as a result of impulse invariance? If $r(n)$ and $r_a(t)$ denote the step responses for the digital and analog filters, respectively, then

$$\begin{aligned} R_a(s) &= \frac{1}{s}H_a(s) = \sum_{i=1}^{N} \frac{A_i}{s(s - p_i)} \\ &= \sum_{i=1}^{N} \frac{A_i}{p_i}\left(\frac{1}{s - p_i} - \frac{1}{s}\right) \end{aligned}$$

and the resulting step response is

$$r_a(t) = \sum_{i=1}^{N} \frac{A_i}{p_i}(e^{p_i t} - 1)u_a(t)$$

If we now sample this analog step response, we obtain

$$r_a(nT) = \sum_{i=1}^{N} \frac{A_i}{p_i}(e^{p_i nT} - 1)u_a(nT) \tag{5.10}$$

To find the unit step response of the digital filter, we first find its z-transform. Using the expression of Eq. (5.5) for $H(z)$, we obtain

$$R(z) = \frac{1}{1 - z^{-1}} H(z) = \sum_{i=1}^{N} \frac{A_i}{(1 - z^{-1})(1 - e^{p_i T} z^{-1})}$$

$$= \sum_{i=1}^{N} \frac{A_i}{1 - e^{p_i T}} \left(\frac{1}{1 - z^{-1}} - \frac{e^{p_i T}}{1 - e^{p_i T} z^{-1}} \right)$$

and, consequently, the resulting digital step response is

$$r(n) = \sum_{i=1}^{N} \frac{A_i}{1 - e^{p_i T}} (1 - e^{p_i T(n+1)}) u(n)$$

Clearly, we can see by comparing $r(n)$ with $r_a(nT)$ in Eq. (5.10) that $r(n) \neq r_a(nT)$.

For the analog function of Eq. (5.8), we find that

$$r_a(nT) = (1 - e^{-t})u_a(t)|_{t=nT}$$

$$= (1 - e^{-nT})u_a(nT)$$

and

$$r(n) = \frac{1 - e^{-(n+1)T}}{1 - e^{-T}} u(n)$$

In particular, $r_a(nT)|_{n=0} = 0$ and $r_a(T) = 1 - e^{-T}$ and $r(0) = 1$ and $r(1) = 1 + e^{-T}$, so that there is not close agreement for small values of n. If we were interested in having $r_a(nT) = r(n)$, then we could have developed a *step-invariant transformation*. The reader is asked to consider this in Prob. 5.4.

Frequency Response

One of the most important properties of a filter is its frequency response. In deriving a digital filter from an analog filter, we generally choose an analog filter that has a frequency response approximately the same as the desired digital-filter frequency response. Since the impulse-invariant transformation involves sampling the analog-filter impulse response, we know from the discussion in Sec. 1.5 that the frequency responses satisfy the relation of Eq. (1.47), which for our case is

$$H(e^{j\omega}) = \frac{1}{T} \sum_{k=-\infty}^{\infty} H_a\left(j\frac{\omega}{T} + j\frac{2\pi k}{T}\right) \qquad \omega = \Omega T \tag{5.11}$$

In the case of a band-limited analog filter, we can obtain

$$H_a(j\omega/T) = 0 \qquad |\omega/T| = |\Omega| \geq \pi/T \geq \Omega_a$$

and, consequently, the corresponding digital-filter frequency response has the form

$$H(e^{j\omega}) = \frac{1}{T}H_a\left(j\frac{\omega}{T}\right) \qquad |\omega| \leq \pi \tag{5.12}$$

For these conditions, it is a simple matter to obtain the desired digital filter from the analog filter. Unfortunately, no practical analog filter is band-limited, so the best that we can hope for is to select an analog filter that results in an aliasing effect that is acceptable.

Whenever the sampling interval T is small, the filter gain at $\omega = 0$ is large because of the factor $1/T$ in Eqs. (5.11) and (5.12). This is generally an undesirable property and can be removed by introducing the multiplication factor T into the impulse-invariant transformation. This modification results if we use the transformation

$$h(n) = Th_a(nT)$$

Then $H(z)$ is obtained by replacing the partial-fraction expansion of Eq. (5.5) by

$$H(z) = \sum_{i=1}^{N} \frac{TA_i}{1 - e^{p_i T}z^{-1}} \tag{5.13}$$

The frequency-response relation of Eq. (5.12) then becomes

$$H(e^{j\omega}) = H_a(j\omega/T) \qquad |\omega| \leq \pi \tag{5.14}$$

so that for this band-limited case, the dc values are related by

$$H(e^{j0}) = H(1) = H_a(j0)$$

Useful Impulse-Invariant Transformations

Some further properties of the impulse-invariant transformation of Eq. (5.6), which the reader is asked to derive in Prob. 5.7, are

$$\frac{1}{(s + s_i)^m} \longrightarrow \frac{(-1)^{m-1}}{(m-1)!}\frac{d^{m-1}}{ds_i^{m-1}}\frac{1}{1 - e^{-s_i T}z^{-1}} \tag{5.15}$$

$$\frac{s + a}{(s + a)^2 + b^2} \longrightarrow \frac{1 - e^{-aT}(\cos bT)z^{-1}}{1 - 2e^{-aT}(\cos bT)z^{-1} + e^{-2aT}z^{-2}} \tag{5.16}$$

$$\frac{b}{(s + a)^2 + b^2} \longrightarrow \frac{e^{-aT}(\sin bT)z^{-1}}{1 - 2e^{-aT}(\cos bT)z^{-1} + e^{-2aT}z^{-2}} \tag{5.17}$$

Note in particular in Eq. (5.16) that the zeros of the analog function and the digital function do not satisfy Eq. (5.7), as mentioned earlier.

Example of a Third-Order Butterworth Filter

Consider as an example the third-order Butterworth analog filter with transfer function given by

$$\begin{aligned} H_a(s) &= \frac{1}{(s+1)(s^2+s+1)} \\ &= \frac{1}{s+1} - \frac{s+\frac{1}{2}}{(s+\frac{1}{2})^2+\frac{3}{4}} + \frac{1}{3^{1/2}}\frac{3^{1/2}/2}{(s+\frac{1}{2})^2+\frac{3}{4}} \end{aligned} \tag{5.18}$$

Using the transformations of Eqs. (5.6), (5.16), and (5.17), we obtain the digital-filter transfer function

$$H(z) = \frac{1}{1-e^{-T}z^{-1}} - \frac{1 - e^{-T/2}\left(\cos\frac{3^{1/2}}{2}T - \frac{1}{3^{1/2}}\sin\frac{3^{1/2}}{2}T\right)z^{-1}}{1 - 2e^{-T/2}\left(\cos\frac{3^{1/2}}{2}T\right)z^{-1} + e^{-T}z^{-2}}$$

It should be noted that the impulse-invariant technique requires a partial-fraction expansion of $H_a(s)$. The resulting digital function is a sum of terms and, therefore, can be easily realized using a parallel realization, which is discussed in Sec. 4.7.

In later sections of this chapter, we give examples of filter designs for Butterworth, Chebyshev, inverse Chebyshev, and elliptic filters using the impulse-invariant transformation.

EXERCISES

5.3.1 For the analog transfer function

$$H_a(s) = \frac{2}{(s+1)(s+2)}$$

determine $H(z)$ using both Eqs. (5.5) and (5.13) if **(a)** $T = 1$ s and **(b)** $T = 0.1$ s. In each case, determine $|H(e^{j0})| = |H(1)|$.

5.3.2 Using impulse invariance with $T = 1$ s, determine $H(z)$ if

$$H_a(s) = \frac{1}{s^2 + 2^{1/2}s + 1}$$

5.3.3 Repeat Ex. 5.3.2 for

$$H_a(s) = \frac{1}{(s+0.5)(s^2+0.5s+2)}$$

5.3.4 Repeat Ex. 5.3.2 for

$$H_a(s) = \frac{2s}{s^2+0.2s+1}$$

5.3.5 If

$$H_a(s) = \frac{\Omega_c^2}{s^2 + b_k\Omega_c s + c_k\Omega_c^2}$$

show that

$$H_a(s) = \frac{\Omega_c^2}{b}\frac{b}{(s+a)^2 + b^2}$$

where

$$a = b_k\Omega_c/2 \qquad b = \Omega_c[c_k - (b_k/2)^2]^{1/2}$$

Determine $H(z)$ using the impulse-invariant transformation given by Eq. (5.13) and the result of Eq. (5.17).
Answer:

$$(\Omega_c T)\frac{\Omega_c}{b}\frac{e^{-aT}(\sin bT)z^{-1}}{1 - 2e^{-aT}(\cos bT)z^{-1} + e^{-2aT}z^{-2}}$$

5.4 BILINEAR TRANSFORMATION

Development of the Transformation

A very simple transformation based on numerical integration that is widely used to derive a digital-filter function from an analog-filter function is the *bilinear transformation,* or, as it is commonly called in control theory, the *Tustin transformation*. It is developed from the trapezoidal rule for integration (see Prob. 1.7). Suppose we begin with

$$\frac{dy_a(t)}{dt} = w_a(t) \tag{5.19}$$

and integrate both sides, obtaining as a result

$$\int_{(n-1)T}^{nT} \frac{dy_a(t)}{dt}\,dt = y_a(nT) - y_a[(n-1)T]$$

$$= \int_{(n-1)T}^{nT} w_a(t)\,dt$$

The integral can be approximated by the trapezoidal rule so that

$$y_a(nT) - y_a[(n-1)T] = (T/2)\{w_a(nT) + w_a[(n-1)T]\}$$

Defining the discrete-time variables in the usual way, we obtain the relation

$$y(n) - y(n-1) = (T/2)[w(n) + w(n-1)]$$

The z-transform of this equation is now written

$$\frac{2}{T}\frac{1 - z^{-1}}{1 + z^{-1}}Y(z) = W(z)$$

so that when it is compared with its analog counterpart

$$sY_a(s) = W_a(s)$$

obtained by taking the Laplace transform of Eq. (5.19) with $y_a(0) = 0$, we see that the analog-to-digital transformation is effected by making the replacements

$$sY_a(s) \longrightarrow \frac{2}{T}\frac{1 - z^{-1}}{1 + z^{-1}}Y(z)$$

$$W_a(s) \longrightarrow W(z)$$

In terms of the transfer functions, this corresponds to the substitution in the analog transfer function of

$$s = \frac{2}{T}\frac{1 - z^{-1}}{1 + z^{-1}} \tag{5.20}$$

This transformation is known as the *bilinear,* or *Tustin, transformation*. The Laplace transforms in the filter expression are replaced by the corresponding z-transforms.

First- and Second-Order Transfer Functions

The digital-filter system function $H(z)$ is obtained quite easily from the analog-filter system function $H_a(s)$ via the transformation of Eq. (5.20). This results in

$$H(z) = H_a\left(\frac{2}{T}\frac{1 - z^{-1}}{1 + z^{-1}}\right) \tag{5.21}$$

In many cases, $H_a(s)$ is in factored form, and, consequently, it is convenient to obtain Eq. (5.21) by applying the bilinear transformation to each factor. This is easily obtained by considering general first- and second-order analog transfer functions.

Applying the bilinear transformation of Eq. (5.20) to a general first-order analog function,

$$H_a(s) = \frac{\beta_0 s + \gamma_0\Omega_c}{s + c_0\Omega_c}$$

results in the digital function

$$\begin{aligned} H(z) &= \frac{\beta_0\dfrac{2}{T}\dfrac{1 - z^{-1}}{1 + z^{-1}} + \gamma_0\Omega_c}{\dfrac{2}{T}\dfrac{1 - z^{-1}}{1 + z^{-1}} + c_0\Omega_c} \\ &= \frac{b_{00} + b_{01}z^{-1}}{1 + a_{01}z^{-1}} \end{aligned} \tag{5.22}$$

where

$$
\begin{aligned}
b_{00} &= \frac{\gamma_0(\Omega_c T/2) + \beta_0}{c_0(\Omega_c T/2) + 1} \\
b_{01} &= \frac{\gamma_0(\Omega_c T/2) - \beta_0}{c_0(\Omega_c T/2) + 1} \qquad (5.23) \\
a_{01} &= \frac{c_0(\Omega_c T/2) - 1}{c_0(\Omega_c T/2) + 1}
\end{aligned}
$$

For the general second-order analog function,

$$H_a(s) = \frac{\alpha_k s^2 + \beta_k \Omega_c s + \gamma_k \Omega_c^2}{s^2 + b_k \Omega_c s + c_k \Omega_c^2}$$

the bilinear transformation yields

$$
\begin{aligned}
H(z) &= \frac{\alpha_k(4/T^2)(1 - z^{-1})^2 + \beta_k\Omega_c(2/T)(1 - z^{-2}) + \gamma_k\Omega_c^2(1 + z^{-1})^2}{(4/T^2)(1 - z^{-1})^2 + b_k\Omega_c(2/T)(1 - z^{-2}) + c_k\Omega_c^2(1 + z^{-1})^2} \\
&= A_{k0}\frac{1 + b_{k1}z^{-1} + b_{k2}z^{-2}}{1 + a_{k1}z^{-1} + a_{k2}z^{-2}} \qquad (5.24)
\end{aligned}
$$

where

$$
\begin{aligned}
A_{k0} &= b_{k0}/a_{k0} \\
b_{k0} &= \alpha_k + \beta_k(\Omega_c T/2) + \gamma_k(\Omega_c T/2)^2 \\
a_{k0} &= 1 + b_k(\Omega_c T/2) + c_k(\Omega_c T/2)^2 \\
b_{k1} &= 2[-\alpha_k + \gamma_k(\Omega_c T/2)^2]/b_{k0} \qquad (5.25) \\
b_{k2} &= [\alpha_k - \beta_k(\Omega_c T/2) + \gamma_k(\Omega_c T/2)^2]/b_{k0} \\
a_{k1} &= 2[-1 + c_k(\Omega_c T/2)^2]/a_{k0} \\
a_{k2} &= [1 - b_k(\Omega_c T/2) + c_k(\Omega_c T/2)^2]/a_{k0}
\end{aligned}
$$

As we will see later, these expressions can be used for low-pass, high-pass, band-pass, and band-reject filters, as well as filters used to modify the phase response.

In the case of all-pole analog transfer functions, we have $\beta_0 = \alpha_k = \beta_k = 0$ in the first- and second-order functions considered previously. It can be shown quite readily that $H(z)$ contains the factor $1 + z^{-1}$ to a power that is equal to the order of the analog filter. This can be an undesirable by-product of using the bilinear transformation. In Ex. 5.4.4, the reader is asked to show that for the all-pole cases of the first- and second-order functions, the digital-filter numerator coefficients satisfy the following relations:

$$
\begin{aligned}
b_{00} = b_{01} &= \frac{\gamma_0(\Omega_c T/2)}{c_0(\Omega_c T/2) + 1} \\
b_{k0} &= \gamma_k(\Omega_c T/2)^2
\end{aligned}
$$

$$
\begin{aligned}
b_{k1} &= 2 \\
b_{k2} &= 1 \\
b_{00} + b_{01}z^{-1} &= b_{00}(1 + z^{-1}) \\
1 + b_{k1}z^{-1} + b_{k2}z^{-2} &= (1 + z^{-1})^2
\end{aligned}
\tag{5.26}
$$

Example of a Third-Order Butterworth Function

Taking the third-order Butterworth analog filter of Eq. (5.18) as an example, we have, for the first-order section,

$$c_0 = \gamma_0 = \Omega_c = 1 \qquad \beta_0 = 0$$

and, for the second-order section,

$$
\begin{aligned}
b_1 = c_1 = \gamma_1 &= \Omega_c = 1 \\
\alpha_1 = \beta_1 &= 0
\end{aligned}
$$

Therefore, from Eqs. (5.22) and (5.24), we obtain the digital-system function

$$H(z) = \frac{b_{00} + b_{01}z^{-1}}{1 + a_{01}z^{-1}} \frac{A_{10}(1 + b_{11}z^{-1} + b_{12}z^{-2})}{1 + a_{11}z^{-1} + a_{12}z^{-2}}$$

where, from Eqs. (5.23), (5.25), and (5.26),

$$
\begin{aligned}
b_{00} + b_{01}z^{-1} &= \frac{T/2}{(T/2) + 1}(1 + z^{-1}) \\
a_{01} &= \frac{(T/2) - 1}{(T/2) + 1} \\
b_{10} &= (T/2)^2 \\
1 + b_{11}z^{-1} + b_{12}z^{-2} &= (1 + z^{-1})^2 \\
a_{10} &= 1 + (T/2) + (T/2)^2 \\
a_{11} &= 2[-1 + (T/2)^2]/[1 + (T/2) + (T/2)^2] \\
a_{12} &= [1 - (T/2) + (T/2)^2]/[1 + (T/2) + (T/2)^2] \\
A_{10} &= (T/2)^2/[1 + (T/2) + (T/2)^2]
\end{aligned}
$$

It should be noted that application of the bilinear transformation to an analog function $H_a(s)$ that is in a product form results in a digital system function that is also in a product form. This makes it very convenient to realize the digital filter using a cascade realization.

Properties of the Bilinear Transformation

The mapping of Eq. (5.20) for the bilinear transformation is a one-to-one mapping; that is, for every point z, there is exactly one corresponding point s, and vice versa.

To get a better idea of the mapping, let us express z in terms of s. From Eq. (5.20), we obtain

$$z = \frac{1 + \dfrac{T}{2}s}{1 - \dfrac{T}{2}s}$$

If we let $s = \sigma + j\Omega$ in this relation, it follows that

$$|z| = \left| \frac{1 + \dfrac{T}{2}\sigma + j\dfrac{T}{2}\Omega}{1 - \dfrac{T}{2}\sigma - j\dfrac{T}{2}\Omega} \right|$$

$$= \left[\frac{\left(1 + \dfrac{T}{2}\sigma\right)^2 + \left(\dfrac{T}{2}\Omega\right)^2}{\left(1 - \dfrac{T}{2}\sigma\right)^2 + \left(\dfrac{T}{2}\Omega\right)^2} \right]^{1/2}$$

When Re $s = \sigma > 0$, we see that $|z| > 1$, and when Re $s = \sigma < 0$, we have $|z| < 1$. Similarly, when $\sigma = 0$, we obtain $|z| = 1$. Therefore, the $j\Omega$ axis maps onto the unit circle $|z| = 1$, the left half of the s plane (LHP) maps onto the interior of the unit circle $|z| = 1$, and the right half of the s plane (RHP) maps onto the exterior of the unit circle $|z| = 1$. As a result, we see that a stable analog filter, one that has only LHP poles, is transformed into a stable digital filter, one that has only poles inside the unit circle.

Of particular interest is the manner in which the $j\Omega$ axis is mapped onto the unit circle $|z| = 1$. Taking $s = j\Omega$ and $z = e^{j\omega}$, we obtain from Eq. (5.20) the expression

$$j\Omega = \frac{2}{T}\frac{1 - e^{-j\omega}}{1 + e^{-j\omega}}$$

$$= \frac{2}{T}\frac{e^{j\omega/2} - e^{-j\omega/2}}{e^{j\omega/2} + e^{-j\omega/2}}$$

and, consequently,

$$\Omega = \frac{2}{T}\tan\frac{\omega}{2} \tag{5.27}$$

and

$$\omega = 2\tan^{-1}\Omega T/2 \tag{5.28}$$

From this relationship, it can be seen that the upper imaginary axis in the s plane is mapped onto the upper half of the unit circle in the z plane. Furthermore, the frequencies Ω and ω satisfy a nonlinear relation, which results in a distortion of the frequency axis. The extent of this distortion can be observed from the graph shown in

Fig. 5.4, which represents the relation of Eq. (5.28). The nonlinearity in the relationship between ω and Ω is quite often referred to as *frequency warping*.

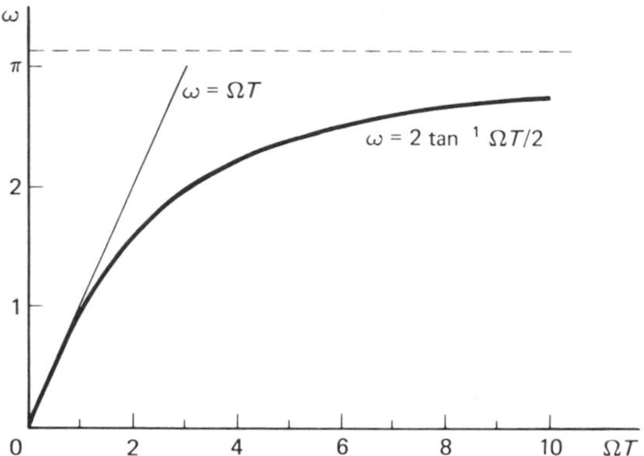

Figure 5.4 Correspondence between analog and digital frequencies resulting from the bilinear transformation.

The bilinear transformation maps the s plane, the analog-filter domain, in a one-to-one fashion onto the z plane, the digital-filter domain. It has the property that rational functions of s are transformed into rational functions of z, and vice versa. Consequently, to every analog filter represented by a rational function of s, there exists a unique digital filter represented by a rational function of z, and vice versa. This concept was illustrated earlier in this section when we transformed analog transfer functions into digital transfer functions. It should be noted that the corresponding digital and analog filters have the same order [48, 49].

If a digital filter is described by a frequency-domain specification, we can find an analog filter whose frequency-domain specification is determined by the digital-filter specification. For a given digital frequency ω_1, there is a corresponding analog frequency Ω_1, given by Eq. (5.27), such that

$$\Omega_1 = \frac{2}{T} \tan \frac{\omega_1}{2}$$

and for which

$$H_a(j\Omega_1) = H(e^{j\omega_1})$$

If

$$H_a(j\Omega_1) = |H_a(j\Omega_1)| \underline{/\phi_a(\Omega_1)}$$
$$H(e^{j\omega_1}) = |H(e^{j\omega_1})| \underline{/\phi(\omega_1)}$$

then we have the corresponding magnitude and phase relations

$$|H_a(j\Omega_1)| = |H(e^{j\omega_1})|$$
$$\phi_a(\Omega_1) = \phi(\omega_1)$$

In Fig. 5.5, the mapping of a typical low-pass analog frequency response to a corresponding digital frequency response is shown. Notice that the magnitude restrictions are similar, differing only in the frequencies at which they change values. We can obtain the proper digital response by properly *prewarping* the critical analog frequencies, that is,

$$\Omega_i = \frac{2}{T} \tan \frac{\omega_i}{2} \qquad i = 1, 2 \tag{5.29}$$

where ω_1, ω_2, Ω_1, and Ω_2 are the frequencies indicated in Fig. 5.5. If the cutoff frequency ω_c is specified, then the cutoff frequency Ω_c must be obtained by prewarping also.

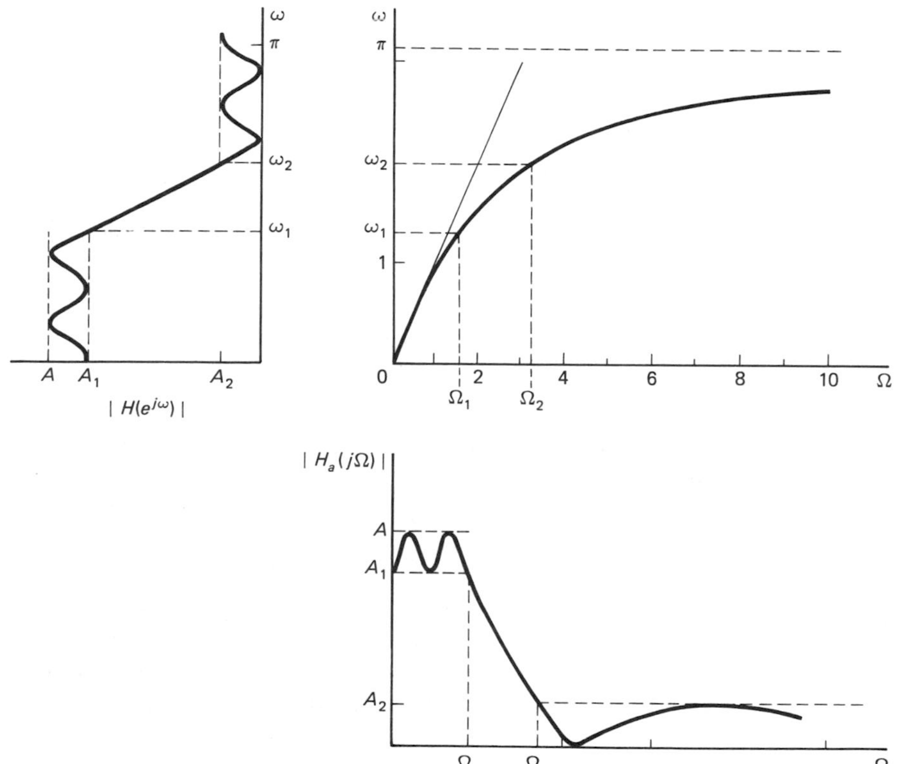

Figure 5.5 Analog frequency response obtained from a digital frequency response by the bilinear transformation.

One distinct advantage of the bilinear-transformation technique is the absence of the aliasing problem that is present, of course, when the impulse-invariant transformation is used. A disadvantage of the bilinear transformation is the introduction of a distortion in the frequency axis.

In the following sections, we give examples of filter designs using the bilinear transformation.

EXERCISES

5.4.1 Apply the bilinear transformation to

$$H_a(s) = \frac{2}{(s+1)(s+2)}$$

with $T = 1$ s and find $H(z)$.

5.4.2 Repeat Ex. 5.4.1 for

$$H_a(s) = \frac{2s}{s^2 + 0.2s + 1}$$

5.4.3 Obtain $H(z)$ from

$$H_a(s) = \frac{s^3}{(s+1)(s^2+s+1)}$$

using the bilinear transformation with $T = 1$ s. Determine $|H(1)|$ and $|H(e^{j\pi})| = |H(-1)|$.

5.4.4 Obtain Eq. (5.26) for all-pole first- and second-order filters.

5.5 ALL-POLE ANALOG FILTERS: BUTTERWORTH AND CHEBYSHEV

In this section, we discuss all-pole analog filters, which are described by transfer functions having the form

$$H_a(s) = \frac{A'}{P(s)} \tag{5.30}$$

where $P(s)$ is a polynomial of degree N, and A' is a constant. The most important low-pass filters of this type are the Butterworth and Chebyshev filters, which are the subject of this section. These filters are most often realized in cascaded form, so that the transfer function is usually written in the factored form

$$H_a(s) = \prod_{k=1}^{N/2} \frac{B_k\Omega_c^2}{s^2 + b_k\Omega_c s + c_k\Omega_c^2} \qquad N = 2, 4, 6, \ldots \tag{5.31}$$

for N even and

$$H_a(s) = \frac{B_0\Omega_c}{s + c_0\Omega_c} \prod_{k=1}^{(N-1)/2} \frac{B_k\Omega_c^2}{s^2 + b_k\Omega_c s + c_k\Omega_c^2} \qquad N = 3, 5, 7, \ldots \tag{5.32}$$

for N odd. The frequency Ω_c is the *cutoff frequency*, and in the normalized case, we have $\Omega_c = 1$ rad/s.

There is a wealth of information available for Butterworth and Chebyshev filters [77, 181, 183]. In many cases, tables are available that give the coefficients of

$H_a(s)$ for the factored forms of Eqs. (5.31) and (5.32) as well as the coefficients of the polynomial $P(s)$ in Eq. (5.30) [79].

In most cases, the filter specifications are given in terms of the magnitude response $|H_a(j\Omega)|$. It is then a straightforward procedure to obtain the transfer function $H_a(s)$. We now examine the magnitude responses for the Butterworth and Chebyshev filters and give the transfer-function coefficients.

Butterworth Filters

The simplest low-pass filter magnitude function is that of the Butterworth filter given by

$$|H_a(j\Omega)| = \frac{A}{[1 + (\Omega/\Omega_c)^{2N}]^{1/2}} \qquad N = 1, 2, 3, \ldots \tag{5.33}$$

where A is the filter gain, and Ω_c is $\Omega_{3\text{dB}}$. The Butterworth magnitude response decreases monotonically as the frequency increases. A number of Butterworth responses are shown in Fig. 5.6. The ideal response is shown by the dashed line. It can be seen that the Butterworth response more closely approximates the ideal response as the order N increases. The Butterworth filter is said to have *maximally flat* magnitude response since the first $2N - 1$ derivatives of $|H_a(j\Omega)|^2$ are zero at $\Omega = 0$.

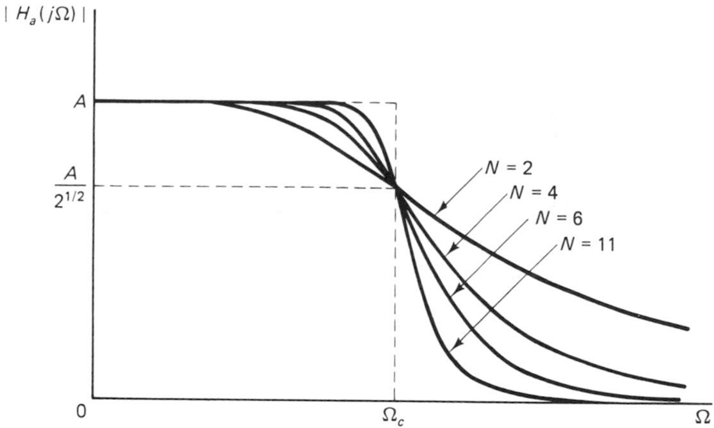

Figure 5.6 Low-pass Butterworth magnitude responses.

The phase responses of Butterworth filters are shown in Fig. 5.7 for several values of N. It can be seen that the phase response becomes more nonlinear as N increases.

The transfer function for a Butterworth filter is given by Eq. (5.31) for N even and by Eq. (5.32) for N odd. For both cases, the coefficients b_k and c_k are given by

$$b_k = 2 \sin [(2k - 1)\pi/2N] \qquad c_k = 1 \tag{5.34}$$

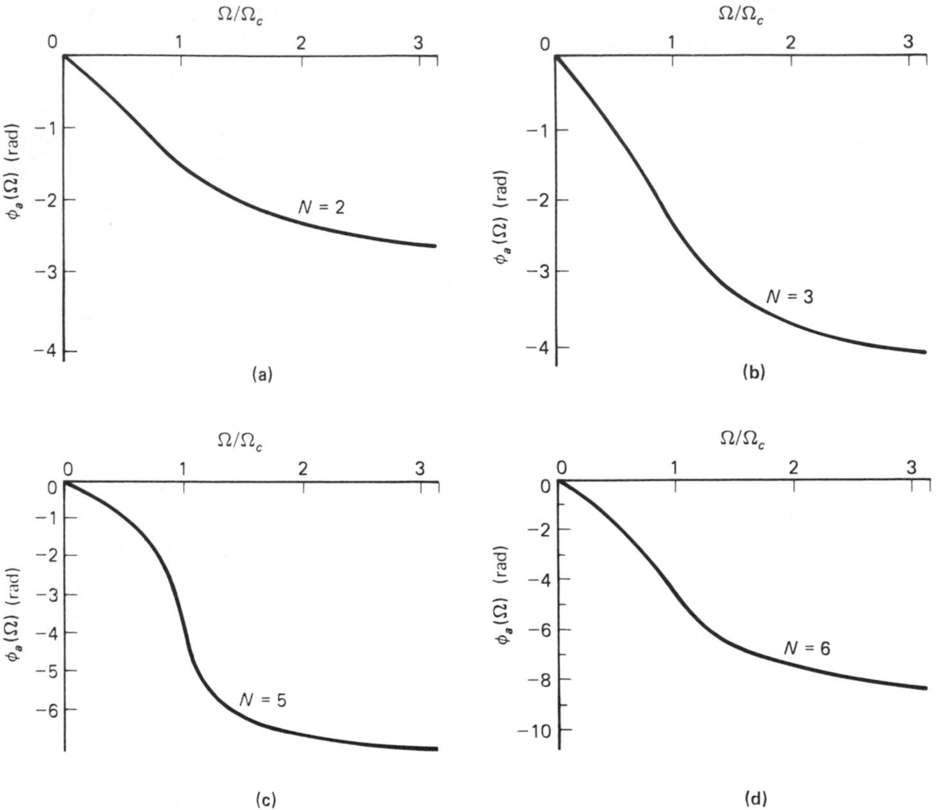

Figure 5.7 Butterworth phase responses for (a) $N = 2$, (b) $N = 3$, (c) $N = 5$, and (d) $N = 6$.

The parameters B_k in Eqs. (5.31) and (5.32), A' in Eq. (5.30), and A in Eq. (5.33) are related. The particular relation can be found by comparing the appropriate expressions. For example, comparing Eqs. (5.31) and (5.32) with Eq. (5.33), we see that

$$A = \prod_{k=1}^{N/2} B_k \qquad N \text{ even}$$

$$= \prod_{k=0}^{(N-1)/2} B_k \qquad N \text{ odd}$$

Example

As an example, when $N = 4$ for a Butterworth filter, we have

$$b_1 = 2 \sin (\pi/8) = 0.76537$$

$$b_2 = 2 \sin (3\pi/8) = 1.84776$$

Chebyshev Filters

The analog low-pass Chebyshev filter is the optimum all-pole filter in that for a given order N and given passband and stopband constraints, no other all-pole filter has a narrower transition bandwidth. Its magnitude response is given by

$$|H_a(j\Omega)| = \frac{A}{[1 + \epsilon^2 C_N^2(\Omega/\Omega_c)]^{1/2}} \tag{5.35}$$

The quantities ϵ and A are constants, and $C_N(x)$ is the Chebyshev polynomial of the first kind of degree N, given by

$$C_N(x) = \cos(N \cos^{-1} x) \tag{5.36}$$

The normalizing frequency Ω_c is the passband frequency Ω_1. For example,

$$\begin{aligned} C_2(x) &= \cos 2\theta \qquad \theta = \cos^{-1} x \\ &= 2\cos^2\theta - 1 \\ &= 2x^2 - 1 \end{aligned}$$

The relation of Eq. (5.36) can be evaluated quite easily when $|x| \leq 1$. However, if $|x| > 1$, then $\cos^{-1} x$ is not a real number. It can be shown that an equivalent form for this case is

$$\begin{aligned} C_N(x) &= \cosh(N \cosh^{-1} x) \\ &= \cosh\{N \ln[x + (x^2 - 1)^{1/2}]\} \end{aligned} \tag{5.37}$$

where "ln" denotes the natural logarithm.

As the value of N increases, it becomes more difficult to use Eq. (5.36) to obtain the polynomial expression for $C_N(x)$. It is frequently more convenient to use the recurrence relation

$$C_N(x) = 2xC_{N-1}(x) - C_{N-2}(x) \tag{5.38}$$

where

$$C_0(x) = 1 \qquad C_1(x) = x \tag{5.39}$$

The reader is asked to obtain Eq. (5.38) in Ex. 5.5.3.

As an example of the use of Eq. (5.38), we see that

$$\begin{aligned} C_3(x) &= 2xC_2(x) - C_1(x) \\ &= 2x(2x^2 - 1) - x \\ &= 4x^3 - 3x \end{aligned}$$

The amplitude response is determined by the manner in which $C_N(\Omega/\Omega_c)$ varies. For $-1 \leq \Omega/\Omega_c \leq 1$, we see from Eq. (5.36) that $-1 \leq C_N(\Omega/\Omega_c) \leq 1$ and, consequently, the amplitude response satisfies the inequality

$$\frac{A}{(1 + \epsilon^2)^{1/2}} \leq |H_a(j\Omega)| \leq A \qquad 0 \leq \Omega/\Omega_c \leq 1$$

The maximum value A is achieved at the zeros of $C_N(\Omega/\Omega_c)$, which are found from Eq. (5.36) to be

$$x_k = \Omega_k/\Omega_c = \cos\,[(2k-1)\pi/2N] \qquad k = 1, 2, \ldots, N \tag{5.40}$$

The minimum value $A/(1+\epsilon^2)^{1/2}$ is achieved when $C_N(\Omega/\Omega_c) = \pm 1$, which occurs when

$$x_m = \Omega_m/\Omega_c = \cos\,(m\pi/N) \qquad m = 0, 1, \ldots, N-1 \tag{5.41}$$

The variation of $C_N(x)$ is shown in Fig. 5.8 for several values of N. The equal ripples of $C_N(\Omega/\Omega_c)$ in the interval $-1 \leq \Omega/\Omega_c \leq 1$ cause equal ripples in $|H_a(j\Omega)|$ with ripple width RW given by

$$RW = A\left[1 - \frac{1}{(1+\epsilon^2)^{1/2}}\right] \tag{5.42}$$

For $x > 1$, $C_N(x)$ increases monotonically, and, hence, for $\Omega > \Omega_c$, the magnitude response decreases monotonically. A number of Chebyshev magnitude responses are shown in Fig. 5.9 for $A = 1$ and $\epsilon = 1$.

The frequency Ω_c rad/s is the end of the *ripple channel,* as shown in Fig. 5.9. In general, this is not the cutoff frequency, or 3-dB frequency, which is given by

$$\Omega_{3\,\mathrm{dB}} = \Omega_c \cosh\left(\frac{1}{N}\cosh^{-1}\frac{1}{\epsilon}\right) \tag{5.43}$$

This form is used since $1/\epsilon > 1$, usually.

The amplitude response of the Chebyshev filter for a given order N is much better than that of the Butterworth filter. However, the phase response of the Chebyshev filter is inferior to that of the Butterworth filter. Some Chebyshev phase responses and one Butterworth phase response are shown in Fig. 5.10.

The parameters in the transfer functions of Eqs. (5.31) and (5.32) for the Chebyshev filter can be computed from the formulas

$$\begin{aligned} b_k &= 2y_N \sin\frac{(2k-1)\pi}{2N} \\ c_k &= y_N^2 + \cos^2\frac{(2k-1)\pi}{2N} \\ c_0 &= y_N \end{aligned} \tag{5.44}$$

where

$$y_N = \tfrac{1}{2}\{[(1/\epsilon^2+1)^{1/2} + 1/\epsilon]^{1/N} - [(1/\epsilon^2+1)^{1/2} + 1/\epsilon]^{-1/N}\} \tag{5.45}$$

The values of B_k are chosen so that $H_a(0)$ has the desired value. This determination varies as to whether N is even or odd. When N is even, the amplitude response at $\Omega = 0$ is at the bottom of the ripple channel and $H_a(0) = A/(1+\epsilon^2)^{1/2}$. On the other hand, when N is odd, the response is at the top of the ripple channel and $H_a(0) = A$. This follows from the fact that $C_N(x)$ is even or odd as N is even or odd.

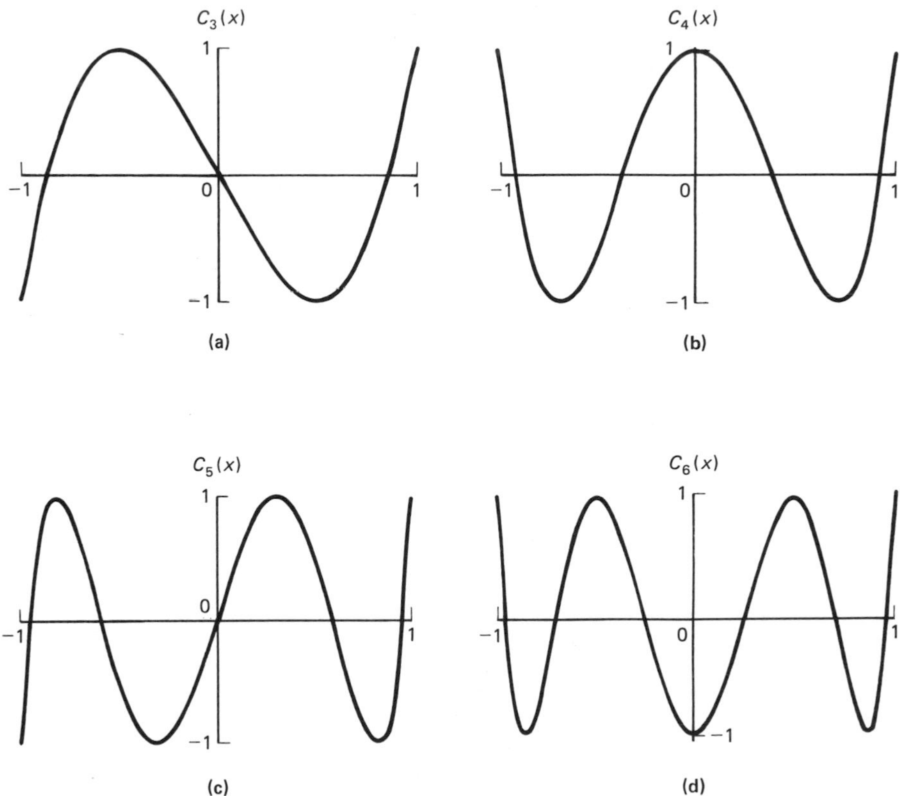

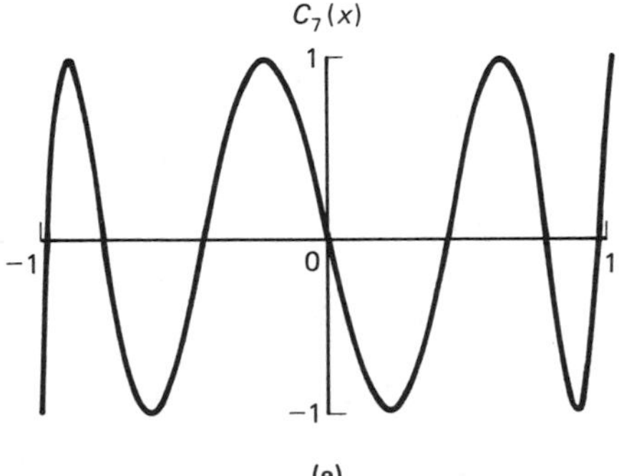

Figure 5.8 Graphs of Chebyshev polynomials $C_N(x)$ for (a) $N = 3$, (b) $N = 4$, (c) $N = 5$, (d) $N = 6$, and (e) $N = 7$.

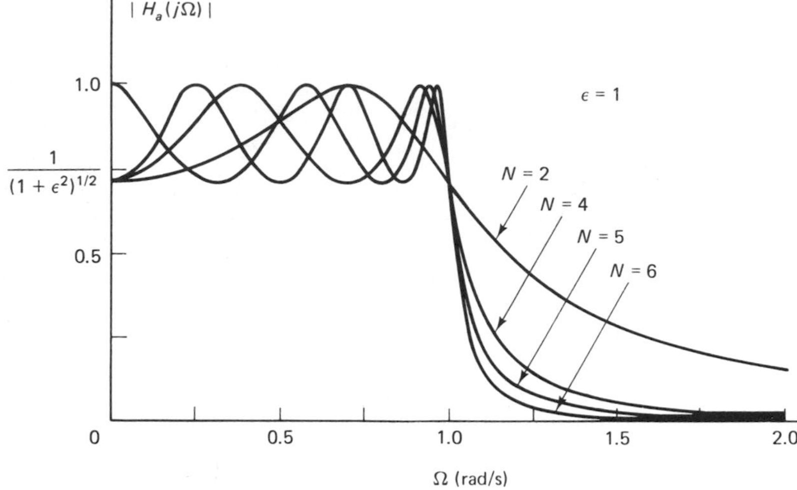

Figure 5.9 Low-pass Chebyshev magnitude responses for $\epsilon = 1$ and $\Omega_1 = 1$.

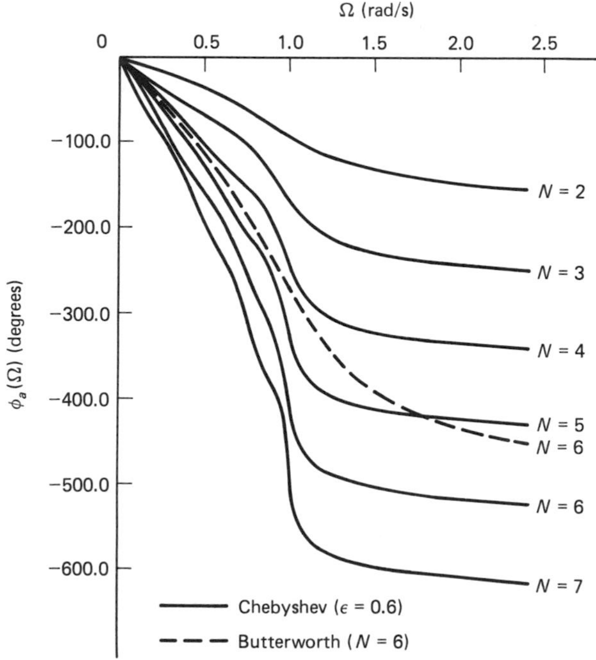

Figure 5.10 Chebyshev and Butterworth phase responses for $\Omega_c = 1$.

Example

If we take $N = 4$ and $\epsilon = 0.8$, we find that

$$y_4 = \frac{1}{2}\left\{\left[\left(\frac{1}{(0.8)^2} + 1\right)^{1/2} + \frac{1}{0.8}\right]^{1/4} - \left[\left(\frac{1}{(0.8)^2} + 1\right)^{1/2} + \frac{1}{0.8}\right]^{-1/4}\right\}$$

$$= 0.26490$$

$$b_1 = 2y_4 \sin \pi/8 = 0.20275$$

$$c_1 = y_4^2 + \cos^2 \pi/8 = 0.92373$$

$$b_2 = 2y_4 \sin 3\pi/8 = 0.48947$$

$$c_2 = y_4^2 + \cos^2 3\pi/8 = 0.21662$$

EXERCISES

5.5.1 Determine $H_a(s)$ for a third-order Butterworth filter. Compare the result with Eq. (5.18).

5.5.2 Repeat Ex. 5.5.1 for $N = 5$.

5.5.3 Using the trigonometric identity

$$\cos(\alpha + \beta) + \cos(\alpha - \beta) = 2\cos\alpha\cos\beta$$

with $\theta = \cos^{-1} x$, $\alpha = (N - 1)\theta$, and $\beta = \theta$, obtain Eq. (5.38).

5.5.4 Show that
(a) $C_N(-x) = (-1)^N C_N(x)$
(b) $C_N(1) = 1$

5.5.5 Determine $H_a(s)$ for a Chebyshev filter with $N = 5$ and $\epsilon = 0.8$.

5.6 DESIGN OF DIGITAL BUTTERWORTH AND CHEBYSHEV FILTERS

In this section, we consider the design of digital Butterworth and Chebyshev filters. We begin with the magnitude constraints on the digital filter and transform them into analog-filter constraints. Using these constraints, we determine the parameters of the magnitude-response function and then the analog transfer function. The magnitude response is given by Eq. (5.33) for a Butterworth filter and by Eq. (5.35) for a Chebyshev filter. Once the analog transfer function is known, the digital transfer function is obtained by using the desired transformation.

Determination of the Filter Parameters

The low-pass digital filters that we consider in this chapter satisfy the magnitude constraints

$$\begin{aligned} A_1 \le |H(e^{j\omega})| &\le 1 \qquad 0 \le \omega \le \omega_1 \\ |H(e^{j\omega})| &\le A_2 \qquad \omega_2 \le \omega \le \pi \end{aligned} \tag{5.46}$$

In the filter-design process, we obtain the corresponding analog magnitude constraints

$$\begin{aligned} A_1 \le |H_a(j\Omega)| &\le 1 \qquad 0 \le \Omega \le \Omega_1 \\ |H_a(j\Omega)| &\le A_2 \qquad \Omega_2 \le \Omega \end{aligned} \tag{5.47}$$

where Ω_1 and Ω_2 are determined from ω_1 and ω_2.

The values of Ω_1 and Ω_2 depend on the design method used. If we use the bilinear transformation, then, from Eq. (5.27), the frequencies are related by

$$\Omega = \frac{2}{T}\tan\frac{\omega}{2}$$

If the impulse-invariant transformation is used, then, from Eq. (5.11), the relation is

$$\Omega = \omega/T$$

As we will see later, determination of the analog-filter parameters requires ratios of analog frequencies such as Ω_2/Ω_1. For the bilinear transformation, we obtain

$$\frac{\Omega_2}{\Omega_1} = \frac{(2/T)\tan(\omega_2/2)}{(2/T)\tan(\omega_1/2)} = \frac{\tan(\omega_2/2)}{\tan(\omega_1/2)} \tag{5.48}$$

and for the impulse-invariant transformation

$$\Omega_2/\Omega_1 = \omega_2/\omega_1 \tag{5.49}$$

These ratios are the same regardless of the type of filter that is being designed. The magnitude-response parameters are then determined by imposing the constraints of Eq. (5.47) at Ω_1 and Ω_2 on the particular $H_a(j\Omega)$ that is being used.

Parameters for the Butterworth Filter

In the case of a Butterworth filter, we can replace $|H_a(j\Omega)|$ in Eq. (5.47) by Eq. (5.33), obtaining for the squared magnitude response the inequalities

$$A_1^2 \le \frac{1}{1 + (\Omega_1/\Omega_c)^{2N}} \le 1$$

$$\frac{1}{1 + (\Omega_2/\Omega_c)^{2N}} \le A_2^2$$

which can be written

$$(\Omega_1/\Omega_c)^{2N} \le (1/A_1^2) - 1$$
$$(\Omega_2/\Omega_c)^{2N} \ge (1/A_2^2) - 1 \tag{5.50}$$

In order to obtain the parameters N and Ω_c, we assume equality in the above inequalities and by division obtain the ratio

$$(\Omega_2/\Omega_1)^{2N_1} = (1/A_2^2 - 1)/(1/A_1^2 - 1) \tag{5.51}$$

where we have replaced N by N_1. This equation can then be solved to give

$$N_1 = \frac{1}{2}\frac{\log\{[(1/A_2)^2 - 1]/[(1/A_1)^2 - 1]\}}{\log(\Omega_2/\Omega_1)}$$

Since this expression normally does not result in an integer value for N_1, we therefore select N to be the smallest integer such that $N \ge N_1$. Since $|H_a(j\Omega_2)|$ decreases as N is increased, the constraint at $\Omega = \Omega_2$ becomes an inequality for $N \ge N_1$. (Recall that N_1 was determined for equality constraints.) Therefore, the constraint at $\Omega = \Omega_2$ is satisfied if we choose N to be the smallest integer such that

$$N \ge N_1 = \frac{1}{2}\frac{\log\{[(1/A_2)^2 - 1]/[(1/A_1)^2 - 1]\}}{\log(\Omega_2/\Omega_1)} \tag{5.52}$$

Since we have a ratio of logarithms, it doesn't matter which base is used.

Once we have obtained N from Eq. (5.52), we can then determine Ω_c by requiring that the constraint at $\Omega = \Omega_1$ be an equality constraint. From Eq. (5.50), using equality, we find that

$$\Omega_c = \frac{\Omega_1}{[(1/A_1)^2 - 1]^{1/2N}} \tag{5.53}$$

For the bilinear transformation, we obtain

$$\frac{\Omega_c T}{2} = \frac{\tan(\omega_1/2)}{[(1/A_1)^2 - 1]^{1/2N}} \tag{5.54}$$

and for the impulse-invariant transformation

$$\Omega_c = \frac{\omega_1/T}{[(1/A_1)^2 - 1]^{1/2N}} \tag{5.55}$$

Parameters for the Chebyshev Filter

For the Chebyshev filter, the constraints on the squared magnitude response, obtained from Eqs. (5.35) and (5.47), can be written

$$A_1^2 \le \frac{1}{1 + \epsilon^2 C_N^2(\Omega_1/\Omega_c)} \le 1 \tag{5.56}$$

$$\frac{1}{1+\epsilon^2 C_N^2(\Omega_2/\Omega_c)} \leq A_2^2 \tag{5.57}$$

The simplest way to satisfy the constraint of Eq. (5.56) is to take

$$\Omega_c = \Omega_1 \tag{5.58}$$

and then choose ϵ so that

$$\frac{1}{1+\epsilon^2} = A_1^2$$

This follows since $C_N(\Omega_1/\Omega_c) = C_N(1) = 1$. The required value of ϵ is then

$$\epsilon = [(1/A_1)^2 - 1]^{1/2} \tag{5.59}$$

The value of N is determined from Eq. (5.57), the constraint at $\Omega = \Omega_2$. When this inequality is solved for $C_N(\Omega_2/\Omega_1)$, we obtain

$$C_N(\Omega_2/\Omega_1) \geq \frac{1}{\epsilon}[(1/A_2)^2 - 1]^{1/2} \tag{5.60}$$

where the ratio Ω_2/Ω_1 is given by Eq. (5.48) for the bilinear transformation and by Eq. (5.49) for the impulse-invariant transformation. The value of N can be determined by using the expression for $C_N(x)$ given in Eq. (5.37) or by successively using the recurrence relation of Eq. (5.38). Using the expression of Eq. (5.37) in Eq. (5.60), we obtain

$$\cosh[N \cosh^{-1}(\Omega_2/\Omega_1)] \geq \frac{1}{\epsilon}[(1/A_2)^2 - 1]^{1/2}$$

which results in N being the smallest integer such that

$$N \geq \frac{\cosh^{-1}\{(1/\epsilon)[(1/A_2)^2 - 1]^{1/2}\}}{\cosh^{-1}(\Omega_2/\Omega_1)} \tag{5.61}$$

In this section, we have considered only the design of low-pass filters. The design of high-pass, bandpass, and band-reject filters can be accomplished by first designing a low-pass filter and then using a frequency transformation. These transformations are discussed in Sec. 5.9.

EXERCISES

5.6.1 Determine N and Ω_c for a Butterworth filter for which $A_1 = 1/2^{1/2}$, $A_2 = 0.1$, $\Omega_1 = 2$ rad/s and $\Omega_2 = 4$ rad/s.

5.6.2 Determine the parameters of a Chebyshev filter that satisfies the constraints given in Ex. 5.6.1.

5.7 DESIGN EXAMPLES: BUTTERWORTH AND CHEBYSHEV FILTERS

In this section, we give several examples illustrating the design of digital Butterworth and Chebyshev filters using both the bilinear and the impulse-invariant transformation.

Butterworth Filter Using the Bilinear Transformation

As a first example, we use the bilinear transformation to design a Butterworth filter satisfying the constraints

$$\begin{aligned} 0.8 \le |H(e^{j\omega})| \le 1 \qquad & 0 \le \omega \le 0.2\pi \\ |H(e^{j\omega})| \le 0.2 \qquad & 0.6\pi \le \omega \le \pi \end{aligned} \tag{5.62}$$

Comparing these constraints with those of Eq. (5.46), we see that

$$\omega_1 = 0.2\pi \qquad \omega_2 = 0.6\pi$$

$$A_1 = 0.8 \qquad A_2 = 0.2$$

In order to obtain $H_a(j\Omega)$, we must determine the values of the parameters N and Ω_c.

From Eq. (5.48), we see that the analog frequency ratio is

$$\frac{\Omega_2}{\Omega_1} = \frac{\tan(\omega_2/2)}{\tan(\omega_1/2)} = \frac{1.376}{0.3249} = 4.235 \tag{5.63}$$

and from Eq. (5.52), the value of N satisfies the relation

$$\begin{aligned} N &\ge \frac{1}{2} \frac{\log[(1/A_2)^2 - 1]/[(1/A_1)^2 - 1]}{\log(\Omega_2/\Omega_1)} \\ &\ge \frac{1}{2} \frac{\log(1/0.04 - 1)/(1/0.64 - 1)}{\log 4.235} \\ &\ge 1.3 \end{aligned}$$

The required filter order is then $N = 2$.

The analog cutoff frequency Ω_c, given by Eq. (5.54), is

$$\begin{aligned} \Omega_c &= \frac{(2/T)\tan(\omega_1/2)}{[(1/A_1)^2 - 1]^{1/2N}} \\ &= \frac{2}{T} \frac{0.3249}{(1/0.64 - 1)^{1/4}} \\ &= \frac{2}{T}(0.3752) \end{aligned}$$

The required analog-filter function can now be obtained from Eq. (5.31) by taking $N = 2$. For unity gain, we take $B_1 = 1$, so that

$$H_a(s) = \frac{\Omega_c^2}{s^2 + b_1\Omega_c s + \Omega_c^2}$$

where $b_1 = 2 \sin(\pi/4) = 1.4142$. When the bilinear transformation is applied to this function, the digital transfer function, obtained using Eqs. (5.24)–(5.26), becomes

$$\begin{aligned} H(z) &= \frac{A_{10}(1 + z^{-1})^2}{1 + a_{11}z^{-1} + a_{12}z^{-2}} \\ &= \frac{0.0842(1 + z^{-1})^2}{1 - 1.0281z^{-1} + 0.3651z^{-2}} \end{aligned}$$

since

$$\begin{aligned} A_{10} &= (\Omega_c T/2)^2[(\Omega_c T/2)^2 c_1 + (\Omega_c T/2)b_1 + 1]^{-1} \\ &= (0.3752)^2[(0.3752)^2 + (0.3752)(1.4142) + 1]^{-1} \\ &= (0.3752)^2(0.5983) = 0.0842 \\ a_{11} &= 2[(\Omega_c T/2)^2 c_1 - 1][(\Omega_c T/2)^2 c_1 + (\Omega_c T/2)b_1 + 1]^{-1} \\ &= -1.0281 \\ a_{12} &= [(\Omega_c T/2)^2 c_1 - (\Omega_c T/2)b_1 + 1][(\Omega_c T/2)^2 c_1 + (\Omega_c T/2)b_1 + 1]^{-1} \\ &= 0.3651 \end{aligned}$$

Graphs of the magnitude and phase responses are shown in Fig. 5.11.

Butterworth Filter Using the Impulse-Invariant Transformation

Next, we use the impulse-invariant transformation to design a digital filter satisfying the constraints of Eq. (5.62). In order to avoid a large gain at $\omega = 0$, we use Eq. (5.14), so that

$$H(e^{j\omega}) = H_a(j\omega/T) \qquad |\omega| \le \pi$$

The analog and digital frequencies are related by $\Omega = \omega/T$, and so, from Eq. (5.49), we have

$$\Omega_2/\Omega_1 = \omega_2/\omega_1 = 0.6/0.2 = 3$$

The required value of N must satisfy the inequality of Eq. (5.52), which becomes

$$N \ge \frac{1}{2}\frac{\log(1/0.04 - 1)/(1/0.64 - 1)}{\log 3} = 1.708$$

and so $N = 2$. The cutoff frequency, determined from Eq. (5.55), is

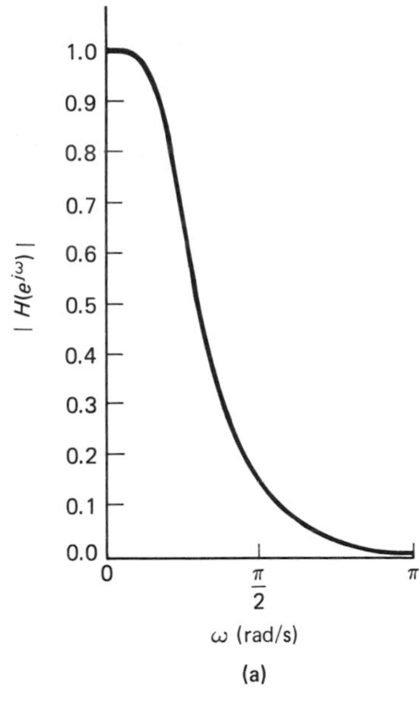

(a)

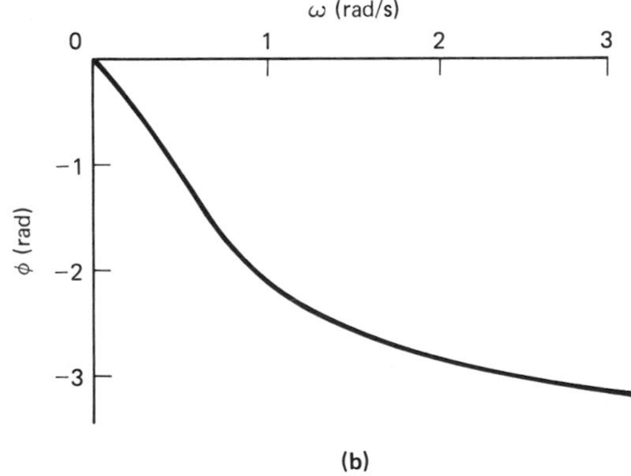

(b)

Figure 5.11 Butterworth (a) magnitude and (b) phase responses obtained for the constraints of Eq. (5.62) using the bilinear transformation.

$$\Omega_c = \frac{\omega_1/T}{[(1/A_1)^2 - 1]^{1/2N}}$$

$$= \frac{0.2\pi/T}{(1/0.64 - 1)^{1/4}} = \frac{0.7255}{T}$$

We can obtain $H(z)$ by using the results of Eq. (5.31) and Ex. 5.3.5. For our case, we have $B_1 = 1$ and

$$a = b_1\,\Omega_c/2 = \Omega_c/2^{1/2}$$

$$b = \Omega_c[c_1 - (b_1/2)^2]^{1/2} = \Omega_c/2^{1/2}$$

Using the result of Ex. 5.3.5, we find that

$$H(z) = \Omega_c T\left(\frac{\Omega_c}{b}\right)\frac{e^{-aT}(\sin bT)z^{-1}}{1 - 2e^{-aT}(\cos bT)z^{-1} + e^{-2aT}z^{-2}}$$

$$= 0.7255(2^{1/2})\frac{e^{-0.7255/2^{1/2}}[\sin (0.7255/2^{1/2})]z^{-1}}{1 - 2e^{-0.7255/2^{1/2}}[\cos (0.7255/2^{1/2})]z^{-1} + e^{-0.7255(2^{1/2})}z^{-2}}$$

$$= \frac{0.3015z^{-1}}{1 - 1.0433z^{-1} + 0.3584z^{-2}}$$

Note that the gain at $\omega = 0$ is the value $H(1) = 0.9568$.

Chebyshev Filter Using the Bilinear Transformation

As an example of the design of a Chebshev digital filter, we consider the same magnitude constraints of Eq. (5.62) that were used for the Butterworth filter designs. If we use the bilinear transformation, then the ratio of frequencies Ω_2/Ω_1 is the same as we obtained in Eq. (5.63) for the Butterworth example. Therefore, we have

$$\frac{\Omega_2}{\Omega_1} = \frac{\tan (\omega_2/2)}{\tan (\omega_1/2)} = 4.235$$

The magnitude response is given by Eq. (5.35) with $A = 1$, as for the Butterworth design problem. To make the response leave the ripple channel at Ω_1, we take $\Omega_1 = \Omega_c$. Comparing the magnitude constraints given in Eq. (5.62) with the Chebyshev responses of Fig. 5.9, we see that equating the minimum values in the passbands results in the relation

$$\frac{1}{(1 + \epsilon^2)^{1/2}} = 0.8$$

Consequently, we have $\epsilon = \frac{3}{4}$.

We still must determine the order N of the filter and it must be such that the constraint at Ω_2 is satisfied. This requires that

$$|H_a(j\Omega_2)| = \frac{1}{[1 + \epsilon^2 C_N^2(\Omega_2/\Omega_1)]^{1/2}} \le 0.2$$

which is equivalent to

$$1 + \epsilon^2 C_N^2(\Omega_2/\Omega_c) \ge 1/(0.2)^2 = 25$$

Solving this inequality for $C_N(\Omega_2/\Omega_c)$ and setting $\epsilon = \frac{3}{4}$ results in

$$C_N(\Omega_2/\Omega_c) = C_N(\Omega_2/\Omega_1) \ge (24)^{1/2}/\epsilon = 8(6)^{1/2}/3$$

The problem now is to determine the smallest integer N such that

$$C_N(4.235) \ge 8(6)^{1/2}/3 = 6.532$$

The value of N that satisfies this relation can be determined from the expression

$$N \ge N_1 = \frac{\cosh^{-1}[C_N(x)]}{\cosh^{-1} x} \tag{5.64}$$

which can be derived from Eq. (5.37) for $x > 1$. Since we know $x = 4.235$ and only that $C_N(4.235) \geq 6.532$, the relation that N must satisfy is

$$N \geq \frac{\cosh^{-1} 6.532}{\cosh^{-1} 4.235} = 1.208$$

Consequently, we must take $N = 2$, which is the same value that was required for the digital Butterworth filter.

The values needed in Eq. (5.64) are easily calculated on most handheld calculators. Another procedure for determining N is to evaluate $C_N(4.235)$ starting with $N = 0$ and continuing until a value is reached for which $C_N(4.235) \geq 6.532$. These calculations can best be carried out using the recurrence relation of Eq. (5.38). In this particular case, we have $C_0(4.235) = 1$, $C_1(4.235) = 4.235$, and from

$$C_2(x) = 2xC_1(x) - C_0(x)$$

we have

$$C_2(4.235) = 2(4.235)(4.235) - 1 = 34.87$$

Consequently, $N = 2$ is the smallest value of N satisfying the given inequality.

The transfer function for the analog filter, given by Eq. (5.31) with $N = 2$, is

$$H_a(s) = \frac{B_1 \Omega_c^2}{s^2 + b_1 \Omega_c s + c_1 \Omega_c^2}$$

where

$$\begin{aligned} y_2 &= \tfrac{1}{2}\{[(16/9 + 1)^{1/2} + 4/3]^{1/2} - [(16/9 + 1)^{1/2} + 4/3]^{-1/2}\} \\ &= 1/(3)^{1/2} \\ b_1 &= 2\left(\frac{1}{3^{1/2}}\right) \sin \frac{\pi}{4} = \left(\frac{2}{3}\right)^{1/2} \\ c_1 &= \left(\frac{1}{3^{1/2}}\right)^2 + \cos^2 \frac{\pi}{4} = \frac{5}{6} \end{aligned}$$

In order to make the amplitude response satisfy the constraints, we must choose B_1 so that $H_a(0) = 1/(1 + \epsilon^2)^{1/2} = \frac{4}{5}$. This is because for N even, the amplitude response at $\Omega = 0$ is at the bottom of the ripple channel. Therefore, we require that

$$H_a(0) = \tfrac{4}{5} = \mathrm{B}_1/\mathrm{c}_1$$

and, as a result,

$$B_1 = \tfrac{4}{5} c_1 = \tfrac{2}{3}$$

Using Eqs. (5.24)–(5.26), we find that

$$\begin{aligned} H(z) &= \frac{B_1 A_{10}(1 + z^{-1})^2}{1 + a_{11} z^{-1} + a_{12} z^{-2}} \\ &= \frac{0.052(1 + z^{-1})^2}{1 - 1.3480z^{-1} + 0.6080z^{-2}} \end{aligned}$$

Since $\Omega_c = \Omega_1 = (2/T)(0.3249)$, the same as for the Butterworth example, we have

$$A_{10} = (0.3249)^2[(0.3249)^2(\tfrac{5}{6}) + (0.3249)(\tfrac{2}{3})^{1/2} + 1]^{-1}$$
$$= 0.078$$
$$a_{11} = 2[(0.3249)^2(\tfrac{5}{6}) - 1]$$
$$= -1.3480$$
$$a_{12} = (0.3249)^2(\tfrac{5}{6}) - (0.3249)(\tfrac{2}{3})^{1/2} + 1$$
$$= 0.6080$$

Graphs of the magnitude and phase responses are shown in Fig. 5.12.

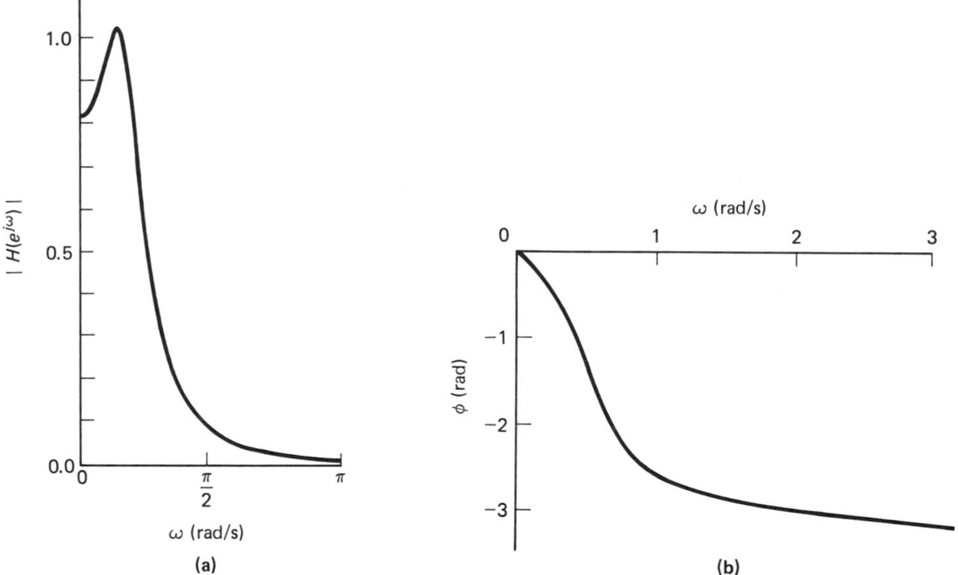

Figure 5.12 Chebyshev (a) magnitude and (b) phase responses obtained for the constraints of Eq. (5.62) by using the bilinear transformation.

Chebyshev Filter Using the Impulse-Invariant Transformation

We next use the impulse-invariant transformation to design a Chebyshev filter satisfying Eq. (5.62). The value of Ω_2/Ω_1 is the same as determined for the previous Chebyshev filter design. Consequently, we have

$$\Omega_2/\Omega_c = \Omega_2/\Omega_1 = 3$$

and

$$C_N(\Omega_2/\Omega_1) \geq 6.532$$

Since $C_1(3) = 3$ and $C_2(3) = 17$, we see again that $N = 2$. The parameters in the analog transfer function are the same as in the preceding Chebyshev filter design.

By using the results of Ex. 5.3.5, it can be shown that

$$H(z) = \frac{0.1948z^{-1}}{1 - 1.3483z^{-1} + 0.5987z^{-2}}$$

Graphs of the magnitude and phase responses are shown in Fig. 5.13.

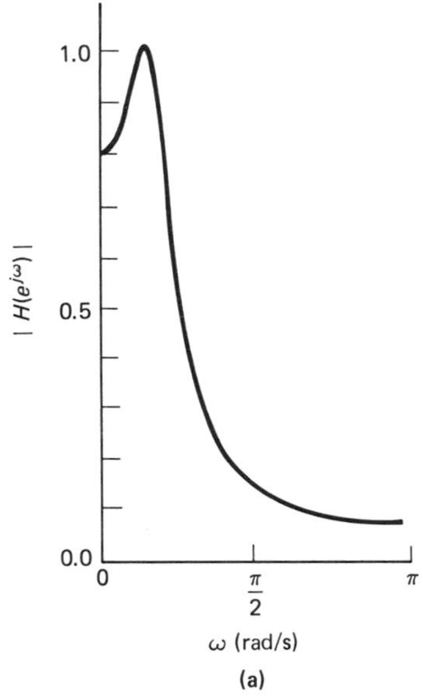

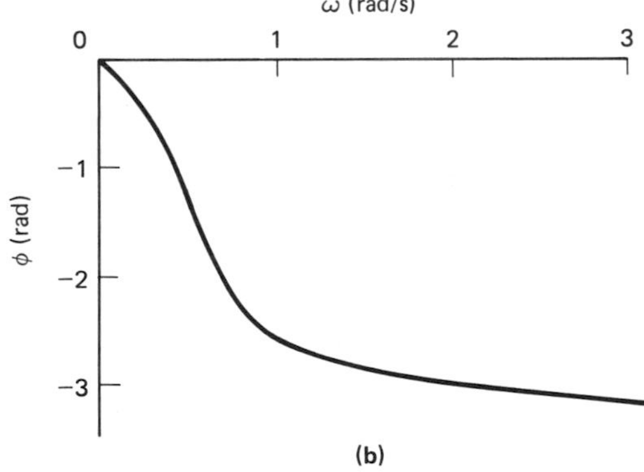

Figure 5.13 Chebyshev (a) magnitude and (b) phase responses obtained for the constraints of Eq. (5.62) by using the impulse-invariant transformation.

Another Design Example

As a final example in this section, let us use the bilinear transformation to design a Chebyshev filter that satisfies the constraints

$$\begin{aligned} 0.8 \le |H(e^{j\omega})| &\le 1 \qquad 0 \le \omega \le 0.2\pi \\ |H(e^{j\omega})| &\le 0.2 \qquad 0.26\pi \le \omega \le \pi \end{aligned} \tag{5.65}$$

First, as before, we take $\Omega_c = \Omega_1 = (2/T)\tan(\omega_1/2) = (2/T)(0.32492)$. The value of N is obtained from the constraint at $\Omega = \Omega_2$, which is the same as in the previous example. Therefore,

$$N \geq \frac{\cosh^{-1}[8(6)^{1/2}/3]}{\cosh^{-1}(\Omega_2/\Omega_1)}$$

and since

$$\frac{\Omega_2}{\Omega_1} = \frac{\tan(\omega_2/2)}{\tan(\omega_1/2)} = \frac{\tan 0.13\pi}{\tan 0.1\pi} = 1.3319$$

we obtain $N = 4$. Since $\epsilon = 0.75$ from the previous example, we obtain, from Eq. (5.45),

$$\begin{aligned} y_4 &= \tfrac{1}{2}\{[(16/9 + 1)^{1/2} + \tfrac{4}{3}]^{1/4} - [(16/9 + 1)^{1/2} + \tfrac{4}{3}]^{-1/4}\} \\ &= 0.278119 \end{aligned}$$

The filter coefficients, as determined from Eq. (5.44), are

$$\begin{aligned} b_1 &= 2(0.278119)\sin(\pi/8) = 0.212863 \\ b_2 &= 2(0.278119)\sin(3\pi/8) = 0.513897 \\ c_1 &= (0.278119)^2 + \cos^2(\pi/8) = 0.930904 \\ c_2 &= (0.278119)^2 + \cos^2(3\pi/8) = 0.223797 \end{aligned}$$

The gains B_1 and B_2 satisfy the relation

$$B_1 B_2/c_1 c_2 = 0.8$$

and so we take

$$\begin{aligned} B_1 = B_2 &= [0.8(0.930904)(0.223797)]^{1/2} \\ &= 0.408248 \end{aligned}$$

Substituting the above values into Eqs. (5.24)–(5.26) and (5.31), we obtain for the transfer function

$$H(z) = H_1(z)H_2(z)$$

where

$$H_1(z) = \frac{0.036919(1 + z^{-1})^2}{1 - 1.544783z^{-1} + 0.881513z^{-2}}$$

$$H_2(z) = \frac{0.036201(1 + z^{-1})^2}{1 - 1.640133z^{-1} + 0.719511z^{-2}}$$

The magnitude response is shown in Fig. 5.14.

To illustrate the superiority of the Chebyshev filter over the Butterworth filter, it is seen that the order of a Butterworth filter obtained using the bilinear transformation and satisfying the constraints of Eq. (5.65) is the smallest integer N such that

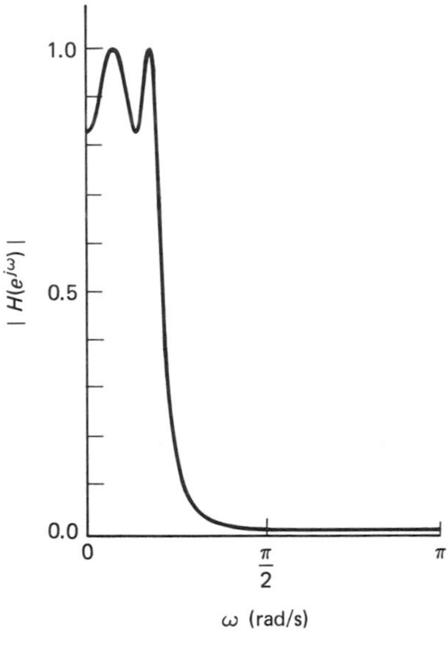

Figure 5.14 Chebyshev magnitude response obtained for the constraints of Eq. (5.65) by using the bilinear transformation.

$$N \geq \frac{\log [24/(9/16)]}{2 \log 1.3319} = 6.55$$

Consequently, the order of the Butterworth filter is $N = 7$. The reader is asked to design this filter in Prob. 5.22.

Design Tables

Tables are available that give the transfer-function coefficients b_k and c_k for the Butterworth and Chebyshev filters for certain values of N and ϵ. For example, in [79] tables are given for both filters for orders $N = 2, 3, \ldots, 10$, with ripple widths RW for the Chebyshev filter of 0.1, 0.5, 1, 2, and 3 dB, where

$$RW \text{ (dB)} = 10 \log_{10} (1 + \epsilon^2)$$

In the next section, we consider inverse Chebyshev and elliptic filters. The transfer functions for these filters have zeros as well as poles and, in the case of the elliptic filter, much faster cutoff can be obtained.

EXERCISES

5.7.1 Obtain an analog Chebyshev-filter transfer function that satisfies the constraints

$$1/2^{1/2} \leq |H_a(j\Omega)| \leq 1 \qquad 0 \leq \Omega \leq 2$$
$$|H_a(j\Omega)| \leq 0.1 \qquad \Omega \geq 4$$

5.7.2 For the constraints

$$1/2^{1/2} \leq |H(e^{j\omega})| \leq 1 \qquad 0 \leq \omega \leq 0.2\pi$$
$$|H(e^{j\omega})| \leq 0.2 \qquad 0.6\pi \leq \omega \leq \pi$$

with $T = 1$ s determine $H(z)$ for a Butterworth filter using **(a)** the bilinear transformation, and **(b)** the impulse-invariant transformation.

5.7.3 Repeat Ex. 5.7.2 if the stopband constraint is

$$|H(e^{j\omega})| \le 0.1 \qquad 0.4\pi \le \omega \le \pi$$

5.7.4 Repeat Ex. 5.7.2 for a Chebyshev filter for the constraints

$$0.9 \le |H(e^{j\omega})| \le 1 \qquad 0 \le \omega \le 0.2\pi$$

$$|H(e^{j\omega})| \le 0.2 \qquad 0.4\pi \le \omega \le \pi$$

5.8 DIGITAL INVERSE CHEBYSHEV AND ELLIPTIC FILTERS

In previous sections, we considered all-pole analog filters of the Butterworth and Chebyshev type. In this section, we consider two types of rational filters (their transfer functions have finite zeros). These are the inverse Chebyshev and the elliptic filters. It is possible to improve the magnitude response by adding finite zeros. For example, the elliptic filter is optimum in the sense that for a given order N and for given passband and stopband magnitude constraints, no other filter achieves a faster transition between the passband and stopband. In other words, it has a narrower transition bandwidth.

Inverse Chebyshev Magnitude Response

The magnitude response of an inverse Chebyshev filter is described by the expression

$$|H_a(j\Omega)| = \frac{\epsilon C_N(\Omega_2/\Omega)}{[1 + \epsilon^2 C_N^2(\Omega_2/\Omega)]^{1/2}} \tag{5.66}$$

where $C_N(x)$ is the Chebyshev polynomial defined by Eq. (5.36). The graph of a typical magnitude function for an inverse Chebyshev filter is shown in Fig. 5.15. In the passband, $0 \le \Omega \le \Omega_c = \Omega_1$, the magnitude is maximally flat, and in the stopband, $\Omega_2 \le \Omega$, there are ripples of equal magnitude. The 3-dB cutoff frequency is given by

$$\Omega_c = \frac{\Omega_2}{\cosh\,[(1/N)\cosh^{-1}(1/\epsilon)]} \tag{5.67}$$

The ripple width is seen to be $\epsilon/(1 + \epsilon^2)^{1/2}$. The transition width TW (in rad/s) is $\Omega_2 - \Omega_c$ and can be found from the expression [79]

$$TW = \Omega_c\{\cosh\,[(1/N)\cosh^{-1}(10^{\alpha_2/10} - 1)^{1/2}] - 1\} \tag{5.68}$$

where $\alpha_2 = MSL$, the minimum stopband loss, is given by

$$\alpha_2 = MSL = -20\log A_2 = -10\log \epsilon^2/(1 + \epsilon^2) \tag{5.69}$$

The quantity TW/Ω_c is the *normalized transition* width.

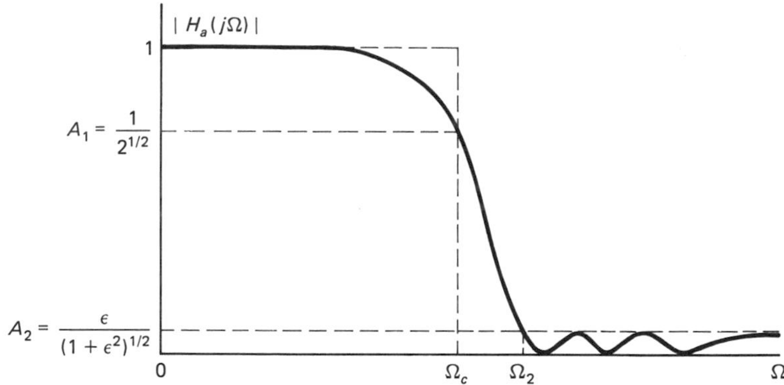

Figure 5.15 $|H_a(j\Omega)|$ for an inverse Chebyshev filter.

The lowest-order N of an inverse Chebyshev filter satisfying the constraints of Fig. 5.15 is the smallest integer satisfying the inequality

$$N \geq \frac{\cosh^{-1}(10^{\alpha_2/10} - 1)^{1/2}}{\cosh^{-1}(\Omega_2/\Omega_c)} \tag{5.70}$$

This is the same order required for a Chebyshev filter that satisfies the same magnitude constraints. This relation should be compared with Eq. (5.61), where $\Omega_c = \Omega_1$, and A_2 and α_2 are related by Eq. (5.69).

Magnitude Response of the Elliptic Filter

The elliptic filter is optimum in that, for a given order and allowable pass-and stopband deviations, it has the shortest transition width. The magnitude response of a fifth-order elliptic filter is shown in Fig. 5.16, where it can be seen that ripples occur in both the pass- and stopbands.

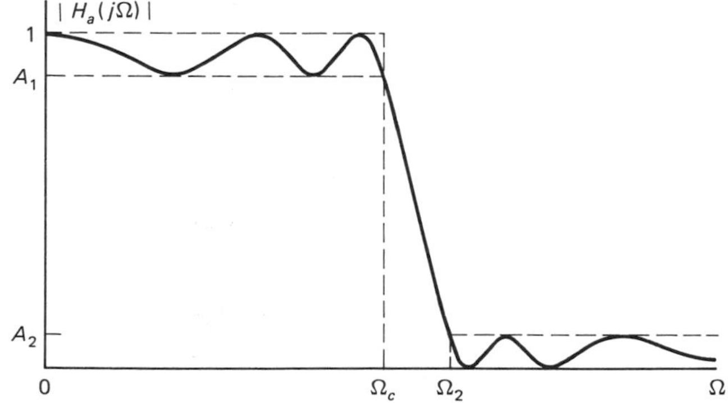

Figure 5.16 Magnitude response of an elliptic filter for $N = 5$.

The passband ripples have equal magnitude and are characterized by the *passband ripple width* (*PRW*), defined by

$$PRW = -20 \log A_1 \text{ (dB)} \tag{5.71}$$

The stopband ripples also have equal magnitude. The *minimum stopband loss* (*MSL*) is the same as for the inverse Chebyshev filter. In terms of the values in Fig. 5.16, we have

$$MSL = -20 \log A_2 \ \text{(dB)} \tag{5.72}$$

The magnitude function is given by

$$|H_a(j\Omega)|^2 = \frac{1}{1 + \epsilon^2 R_N^2(\Omega)} \tag{5.73}$$

where $R_N(\Omega)$ is the *Chebyshev rational function*. As its name implies, it is a rational function and its poles and zeros are related to the *Jacobian elliptic sine functions*. We shall not consider these functions, but the interested reader is referred to [77, 181, 183].

Transfer Functions for Rational Filters

The transfer function for a low-pass analog inverse Chebyshev filter or an elliptic filter with $\Omega_c = 1$ rad/s has the factored form

$$H_a(s) = \prod_{k=1}^{N/2} \frac{B_k(s^2 + a_k)}{s^2 + b_k s + c_k} \tag{5.74}$$

TABLE 5.1 COEFFICIENTS FOR THE TRANSFER FUNCTION OF FOURTH-ORDER INVERSE CHEBYSHEV FILTERS WITH $\Omega_c = 1$ rad/s, AND $N = 4$

MSL	A	B	C
30.0	2.951050	0.630988	1.061509
	17.199978	2.169970	1.512100
35.0	3.719203	0.664072	1.048291
	21.677103	2.091600	1.367632
40.0	4.748478	0.689168	1.037463
	27.676159	2.031494	1.266740
45.0	6.124879	0.708119	1.028812
	35.698408	1.985895	1.195116
50.0	7.963280	0.722391	1.022018
	46.413396	1.951490	1.143607
55.0	10.417060	0.733119	1.016745
	60.715075	1.925605	1.106189
60.0	13.690914	0.741175	1.012691
	79.796493	1.906160	1.078796
65.0	18.057937	0.747222	1.009592
	105.249368	1.891564	1.058624
70.0	23.882409	0.751759	1.007236
	139.196880	1.880612	1.043702
75.0	31.650183	0.755161	1.005450
	184.470785	1.872397	1.032626
80.0	42.009214	0.757714	1.004101
	244.847640	1.866236	1.024385
85.0	55.823606	0.759628	1.003083
	325.363818	1.861615	1.018240

for N even, and

$$H_a(s) = \frac{B_0}{s + c_0} \prod_{k=1}^{(N-1)/2} \frac{B_k(s^2 + a_k)}{s^2 + b_k s + c_k} \tag{5.75}$$

for N odd. The parameters a_k, b_k, and c_k are available in [79] for $N = 2, 3, \ldots, 10$. For the inverse Chebyshev filter, the minimum stopband loss MSL varies from 30 to 100 dB in steps of 5 dB in most cases. For the elliptic filter, the passband ripples that are available are 0.1, 0.5, 1, 2, and 3 dB with minimum stopband losses, in most cases, of 30 to 100 dB in steps of 5 dB. For example, the information for inverse Chebyshev filters for $N = 4$ is shown in Table 5.1 for the normalized case, $\Omega_c = 1$ rad/s. Table 5.2 contains the information for fifth-order elliptic filters with

TABLE 5.2 COEFFICIENTS FOR THE TRANSFER FUNCTION OF FIFTH-ORDER ELLIPTIC FILTERS WITH $\Omega_c = 1$ rad/s, $PRW = 0.5$ dB, AND $N = 5$

MSL	*TW*	*A*	*B*	*C*
30.0	0.1291	1.333847	0.510799	0.732427
		2.421823	0.093582	1.024337
		—	—	0.543429
35.0	0.1929	1.501984	0.535157	0.681055
		2.913728	0.113861	1.026958
		—	—	0.501433
40.0	0.2726	1.722931	0.551409	0.639213
		3.534235	0.132172	1.028998
		—	—	0.470007
45.0	0.3695	2.008726	0.562257	0.605620
		4.316373	0.148211	1.030568
		—	—	0.446162
50.0	0.4847	2.374727	0.569532	0.578860
		5.301775	0.161946	1.031769
		—	—	0.427884
55.0	0.6198	2.840495	0.574452	0.557630
		6.542906	0.173509	1.032685
		—	—	0.413763
60.0	0.7766	3.430868	0.577813	0.540817
		8.105853	0.183118	1.033385
		—	—	0.402789
65.0	0.9572	4.177303	0.580138	0.527513
		10.073844	0.191026	1.033920
		—	—	0.394222
70.0	1.1638	5.119559	0.581766	0.516985
		12.551680	0.197483	1.034331
		—	—	0.387508
75.0	1.3992	6.307818	0.582922	0.508652
		15.671314	0.202725	1.034648
		—	—	0.382232
80.0	1.6664	7.805364	0.583753	0.502053
		19.598876	0.206960	1.034894
		—	—	0.378077
85.0	1.9692	9.691945	0.584359	0.496826
		24.543523	0.210370	1.035085
		—	—	0.374799

$PRW = 0.5$ dB. In the tables A, B, and C represent the values of a_k, b_k, and c_k, respectively, in Eqs. (5.74) and (5.75). In addition to the coefficients for the transfer functions, the normalized transition width TW is given for each case.

The minimum-order N required for an elliptic filter that gives the maximum allowable normalized transition width TW/Ω_c for a specified PRW and MSL has been tabulated [79]. The results for $PRW = 0.5$ dB, MSL from 30 to 100 dB, and $N = 2, 3, \ldots, 10$ are given in Table 5.3.

TABLE 5.3 NORMALIZED TRANSITION WIDTHS TW/Ω_C FOR ELLIPTIC FILTERS WITH $PRW = 0.5$ dB

	N								
MSL	2	3	4	5	6	7	8	9	10
30.0	3.8087	0.9232	0.3244	0.1291	0.0539	0.0230	0.0099		
35.0	5.3829	1.2753	0.4611	0.1929	0.0857	0.0391	0.0180		
40.0	7.4892	1.7115	0.6284	0.2726	0.1270	0.0611	0.0299	0.0147	
45.0		2.2479	0.8298	0.3695	0.1785	0.0897	0.0460	0.0238	
50.0		2.9043	1.0692	0.4847	0.2410	0.1254	0.0668	0.0361	0.0196
55.0		3.7049	1.3518	0.6198	0.3151	0.1687	0.0928	0.0519	0.0292
60.0		4.6793	1.6832	0.7766	0.4014	0.2198	0.1243	0.0715	0.0416
65.0		5.8635	2.0704	0.9572	0.5008	0.2793	0.1615	0.0953	0.0569
70.0		7.3012	2.5213	1.1638	0.6141	0.3476	0.2047	0.1234	0.0754
75.0		9.0454	3.0454	1.3992	0.7425	0.4252	0.2542	0.1560	0.0973
80.0			3.6533	1.6664	0.8870	0.5125	0.3104	0.1934	0.1227
85.0			4.3578	1.9692	1.0490	0.6101	0.3734	0.2357	0.1517
90.0			5.1735	2.3113	1.2299	0.7187	0.4436	0.2831	0.1846
95.0			6.1173	2.6974	1.4314	0.8390	0.5213	0.3359	0.2215
100.0			7.2087	3.1326	1.6553	0.9718	0.6071	0.3941	0.2624

To see how Table 5.1 can be used, let us consider the case where the minimum stopband loss is 40 dB. From Table 5.1, we see that

$$a_1 = 4.748478 \qquad b_1 = 0.689168 \qquad c_1 = 1.037463$$

$$a_2 = 27.676159 \qquad b_2 = 2.031494 \qquad c_2 = 1.266740$$

and, consequently, the analog transfer function for a fourth-order inverse Chebyshev filter, with an arbitrary cutoff frequency Ω_c, is

$$H_a(s) = \frac{B_1(s^2 + 4.748478\Omega_c^2)}{s^2 + 0.689168\Omega_c s + 1.037463\Omega_c^2} \times \frac{B_2(s^2 + 27.676159\Omega_c^2)}{s^2 + 2.031494\Omega_c s + 1.266740\Omega_c^2} \tag{5.76}$$

By using this analog transfer function, a digital transfer function can be obtained using the bilinear transformation or the impulse-invariant technique. The reader is asked to apply the bilinear transformation in Ex. 5.8.2.

EXERCISES

5.8.1 Determine the analog transfer function of a fourth-order inverse Chebyshev filter with $\Omega_c = 1$ rad/s, $MSL = 50$ dB, and B_1 and B_2 in Eq. (5.74) chosen so that $|H_a(j0)| = 1$. Also determine the transition width TW.

5.8.2 Obtain a digital transfer function from Eq. (5.76) using the bilinear transformation with $\Omega_c = 1$ rad/s and $T = 1$ s. Use Eqs. (5.24) and (5.25).

5.8.3 If $A_1 = 0.9$ and $A_2 = 0.1$ in Fig. 5.16, determine PRW and MSL.

5.8.4 Determine the minimum-order N of an elliptic filter if $PRW = 0.5$ dB, $MSL = 60$ dB, and $TW/\Omega_c \leq 0.5$.

5.8.5 Determine the transfer function of an analog elliptic filter with $N = 5$, $PRW = 0.5$ dB, $MSL = 50$ dB, and $\Omega_c = 2$ rad/s.

5.9 FREQUENCY TRANSFORMATIONS

The four basic types of frequency-selective filters are low-pass, high-pass, bandpass, and band-reject. Their frequency responses for the ideal cases are shown as solid lines in Fig. 5.17. These ideal responses are not obtainable in practice because of the sharp corners. Typical realizable responses are shown also in Fig. 5.17 as dashed lines. More complicated filters can be designed having more passbands and stopbands than the responses of Fig. 5.17, but we shall not consider them here. The interested reader should consult [29].

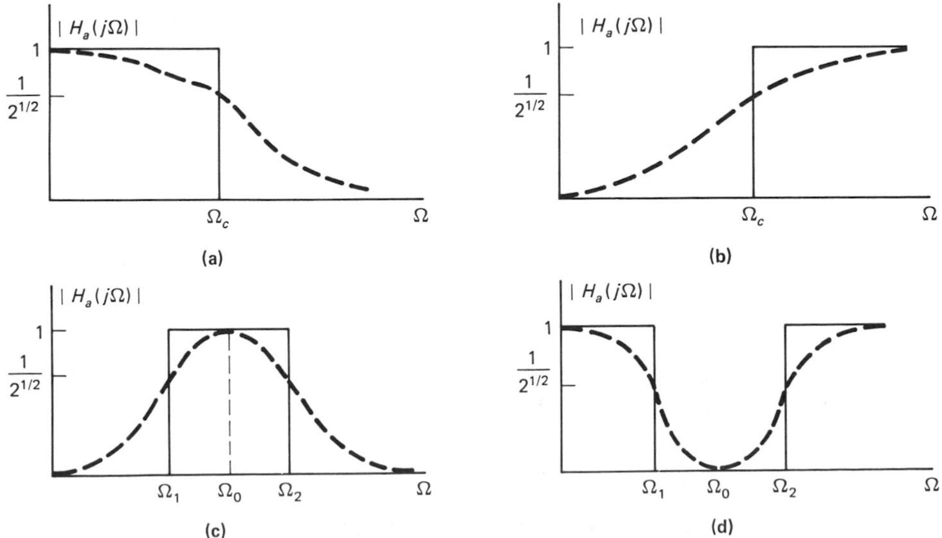

Figure 5.17 Ideal (solid line) and practical (dashed line) magnitude responses of (a) a low-pass, (b) a high-pass, (c) a bandpass, and (d) a band-reject filter.

Many filters are designed by first designing a low-pass filter and then using a frequency transformation to obtain the desired filter. In the case of digital filters, we can first design a suitable low-pass analog filter and then use a frequency transformation to obtain another analog filter of the desired type. The digital filter can then be determined from the final analog filter using the bilinear transformation, impulse invariance, or any other suitable method. Another procedure that could be used is to first design a low-pass digital filter and then use a frequency transformation to obtain the desired digital filter. In the first case, we are transforming the analog frequency, and in the second case, we are transforming the digital frequency.

Analog Frequency Transformations

The frequency transformations that can be used to obtain a low-pass, high-pass, bandpass, or band-reject analog filter from a normalized low-pass analog filter (cutoff or center frequency is 1 rad/s) are shown in Table 5.4. For low-pass and high-pass filters, Ω_c is the cutoff frequency (either the 3-dB frequency or the end or beginning of the passband ripple channel). For bandpass and band-reject filters, Ω_0 is the *center frequency* and Q is the *quality factor* given by

$$\Omega_0^2 = \Omega_1\Omega_2$$

$$Q = \frac{\Omega_0}{\Omega_2 - \Omega_1}$$

where Ω_1 and Ω_2 are the lower and upper cutoff frequencies, respectively. The quantity $\Omega_2 - \Omega_1$ is the *bandwidth*.

TABLE 5.4

Filter Type	Transformation	
Low-pass	$S = s/\Omega_c$	(5.77)
High-pass	$S = \Omega_c/s$	(5.78)
Bandpass	$S = \dfrac{Q(s^2 + \Omega_0^2)}{\Omega_0 s}$	(5.79)
Band-reject	$S = \dfrac{\Omega_0 s}{Q(s^2 + \Omega_0^2)}$	(5.80)

As a simple example, consider a first-order section of a normalized low-pass filter with transfer function

$$H_a(S) = \frac{Bc}{S + c} \tag{5.81}$$

If we desire a bandpass filter with center frequency Ω_0 and quality factor Q, we use the bandpass transformation of Eq. (5.79) of Table 5.4. The resulting transfer function is

$$\begin{aligned} H_{aBP}(s) &= \frac{Bc}{Q(s^2 + \Omega_0^2)/\Omega_0 s + c} \\ &= \frac{Bc(\Omega_0/Q)s}{s^2 + c(\Omega_0/Q)s + \Omega_0^2} \end{aligned} \tag{5.82}$$

If we apply the transformations of Eqs. (5.77), (5.78), and (5.80) to the first-order low-pass transfer function of Eq. (5.81), then we obtain another first-order low-pass function, a first-order high-pass function, and a second-order band-reject function, respectively.

Filter Design Using Frequency Transformations

In order to design a filter using frequency transformations, we start with the transfer function given in Eq. (5.31) or (5.32) for Butterworth and Chebyshev low-pass filters or in Eqs. (5.74) and (5.75) for inverse Chebyshev and elliptic filters. In these transfer functions, there is a first-order section of the form of Eq. (5.81) and two types of second-order sections given by

$$H_a(S) = \frac{Bc}{S^2 + bS + c} \tag{5.83}$$

for Butterworth and Chebyshev filters, and by

$$H_a(S) = \frac{B(S^2 + a)}{S^2 + bS + c} \tag{5.84}$$

for inverse Chebyshev and elliptic filters.

High-pass filters can be obtained rather easily by substituting Eq. (5.78) into Eq. (5.81) for the first-order section and Eq. (5.83) or Eq. (5.84) for the second-order sections. The reader is asked to do this in Ex. 5.9.2. The situation is somewhat more complicated for bandpass and band-reject filters. This complication arises since second-degree polynomials in S are converted into fourth-degree polynomials in s when the transformations of Eqs. (5.79) and (5.80) are used. Consequently, a second-order low-pass section gives rise to two second-order bandpass or band-reject sections. These second-order functions are given in Table 5.5 for Butterworth, Chebyshev, inverse Chebyshev, and elliptic filters. The parameters that appear in these functions are also given there. In each case, we have $H_a(s) = H_{a1}(s)H_{a2}(s)$, where the gains satisfy the relation $B = B_1B_2$.

Examples Utilizing a Third-Order Butterworth Function

To illustrate the use of the frequency transformations, consider the third-order low-pass Butterworth-filter transfer function of Eq. (5.18). We will obtain corresponding high-pass, bandpass, and band-reject filters. Beginning with the low-pass transfer function

$$H_a(S) = \frac{1}{(S + 1)(S^2 + S + 1)}$$

TABLE 5.5 TRANSFER FUNCTIONS

	$H_{a1}(s)$	$H_{a2}(s)$	
Bandpass			
Butterworth Chebyshev	$\dfrac{(B_1\Omega_0 c^{1/2}/Q)\,s}{s^2 + (D\Omega_0/E)s + D^2\Omega_0^2}$	$\dfrac{(B_2\Omega_0 c^{1/2}/Q)\,s}{s^2 + (\Omega_0/DE)s + \Omega_0^2/D^2}$	(5.85)
Inverse Chebyshev Elliptic	$\dfrac{B_1(c/a)^{1/2}(s^2 + A_1\Omega_0^2)}{s^2 + (D\Omega_0/E)\,s + D^2\Omega_0^2}$	$\dfrac{B_2(c/a)^{1/2}(s^2 + \Omega_0^2/A_1)}{s^2 + (\Omega_0/DE)\,s + \Omega_0^2/D^2}$	(5.86)
Band-reject			
Butterworth Chebyshev	$\dfrac{B_1(s^2 + \Omega_0^2)}{s^2 + (D_1\Omega_0/E_1)\,s + D_1^2\Omega_0^2}$	$\dfrac{B_2(s^2 + \Omega_0^2)}{s^2 + (\Omega_0/D_1E_1)\,s + \Omega_0^2/D_1^2}$	(5.87)
Inverse Chebyshev Elliptic	$\dfrac{B_1(s^2 + A_2\Omega_0^2)}{s^2 + (D_1\Omega_0/E_1)\,s + D_1^2\Omega_0^2}$	$\dfrac{B_2(s^2 + \Omega_0^2/A_2)}{s^2 + (\Omega_0/D_1E_1)\,s + \Omega_0^2/D_1^2}$	(5.88)

Parameters

$$E = \frac{1}{b}\left\{\frac{c + 4Q^2 + [(c + 4Q^2)^2 - (2bQ)^2]^{1/2}}{2}\right\}^{1/2} \tag{5.89}$$

$$D = \tfrac{1}{2}\{bE/Q + [(bE/Q)^2 - 4]^{1/2}\} \tag{5.90}$$

$$A_1 = 1 + \frac{1}{2Q^2}[a + (a^2 + 4aQ^2)^{1/2}] \tag{5.91}$$

$$E_1 = \frac{1}{b}\left(\frac{c}{2}\{1 + 4cQ^2 + [(1 + 4cQ^2)^2 - (2bQ)^2]^{1/2}\}\right)^{1/2} \tag{5.92}$$

$$D_1 = \tfrac{1}{2}\{bE_1/cQ + [(bE_1/cQ)^2 - 4]^{1/2}\} \tag{5.93}$$

$$A_2 = 1 + \frac{1}{2aQ^2}[1 + (1 + 4aQ^2)^{1/2}] \tag{5.94}$$

we apply the transformation $S = \Omega_c/s$ of Eq. (5.78) and obtain the high-pass filter function

$$H_{aHP}(s) = \frac{s^3}{(s + \Omega_c)(s^2 + \Omega_c s + \Omega_c^2)} \tag{5.95}$$

The bandpass transfer function is obtained by applying the transformation $S = Q(s^2 + \Omega_0^2)/\Omega_0 s$ of Eq. (5.79). The factored form can be obtained by using the results of Eqs. (5.82) and (5.85). The resulting bandpass transfer function is then

$$H_{aBP}(s) = \frac{(\Omega_0/Q)s}{s^2 + (\Omega_0/Q)s + \Omega_0^2}\,\frac{(\Omega_0/Q)s}{s^2 + (D\Omega_0/E)s + D^2\Omega_0^2} \times \frac{(\Omega_0/Q)s}{s^2 + (\Omega_0/DE)s + \Omega_0^2/D^2} \tag{5.96}$$

where E and D, given by Eqs. (5.89) and (5.90) for $b = c = 1$, are

$$E = \left\{\frac{1 + 4Q^2 + [(1 + 4Q^2)^2 - 4Q^2]^{1/2}}{2}\right\}^{1/2}$$

$$D = \tfrac{1}{2}\{E/Q + [(E/Q)^2 - 4]^{1/2}\}$$

The band-reject filter is obtained from Eq. (5.18) by using the transformation $S = \Omega_0 s/Q(s^2 + \Omega_0^2)$ of Eq. (5.80). Making use of Eq. (5.87), we obtain

$$H_{aBR}(s) = \frac{s^2 + \Omega_0^2}{s^2 + (\Omega_0/Q)s + \Omega_0^2} \frac{s^2 + \Omega_0^2}{s^2 + (D_1\Omega_0/E_1)s + D_1^2\Omega_0^2} \times \frac{s^2 + \Omega_0^2}{s^2 + (\Omega_0/D_1E_1)s + \Omega_0^2/D_1^2} \tag{5.97}$$

where E_1 and D_1 are given by Eqs. (5.92) and (5.93). For the case where $b = c = 1$, it can be seen that $E_1 = E$ and $D_1 = D$, which are the values given for the bandpass filter.

Digital Frequency Transformations

Frequency transformations are available also for transforming a low-pass digital filter into another low-pass digital filter, a high-pass filter, a bandpass filter, or a band-reject filter. These transformations are shown in Table 5.6 along with formulas for the design parameters [24, 126, 136]. In each case, it is assumed that the cutoff frequency of the original low-pass digital filter is ω_c' and that the cutoff frequency of the resulting low-pass or high-pass filter is ω_c. The upper and lower cutoff frequencies for the derived bandpass and band-reject filters are denoted by ω_2 and ω_1, respectively, and the center frequency is ω_0. We assume that Z^{-1} is the variable associated with the original low-pass filter and it is replaced by a function of z^{-1}, $Z^{-1} = f(z^{-1})$, as indicated by the transformations given in Table 5.6.

Derivation of Low-Pass to High-Pass Transformation

To see how these transformations are obtained, we develop the frequency relation for the low-pass to high-pass transformation given in Eq. (5.99) in Table 5.6. Denoting the frequency of the low-pass filter by ω' and the frequency of the high-pass filter by ω, we see from Fig. 5.17 that we should have the following mappings:

$$\omega' = 0 \longrightarrow \omega = \pm\pi$$

$$\omega' = \pm\pi \longrightarrow \omega = 0$$

$$\omega' = \pm\omega_c' \longrightarrow \omega = \mp\omega_c$$

If Z and z are used as the frequency-domain variables for the low-pass and the high-pass filter, respectively, we can obtain a one-to-one mapping for these variables by using the bilinear transformation

$$Z^{-1} = \frac{az^{-1} + b}{cz^{-1} + d}$$

The parameters in this transformation are determined to satisfy the frequency

TABLE 5.6

Filter Type	Transformation	Design Parameters	
Low-pass	$\dfrac{z^{-1}-\alpha}{1-\alpha z^{-1}}$	$\alpha = \dfrac{\sin\left(\dfrac{\omega_c' - \omega_c}{2}\right)}{\sin\left(\dfrac{\omega_c' + \omega_c}{2}\right)}$	(5.98)
High-pass	$-\dfrac{z^{-1}+\alpha}{1+\alpha z^{-1}}$	$\alpha = -\dfrac{\cos\left(\dfrac{\omega_c' + \omega_c}{2}\right)}{\cos\left(\dfrac{\omega_c' - \omega_c}{2}\right)}$	(5.99)
Bandpass	$-\dfrac{z^{-2} - \dfrac{2\alpha k}{k+1} z^{-1} + \dfrac{k-1}{k+1}}{\dfrac{k-1}{k+1} z^{-2} - \dfrac{2\alpha k}{k+1} z^{-1} + 1}$	$\alpha = \dfrac{\cos\left(\dfrac{\omega_2 + \omega_1}{2}\right)}{\cos\left(\dfrac{\omega_2 - \omega_1}{2}\right)} = \cos\omega_0$ $k = \cot\left(\dfrac{\omega_2 - \omega_1}{2}\right)\tan\dfrac{\omega_c'}{2}$	(5.100)
Band-reject	$\dfrac{z^{-2} - \dfrac{2\alpha}{k+1} z^{-1} + \dfrac{1-k}{1+k}}{\dfrac{1-k}{k+1} z^{-2} - \dfrac{2\alpha}{k+1} z^{-1} + 1}$	$\alpha = \dfrac{\cos\left(\dfrac{\omega_2 + \omega_1}{2}\right)}{\cos\left(\dfrac{\omega_2 - \omega_1}{2}\right)} = \cos\omega_0$ $k = \tan\left(\dfrac{\omega_2 - \omega_1}{2}\right)\tan\dfrac{\omega_c'}{2}$	(5.101)

transformations just given. This requires that $Z^{-1} = e^{-j\omega'} = e^{-j0} = 1$ maps onto $z^{-1} = e^{-j\omega} = e^{j\pi} = -1$, $Z^{-1} = e^{j\pi} = -1$ maps onto $z^{-1} = e^{j0} = 1$ and $Z^{-1} = e^{j\omega_c'}$ maps onto $z^{-1} = e^{-j\omega_c}$. The corresponding relations between parameters a, b, c, and d for these mappings are

$$1 = \frac{-a+b}{-c+d}$$

$$-1 = \frac{a+b}{c+d}$$

$$e^{j\omega_c'} = \frac{ae^{-j\omega_c}+b}{ce^{-j\omega_c}+d}$$

The first two relations can be written

$$-c + d = -a + b$$

$$-c - d = a + b$$

By adding these equations, we obtain $c = -b$, and by subtracting, we obtain $d = -a$. Substituting these values into the last of the mapping relations, we obtain

$$e^{j\omega_c'} = -\frac{ae^{-j\omega_c} + b}{be^{-j\omega_c} + a}$$

$$= -\frac{e^{-j\omega_c} + \alpha}{\alpha e^{-j\omega_c} + 1}$$

where $\alpha = b/a$. Clearing this expression of fractions and solving for α, we find that

$$\alpha = -\frac{e^{j\omega_c'} + e^{-j\omega_c}}{1 + e^{j(\omega_c' - \omega_c)}}$$

$$= -\frac{e^{j(\omega_c'-\omega_c)/2}\, e^{j(\omega_c'+\omega_c)/2} + e^{-j(\omega_c'+\omega_c)/2}}{e^{j(\omega_c'-\omega_c)/2}\, e^{j(\omega_c'-\omega_c)/2} + e^{-j(\omega_c'-\omega_c)/2}}$$

$$= -\frac{\cos\,[(\omega_c' + \omega_c/2]}{\cos\,[(\omega_c' - \omega_c)/2]}$$

Using the parameter values just obtained results in the transformation of Eq. (5.99) given in Table 5.6.

Examples

We now consider some simple examples illustrating these transformations. The first-order normalized analog function

$$H_a(S) = \frac{1}{S + 1} \tag{5.102}$$

is used to obtain other analog filters as well as digital filters. Using the bilinear transformation with $T = 1$, we obtain the digital transfer function

$$H(Z) = \frac{1 + Z^{-1}}{3 - Z^{-1}} \tag{5.103}$$

with cutoff frequency $\omega_c' = 2\tan^{-1}(\Omega_c T/2) = 2\tan^{-1} 0.5 = 0.9273$ rad/s.

First, let us apply the digital frequency transformation of Eq. (5.99) to Eq. (5.103) to obtain a high-pass filter. Substituting

$$Z^{-1} = -\frac{z^{-1} + \alpha}{1 + \alpha z^{-1}}$$

into (5.103) results in

$$H_{HP}(z) = \frac{1 - \alpha}{3 + \alpha}\,\frac{1 - z^{-1}}{1 + \dfrac{1 + 3\alpha}{3 + \alpha} z^{-1}} \tag{5.104}$$

where

$$\alpha = -\frac{\cos\left(\dfrac{\omega_c' + \omega_c}{2}\right)}{\cos\left(\dfrac{\omega_c' - \omega_c}{2}\right)}$$

If we take $\omega_c = 2$ rad/s as the cutoff frequency, then

$$\alpha = -\frac{\cos\left(\dfrac{0.9273 + 2}{2}\right)}{\cos\left(\dfrac{0.9273 - 2}{2}\right)} = -0.1244$$

and

$$H_{HP}(z) = 0.391\frac{1 - z^{-1}}{1 + 0.218z^{-1}}$$

It should be noted that $H_{HP}(e^{j0}) = H_{HP}(1) = 0$ and $H_{HP}(e^{j\pi}) = H_{HP}(-1) = 1$ from Eq. (5.104), as is expected for a high-pass filter.

To illustrate the other procedure for finding a high-pass digital filter, we start with the low-pass analog filter of Eq. (5.102) and first apply the transformation of Eq. (5.78) to obtain a high-pass analog filter. We then apply the bilinear transformation to the high-pass analog transfer function and obtain the desired high-pass digital-filter transfer function. Applying Eq. (5.78) to Eq. (5.102) results in the high-pass function

$$H_{aHP}(s) = \frac{1}{\Omega_c/s+1} = \frac{s}{s + \Omega_c}$$

The resulting high-pass digital transfer function is given by

$$H_{HP}(z) = \frac{\dfrac{2}{T}\dfrac{1 - z^{-1}}{1 + z^{-1}}}{\dfrac{2}{T}\dfrac{1 - z^{-1}}{1 + z^{-1}} + \Omega_c}$$

$$= \frac{1 - z^{-1}}{1 + \Omega_c T/2 - (1 - \Omega_c T/2)z^{-1}}$$

where the digital cutoff frequency satisfies the relation $\Omega_c T/2 = \tan(\omega_c/2)$. If we choose $\omega_c = 2$ rad/s, then $\Omega_c T/2 = \tan 1 = 1.5574$, and it follows that

$$H_{HP}(z) = \frac{1 - z^{-1}}{2.5574 + 0.5574z^{-1}}$$

$$= 0.391\frac{1 - z^{-1}}{1 + 0.218z^{-1}}$$

This is the same transfer function that we obtained earlier.

As a final example in this section, let us obtain a band-reject digital-filter function from the low-pass digital transfer function of Eq. (5.103) by using the frequency transformation of Eq. (5.101). The band-reject digital transfer function is then

$$H_{BR}(z) = \frac{1 + \dfrac{z^{-2} - \dfrac{2\alpha}{k+1}z^{-1} + \dfrac{1-k}{k+1}}{\dfrac{1-k}{k+1}z^{-2} - \dfrac{2\alpha}{k+1}z^{-1} + 1}}{3 - \dfrac{z^{-2} - \dfrac{2\alpha}{k+1}z^{-1} + \dfrac{1-k}{k+1}}{\dfrac{1-k}{k+1}z^{-2} - \dfrac{2\alpha}{k+1}z^{-1} + 1}}$$

If we take the cutoff frequencies to be $\omega_1 = \pi/3$ and $\omega_2 = 2\pi/3$, then from Eq. (5.101) we have

$$\begin{aligned} k &= \tan\left(\frac{\omega_2 - \omega_1}{2}\right)\tan\frac{\omega_c'}{2} = \tan\frac{\pi}{6}\tan\frac{0.9273}{2} \\ &= 0.2887 \end{aligned}$$

$$\alpha = \frac{\cos\left(\dfrac{\omega_2 + \omega_1}{2}\right)}{\cos\left(\dfrac{\omega_2 - \omega_1}{2}\right)} = 0$$

Substitution of these values into the expression for $H_{BR}(z)$ results in

$$H_{BR}(z) = \frac{0.634(1 + z^{-2})}{1 + 0.268z^{-2}}$$

The values for $\omega = 0$ and $\omega = \pi$ are given by $H_{BR}(1) = H_{BR}(-1) = 1$, as they should be for a band-reject filter. The center frequency is at $\omega = \pi/2$, as can be seen from $H_{BR}(j1) = 0$.

EXERCISES

5.9.1 A normalized low-pass 1-dB Chebyshev filter has transfer function

$$H_a(s) = \frac{1.102510}{s^2 + 1.097734s + 1.102510}$$

Obtain a high-pass filter with $\Omega_c = 2$ rad/s.

5.9.2 Obtain the high-pass filter sections from Eqs. (5.81), (5.83), and (5.84).

5.9.3 Given the low-pass digital-filter function

$$H(z) = \frac{1}{2}\frac{1 + z^{-1}}{2 - z^{-1}}$$

show that the cutoff frequency is $\omega_c' = 0.6435$ rad/s. Use a low-pass transformation to obtain another low-pass filter with $\omega_c = 1$ rad/s.

5.9.4 Apply a transformation to $H(z)$ of Ex. 5.9.3 to obtain a high-pass filter with $\omega_c = 2$ rad/s.

5.9.5 Obtain a bandpass filter from $H(z)$ in Ex. 5.9.3 with $\omega_1 = \pi/4$ rad/s and $\omega_2 = 3\pi/4$ rad/s.

5.10 PHASE MODIFICATION

In this section, we consider filters that are designed specifically to modify the phase response of a digital system. We begin the discussion with all-pass filters and then consider filters whose time delay is approximately constant.

All-Pass Filters: First-Order Functions

An *all-pass* filter is one whose amplitude response is constant (flat) for all frequencies. These filters are frequently used for phase and time-delay improvement. We begin the discussion with the first-order function

$$H(z) = K_1 \frac{1 + bz^{-1}}{1 + az^{-1}} \qquad a \neq b \tag{5.105}$$

The square of the amplitude response can be written

$$\begin{aligned} |H(e^{j\omega})|^2 &= |K_1|^2 \frac{1 + be^{-j\omega}}{1 + ae^{-j\omega}} \frac{1 + b^*e^{j\omega}}{1 + a^*e^{j\omega}} \\ &= |K_1 b|^2 \frac{1 + (b^*)^{-1}e^{-j\omega} + b^{-1}e^{j\omega} + (bb^*)^{-1}}{1 + ae^{-j\omega} + a^*e^{j\omega} + aa^*} \end{aligned}$$

It can be seen from this relation that $|H(e^{j\omega})|^2 = |K_1 b|^2$ if $a = 1/b^*$, or $b = 1/a^*$. In terms of the pole and zero of $H(z)$, we see that the zero at $z = -b$ is the reciprocal of the complex conjugate of the pole at $z = -a$.

Phase and Time Delay

An all-pass filter can be used to modify the phase response without affecting the magnitude response. To see how this can be accomplished, we determine the phase response and time delay for the first-order case of Eq. (5.105) for a real and $b = 1/a$. If we write

$$H(e^{j\omega}) = \frac{K_1}{a} \frac{a + e^{-j\omega}}{1 + ae^{-j\omega}} \tag{5.106}$$

and take $K = K_1/a > 0$, then the phase response is given by

$$\phi(\omega) = \tan^{-1} \frac{a \sin \omega}{1 + a \cos \omega} - \tan^{-1} \frac{\sin \omega}{a + \cos \omega} \tag{5.107}$$

A simpler form for the phase is obtained if we write

$$H(e^{j\omega}) = Ke^{-j\omega}\frac{1 + ae^{j\omega}}{1 + ae^{-j\omega}} \tag{5.108}$$

From this representation, the phase has the form

$$\phi(\omega) = -\omega + 2\tan^{-1}\frac{a\sin\omega}{1 + a\cos\omega} \tag{5.109}$$

The time delay, computed from either Eq. (5.107) or (5.109), is given by

$$\tau(\omega) = -\frac{d\phi}{d\omega} = \frac{1 - a^2}{1 + a^2 + 2a\cos\omega} \tag{5.110}$$

Graphs of phase and time-delay responses of $H(e^{j\omega})$ for $K > 0$ are shown in Fig. 5.18 for several values of a.

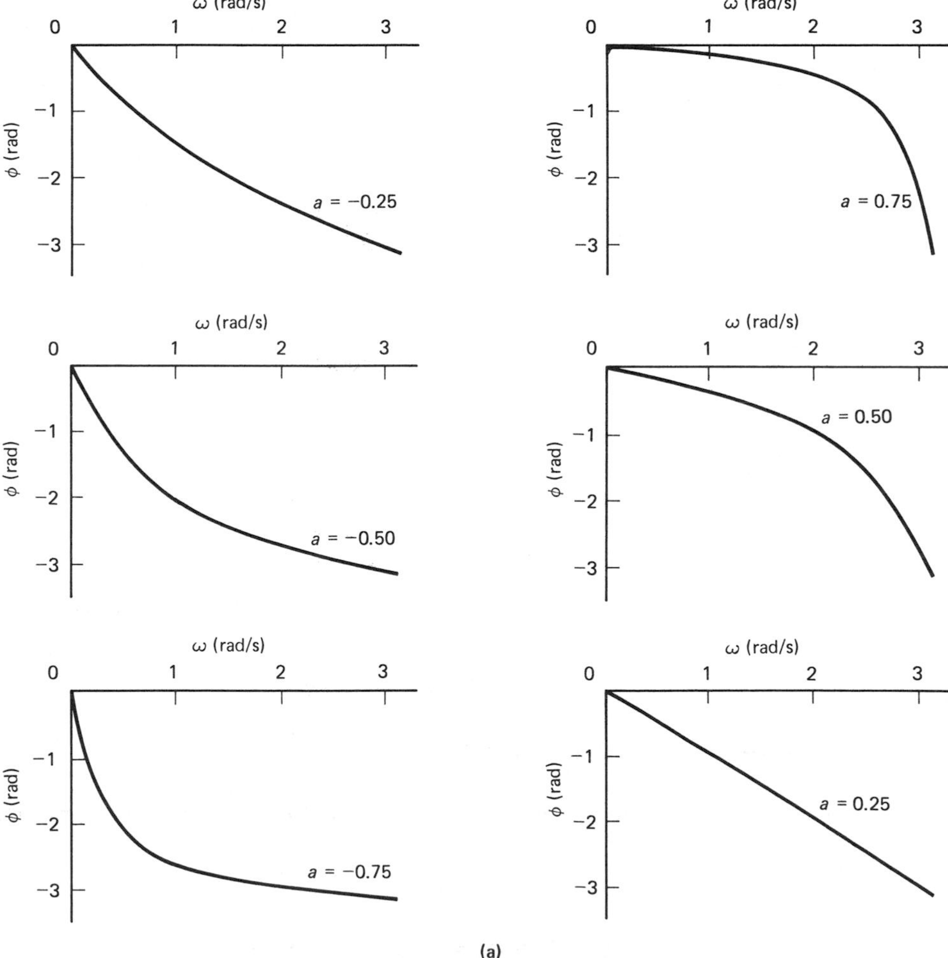

Figure 5.18 (a) Phase and (b) time-delay responses for $H(e^{j\omega})$ in Eq. (5.108) for various values of a.

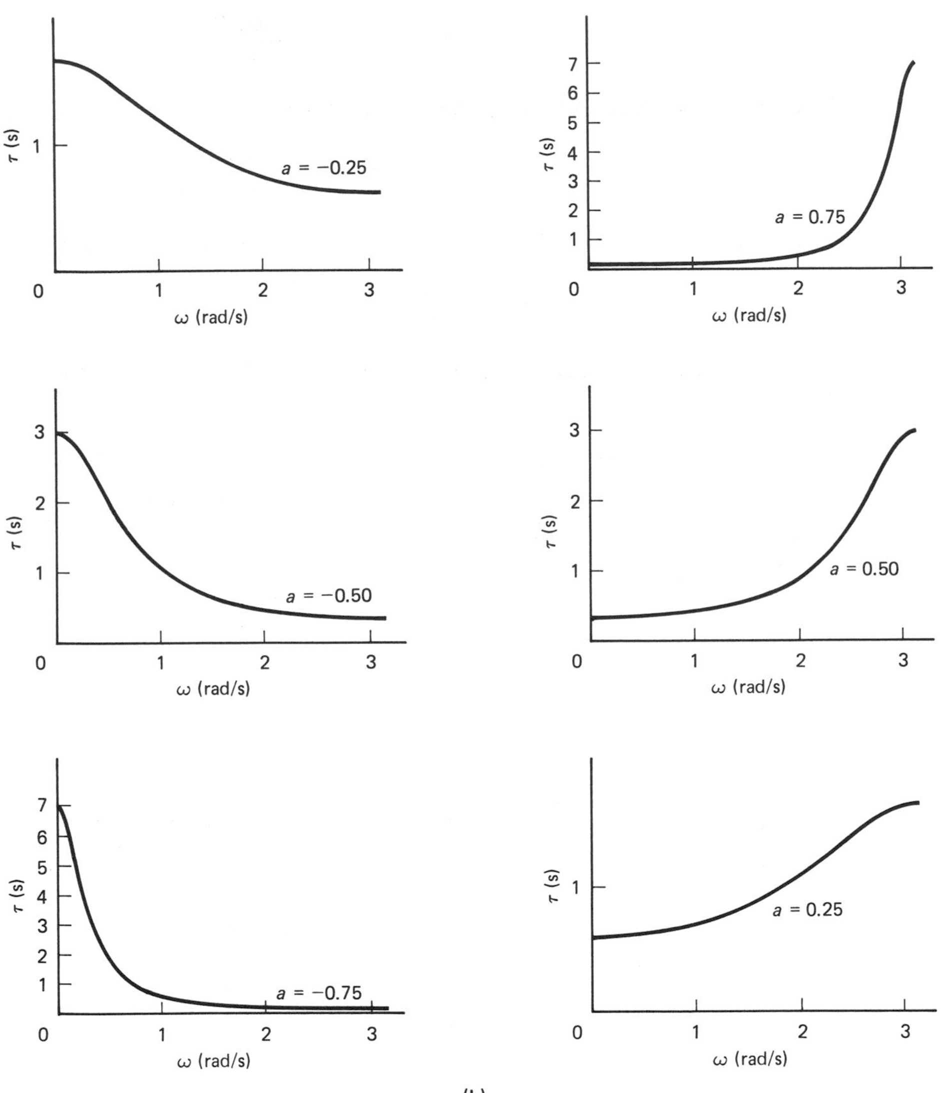

Figure 5.18 *(continued)*

Determination of the Parameter *a*

We may be interested in determining a value of a that will give a particular phase at a given frequency. For example, if we want $\phi(\omega_0) = \phi_0$, then we must have, from Eq. (5.109),

$$\phi(\omega_0) = \phi_0 = -\omega_0 + 2 \tan^{-1} \frac{a \sin \omega_0}{1 + a \cos \omega_0}$$

When this equation is solved for a, we obtain

$$a = \frac{\tan\left(\dfrac{\phi_0 + \omega_0}{2}\right)}{\sin\omega_0 - \cos\omega_0 \tan\left(\dfrac{\phi_0 + \omega_0}{2}\right)} \tag{5.111}$$

General All-Pass Filters

In the general case, the transfer function of an all-pass filter has the form

$$H(z) = K\frac{z^{-N}Q(z^{-1})}{Q(z)} \tag{5.112}$$

where

$$Q(z) = 1 + a_1 z^{-1} + a_2 z^{-2} + \cdots + a_N z^{-N}$$

To see that this is an all-pass function, we note that

$$|H(e^{j\omega})| = |K|\frac{|e^{-j\omega}|^N |Q(e^{-j\omega})|}{|Q(e^{j\omega})|}$$

Since $|Q(e^{-j\omega})| = |Q^*(e^{j\omega})| = |Q(e^{j\omega})|$, it follows that $|H(e^{j\omega})| = |K|$ and Eq. (5.112) is an all-pass function. To get a better idea of the form of Eq. (5.112), we note that

$$\begin{aligned} z^{-N}Q(z^{-1}) &= z^{-N}(1 + a_1 z + a_2 z^2 + \cdots + a_N z^N) \\ &= a_N + a_{N-1}z^{-1} + \cdots + a_1 z^{-(N-1)} + z^{-N} \end{aligned}$$

and thus has the same coefficients as $Q(z)$ but in reverse order. If $K > 0$, then the phase response of Eq. (5.112) has the form

$$\phi(\omega) = -N\omega - 2\arg Q(e^{j\omega}) \tag{5.113}$$

(See [76] for additional design information.)

Bessel Filters

Linear phase is a desirable property for a filter. This cannot be achieved with an analog filter obtained from resistors, capacitors, inductors, and op amps, but, as we have seen, it can be achieved for a digital filter. For linear phase, we have

$$\phi(\omega) = -\omega\tau_0$$

where τ_0 is a constant. For this case, the time delay is given by

$$\tau(\omega) = -d\phi/d\omega = \tau_0$$

so that linear phase is accompanied by constant time delay. (Here ω could be either the digital- or analog-system frequency.)

There have been several techniques developed for obtaining analog filters in which the phase is approximately linear or the time delay is approximately constant. The Bessel filter [77, 181, 183] is an all-pole filter that provides a maximally flat time delay; that is, for a filter of order N, the first $2N - 1$ derivatives of $\tau(\omega)$ are zero when $\omega = 0$. The transfer function of an analog Bessel filter of order N is given by

$$H_a(s) = \frac{B_N(0)}{B_N(s/\Omega_c)} \tag{5.114}$$

where $B_N(s)$ is the Bessel polynomial of degree N defined by

$$B_N(s) = \sum_{k=0}^{N} \frac{(2N - k)!\, s^k}{2^{N-k} k!\, (N - k)!} \tag{5.115}$$

The amplitude response of a Bessel filter decreases monotonically from its maximum value at $\Omega = 0$. For $N \geq 3$, the cutoff frequency is given approximately by

$$\Omega_{3\,\text{dB}} = [0.69315(2N - 1)]^{1/2} \tag{5.116}$$

If a specific time delay τ_c is desired at Ω_c, $\tau_c = \tau_a(\Omega_c)$, then we have approximately

$$\Omega_c = 1/\tau_c \text{ (rad/s)} \tag{5.117}$$

or

$$f_c = \frac{1}{2\pi\tau_c} = \frac{0.15915}{\tau_c} \text{ Hz}$$

Example of an Analog Bessel Filter

As an example, consider a second-order analog Bessel filter with

$$H_a(s) = \frac{3}{s^2 + 3s + 3} \tag{5.118}$$

The phase response is found to be

$$\phi_a(\Omega) = -\tan^{-1} \frac{3\Omega}{3 - \Omega^2}$$

and the time delay is given by

$$\begin{aligned}\tau_a(\Omega) &= -\frac{d}{d\Omega}\left(-\tan^{-1} \frac{3\Omega}{3 - \Omega^2}\right) \\ &= \frac{9 + 3\Omega^2}{9 + 3\Omega^2 + \Omega^4} = 1 - \frac{1}{9}\Omega^4 + \frac{1}{27}\Omega^6 + \cdots +\end{aligned}$$

For small values of Ω, we see that Ω^4 is small and so $\tau_a(\Omega) \simeq 1$.

In order to see how nearly constant is the time delay, we find that

$$N = 2\text{: } \Omega_c \tau_a(\Omega_c) = 12/13 = 0.92308$$

$$N = 3\text{: } \Omega_c \tau_a(\Omega_c) = 276/277 = 0.99639$$

$$N = 4\text{: } \Omega_c \tau_a(\Omega_c) = 12{,}745/12{,}746 = 0.99992$$

This is also illustrated in the graphs in Fig. 5.19, showing the Bessel phase responses for $N = 2$, 3, 4, and 5.

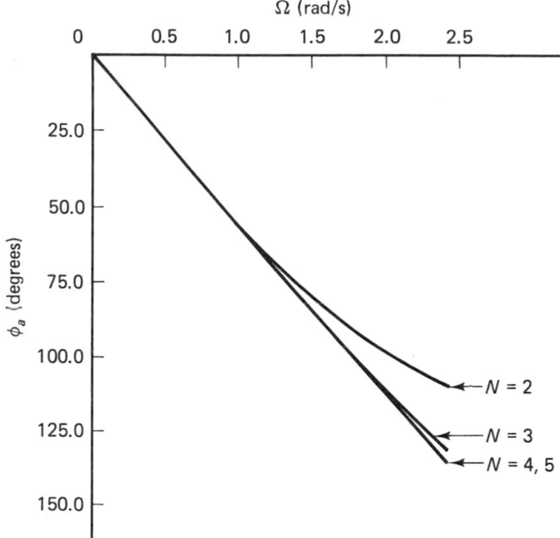

Figure 5.19 Analog Bessel phase responses.

Example of a Digital Bessel Filter

Now that we are familiar with analog Bessel filters, let us now obtain digital filters by applying the bilinear transformation to the Bessel function of Eq. (5.114). Factorizations of the Bessel polynomials are available (see for example, [79]) and can be used to obtain $H_a(s)$ in the same form as Eq. (5.31) or (5.32). Then we can use Eqs. (5.22)–(5.26) to obtain $H(z)$.

To illustrate the above procedure, let us consider a second-order Bessel filter given by Eq. (5.118) with digital cutoff frequency $\omega_c = 1$ rad/s. Using the bilinear transformation results in the digital function

$$H(z) = \frac{0.25333(1 + z^{-1})^2}{1 - 0.059194z^{-1} + 0.072559z^{-2}}$$

This last expression for $H(z)$ is obtained from Eqs. (5.24)–(5.26), where $\Omega_c T/2 = \tan(\omega_c/2) = \tan 0.5 = 0.546302$ and

$$\begin{aligned} b_{10} &= \gamma_1(\Omega_c T/2)^2 = 3(0.546302)^2 \\ &= 0.895338 \\ a_{10} &= 1 + b_1(\Omega_c T/2) + c_1(\Omega_c T/2)^2 \\ &= 1 + 3(0.546302) + 3(0.546302)^2 \\ &= 3.534243 \\ A_{10} &= b_{10}/a_{10} = 0.253333 \\ a_{11} &= 2[-1 + c_1(\Omega_c T/2)^2]/a_{10} \end{aligned}$$

$$= 2[-1 + 3(0.546302)^2]/(3.534243)$$
$$= -0.059228$$
$$a_{12} = [1 - b_1(\Omega_c T/2) + c_1(\Omega_c T/2)^2]/a_{10}$$
$$= [1 - 3(0.546302) + 3(0.546302)^2]/(3.534243)$$
$$= 0.072556$$

The frequency response is

$$H(e^{j\omega}) = \frac{1.01332e^{-j\omega}\cos^2\omega/2}{1 - 0.059228e^{-j\omega} + 0.072556e^{-j2\omega}}$$

and the phase response is

$$\phi(\omega) = -\omega - \tan^{-1}\frac{0.059194 \sin\omega - 0.072559 \sin 2\omega}{1 - 0.059228 \cos\omega + 0.072556 \cos 2\omega} \tag{5.119}$$

The graph of this phase response is shown in Fig. 5.20.

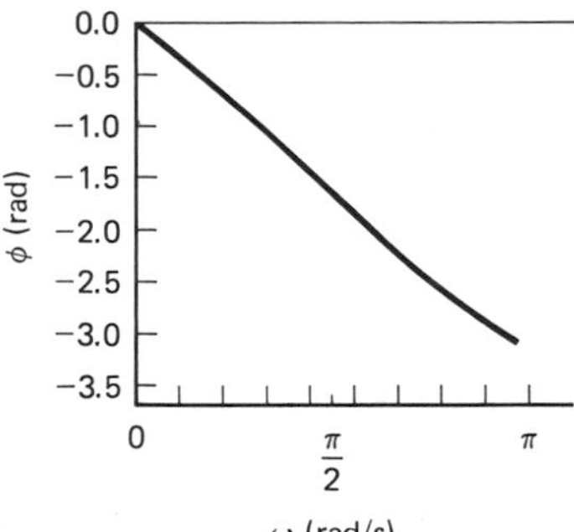

Figure 5.20 Phase response of Eq. (5.119).

The maximally flat time-delay property of the analog Bessel filter is not preserved in general by the transformations we have considered. For example, when the bilinear transformation is used, the analog and digital time delays are related by the expression

$$\tau(\omega) = -\frac{d}{d\omega}\phi(\omega) = -\frac{d}{d\omega}\phi_a(\Omega) = -\frac{d\Omega}{d\omega}\frac{d}{d\Omega}\phi_a(\Omega)$$
$$= \frac{d\Omega}{d\omega}\tau_a(\Omega)$$

where

$$\frac{d\Omega}{d\omega} = \frac{d}{d\omega}\frac{2}{T}\tan\frac{\omega}{2} = \frac{1}{T}\sec^2\frac{\omega}{2}$$
$$= \frac{1}{T}\left(1 + \tan^2\frac{\omega}{2}\right) = \frac{1}{T}\left(1 + \frac{T^2\Omega^2}{4}\right)$$

Substituting this result into the previous relation, we obtain

$$\tau(\omega) = \frac{1}{T}\left(1 + \frac{T^2\Omega^2}{4}\right)\tau_a(\Omega)$$

Methods are available for designing digital filters with maximally flat time delay. The reader is referred to the work of Thiran [178], Fettweis [38], and Thajchayapong and Lomtong [176]. Thiran [177] has also given a method for obtaining an equal-ripple time delay. We will give an example of a second-order digital filter with maximally flat time delay using the results of Fettweis. He begins by defining a function F_N by a continued fraction

$$F_N(s) = \cfrac{\mu}{\cfrac{1}{s} + \cfrac{\mu^2 - 1}{\cfrac{3}{s} + \cfrac{\mu^2 - 4}{\cfrac{5}{s} + \cdots + \cfrac{\mu^2 - (N-1)^2}{(2N-1)/s}}}}$$

Simplifying the continued fraction, we can write $F_N(s) = P(s)/Q(s)$, where $P(s)$ and $Q(s)$ are polynomials. The analog-filter transfer function is defined to be

$$H_a(s) = \frac{(1 + s)^N}{P(s) + Q(s)}$$

The digital filter is obtained by applying the transformation

$$s = \frac{z - 1}{z + 1}$$

which is the bilinear transformation of Eq. (5.20) with $T = 2$. The parameter μ is given by

$$\mu = N + 2\tau_0/T$$

where τ_0 is the delay to be approximated. A stable filter is obtained if $\mu > N - 1$ and $\tau_0 > -T/2$.

To illustrate the method of Fettweis, we take $N = 2$. Then

$$F_2(s) = \cfrac{\mu}{\cfrac{1}{s} + \cfrac{\mu^2 - 1}{3/s}} = \frac{3\mu s}{3 + (\mu^2 - 1)s^2}$$

The analog transfer function is then

$$H_a(s) = \frac{(s + 1)^2}{(\mu^2 - 1)s^2 + 3\mu s + 3}$$

Now applying the bilinear transformation, we obtain the digital transfer function

$$H(z) = \frac{4z^2}{(\mu^2 + 3\mu + 2)z^2 + (8 - 2\mu^2)z + \mu^2 - 3\mu + 2}$$

For convenience, we write

$$H(e^{j\omega}) = \frac{4e^{j2\omega}}{a_2 e^{j2\omega} + a_1 e^{j\omega} + a_0}$$

from which we obtain

$$\phi(\omega) = 2\omega - \tan^{-1} \frac{a_2 \sin 2\omega + a_1 \sin \omega}{a_2 \cos 2\omega + a_1 \cos \omega + a_0}$$

and

$$\tau(\omega) = -2 + \frac{b_0 + b_1 \cos \omega + b_2 \cos 2\omega}{c_0 + c_1 \cos \omega + c_2 \cos 2\omega}$$

where

$$b_0 = 2a_2^2 + a_1^2 \qquad b_1 = a_1(3a_2 + a_0) \qquad b_2 = 2a_2 a_0$$

$$c_0 = a_2^2 + a_1^2 + a_0^2 \qquad c_1 = 2a_1(a_2 + a_0) \qquad c_2 = 2a_2 a_0$$

If we take $\mu = 3$, then $a_2 = \mu^2 + 3\mu + 2 = 20$, $a_1 = 8 - 2\mu^2 = -10$, and $a_0 = \mu^2 - 3\mu + 2 = 2$. After using these values to calculate b_0, b_1, b_2, c_0, c_1, and c_2, we obtain

$$\tau(\omega) = -2 + \frac{5}{2} \frac{45 - 31 \cos \omega + 4 \cos 2\omega}{63 - 55 \cos \omega + 10 \cos 2\omega}$$

Using the series expansions for $\cos \omega$ and $\cos 2\omega$, we obtain

$$\begin{aligned}\tau(\omega) &= -2 + \frac{5}{2} \frac{18 + 15(\omega^2/2) + 33(\omega^4/24) + \cdots +}{18 + 15(\omega^2/2) + 105(\omega^4/24) + \cdots +} \\ &= \frac{1}{2} - \frac{5}{12}\omega^4 + \cdots +\end{aligned}$$

This is a maximally flat time delay for $N = 2$.

To get a comparison between this time delay and that obtained from an analog Bessel filter, let us replace s in $H_a(s)$ by $s = (z - 1)/(z + 1)$. This results in

$$H(z) = \frac{3(z + 1)^2}{7z^2 + 4z + 1}$$

The reader is asked in Ex. 5.10.5 to show that the time delay is

$$\tau(\omega) = \frac{1}{2} + \frac{1}{4}\frac{\omega^2}{2} + \frac{5}{12}\frac{\omega^4}{24} + \cdots +$$

which is not maximally flat.

Bessel filters have been used in the design of filter banks for speech processing [144]. A filter bank consists of N bandpass filters chosen with uniformly spaced center frequencies $\omega_k = 2\pi k/N$, $k = 0, 1, 2, \ldots, N - 1$, so that the entire base frequency band is covered. Such a filter bank is shown in Fig. 5.21.

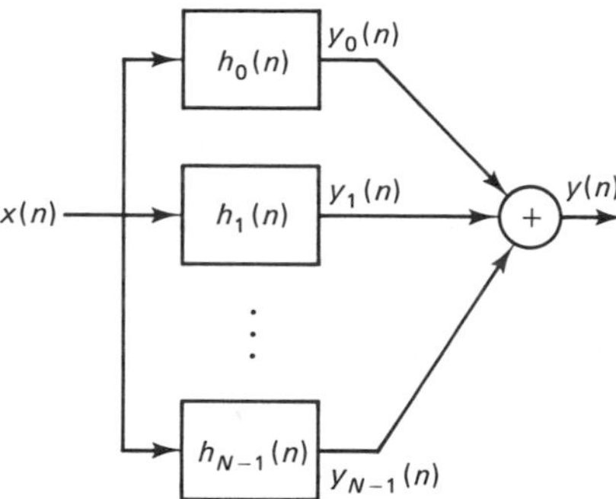

Figure 5.21 A filter bank.

In order to achieve the objective of a flat composite magnitude response and linear phase, Bessel filters were used for the individual filters in the bank. (Of course, this does not result in linear phase, but only *approximates* linear phase.) For one example, the digital filters were obtained from a sixth-order analog Bessel filter using the impulse-invariant transformation [155].

Bessel Rational Filters

An all-pass constant-time-delay filter can be obtained from the Bessel filter [77] by considering the rational transfer function ($\Omega_c = 1$)

$$H_a(s) = \frac{KB_N(-s/2)}{B_N(s/2)} \tag{5.120}$$

Such a filter is called a Bessel rational filter [77]. The frequency range over which the time delay is nearly constant is twice that of the Bessel filter.

Example of a Bessel Rational Filter

The second-order Bessel rational filter with $K = 1$ has the transfer function

$$H_a(s) = \frac{s^2 - 6s + 12}{s^2 + 6s + 12} \tag{5.121}$$

The time delay can be shown to be

$$\tau_a(\Omega) = \frac{144 + 12\Omega^2}{144 + 12\Omega^2 + \Omega^4} \tag{5.122}$$

From this equation, we have $\tau_a(1) = 156/157$ and $\tau_a(2) = 12/13$, which should be compared to the value $12/13$ obtained for the second-order analog Bessel filter when $\Omega = 1$.

To further illustrate the time-delay properties of the analog Bessel rational filter, its time delay is shown in Fig. 5.22 for $N = 1$ to 5. The time delay for the third-order Bessel filter is also shown.

The digital filter obtained from the analog Bessel rational filter using the bilinear transformation does not result in linear phase, just as was the case for the Bessel filter. As an example, the digital filter obtained from the second-order Bessel rational filter of Eq. (5.121) using the transformation $s = (z - 1)/(z + 1)$ is

$$H(z) = \frac{7z^2 + 22z + 19}{19z^2 + 22z + 7}$$

Comparing this with Eq. (5.112), we see that it is an all-pass function, as it should be. The phase, using Eq. (5.113), is given by

$$\phi(\omega) = 2\omega - 2 \arg (19e^{j2\omega} + 22e^{j\omega} + 7)$$

and the time delay is

$$\tau(\omega) = \frac{1}{2} + \frac{1}{4}\frac{\omega^2}{2} + \cdots +$$

which is not maximally flat.

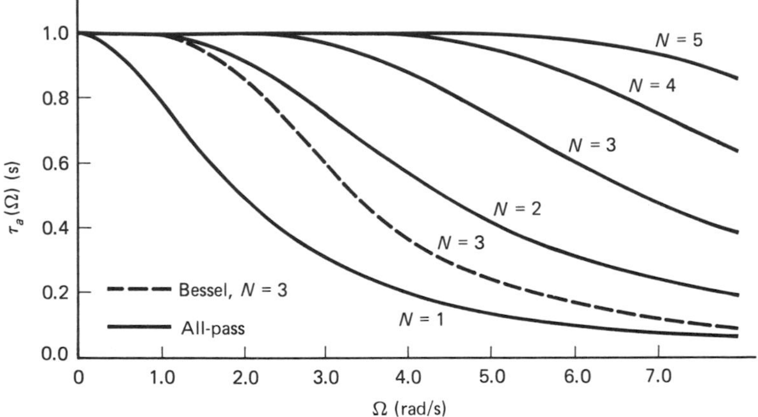

Figure 5.22 Time-delay graphs of analog Bessel rational filters.

One method that can be used to obtain an all-pass function that has maximally flat time delay is to use the denominator obtained from the method of Fettweis as $z^N Q(z)$ in Eq. (5.112). For example, using the denominator

$$(\mu^2 + 3\mu + 2)z^2 + (8 - 2\mu^2)z + \mu^2 - 3\mu + 2$$

we obtain the all-pass function

$$H(z) = \frac{(\mu^2 - 3\mu + 2)z^2 + (8 - 2\mu^2)z + \mu^2 + 3\mu + 2}{(\mu^2 + 3\mu + 2)z^2 + (8 - 2\mu^2)z + \mu^2 - 3\mu + 2}$$

For $\mu = 3$, as before, this becomes

$$H(z) = \frac{2z^2 - 10z + 20}{20z^2 - 10z + 2} = \frac{z^2 - 5z + 10}{10z^2 - 5z + 1}$$

with phase response

$$\begin{aligned}\phi(\omega) &= 2\omega - 2 \arg (10e^{j2\omega} - 5e^{j\omega} + 1) \\ &= 2\omega - 2 \tan^{-1} \frac{10 \sin 2\omega - 5 \sin \omega}{10 \cos 2\omega - 5 \cos \omega + 1}\end{aligned}$$

and time delay

$$\tau(\omega) = -2 + 5\frac{18 + 15(\omega^2/2) + 33(\omega^4/24) + \cdots +}{18 + 15(\omega^2/2) + 105(\omega^4/24) + \cdots +}$$

$$= -2 + 5\left(1 - \frac{\omega^4}{6} + \cdots +\right) = 3 - \frac{5}{6}\omega^4 + \cdots +$$

which is maximally flat.

Summary of Time Delay

Summarizing our results for the time delays of the four filters considered in this section, we have for the second-order Bessel filter,

$$H(z) = \frac{3(z+1)^2}{7z^2 + 4z + 1}$$

$$\tau(\omega) = -1 + \frac{57 + 44\cos\omega + 7\cos 2\omega}{33 + 32\cos\omega + 7\cos 2\omega}$$

for the second-order Fettweis filter,

$$H(z) = \frac{4z^2}{20z^2 - 10z + 2} = \frac{2z^2}{10z^2 - 5z + 1}$$

$$\tau(\omega) = -2 + \frac{5}{2}\frac{45 - 31\cos\omega + 4\cos 2\omega}{63 - 55\cos\omega + 10\cos 2\omega}$$

for the second-order Bessel rational filter,

$$H(z) = \frac{7z^2 + 22z + 19}{19z^2 + 22z + 7}$$

$$\tau(\omega) = -2 + 2\frac{603 + 704\cos\omega + 133\cos 2\omega}{447 + 572\cos\omega + 133\cos 2\omega}$$

and for the second-order Fettweis rational filter,

$$H(z) = \frac{2z^2 - 10z + 20}{20z^2 - 10z + 2} = \frac{z^2 - 5z + 10}{10z^2 - 5z + 1}$$

$$\tau(\omega) = -2 + 5\frac{45 - 31\cos\omega + 4\cos 2\omega}{63 - 55\cos\omega + 10\cos 2\omega}$$

Finally, for the second-order Bessel filter obtained using the impulse-invariant transformation with $T = 1$ s, considered in Ex. 5.10.3, we obtain

$$\tau(\omega) = -1 + \frac{2.0836 - 0.8817\cos\omega + 0.0996\cos 2\omega}{1.0861 - 0.607\cos\omega + 0.0996\cos 2\omega}$$

The graphs of these time delays are shown in Fig. 5.23.

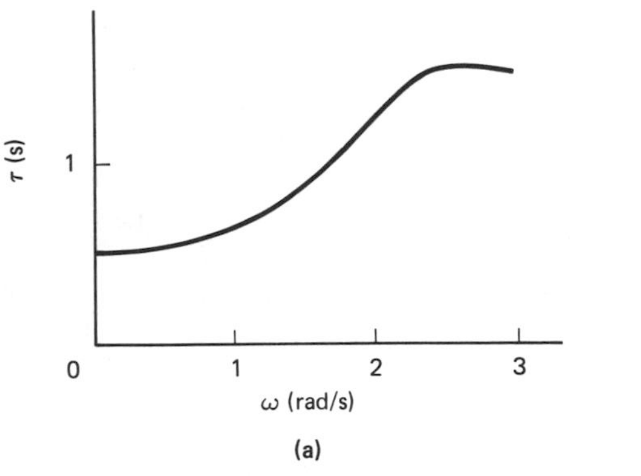

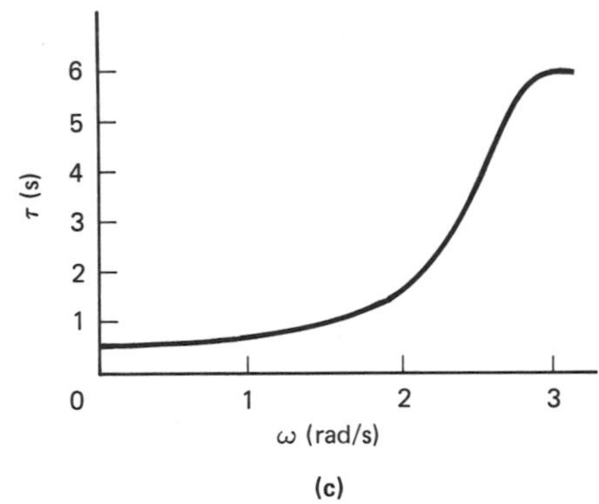

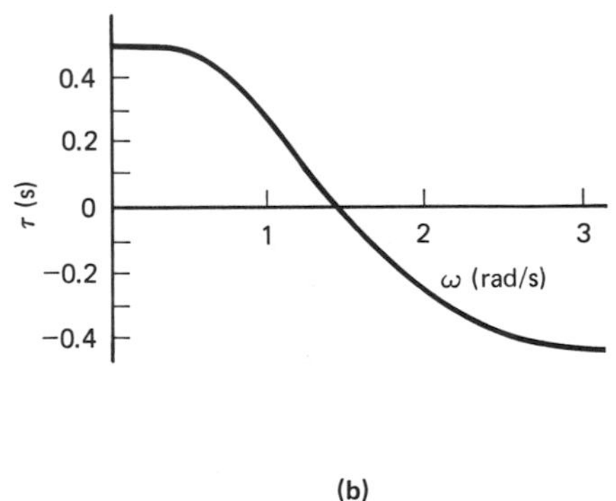

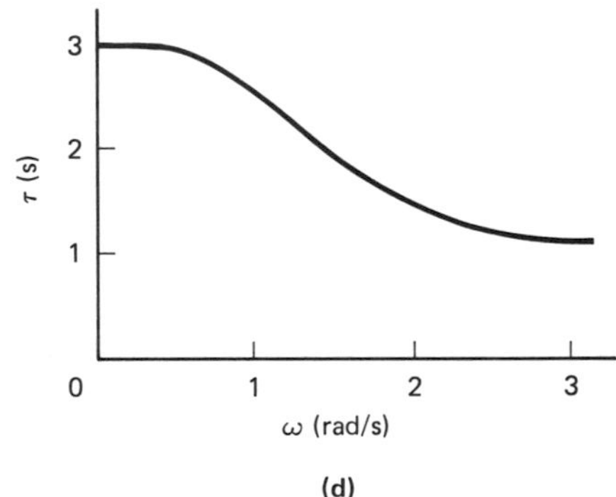

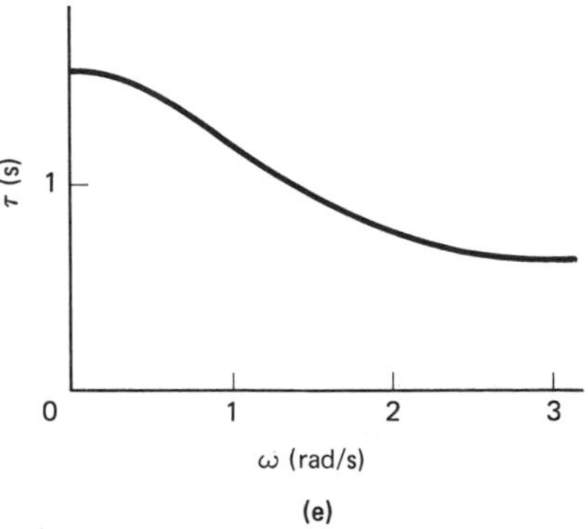

Figure 5.23 Graphs of the time delay for (a) a Bessel filter, (b) a Fettweis filter, (c) a Bessel rational filter, (d) a Fettweis rational filter obtained by using the bilinear transformation, and (e) a Bessel filter obtained by using the impulse-invariant transformation.

EXERCISES

5.10.1 Given

$$H(z) = K_1 \frac{1 + b_1 z^{-1} + b_2 z^{-2}}{1 + a_1 z^{-1} + a_2 z^{-2}}$$

where a_1, a_2, b_1, b_2, and K_1 are real. If $H(z)$ has a complex pole at $z = p$, determine b_1 and b_2 in terms of a_1 and a_2 so that $H(z)$ is all-pass. Use the result for an all-pass first-order filter. Determine $|H(e^{j\omega})|$ and $\phi(\omega)$. *Hint:* Factor $e^{-j2\omega}$ out of the numerator to determine $\phi(\omega)$.

5.10.2 Apply the step-invariant transformation of Prob. 5.4 to the Bessel-filter function

$$H_a(s) = \frac{1}{s^2 + 3s + 3}$$

and determine the phase response $\phi(\omega)$ for $T = 1$ s.

5.10.3 Use the impulse-invariant transformation to obtain a low-pass digital filter from the second-order analog Bessel filter. Compare the time delay with that obtained using the bilinear transformation.

5.10.4 Use both impulse and step invariance to obtain a digital-filter function from the Bessel rational function

$$H_a(s) = \frac{2 - s}{2 + s}$$

with **(a)** T = 1 s and **(b)** T = 0.5 s. Determine the magnitude and phase responses for each digital filter and compare the results.

5.10.5 Verify that the time delay of the digital filter with

$$H(z) = \frac{3(z + 1)^2}{7z^2 + 4z + 1}$$

is

$$\tau(\omega) = \frac{1}{2} + \frac{1}{4}\frac{\omega^2}{2} + \frac{5}{12}\frac{\omega^4}{24} + \cdots$$

5.10.6 Verify Eq. (5.122).

5.11 COMPUTER-AIDED DESIGN TECHNIQUES

Introduction to Optimization

In this section, we give a brief discussion of computer-aided design techniques for IIR filters. In most cases, a particular filter structure (see Chapter 4) is assumed and the filter response is then determined in terms of the filter coefficients, some of which may be fixed. The variable coefficients are referred to as the *designable parameters*. Normally, it is impossible to choose these parameters so that a filter has the desired response. Consequently, design procedures are sought that yield filter parameters resulting in a response that is as close to the desired response as possible. In order to obtain a mathematical design algorithm, it is convenient to introduce a *performance function,* or *index,* which is a function of the designable parameters and whose value measures how well the desired response is approximated. The parameters that minimize the performance index give the filter response that best approximates the desired response for the particular structure chosen. (For an overview of optimization techniques, the reader is referred to [22, 36, 175].)

The designable parameters can be represented as entries of the *parameter vector*

$$\mathbf{c} = [c_1 \quad c_2 \quad \cdots \quad c_m]^T$$

The performance function is a scalar function of the parameter vector and is denoted by $E(c_1, c_2, \ldots, c_m)$ or $E(\mathbf{c})$. The optimization problem is to determine $\mathbf{c}$ so that $E(\mathbf{c})$ is minimized. Normally, $E(\mathbf{c}) \geq 0$ and ideally its minimum value is $E(\mathbf{c}) = 0$.

A rather general performance function is given by

$$E(\mathbf{c}) = \int_\Theta w(\theta)|e(\mathbf{c}, \theta)|^p \, d\theta \tag{5.123}$$

where θ is the independent variable (time, frequency, temperature, etc.), p is a positive integer, and Θ is the range of allowable values of θ. The *error* function $e(\mathbf{c}, \theta)$ is defined by

$$e(\mathbf{c}, \theta) = F_d(\theta) - F(\mathbf{c}, \theta)$$

where $F_d(\theta)$ is the desired response, and $F(\mathbf{c}, \theta)$ is the actual response. The function $w(\theta)$ is a nonnegative weighting function that can be chosen to give more or less weight to the error for a particular range of θ. *Least-squares* or *mean-square* minimization is obtained when $p = 2$. *Chebyshev* or *minimax* optimization can be achieved by letting $p \to \infty$. It is known that

$$\lim_{p\to\infty} \left[\int_\Theta w(\theta)|e(\mathbf{c}, \theta)|^p \, d\theta \right]^{1/p} = \max_\Theta w(\theta)|e(\mathbf{c}, \theta)|$$

and so the value of $\mathbf{c}$ that minimizes Eq. (5.123) also minimizes the pth root of the integral and the *Chebyshev* performance function

$$E(\mathbf{c}) = \max_\Theta w(\theta)|e(\mathbf{c}, \theta)|$$

as well.

In most cases, the computations required to minimize $E(\mathbf{c})$ are done on a digital computer. It is then more convenient to use the simpler function

$$E(\mathbf{c}) = \sum_{k=1}^{K} w(\theta_k)|e(\mathbf{c}, \theta_k)|^p$$

Frequently, a value of $p = 40$ yields a reasonable approximation to the minimax problem.

Example of a Performance Index

To give the reader a flavor for computer-aided design procedures, we give a brief summary of a method considered by Deczky [30] and by Stieglitz [170]. The structure chosen for the filter is the cascade form. One reason for this choice is that the quantization effects due to finite register lengths are not as severe as for the direct form. Second, it is very easy to test the first- and second-degree factors of the factored denominator for stability.

If we express the system function $H(z)$ in the form

$$H(z) = A_0 \prod_{i=1}^{N} \frac{1 + b_{1i}z^{-1} + b_{2i}z^{-2}}{1 + a_{1i}z^{-1} + a_{2i}z^{-2}} = A_0 \prod_{i=1}^{N} H_i(z)$$

or

$$H(z) = A_0 \prod_{i=1}^{N} \frac{(1 - r_i e^{j\phi_i} z^{-1})(1 - r_i e^{-j\phi_i} z^{-1})}{(1 - \gamma_i e^{j\theta_i} z^{-1})(1 - \gamma_i e^{-j\theta_i} z^{-1})}$$

then the vector of designable parameters is

$$\mathbf{c} = [b_{11}, b_{21}, a_{11}, a_{21}, b_{12}, b_{22}, a_{12}, a_{22}, \ldots, b_{1N}, b_{2N}, a_{1N}, a_{2N}, A_0]^T$$

or

$$\mathbf{c} = [r_1, \phi_1, \gamma_1, \theta_1, r_2, \phi_2, \gamma_2, \theta_2, \ldots, r_N, \phi_N, \gamma_N, \theta_N, A_0]^T$$

The magnitude, phase, and time delay for each second-order section with system function $H_i(z)$ can be obtained in terms of the design parameters.

For example, if we desire the magnitude, we can first determine the section magnitudes, given by

$$|H_i(e^{j\omega})|^2 = \frac{M_i M_i'}{N_i N_i'}$$

where

$$M_i = 1 - 2r_i \cos(\omega - \phi_i) + r_i^2$$

$$N_i = 1 - 2\gamma_i \cos(\omega - \theta_i) + \gamma_i^2$$

and $M_i'(N_i')$ is obtained from $M_i(N_i)$ by replacing ϕ_i by $-\phi_i(\theta_i$ by $-\theta_i)$. It then follows that the squared magnitude is

$$|H(e^{j\omega})|^2 = A_0^2 \prod_{i=1}^{N} |H_i(e^{j\omega})|^2 = A_0^2 \prod_{i=1}^{N} \frac{M_i M_i'}{N_i N_i'}$$

In a similar manner, the phase and time delay can be obtained.

A suitable performance index for this problem is of the form

$$E(\mathbf{c}) = \sum_{k=1}^{K} w(\omega_k)\,|F(\mathbf{c}, \omega_k) - F_d(\omega_k)|^{2p}$$

where $F(\mathbf{c}, \omega)$ can be taken as $|H(e^{j\omega})|$ or as the time delay $\tau(\omega)$.

Minimization of the Performance Index

The requirements for an optimal solution $\mathbf{c}$, that which minimizes $E(\mathbf{c})$, can be obtained from the Taylor's expansion of $E(\mathbf{c})$. For the case where $\mathbf{c} = [c_1 \quad c_2]^T$, the transpose of the row vector, this expansion is given by

$$E(\mathbf{c} + \Delta\mathbf{c}) = E(\mathbf{c}) + \frac{\partial E(\mathbf{c})}{\partial c_1}\Delta c_1 + \frac{\partial E(\mathbf{c})}{\partial c_2}\Delta c_2 + \frac{1}{2}\left[\frac{\partial^2 E(\mathbf{c})}{\partial c_1^2}(\Delta c_1)^2 + 2\frac{\partial^2 E(\mathbf{c})}{\partial c_1 \partial c_2}\Delta c_1 \Delta c_2 + \frac{\partial^2 E(\mathbf{c})}{\partial c_2^2}(\Delta c_2)^2\right] + \cdots +$$

Denoting the change in $\mathbf{c}$ by

$$\Delta\mathbf{c} = [\Delta c_1 \quad \Delta c_2]^T$$

the gradient of $E(\mathbf{c})$ by

$$\nabla E(\mathbf{c}) = \left[\frac{\partial E}{\partial c_1} \quad \frac{\partial E}{\partial c_2}\right]^T$$

and the *Hessian* matrix of $E(\mathbf{c})$ by $\mathbf{H}(\mathbf{c})$, the matrix whose ijth element is $\partial^2 E/\partial c_i \partial c_j$, then, by ignoring the higher-order terms, we obtain

$$E(\mathbf{c} + \Delta\mathbf{c}) \approx E(\mathbf{c}) + [\nabla E(\mathbf{c})]^T \Delta\mathbf{c} + \tfrac{1}{2}(\Delta\mathbf{c})^T \mathbf{H}(\mathbf{c})\Delta\mathbf{c}$$

The conditions to be satisfied in order for E to have a local minimum at $\mathbf{c}$ are

$$\nabla E(\mathbf{c}) = \mathbf{0}$$

$$(\Delta\mathbf{c})^T \mathbf{H}(\mathbf{c})\Delta\mathbf{c} > 0$$

In other words, the components of the gradient must vanish, $\partial E/\partial c_i = 0$, and the Hessian matrix must be positive definite. When these conditions are satisfied, it follows that

$$E(\mathbf{c} + \Delta\mathbf{c}) > E(\mathbf{c})$$

for any $\Delta\mathbf{c} \neq 0$, but having components whose magnitudes are sufficiently small to justify the omission of higher-order terms. The vanishing of the gradient is a necessary condition for a minimum, but is usually the first step taken in finding either a maximum or a minimum value of a function.

There are numerous methods available for finding the minimum value of $E(\mathbf{c})$, most of which utilize numerical algorithms that make the gradient $\nabla E(\mathbf{c} + \Delta\mathbf{c}) \approx 0$ at each step in the iteration. For example, using the first-order approximation of Taylor's expansion of the gradient $\nabla E(\mathbf{c})$, we can write (see Prob. 5.43)

$$\nabla E(\mathbf{c} + \Delta\mathbf{c}) = \nabla E(\mathbf{c}) + \mathbf{H}(\mathbf{c})\Delta\mathbf{c}$$

If we let $\mathbf{c} = \mathbf{c}^j$ and $\mathbf{c} + \Delta\mathbf{c} = \mathbf{c}^{j+1}$, then $\Delta\mathbf{c} = \mathbf{c}^{j+1} - \mathbf{c}^j$. Now setting $\nabla E(\mathbf{c} + \Delta\mathbf{c}) = 0$ and solving for $\mathbf{c}^{j+1}$, we obtain the iteration scheme

$$\mathbf{c}^{j+1} = \mathbf{c}^j - \mathbf{H}^{-1}(\mathbf{c}^j)\nabla E(\mathbf{c}^j)$$

By using this iteration procedure, $\nabla E(\mathbf{c}^{j+1})$ is calculated and the process continued until the components of the gradient are sufficiently small. This procedure requires that the inverse of the Hessian matrix be computed at each iteration and, consequently, requires considerable computational effort. There are numerous methods available, notably that of Fletcher and Powell [41], that avoid this problem by estimating this inverse and updating it at each iteration.

Minimization Using Linear Programming

Another procedure that is sometimes used is *linear programming*. In this case, the constraints on the system function are ultimately expressed as a system of linear equations in the design parameters and *slack variables* with a linear performance function [44, 136, 137].

To illustrate the procedure, let us consider obtaining a magnitude response that satisfies the constraints shown in Fig. 5.5. For example, if ω_1 is in the passband, then we have

$$A_1^2 \le |H(e^{j\omega})|^2 \le A^2 \qquad 0 \le \omega_i \le \omega_1 \tag{5.124}$$

If we assume that the system function has the form of Eq. (4.2), which we repeat as

$$H(z) = \frac{\sum_{k=0}^{M} b_k z^{-k}}{\sum_{k=0}^{N} a_k z^{-k}}$$

then it can be shown that

$$|H(e^{j\omega_i})|^2 = \frac{Q(\omega_i)}{P(\omega_i)} = \frac{c_0 + \sum_{k=1}^{M} 2c_k \cos k\omega_i}{d_0 + \sum_{k=1}^{N} 2d_k \cos k\omega_i}$$

Substitution of this expression into Eq. (5.124) results in the two linear inequalities

$$\begin{aligned} Q(\omega_i) - A^2 P(\omega_i) &\le 0 \\ -Q(\omega_i) + A_1^2 P(\omega_i) &\le 0 \end{aligned} \tag{5.125}$$

The design parameters can be taken as $c_0, c_1, \ldots, c_M, d_0, d_1, \ldots, d_N$. The parameters $a_0, a_1, \ldots, a_N, b_0, b_1, \ldots, b_M$ can be determined from the c_k's and d_k's. Without loss of generality, we can take $a_0 = 1$ and this results in $d_0 = 1$. For frequencies in the stopband, we obtain constraints similar to Eq. (5.125). In addition to these constraints, we must also have

$$\begin{aligned} -Q(\omega_i) &\le 0 \\ -P(\omega_i) &\le 0 \end{aligned} \tag{5.126}$$

To complete the formulation of the design constraints as a linear programming problem, we introduce an auxiliary parameter ν that is subtracted from the left-hand side of each constraint. In the case of Eqs. (5.125) and (5.126), we obtain

$$Q(\omega_i) - A^2 P(\omega_i) - \nu \leq 0$$
$$-Q(\omega_i) + A_1^2 P(\omega_i) - \nu \leq 0$$
$$-Q(\omega_i) - \nu \leq 0$$
$$-P(\omega_i) - \nu \leq 0$$

and minimize the performance function

$$E(\mathbf{c}) = \nu$$

Once the linear programming problem is formulated, it can be solved rather easily using a procedure known as the *simplex method* [44].

PROBLEMS

5.1 From

$$H_a(s) = \frac{3}{(s+1)(s+3)}$$

determine $H(z)$ using impulse invariance and using both Eqs. (5.5) and (5.13) for **(a)** $T = 1$ s and **(b)** $T = 0.1$ s. In each case, determine $|H(1)|$ and $|H(-1)|$.

5.2 Repeat Prob. 5.1 for

(a) $H_a(s) = \dfrac{s+2}{(s+1)(s+3)}$

(b) $H_a(s) = \dfrac{2s^2 + 3s + 3}{(s+1)(s^2 + 2s + 2)}$

5.3 Determine the step response $r(n)$ for the digital systems obtained in Probs. 5.1 and 5.2 for $T = 1$ s.

5.4 **(a)** Develop a step-invariant transformation starting with the system function of Eq. (5.2).

(b) Use this transformation to obtain $H(z)$ for the function of Prob. 5.1 with $T = 1$ s.

(c) Find the step response $r(n)$.

Answer:

(a) $\dfrac{1}{s + p_k} \longrightarrow \dfrac{(1 - e^{-p_k T})z^{-1}}{p_k(1 - e^{-p_k T} z^{-1})}$

5.5 Apply both impulse and step invariance to the analog function

$$H_a(s) = \frac{2}{(s+1)(s+2)}$$

with $T = 1$ s to obtain $H(z)$. Find the step responses for the two digital functions and compare the results.

5.6 Obtain $H(z)$ from the $H_a(s)$ in Prob. 5.2(a) using step invariance with $T = 1$ s.

5.7 Develop the relations given in Eqs. (5.15)–(5.17). Note that $H(z) = Z\{h_a(nT)\}$. The inverse Laplace transform

$$\mathcal{L}^{-1}\left\{\frac{1}{(s + s_i)^m}\right\} = \frac{1}{(m - 1)!}t^{m-1}e^{-s_i t}$$

should be used to obtain Eq. (5.15). Partial fractions should be used to obtain Eqs. (5.16) and (5.17).

5.8 Use impulse invariance to obtain $H(z)$ if $T = 1$ s and $H_a(s)$ is

(a) $\dfrac{1}{(s + 1)^2}$

(b) $\dfrac{1}{s^2 + (2^{1/2})s + 1}$

(c) $\dfrac{s}{s^2 + s + 1}$

5.9 Determine $H(z)$, which results when impulse invariance is applied to

$$H_a(s) = \frac{s^2 + c}{(s + a)^2 + b^2}$$

Apply this result to the function

$$H_a(s) = \frac{s^2 + 4.525}{s^2 + 0.692s + 0.504}$$

with $T = 1$ s. Determine $|H(1)|$ and $|H(-1)|$.

5.10 Find $H(z)$ using impulse invariance with $T = 1$ s if

$$H_a(s) = \frac{3s^2 + 4}{(s + 2)(s^2 + 0.5s + 1)}$$

5.11 Sketch $|H_a(j\Omega)|^2$ for $H_a(s)$ given in Prob. 5.1 and determine $\Omega_{3\,\mathrm{dB}}$. Use the bilinear transformation with $T = 1$ s and obtain $H(z)$. What is $\omega_{3\,\mathrm{dB}}$ for the digital filter? Determine $|H(1)|$ and $|H(-1)|$. Compare the results with those obtained in Prob. 5.1 for $T = 1$ s.

5.12 Obtain $H(z)$ using the bilinear transformation with $T = 1$ s for $H_a(s)$ given in Prob. 5.8.

5.13 Determine $H(z)$ that results when the bilinear transformation is applied to

$$H_a(s) = \frac{s^2 + a}{s^2 + bs + c}$$

Apply the result to

$$H_a(s) = \frac{s^2 + 4.525}{s^2 + 0.692s + 0.504}$$

and determine $|H(1)|$ and $|H(-1)|$. Compare these results with those obtained in Prob. 5.9.

5.14 Apply the results of Prob. 5.13 to $H_a(s)$ given in Prob. 5.10.

5.15 Design a digital Butterworth filter satisfying the constraints

$$1/2^{1/2} \le |H(e^{j\omega})| \le 1 \qquad 0 \le \omega \le \pi/2$$

$$|H(e^{j\omega})| \le 0.2 \qquad 3\pi/4 \le \omega \le \pi$$

with $T = 1$ s using **(a)** the bilinear transformation and **(b)** impulse invariance. Realize the filter in each case using the most convenient realization form.

5.16 Repeat Prob. 5.15 for the constraints

$$0.9 \le |H(e^{j\omega})| \le 1 \qquad 0 \le \omega \le \pi/2$$
$$|H(e^{j\omega})| \le 0.2 \qquad 3\pi/4 \le \omega \le \pi$$

5.17 Repeat Prob. 5.15 for the constraints

$$0.9 \le |H(e^{j\omega})| \le 1 \qquad 0 \le \omega \le \pi/4$$
$$|H(e^{j\omega})| \le 0.2 \qquad 3\pi/4 \le \omega \le \pi$$

5.18 Repeat Prob. 5.15 using a digital Chebyshev filter.

5.19 Repeat Prob. 5.16 using a ditigal Chebyshev filter.

5.20 Repeat Prob. 5.17 using a digital Chebyshev filter.

5.21 For the constraints

$$1/2^{1/2} \le |H(e^{j\omega})| \le 1 \qquad 0 \le \omega \le 0.2\pi$$
$$|H(e^{j\omega})| \le 0.1 \qquad 0.5\pi \le \omega \le \pi$$

design **(a)** a digital Butterworth filter and **(b)** a digital Chebyshev filter using $T = 1$ s and the bilinear transformation.

5.22 Use the bilinear transformation to obtain $H(z)$ for a Butterworth filter that satisfies the constraints of Eq. (5.65). Compare the results with the Chebyshev filter obtained in Sec. 5.7.

5.23 Use the forward Euler formula to approximate the integral obtained by integrating both sides of Eq. (5.19) and obtain a relation between s and z. (See Prob. 1.7.)

5.24 Repeat Prob. 5.23 using the backward Euler formula.

5.25 Obtain a digital filter from the analog function $H_a(s) = 1/(s + 1)$ using $T = 1$ s by applying **(a)** the bilinear transformation, **(b)** the transformation of Prob. 5.23, and **(c)** the transformation of Prob. 5.24. Compare the results obtained.

5.26 Apply the impulse-invariant technique to the analog transfer function given in Prob. 5.13 with $T = 1$ s.

5.27 Determine $H_a(s)$ for an elliptic filter with $PRW = 0.5$ dB, $MSL = 60$ dB, and $\Omega_c = 2$ rad/s. Find the transition width.

5.28 Determine the minimum order N of an analog elliptic filter for a cutoff frequency of $\Omega_c = 2000\pi$ rad/s, $PRW = 0.5$ dB, $MSL = 60$ dB, and $TW \le 200\pi$ rad/s.

5.29 A normalized low-pass elliptic filter with $PRW = 1$ dB and $MSL = 60$ dB has a transfer function

$$H_a(s) = \frac{s^2 + 33.448141}{(s + 0.500423)(s^2 + 0.485510s + 0.996740)}$$

Obtain a high-pass filter with $\Omega_c = 2$ rad/s.

5.30 A normalized low-pass 1-dB Chebyshev filter has a transfer function

$$H_a(s) = \frac{1.102510}{s^2 + 1.097734s + 1.102510}$$

Obtain a bandpass filter with $\Omega_0 - 2$ rad/s and $Q = 10$.

5.31 Given the low-pass digital-filter function

$$H(z) = \frac{1 + 2z^{-1}}{4 - z^{-1}}$$

determine the following:

(a) the cutoff frequency ω_c'

(b) another low-pass filter with $\omega_c = 2$ rad/s

(c) a high-pass filter with $\omega_c = 2$ rad/s

(d) a bandpass filter with $\omega_1 = \pi/4$ rad/s and $\omega_2 = 3\pi/4$ rad/s

(e) a band-reject filter with $\omega_1 = \pi/4$ rad/s and $\omega_2 = 3\pi/4$ rad/s

5.32 The digital low-pass filter obtained from a second-order low-pass Butterworth filter is

$$H(z) = \frac{(1 + z^{-1})^2}{(5 - 2^{1/2})z^{-2} - 6z^{-1} + (5 + 2^{1/2})}$$

The cutoff frequency is $\omega_c' = 2\tan^{-1} 0.5 = 0.927295$ rad/s. Obtain the following:

(a) a high-pass filter with $\omega_c = 1.5$ rad/s

(b) another low-pass filter with $\omega_c = 1.5$ rad/s

(c) a bandpass filter with $\omega_1 = 1$ rad/s and $\omega_2 = 2$ rad/s

5.33 Determine $H_a(s)$ for a third-order Bessel filter with $\Omega_c = 1$ rad/s. Show that

$$\tau_a(\Omega) = \frac{225 + 45\Omega^2 + 6\Omega^4}{225 + 45\Omega^2 + 6\Omega^4 + \Omega^6}$$

5.34 It is known that [79]

$$B_3(s) = (s + 2.322185)(s^2 + 3.677815s + 6.459433)$$

Write $\phi(\Omega)$ as a sum $\phi_1(\Omega) + \phi_2(\Omega)$, where $\phi_1(\Omega)$ is the phase of the first-order section, and ϕ_2 is the phase of the second-order section. Determine $\tau_a(\Omega)$ from this sum.

5.35 Show that

$$|H(e^{j\omega})|^2 = \frac{1}{1 + \left(\dfrac{\tan \omega/2}{\tan \omega_c/2}\right)^{2N}}$$

represents the squared magnitude function of a low-pass filter with cutoff frequency ω_c. If $z = e^{j\omega}$, show that

$$\tan\frac{\omega}{2} = -j\frac{z - 1}{z + 1} = -jw$$

and consequently that

$$H(z)H(z^{-1}) = \frac{1}{1 + (-1)^N(w \cot \omega_c/2)^{2N}}$$

The system function $H(z)$ is then given by

$$H(z) = \frac{B(z + 1)^N}{\prod_{k=1}^{N} (z - z_k)}$$

where z_k, $k = 1, 2, \ldots, N$, are the poles of $H(z)H(z^{-1})$ for which $|z_k| < 1$. Determine these poles from the equation

$$(-1)^N w_k^{2N} = -(\tan \omega_c/2)^{2N} = (\tan \omega_c/2)^{2N} e^{j(2k-1)\pi} \qquad k = 1, 2, \ldots, 2N$$

(Note that these functions resemble Butterworth functions.) Find $H(z)$ for the special cases $N = 2$ and 3 and $\omega_c = \pi/2$. (See [16, 51].)

5.36 Show that

$$|H(e^{j\omega})|^2 = \frac{1}{1 + \left(\dfrac{\cot \omega/2}{\cot \omega_c/2}\right)^{2N}}$$

represents the squared magnitude function of high-pass filter. Repeat the remainder of Prob. 5.35 for this function. (See [16].)

5.37 Repeat the procedure of Prob. 5.35 for the function

$$|H(e^{j\omega})|^2 = \frac{1}{1 + \left(\dfrac{\sin \omega/2}{\sin \omega_c/2}\right)^{2N}}$$

5.38 Show that the squared magnitude function given by

$$|H(e^{j\omega})|^2 = \frac{1}{1 + \epsilon^2 C_N^2\left(\dfrac{\tan \omega/2}{\tan \omega_c/2}\right)}$$

represents a low-pass filter. By using the transformation of Prob. 5.35, it can be shown that the poles of $H(z)H(z^{-1})$ are

$$w_k = \tan \frac{\omega_c}{2}\left[-\sin \frac{(2k-1)\pi}{2N} \sinh v + j \cos \frac{(2k-1)\pi}{2N} \cosh v\right] \qquad k = 1, 2, \ldots, 2N$$

where

$$\sinh v = \tfrac{1}{2}\{[(1 + 1/\epsilon^2)^{1/2} + 1/\epsilon]^{1/N} - [(1 + 1/\epsilon^2)^{1/2} + 1/\epsilon]^{-1/N}\}$$

$$\cosh v = \tfrac{1}{2}\{[(1 + 1/\epsilon^2)^{1/2} + 1/\epsilon]^{1/N} + [(1 + 1/\epsilon^2)^{1/2} + 1/\epsilon]^{-1/N}\}$$

Use these results to find a realizable $H(z)$ for $\epsilon = 1$, $\omega_c = \pi/2$, and $N = 2$ and 3.

5.39 Use (a) the impulse-invariant transformation and (b) the step-invariant transformation to obtain a low-pass digital filter from the second-order analog Bessel rational filter. Determine the phase and time delay and compare them with those obtained using the bilinear transformation.

5.40 Use the impulse-invariant transformation to obtain a low-pass digital filter from the third-order analog Bessel filter. Compare the time delay with that obtained using the bilinear transformation.

5.41 Use the method of Fettweis to obtain a low-pass digital filter from the third-order analog Bessel filter with $\mu = 3$. Determine the phase and the time delay.

5.42 Use the modified method of Fettweis given in the text to obtain a digital filter from the third-order analog Bessel rational filter with $\mu = 3$. Determine the phase and the time delay.

5.43 Obtain the first-order approximation of Taylor's expansion of the gradient $\nabla E(\mathbf{c})$ for the two-dimensional case and then extend the result to n dimensions.

5.44 Find the minimum of

$$E(c_1, c_2, c_3) = 2(c_1 - 1)^2 + (c_2 + 1)^2 + 2(c_3 - 2)^2 + 4$$

5.45 Given the analog-filter function

$$H_a(s) = \frac{s^2 + b^2}{(s + b)^2}$$

and the digital-filter function

$$G(z) = \frac{A[1 - 2(\cos bT)z^{-1} + z^{-2}]}{(1 - az^{-1})^2}$$

where a and b are real numbers.

(a) Determine what type filter $H_a(s)$ represents and determine its distinguishing characteristics.

(b) Determine the value of a (in terms of b and T) such that the phase of $H_a(j\Omega')$ is the same as the phase of $G(e^{j\Omega' T})$.

Answer: $a = \cos \Omega' T - \sin \Omega' T / \tan [(\Omega' T/2) + \tan^{-1} (\Omega'/b)]$.

5.46 For the analog bandpass and band-reject transfer functions

$$H_{aBP}(s) = \frac{K\Omega_0 s}{s^2 + a_1 \Omega_0 s + \Omega_0^2}$$

$$H_{aBR}(s) = \frac{K(s^2 + \Omega_0^2)}{s^2 + a_1 \Omega_0 s + \Omega_0^2}$$

show that if the bilinear transformation is used, the analog bandwidth $ABW = a_1 \Omega_0$ is related to the digital bandwidth DBW in each case by

$$\tan (DBW/2) = \frac{(T/2)ABW}{1 + (\Omega_0 T/2)^2}$$

Express the digital transfer functions in terms of DBW and the center frequency ω_0 for each case.

5.47 The following function is given:

$$H(z) = \frac{K(1 - z_n z^{-1})(1 - z_n^* z^{-1})}{(1 - z_d z^{-1})(1 - z_d^* z^{-1})}$$

If $z_n = r_n e^{j\phi_n}$, determine $z_d = r_d e^{j\phi_d}$ so that the phase of $H(e^{j\omega_0 T})$ is zero for $\omega_0 T < \phi_n$. If $z_d = x + jy$, determine the locus of these points that give zero phase at $z = e^{j\omega_0 T}$ [42].

Answer:

$$\cos \phi_d = \frac{(r_d^2 - r_n^2) \cos \omega_0 T + (1 - r_d^2) r_n \cos \phi_n}{r_d(1 - r_n^2)}$$

$$x^2 - ax + y^2 = b$$

$$a = \frac{1 - r_n^2}{\cos \omega_0 T - r_n \cos \phi_n}$$

$$b = \frac{r_n(r_n \cos \omega_0 T - \cos \phi_n)}{\cos \omega_0 T - r_n \cos \phi_n}$$

5.48 **(a)** Given the block diagram representation of an analog system shown in the figure, obtain the transfer functions $Y_a(s)/X_a(s)$ and $V_a(s)/X_a(s)$ and use the bilinear transformation to obtain a digital system with outputs $Y(z)$ and $V(z)$ and input $X(z)$. Obtain a block diagram realization using one delay.

(b) Obtain a digital system by first applying the bilinear transformation to the transfer function K/s. Complete the signal-flow graph realization and discuss its physical realization.

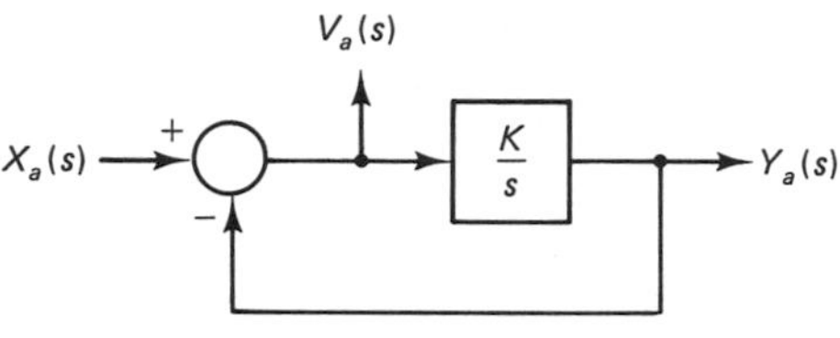

Problem 5.48

6

FINITE IMPULSE-RESPONSE FILTER DESIGN

6.1 INTRODUCTION

As we have seen in the previous chapters, finite impulse-response (FIR) filters possess certain very desirable properties. First of all, we know that an FIR filter is stable. Moreover, a realizable filter can always be obtained. If the sequence $h(n)$ is noncausal, then it can be sufficiently delayed so as to obtain a causal sequence representing a realizable filter. Consequently, the questions of stability and realizability never arise for FIR filters.

Another very important feature of FIR filters, which is not shared by IIR filters, is the possibility of obtaining exactly linear phase, which is accompanied by constant time or group delay. As we saw in Sec. 2.11, this is achieved by requiring that the impulse response $h(n)$ be symmetric, $h(n) = h(N - 1 - n)$, or be antisymmetric, $h(n) = -h(N - 1 - n)$. For IIR filters, the best we can hope for is a maximally flat time delay.

Attainment of linear phase is quite desirable in many signal-processing applications. For example, it makes possible precise time alignment in speech-processing applications. In data-transmission applications, linear phase prevents pulse dispersion, since the output signal is simply the input signal delayed in time and possibly amplified. The delay obtained is not always an integral number of samples and this can sometimes be a problem. Since linear phase is attainable, the approximation problem is often simplified by using a FIR filter, since the designer can concentrate on obtaining a desired magnitude response.

One disadvantage of FIR filters is that long sequences for $h(n)$ are generally required to adequately approximate sharp cutoff filters. A large amount of processing is required to realize the filter if slow convolution is used. By using the fast Fourier-transform (FFT) algorithm, these filters can be designed more efficiently. (The FFT algorithm and its use in designing FIR filters are discussed in Chapter 8.) If linear phase is desired, it may be more efficient to use a FIR filter, since the phase response of an IIR filter can be too nonlinear and require an all-pass section to improve it. (For a discussion comparing FIR and IIR filters, see [139].)

There are three well-known design methods for designing FIR filters with linear phase. These are the window method, frequency-sampling method, and optimal or minimax design. The window method involves a straightforward analytical procedure; however, in some cases, iteration is necessary to obtain the desired result. Several windows are considered. The Kaiser family of windows provides the designer considerable flexibility in meeting the filter specifications. Kaiser has obtained empirical equations that allow selection of design parameters that give a filter satisfying the given specifications.

The second method of design that is considered is frequency-sampling. A desired frequency response is uniformly sampled and filter coefficients are then determined from these samples using the inverse discrete Fourier transform.

Optimal FIR filters are considered based on the representations given in Sec. 2.11 for the frequency response. An efficient computer program is available for designing optimal filters, so that this technique is very attractive. Analytical solutions can be found for a limited class of FIR filters and these help the reader to understand this design procedure.

The terminology used in describing FIR filters is generally not the same as that used for describing IIR filters. For the FIR filter, the magnitude specifications are

$$\begin{aligned} 1 - \delta_1 \le |H(e^{j\omega})| &\le 1 + \delta_1 \qquad 0 \le \omega \le \omega_p \qquad 0 \le f \le F_p \\ 0 \le |H(e^{j\omega})| &\le \delta_2 \qquad \omega_s \le \omega \le \pi \qquad F_s \le f \le 0.5 \end{aligned}$$

where $f = \omega/2\pi$ is the frequency in hertz. The passband is the range of frequencies $0 \le f \le F_p$, and the stopband is the range $F_s \le f \le 0.5$. Usually, the magnitude is described by a pseudomagnitude $H_1(\omega)$ for linear-phase filters that oscillates between $1 - \delta_1$ and $1 + \delta_1$ in the passband and between $-\delta_2$ and $+\delta_2$ in the stopband. For the general case of optimal filters, there does not exist any simple analytical relationships between the filter parameters δ_1, δ_2, F_p, F_s, and N, the filter length. The exceptions are the cases where there is either one passband ripple or one stopband ripple, which are considered in Sec. 6.8.

Approximate empirical relations have been obtained [61, 135, 136] that give satisfactory relations between the filter parameters. These relations are

$$N = 1 + \frac{D_\infty(\delta_1, \delta_2)}{\Delta F} - f(\delta_1, \delta_2)\Delta F$$

where

$$\Delta F = F_s - F_p = \text{relative transition width}$$

$$D_\infty(\delta_1, \delta_2) = [0.005309(\log_{10} \delta_1)^2 + 0.07114 \log_{10} \delta_1 - 0.4761] \log_{10} \delta_2$$
$$-[0.00266(\log_{10} \delta_1)^2 + 0.5941 \log_{10} \delta_1 + 0.4278]$$

$$f(\delta_1, \delta_2) = 0.51244 \log_{10} (\delta_1/\delta_2) + 11.01$$

A simpler but less accurate expression for N, obtained as a modification of Kaiser's relations for window designs [86, 87], is

$$N = \frac{-10 \log_{10} (\delta_1 \cdot \delta_2) - 15}{14\Delta F} + 1$$

EXERCISES

6.1.1 Calculate N from both empirical relations for $\delta_1 = \delta_2 = 0.05$ and $\Delta F = 0.1$.

6.1.2 Repeat Ex. 6.1.1 for $\Delta F = 0.05$.

6.2 WINDOWING AND THE RECTANGULAR WINDOW

Obtaining Finite-Length Impulse Responses by Truncation

One way to obtain a finite-length impulse response is to simply truncate an infinite impulse response. Suppose that the desired frequency response is $H_d(e^{j\omega})$ with a Fourier series representation

$$H_d(e^{j\omega}) = \sum_{n=-\infty}^{\infty} h_d(n)e^{-j\omega n} \tag{6.1}$$

The impulse-response sequence $h_d(n)$ has infinite length and this frequency repsonse is that of an IIR system. Also, the system is not realizable since $h_d(n) \neq 0$ for $n < 0$. The realizability problem can be solved by simply shifting the sequence to the right if it has only a finite number of nonzero terms for $n < 0$.

We examine some ways to obtain a new frequency response $H(e^{j\omega})$ from Eq. (6.1) that will represent a FIR system. Once this is done, the realizability problem can be easily solved. One obvious way to achieve this desired representation is to simply truncate the series of Eq. (6.1). This section is devoted to truncation; other methods are described in later sections.

In order to obtain a realizable filter, we can truncate the impulse-response sequence to obtain a new sequence having length N with defining relation

$$h(n) = \begin{cases} h_d(n) & 0 \le n \le N - 1 \\ 0 & \text{otherwise} \end{cases} \tag{6.2}$$

The frequency response corresponding to this finite-duration sequence is

$$\begin{aligned} H(e^{j\omega}) &= \sum_{n=-\infty}^{\infty} h(n)e^{-j\omega n} \\ &= \sum_{n=0}^{N-1} h_d(n)e^{-j\omega n} \end{aligned} \tag{6.3}$$

It can also be obtained from Eq. (6.1) by truncation. Since we use only the N values of $h_d(n)$, $0 \le n \le N-1$, we could also think of obtaining Eq. (6.3) by looking through a window and seeing only these terms of $h_d(n)$. For this reason, the process of obtaining Eq. (6.3) from Eq. (6.1) is called *windowing*.

Rectangular Windows

The relation expressed in Eq. (6.2) can also be obtained by multiplying the sequence $h_d(n)$ by the sequence $w_R(n)$ defined by

$$w_R(n) = \begin{cases} 1 & 0 \le n \le N-1 \\ 0 & \text{otherwise} \end{cases} \tag{6.4}$$

The resulting expression, which is equivalent to Eq. (6.2), is

$$h(n) = w_R(n)h_d(n) \tag{6.5}$$

The graph of the sequence $w_R(n)$ is shown in Fig. 6.1 and, because of its appearance, $w_R(n)$ is called a *rectangular window*.

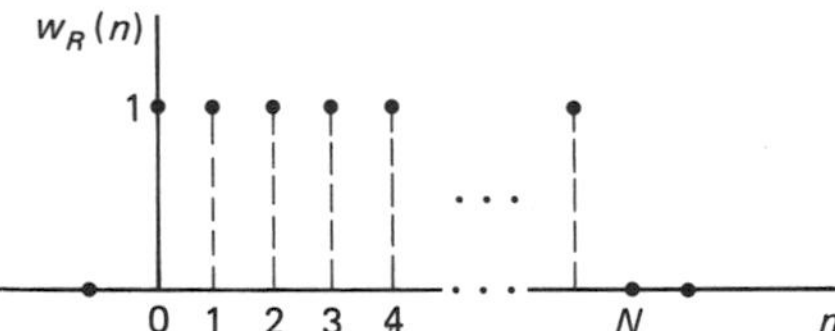

Figure 6.1 Graph of a rectangular window $w_R(n)$.

We could also obtain a FIR approximation by truncating outside the interval $-M \le n \le M$. A realizable filter can be obtained by delaying the resulting sequence by M samples. This type of truncation has certain symmetry properties that make it desirable from an analytical standpoint. It has the disadvantage that there are always an odd number of terms, $2M+1$, in the window sequence. The truncated series has the form

$$H(e^{j\omega}) = \sum_{n=-M}^{M} h_d(n)e^{-j\omega n} \tag{6.6}$$

with corresponding rectangular window

$$w_{RS}(n) = \begin{cases} 1 & -M \le n \le M \\ 0 & \text{otherwise} \end{cases} \tag{6.7}$$

We have added S to the subscript to indicate the symmetry of the window and to distinguish it from the causal window $w_R(n)$ of Eq. (6.4). The relation between the two impulse-response sequences corresponding to Eq. (6.5) is given by

$$h_S(n) = w_{RS}(n)h_d(n) \tag{6.8}$$

The Gibbs Phenomenon

One disadvantage in directly truncating the infinite series of Eq. (6.1) to obtain a finite impulse-response sequence is that it leads directly to the *Gibbs phenomenon,* named for J. Willard Gibbs. (Gibbs was the first to publicize this effect, but he did not publish his results immediately.) The Gibbs phenomenon manifests itself whenever a finite discontinuity is being fitted by using a finite number of terms of the series. This phenomenon appears as a fixed-percentage overshoot and ripple before and after the discontinuity. This is true because it is impossible to obtain an infinite slope using only a finite number of terms. As the number of terms increases, the ripples do not decrease, but are squeezed into a narrower interval about the discontinuity. Even in the infinite sum, this overshooting and undershooting persists and the complete series has *flanges,* as shown by D^+ and D^- in Fig. 6.2 [51, 55, 119].

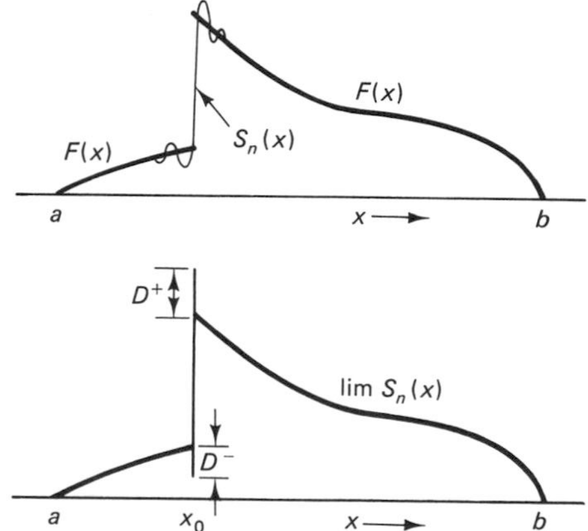

Figure 6.2 Illustration of the Gibbs phenomenon.

As a specific example, consider the function with period 2π defined by

$$f(t) = \begin{cases} +1 & 0 < t < \pi \\ -1 & \pi < t < 2\pi \end{cases} \tag{6.9}$$

A truncated Fourier series has the form

$$S_N(t) = \frac{4}{\pi} \sum_{k=0}^{N} \frac{\sin (2k + 1)t}{2k + 1}$$

and it can be shown that [119]

$$\lim_{N \to \infty} S_N\left(\frac{\pi}{2N + 1}\right) = 1.179$$

Consequently, the series, in trying to follow the discontinuity at $t = 0$, overshoots the mark by about 18 percent over a region of vanishingly small width before settling down to a correct value of unity. A similar situation occurs at $t = \pi$, where the series undershoots the mark by the same percentage as at $t = 0$. The graphs of $S_N(t)$ for $N = 0$, 4, and 9 are shown in Fig. 6.3.

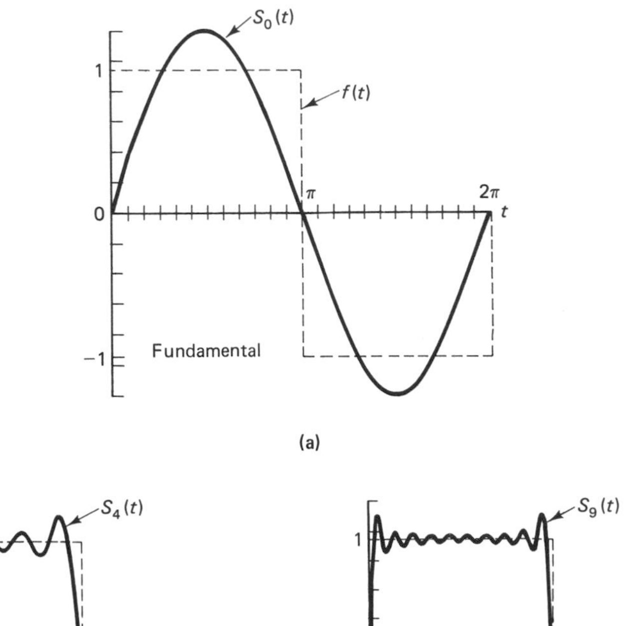

Figure 6.3 Graphs of $S_N(t)$ for (a) $N = 0$, (b) $N = 4$, and (c) $N = 9$ for $f(t)$ given by Eq. (6.9).

The most important standard filters are low-pass, high-pass, bandpass, and band-reject filters, all of which require approximation of a discontinuity in the ideal amplitude response. When the rectangular window is used to truncate the infinite series, the Gibbs phenomenon is present and usually an undesirable approximation results. At a discontinuity, the approximation has a fixed-percentage overshoot with ripples before and after the discontinuity. For this reason, the rectangular window is not of much practical use and other window sequences $w(n)$ are sought that alleviate this problem.

Complex Convolution

In most cases involving the design of FIR filters using windows, we start with a frequency response $H_d(e^{j\omega})$. The problem is then to choose a window sequence $w(n)$ so that the sequence

$$h(n) = w(n)h_d(n) \tag{6.10}$$

has a suitable frequency response $H(e^{j\omega})$. We can determine $H(e^{j\omega})$ directly from the sum

$$\begin{aligned} H(e^{j\omega}) &= \sum_{n=0}^{N-1} h(n)e^{-j\omega n} \\ &= \sum_{n=0}^{N-1} h_d(n)w(n)e^{-j\omega n} \end{aligned} \tag{6.11}$$

but this does not give us much insight concerning how variations in $w(n)$ affect $H(e^{j\omega})$. Since we are interested in comparing $H(e^{j\omega})$ and $H_d(e^{j\omega})$, it would be useful to have a relation giving $H(e^{j\omega})$ in terms of $H_d(e^{j\omega})$. The desired relation is known as the *complex convolution* of the product of the sequences $h_d(n)$ and $w(n)$ and we now develop this expression.

In the development of the complex convolution, we assume that both $h_d(n)$ and $w(n)$ are arbitrary sequences. Then, as before, the Fourier transform $H(e^{j\omega})$ is given by

$$\begin{aligned} H(e^{j\omega}) &= \sum_{n=-\infty}^{\infty} h(n)e^{-j\omega n} \\ &= \sum_{n=-\infty}^{\infty} h_d(n)w(n)e^{-j\omega n} \end{aligned}$$

The first step in the development is to replace $h_d(n)$ by its integral value given in Eq. (1.29). This results in

$$\begin{aligned} H(e^{j\omega}) &= \sum_{n=-\infty}^{\infty} w(n)e^{-j\omega n}\left(\frac{1}{2\pi}\int_{-\pi}^{\pi} H_d(e^{j\theta})e^{jn\theta}\, d\theta\right) \\ &= \frac{1}{2\pi}\int_{-\pi}^{\pi} H_d(e^{j\theta})\left(\sum_{n=-\infty}^{\infty} w(n)e^{-jn(\omega-\theta)}\right) d\theta \end{aligned}$$

where we have assumed the validity of term-by-term integration. The term in square brackets has the form of a Fourier transform in the continuous variable $\omega - \theta$. By making this substitution, the resulting expression, which is the complex convolution of the product of the sequences $h_d(n)$ and $w(n)$, becomes

$$H(e^{j\omega}) = \frac{1}{2\pi}\int_{-\pi}^{\pi} H_d(e^{j\theta})W(e^{j(\omega-\theta)})\, d\theta$$

If we had first replaced $w(n)$ by its value from Eq. (1.29), we would have obtained the equivalent expression

$$H(e^{j\omega}) = \frac{1}{2\pi}\int_{-\pi}^{\pi} W(e^{j\theta})H_d(e^{j(\omega-\theta)})\, d\theta \tag{6.12}$$

This expression could also have been obtained from the previous expression by a change of variable.

We can utilize Eq. (6.12) to obtain some useful information concerning the behavior of $H(e^{j\omega})$ as we vary the characteristics of the window. In order to obtain a successful design, $H(e^{j\omega})$ should be a reasonable approximation of $H_d(e^{j\omega})$. Let us first see if it is possible to choose $W(e^{j\omega})$ so that $H(e^{j\omega}) = H_d(e^{j\omega})$. If we compare this expression to the *sifting property* of the continuous-time impulse function $\delta_a(\omega)$, we see that this equality is attained if

$$W(e^{j\omega}) = 2\pi\delta_a(\omega) \qquad -\pi < \omega < \pi \tag{6.13}$$

This result is not surprising, of course, since for this situation we would have $w(n) = 1$ for all n. It appears that the development leading to Eq. (6.13) has no value since the resulting $w(n)$ has an infinite duration. However, we can see from this development that if we can find a finite-duration sequence $w(n)$ whose Fourier transform approximates the impulse function, then by Eq. (6.12) $H(e^{j\omega})$ approximates $H_d(e^{j\omega})$. We say that the frequency response $H(e^{j\omega})$ is a *smeared* version of $H_d(e^{j\omega})$.

Graphical Interpretation of Complex Convolution

To illustrate the usefulness of Eq. (6.12) in obtaining an idea of how the windowed frequency response looks, let us consider the symmetric rectangular window $w_{RS}(n)$ given in Eq. (6.7). Its frequency response is given by

$$\begin{aligned} W_{RS}(e^{j\omega}) &= \sum_{n=-M}^{M} e^{-j\omega n} = \frac{e^{j\omega M} - e^{-j\omega(M+1)}}{1 - e^{-j\omega}} \\ &= \frac{e^{j\omega(M+\frac{1}{2})} - e^{-j\omega(M+\frac{1}{2})}}{e^{j\omega/2} - e^{-j\omega/2}} = \frac{\sin \dfrac{(2M+1)\omega}{2}}{\sin \omega/2} \end{aligned} \tag{6.14}$$

which is a Chebyshev polynomial of the second kind in the variable $\xi = \cos \omega/2$ [78]. The frequency response is real and its zeros occur when $(2M + 1)\omega/2 = k\pi$ or $\omega = 2k\pi/(2M + 1)$, where k is an integer. The zeros satisfy the inequality $-\pi < \omega < \pi$ whenever $0 < |k| < (2M + 1)/2$. The response for ω between the consecutive zeros $-2\pi/(2M + 1)$ and $2\pi/(2M + 1)$ is called the *main lobe* and the remainder of the response makes up the *side lobes*. There are M ripples or complete oscillations in the interval $-\pi \le \omega \le \pi$. The largest amplitude occurs at $\omega = 0$ and has the value $2M + 1$, obtained in the limit as $\omega \to 0$, while the other ripple amplitudes decrease as ω increases. The graphs of $W_{RS}(e^{j\omega})$ for $M = 3$, 12, and 50 ($N = 7$, 25, and 101) are shown in Fig. 6.4.

In order for the frequency response $W_{RS}(e^{j\omega})$ to approximate an impulse function, the main lobe should be narrow and the side lobes should be insignificant. The main lobe can be narrowed and its peak value $2M + 1$ can be increased by increasing M. The side lobes grow in the same manner with decreased width, and the number of ripples in the response increases. These effects can be observed by comparing the responses shown in Fig. 6.4.

The side lobes of the window frequency response are responsible for the oscillatory behavior of the response $|H(e^{j\omega})|$ at a discontinuity of $|H_d(e^{j\omega})|$. This is the Gibbs phenomenon. To illustrate the origin of these oscillations, consider the desired response

$$H_d(e^{j\omega}) = \begin{cases} 1 & -8\pi/25 < \omega < 8\pi/25 \\ 0 & \text{otherwise} \end{cases}$$

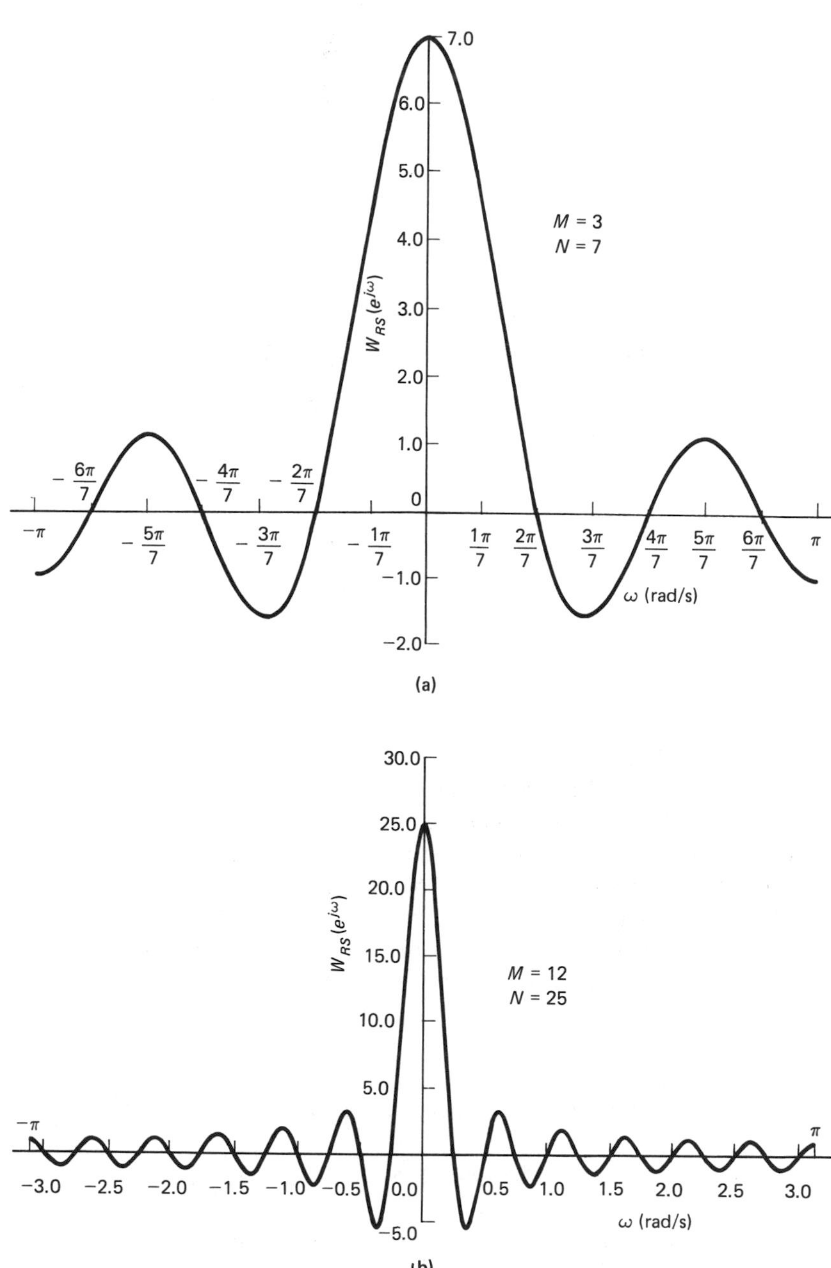

Figure 6.4 Graphs of $W_{RS}(e^{j\omega})$ for (a) $M = 3$, (b) $M = 12$, and (c) $M = 50$.

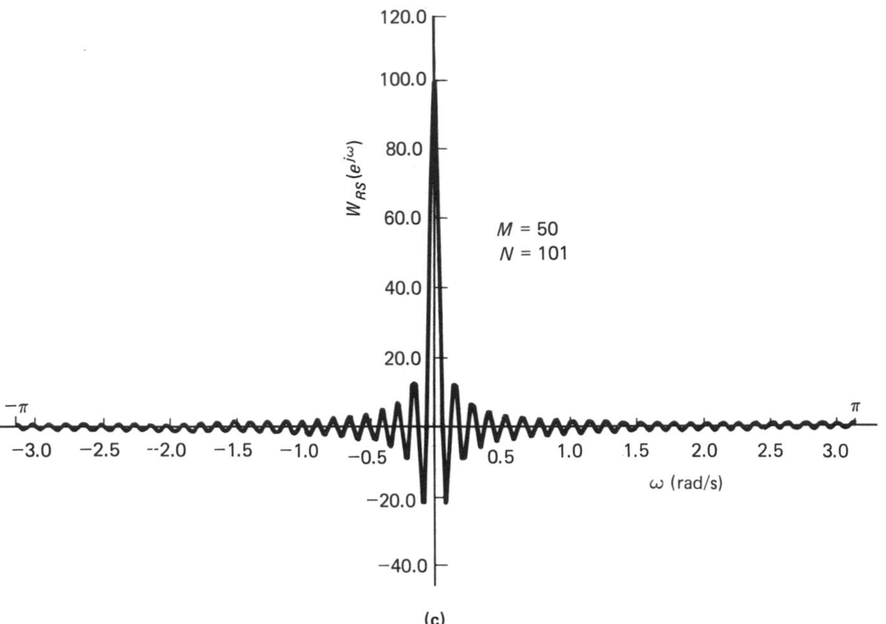

(c)

Figure 6.4 *(continued)*

The function $H_d(e^{j(\omega - \theta)})$ is shown as a rectangle in Fig. 6.5 for $\omega = 0$. The window chosen is $w_{RS}(n)$ with $M = 12$. Its zeros occur at $\omega = 2k\pi/25$ rad/s. As ω increases, the rectangle moves to the right and the area under $W(e^{j\theta})$ corresponding to $H(e^{j\omega})$ given by Eq. (6.12) increases until the right edge of the rectangle reaches $\theta = 10\pi/25$. As ω continues to increase, the area under $W(e^{j\theta})$ between the ends of the rectangle increases as the right edge of the rectangle moves from $8\pi/25$ to $10\pi/25$ (ω goes from 0 to $2\pi/25$), decreases as the right edge moves from $10\pi/25$ to $12\pi/25$ (ω goes from $2\pi/25$ to $4\pi/25$), and alternately increases and decreases until the right edge reaches $14\pi/25(\omega = 6\pi/25)$. The left edge is then at $\theta = -2\pi/25$. Now, as the left edge moves across the main lobe (ω goes from $6\pi/25$ to $10\pi/25$), the value of $H(e^{j\omega})$ decreases to a negative value. This results in the transition region. As ω continues to increase, $H(e^{j\omega})$ has ripples due to the ripples in $W(e^{j\theta})$.

This effect is shown in Fig. 6.5 for $M = 12$ $(N = 25)$ and

$$H_d(e^{j\omega}) = \begin{cases} 1 & -1 \leq \omega \leq 1 \\ 0 & 1 < |\omega| \leq \pi \end{cases}$$

This choice of $H_d(e^{j\omega})$ results in $H(e^{j\omega})$ having linear phase, so that (see Prob. 6.2)

$$H(e^{j\omega}) = H_1(\omega)$$

The graph of $H_1(\omega)$ is shown in Fig. 6.6. Oscillations occur before and after the discontinuity in $|H_d(e^{j\omega})|$ at $\omega = 1$ rad/s. Also, this discontinuity results in a transition band for $|H(e^{j\omega})|$.

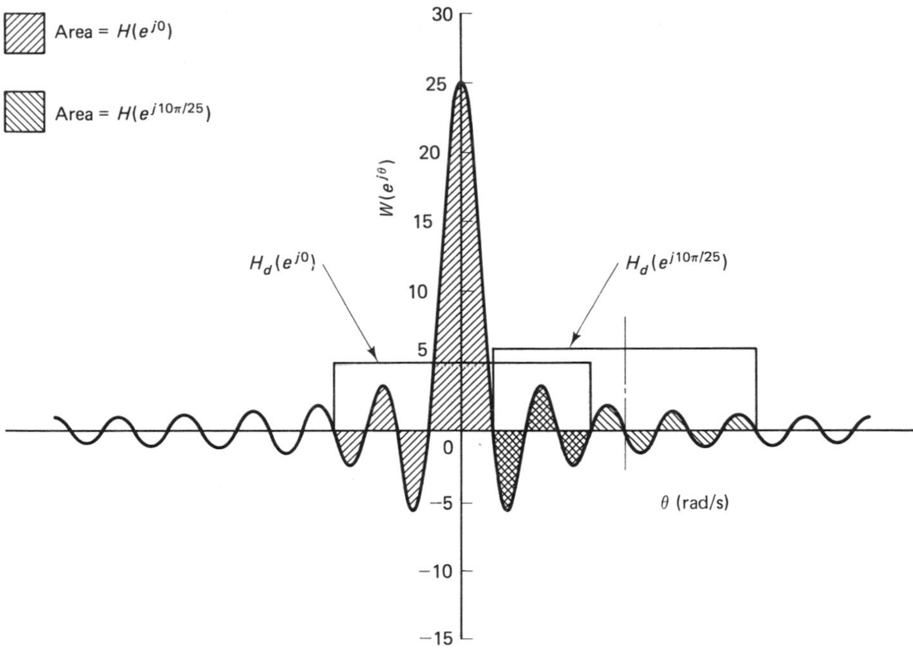

Figure 6.5 Graphical evaluation of the convolution integral of Eq. (6.12).

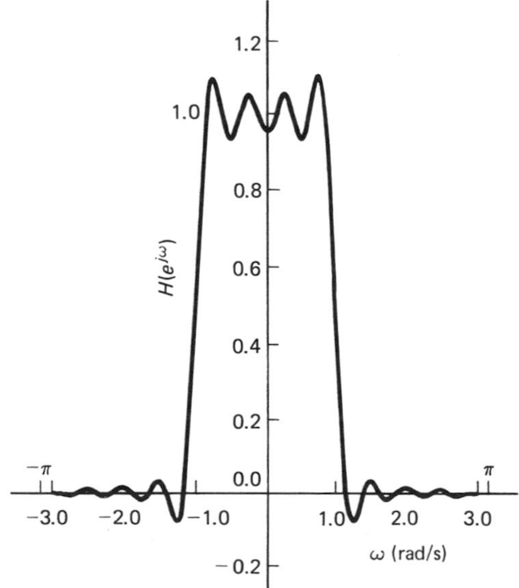

Figure 6.6 Graph of the smeared response $H(e^{j\omega})$ obtained from the convolution integral.

We have used the symmetric window because it has a real Fourier transform and this makes the analysis easier. It can be easily shown that the Fourier transform of the rectangular window $w_R(n)$ is

$$W_R(e^{j\omega}) = e^{-j\omega(N-1)/2} \frac{\sin \omega N/2}{\sin \omega/2} \tag{6.15}$$

The exponential function arises since $w_R(n)$ can be obtained by shifting $w_{RS}(n)$ to the right. This shift corresponds to multiplication by an exponential, as we saw in Ex. 3.7.3. Use of Eq. (6.15) complicates the evaluation of the convolution integral, making it essentially impossible to obtain results similar to those obtained using $w_{RS}(n)$.

The problems caused by the truncation of the Fourier series can be alleviated somewhat by tapering the window sequence smoothly to zero at each end. Windows of this type are considered in the following sections.

Some examples of filter design are considered in the exercises that follow. After we discuss the other windows of interest in Sec. 6.3, we devote Sec. 6.4 to the filter-design problem using windows and give numerous examples.

EXERCISES

6.2.1 The desired frequency response of a low-pass filter is given by

$$H_d(e^{j\omega}) = \begin{cases} e^{-j2\omega} & -\pi/4 \le \omega \le \pi/4 \\ 0 & \pi/4 < |\omega| \le \pi \end{cases}$$

Determine the filter coefficients $h_d(n)$. If we define new filter coefficients by $h(n) = w(n)h_d(n)$, where

$$w(n) = \begin{cases} 1 & 0 \le n \le 4 \\ 0 & \text{otherwise} \end{cases}$$

determine the response $H(e^{j\omega})$ of the new filter and compare $|H(e^{j\omega})|$ with $|H_d(e^{j\omega})|$.

6.2.2 Repeat Ex. 6.2.1 using

$$w(n) = 1 \qquad 1 \le n \le 3$$

$$w(0) = w(4) = \tfrac{1}{2}$$

$$w(n) = 0 \qquad \text{otherwise}$$

6.2.3 Repeat Ex. 6.2.1 if

$$H_d(e^{j\omega}) = \begin{cases} 0 & -\pi/4 \le \omega \le \pi/4 \\ e^{-j2\omega} & \pi/4 < |\omega| \le \pi \end{cases}$$

6.2.4 Find $h_d(n)$ and $H(e^{j\omega})$ if

$$H_d(e^{j\omega}) = \begin{cases} 1 & -\pi/4 \le \omega \le \pi/4 \\ 0 & \pi/4 < |\omega| \le \pi \end{cases}$$

and

$$w(n) = \begin{cases} 1 & -2 \le n \le 2 \\ 0 & \text{otherwise} \end{cases}$$

Is this a realizable filter? Compare this result with that of Ex. 6.2.1. Note that $|H_d(e^{j\omega})|$ is the same for these two cases.

6.3 OTHER COMMONLY USED WINDOWS

As we have seen in the previous discussion, use of the rectangular window, with its uniform weighting of the Fourier coefficients that are kept, is undesirable in many cases because of the resulting Gibbs phenomenon. In order to achieve better results, a more severe weighting of those coefficients that are kept in the truncation process is required. In this section, we look at some of the more commonly used windows.

Modified Rectangular Window

One of the first windows that was suggested as an alternate to the rectangular window was the *modified rectangular window* due to R. W. Hamming. It differs from the rectangular window only at the two end values, where $w(0) = w(N-1) = \frac{1}{2}$. The reader was asked to find the frequency response of a modified rectangular window for $N = 5$ in Ex. 6.2.2. The symmetric modified rectangular window is considered in Ex. 6.3.1.

Hann Window

A window that is more severely tapered but still relatively simple is the *Hann window,* due to J. von Hann. It is known also as the *raised cosine window* and the *hanning window*. The Hann window is defined by

$$w_h(n) = \frac{1}{2}\left[1 - \cos\left(\frac{2\pi n}{N-1}\right)\right] \qquad 0 \le n \le N-1 \tag{6.16}$$

Note that $w(0) = w(N-1) = 0$, so that there are actually only $N-2$ nonzero terms in the window sequence. A graph of the shifted window is shown in Fig. 6.7(a) for $N = 7$ ($M = 3$), and its frequency responses for $M = 3$ and $M = 12$ are shown in Figs. 6.7(b) and (c), respectively.

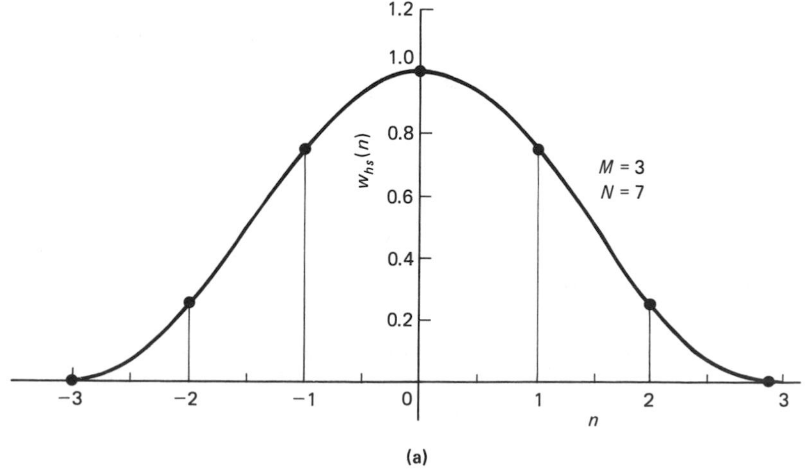

Figure 6.7 Graphs of (a) the shifted Hann window $w_{hs}(n)$ for $M = (N-1)/2 = 3$ and its frequency responses for (b) $M = 3$ and (c) $M = 12$.

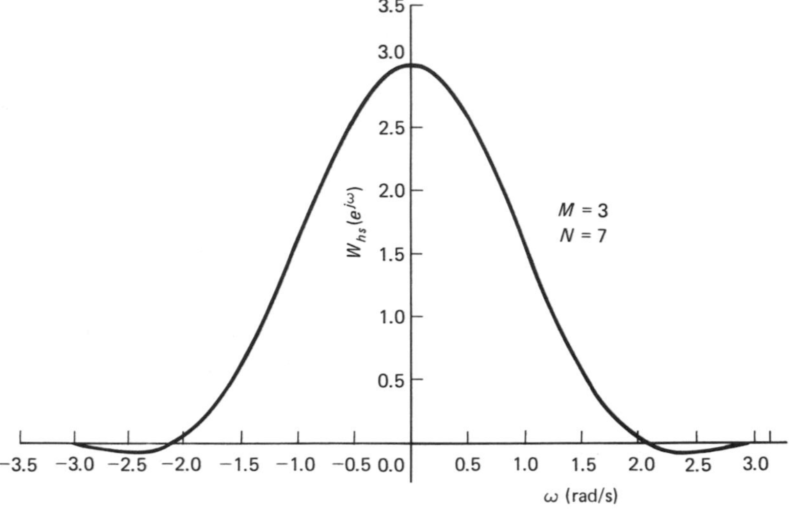

(b)

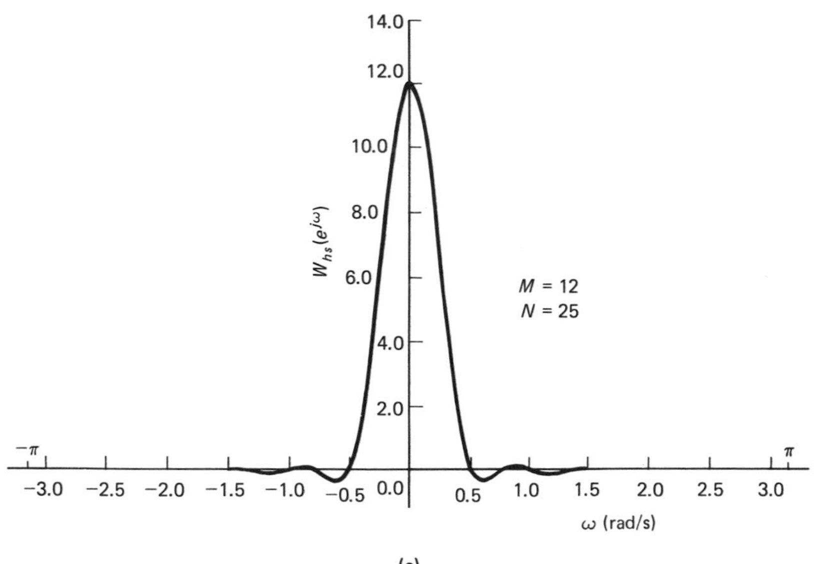

(c)

Figure 6.7 *(continued)*

Hamming Window

A window similar to the Hann window, but which has N nonzero terms is the *Hamming window,* given by R. W. Hamming. It is also called a *raised cosine with a platform* because of the nonzero terms at $n = 0$ and $n = N - 1$. The relation for the nonzero terms is

$$w_H(n) = 0.54 - 0.46 \cos\left(\frac{2\pi n}{N-1}\right) \qquad 0 \leq n \leq N - 1 \tag{6.17}$$

and it can be seen that $w(0) = w(N - 1) = 0.08$. The graph of the shifted window is shown in Fig. 6.8(a) for $N = 7$ ($M = 3$) and its frequency responses for $M = 3$ and $M = 12$ are shown in Figs. 6.8(b) and (c), respectively.

Generalized Hamming Window

The Hann and Hamming windows are special cases of a class of windows, the so-called *generalized Hamming window,* given by Rabiner and Gold [136]. It is de-

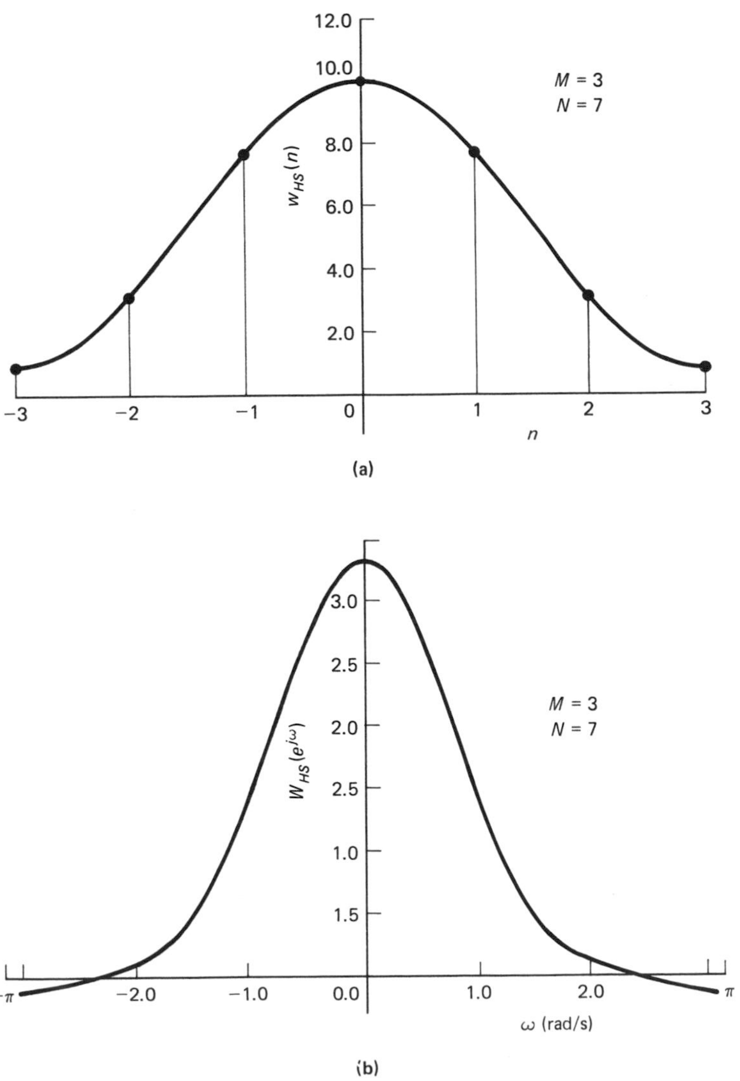

Figure 6.8 Graphs of (a) the shifted Hamming window $w_{HS}(n)$ for $M = (N - 1)/2 = 3$ and its frequency responses for (b) $M = 3$ and (c) $M = 12$.

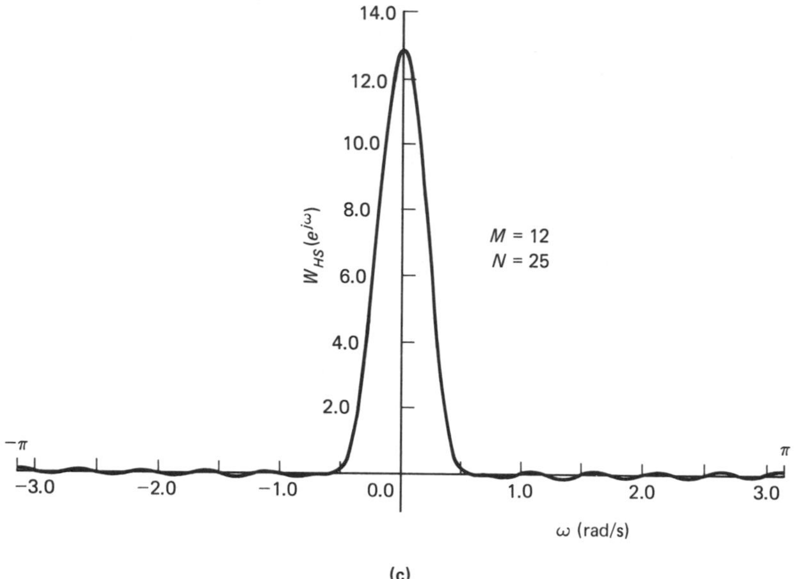

(c)

Figure 6.8 *(continued)*

scribed by

$$w_\alpha(n) = \alpha - (1 - \alpha)\cos\left(\frac{2\pi n}{N-1}\right) \qquad 0 \le n \le N-1 \tag{6.18}$$

where α is in the range $0 \le \alpha \le 1$. The Hann window results when $\alpha = 0.5$, and $\alpha = 0.54$ yields the Hamming window. The frequency response $W_\alpha(e^{j\omega})$ for the generalized Hamming window can be expressed in terms of $W_R(e^{j\omega})$, the frequency response of the rectangular window. By using the relation

$$\cos n\theta = \tfrac{1}{2}(e^{jn\theta} + e^{-jn\theta}) \tag{6.19}$$

it is found that

$$\begin{aligned} W_\alpha(e^{j\omega}) = \alpha W_R(e^{j\omega}) &- \frac{1-\alpha}{2} W_R(e^{j[\omega - 2\pi/(N-1)]}) \\ &- \frac{1-\alpha}{2} W_R(e^{j[\omega + 2\pi/(N-1)]}) \end{aligned} \tag{6.20}$$

The reader is asked to obtain this expression in Prob. 6.7.

It is more informative to consider the symmetric generalized Hamming window,

$$w_{\alpha s}(n) = \begin{cases} \alpha + (1-\alpha)\cos \pi n/M & -M \le n \le M \\ 0 & \text{otherwise} \end{cases} \tag{6.21}$$

and find its frequency response in terms of $W_{RS}(e^{j\omega})$. Using the relation of Eq. (6.19), we can write

$$W_{\alpha s}(e^{j\omega}) = \sum_{n=-M}^{M} \left[\alpha + \frac{1-\alpha}{2}\left(e^{j\pi n/M} + e^{-j\pi n/M}\right) \right] e^{-jn\omega}$$

$$= \alpha \sum_{n=-M}^{M} e^{-jn\omega} + \frac{1-\alpha}{2} \sum_{n=-M}^{M} e^{-jn(\omega-\pi/M)}$$

$$+ \frac{1-\alpha}{2} \sum_{n=-M}^{M} e^{-jn(\omega+\pi/M)}$$

Using the summation for $W_{RS}(e^{j\omega})$ in Eq. (6.14), we see that the previous expression can be written

$$W_{\alpha s}(e^{j\omega}) = \alpha W_{RS}(e^{j\omega}) + \frac{1-\alpha}{2} W_{RS}(e^{j(\omega-\pi/M)}) + \frac{1-\alpha}{2} W_{RS}(e^{j(\omega+\pi/M)}) \tag{6.22}$$

In Fig. 6.9(a), the graphs of the three components of Eq. (6.22) are shown, and the composite graph is shown in Fig. 6.9(b). The width of the main lobe is ap-

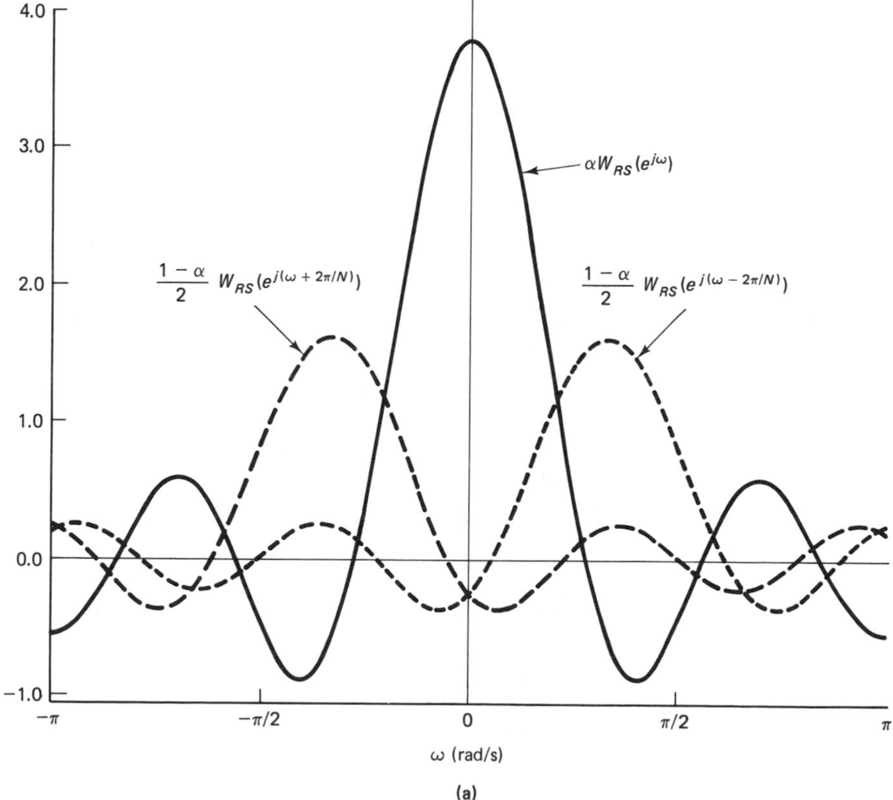

Figure 6.9 Frequency response of a Hamming window. (a) Graphs of the components of Eq. (6.22) and (b) the resulting composite graph.

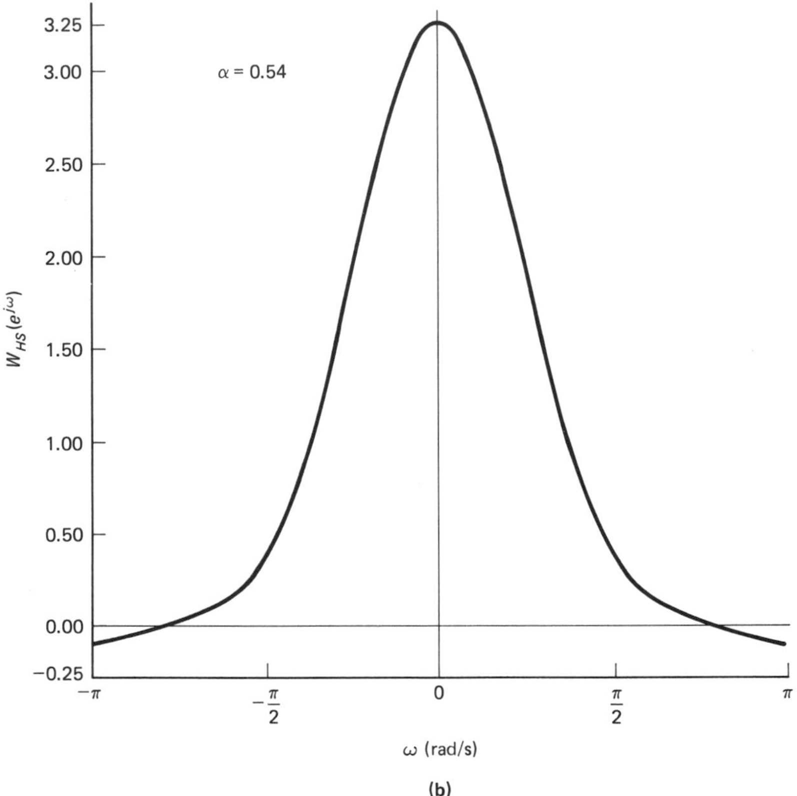

Figure 6.9 *(continued)*

proximately $8\pi/(2M + 1)$, which is about twice that of the rectangular window. This increased main-lobe width makes the transition band wider for the same value of M. The side-lobe amplitudes of the generalized Hamming window are considerably smaller than for the rectangular window. This reduction makes for smaller ripples in $H(e^{j\omega})$. These results can be obtained from a qualitative analysis of the convolution integral of Eq. (6.12). The frequency responses for the generalized Hamming window are shown in Fig. 6.10 for $M = 3$ and $\alpha = 0.3$, 0.4, and 0.9.

Other Windows

Some other windows that have been used are the *Bartlett triangular window*, defined by

$$w(n) = \begin{cases} \dfrac{2n}{N-1} & 0 \le n \le \dfrac{N-1}{2} \\[2ex] 2 - \dfrac{2n}{N-1} & \dfrac{N-1}{2} \le n \le N-1 \end{cases} \tag{6.23}$$

and the *Blackman window*, given by

$$w(n) = 0.42 - 0.5 \cos\left(\frac{2\pi n}{N-1}\right) + 0.08 \cos\left(\frac{4\pi n}{N-1}\right) \qquad 0 \le n \le N-1 \tag{6.24}$$

Other windows are the *Kaiser window* and the *Dolph–Chebyshev window,* which are essentially optimum windows. The Bartlett, Blackman, and Dolph–Chebyshev windows are considered in the problems at the end of the chapter. Graphs of $w(n)$ for $M = (N - 1)/2 = 3$ and $W(e^{j\omega})$ for $M = 3$ and $M = 10$ for the Bartlett and

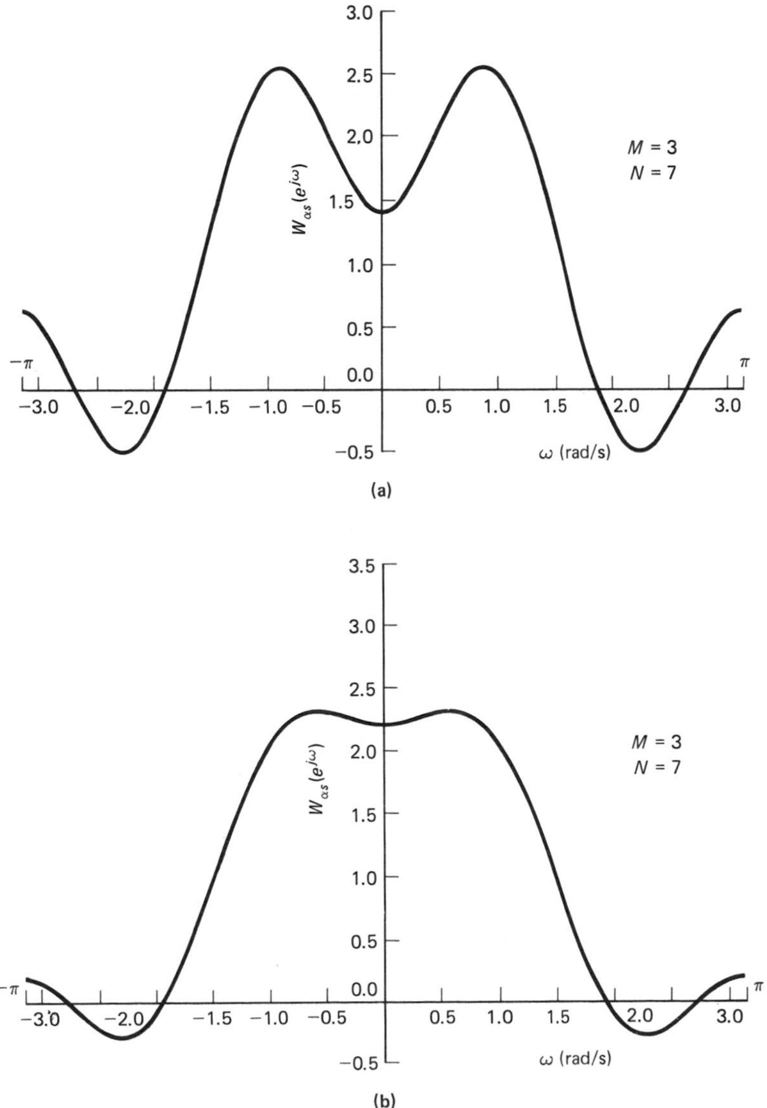

Figure 6.10 Graphs of the frequency responses of the generalized Hamming window for $M = 3$ and (a) $\alpha = 0.3$, (b) $\alpha = 0.4$, and (c) $\alpha = 0.9$.

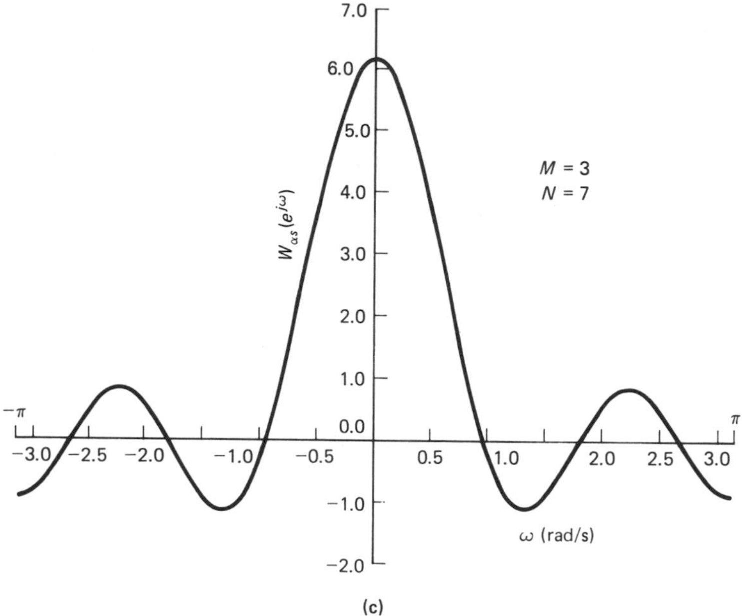

Figure 6.10 *(continued)*

Blackman windows are shown in Figs. 6.11 and 6.12, respectively. The Kaiser window, because of its importance, is the subject of Sec. 6.5. Actually, the Kaiser and Dolph–Chebyshev windows are classes of windows like the generalized Hamming window, since they have variable parameters.

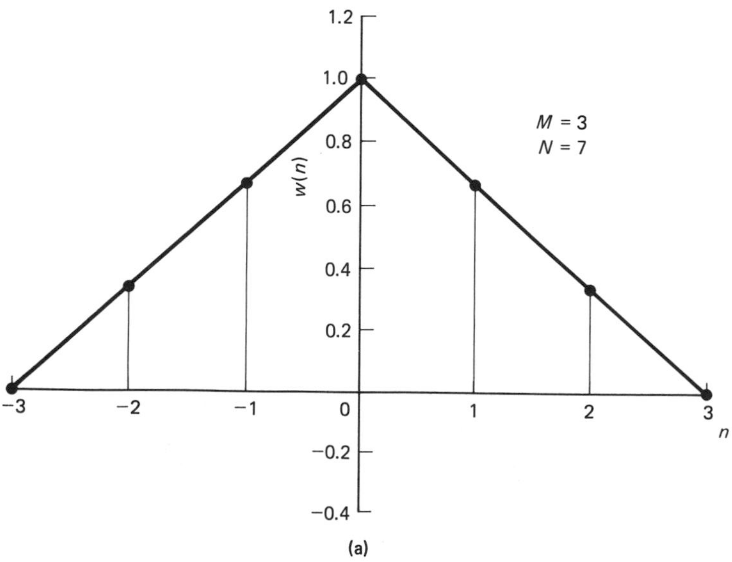

Figure 6.11 Graphs of (a) the Bartlett window $w(n)$ for $M = (N - 1)/2 = 3$ and the frequency responses for (b) $M = 3$ and (c) $M = 10$.

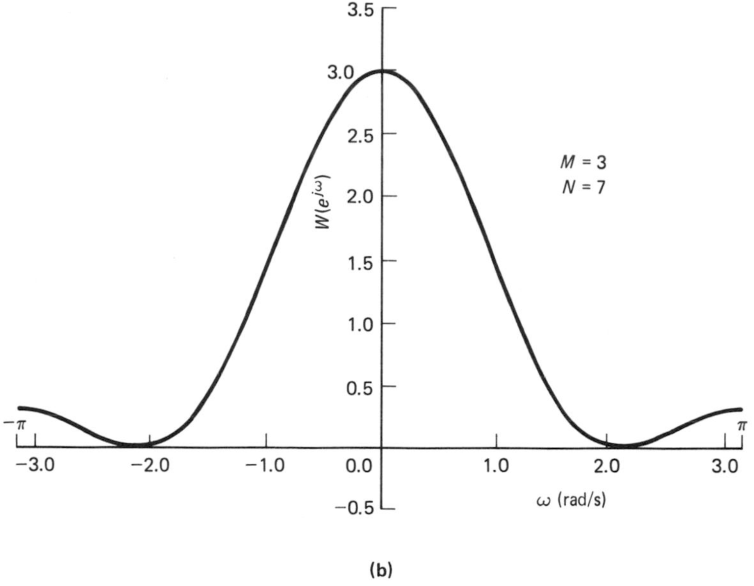

(b)

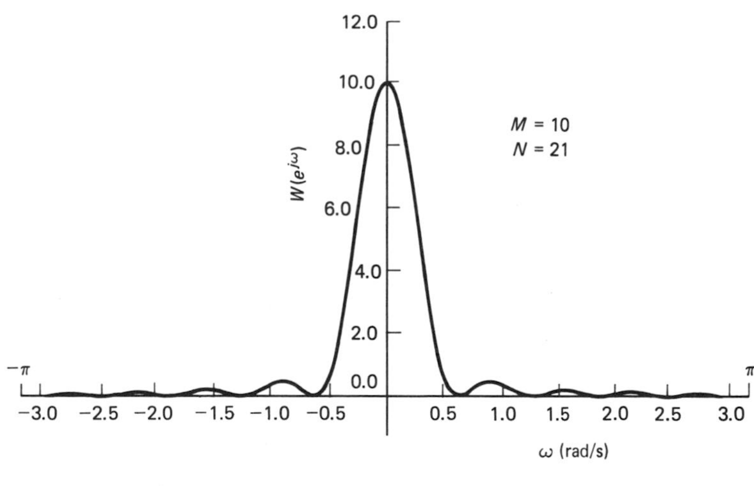

(c)

Figure 6.11 *(continued)*

Symmetry Properties

It should be noted that all windows described thus far have the symmetry property

$$w(n) = w(N - 1 - n)$$

or

$$w(n) = w(-n)$$

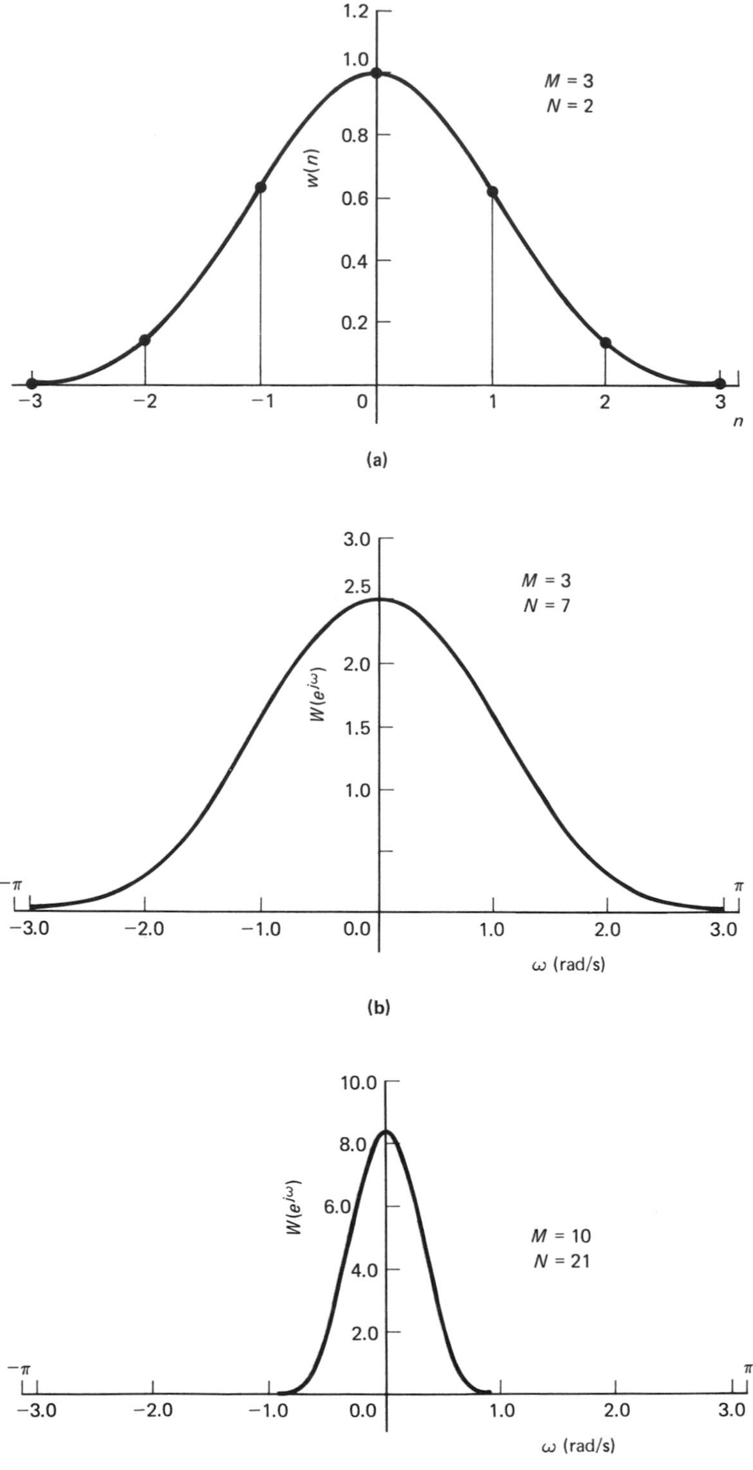

Figure 6.12 Graphs of (a) the Blackman window $w(n)$ for $M = (N - 1)/2 = 3$ and the frequency responses for (b) $M = 3$ and (c) $M = 10$.

depending on whether the range is $0 \leq n \leq N - 1$ or $-(N - 1)/2 \leq n \leq (N - 1)/2 = M$. If we require that $h_d(n)$ have either of these symmetry properties, too, then so will $h(n)$. The resulting FIR filter then has linear phase.

For more information on windows, the reader should see also [34, 55, 56, 57, 87, 88, 103, 121, 122, 126, 136, 141].

EXERCISES

6.3.1 Determine the frequency response for the symmetric modified rectangular window given by

$$\begin{aligned} w(n) &= 1 \qquad -M + 1 \leq n \leq M - 1 \\ w(-M) &= w(M) = \tfrac{1}{2} \\ w(n) &= 0 \qquad \text{otherwise} \end{aligned}$$

Draw its graph for $M = 3$.

6.3.2 Show that the substitution $n' = n + (N - 1)/2$, N odd, transforms a window $w(n')$, $0 \leq n' \leq N - 1$, into a window with $-(N - 1)/2 \leq n \leq (N - 1)/2$. Apply this substitution to the Hann, Hamming, and generalized Hamming windows and obtain windows symmetric about the origin.

6.3.3 Determine the frequency response for the symmetric (shifted) Hann window given by

$$w(n) = \begin{cases} \frac{1}{2}(1 + \cos \pi n/M) & -M \leq n \leq M \\ 0 & \text{otherwise} \end{cases}$$

Consider the special case where $M = 1$ and draw the graph.

6.3.4 Repeat Ex. 6.3.3 for the symmetric (shifted) Hamming window given by

$$w(n) = \begin{cases} 0.54 + 0.46 \cos \pi n/M & -M \leq n \leq M \\ 0 & \text{otherwise} \end{cases}$$

6.4 EXAMPLES OF FILTER DESIGNS USING WINDOWS

Low-Pass Filter Coefficients

As an example illustrating the use of windows in the design of a low-pass FIR filter, consider the desired frequency response given by

$$H_d(e^{j\omega}) = \begin{cases} e^{-j\omega\tau} & |\omega| \leq \omega_c < \pi \\ 0 & \text{otherwise} \end{cases} \tag{6.25}$$

for $|\omega| \leq \pi$. The impulse-response sequence determined by this frequency response is obtained from

$$h_d(n) = \frac{1}{2\pi}\int_{-\pi}^{\pi} H_d(e^{j\omega})e^{j\omega n}\, d\omega$$

and for the special case of Eq. (6.25) is

$$\begin{aligned} h_d(n) &= \frac{1}{2\pi}\int_{-\omega_c}^{\omega_c} e^{-j\omega(\tau-n)}\, d\omega \\ &= \frac{1}{2\pi}\frac{e^{j\omega(n-\tau)}}{j(n-\tau)}\bigg|_{-\omega_c}^{\omega_c} \\ &= \frac{\sin \omega_c(n-\tau)}{\pi(n-\tau)} \qquad n \neq \tau \\ &= \omega_c/\pi \qquad n = \tau \end{aligned} \tag{6.26}$$

If τ is an integer, then $h_d(n)$ for $n = \tau$ is computed from the integral or is obtained from the first of Eqs. (6.26) by a limiting process. To obtain a finite-duration causal sequence of length N, we take

$$h(n) = h_d(n)w(n) \tag{6.27}$$

where $w(n)$ is any of the causal windows that we have discussed.

If we desire a linear-phase filter, then we must choose τ to satisfy the symmetry condition $h(n) = h(N-1-n)$, which for our case is

$$\frac{\sin \omega_c(N-1-n-\tau)}{N-1-n-\tau} = \frac{\sin \omega_c(n-\tau)}{n-\tau}$$

This can be accomplished only if

$$N-1-n-\tau = -(n-\tau)$$

which requires that

$$\tau = \frac{N-1}{2}$$

It is shown in Prob. 6.2 that $\tau = 0$ for symmetry about $n = 0$.

Examples of Low-Pass Filter Designs

We give some simple design examples illustrating the use of rectangular, Hann, and Hamming windows. In particular, examples of low-pass, high-pass, bandpass, and band-reject filters are given for $N = 7$ and $N = 25$. The frequency response for linear-phase FIR filters is given in Eqs. (2.85) and (2.86) for N odd and is repeated here for convenience as

$$H(e^{j\omega}) = e^{-j(N-1)\omega/2}H_1(\omega)$$

where the pseudomagnitude $H_1(\omega)$ is given by

$$H_1(\omega) = h\left(\frac{N-1}{2}\right) + 2\sum_{n=1}^{(N-1)/2} h\left(\frac{N-1}{2} - n\right)\cos n\omega \tag{6.28}$$

In the design examples to follow, we take $N = 7$ and, consequently, we require the desired filter coefficients $h_d(n)$ for $0 \le n \le 6$. For a low-pass filter, these coefficients are given by Eq. (6.26). By taking $N = 7$, $\tau = 3$, *and* $\omega_c = 1$ rad/s, these coefficients are

$$\begin{aligned} h_d(0) &= h_d(6) = \frac{\sin 3}{3\pi} = 0.01497 \\ h_d(1) &= h_d(5) = \frac{\sin 2}{2\pi} = 0.14472 \\ h_d(2) &= h_d(4) = \frac{\sin 1}{\pi} = 0.26785 \\ h_d(3) &= \frac{1}{\pi} = 0.31831 \end{aligned} \tag{6.29}$$

For $N = 7$ the pseudomagnitude of Eq. (6.28) becomes

$$H_1(\omega) = h(3) + 2h(2)\cos\omega + 2h(1)\cos 2\omega + 2h(0)\cos 3\omega \tag{6.30}$$

Using a rectangular window, we have $w(n) = 1$ and $h(n) = h_d(n)$, $0 \le n \le 6$. For the coefficients given in Eq. (6.29), the pseudomagnitude $H_1(\omega)$ of Eq. (6.30) has the form

$$H_1(\omega) = 0.31831 + 0.5357\cos\omega + 0.28944\cos 2\omega + 0.02994\cos 3\omega \tag{6.31}$$

The graph of $H_1(\omega)$ is shown in Fig. 6.13(a).

The next type of window to be considered is the Hann window described by Eq. (6.16). For our case, this becomes

$$w(n) = \tfrac{1}{2}(1 - \cos n\pi/3) \qquad 0 \le n \le 6$$

with the particular values

$$\begin{aligned} w(0) &= w(6) = 0 \\ w(1) &= w(5) = \tfrac{1}{2}(1 - \cos\pi/3) = \tfrac{1}{4} \\ w(2) &= w(4) = \tfrac{1}{2}(1 - \cos 2\pi/3) = \tfrac{3}{4} \\ w(3) &= \tfrac{1}{2}(1 - \cos\pi) = 1 \end{aligned} \tag{6.32}$$

The FIR filter coefficients, using Eqs. (6.27), (6.29), and (6.32), are

$$\begin{aligned} h(0) &= h(6) = 0 \\ h(1) &= h(5) = (0.14472)(0.25) = 0.03618 \\ h(2) &= h(4) = (0.26785)(0.75) = 0.20089 \\ h(3) &= (0.31831)(1) = 0.31831 \end{aligned}$$

and the resulting pseudomagnitude is

$$H_1(\omega) = 0.31831 + 0.40178\cos\omega + 0.07236\cos 2\omega \tag{6.33}$$

The graph of $H_1(\omega)$ is shown in Fig. 6.13(b).

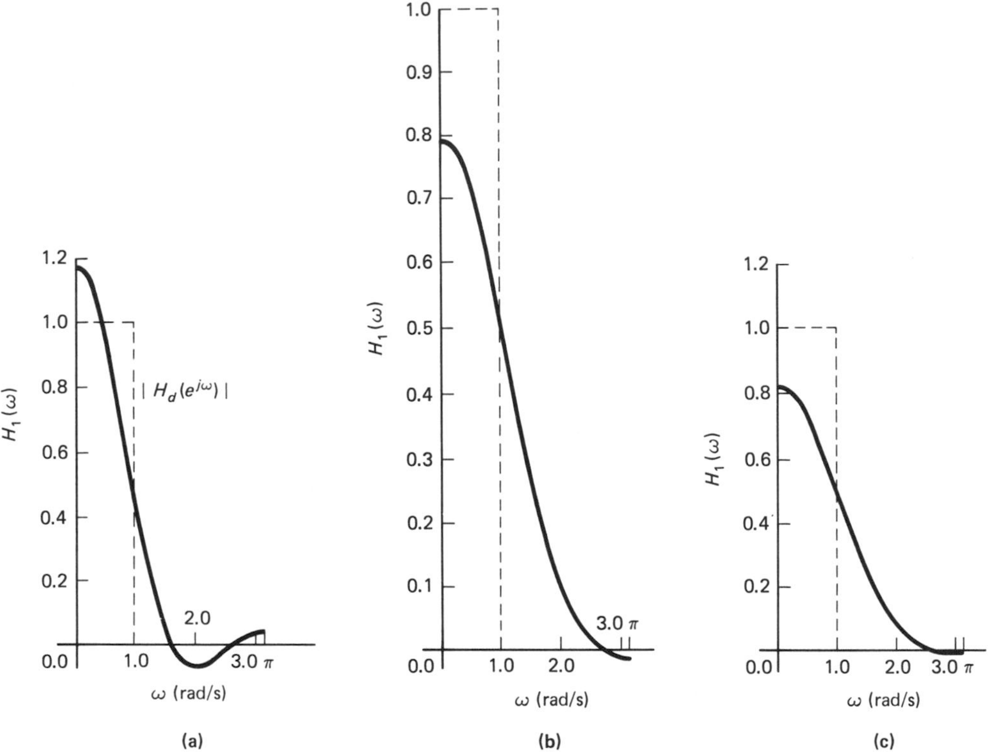

Figure 6.13 Graphs of the low-pass function $H_1(\omega)$ for (a) a rectangular window, (b) a Hann window, and (c) a Hamming window, all for $M = 3$. The desired responses are shown by the dashed lines.

Now using a Hamming window, we have, from Eq. (6.17),

$$w(n) = 0.54 - 0.46 \cos 2\pi n/6 \qquad 0 \leq n \leq 6$$

and, in particular,

$$\begin{aligned} w(0) &= w(6) = 0.08 \\ w(1) &= w(5) = 0.54 - 0.46 \cos \pi/3 = 0.31 \\ w(2) &= w(4) = 0.54 - 0.46 \cos 2\pi/3 = 0.77 \\ w(3) &= 1 \end{aligned} \tag{6.34}$$

Using the coefficients $h_d(n)$ given in Eq. (6.29), we find that the FIR filter has pseudomagnitude

$$H_1(\omega) = 0.31831 + 0.41249 \cos \omega + 0.08973 \cos 2\omega + 0.0024 \cos 3\omega \tag{6.35}$$

The corresponding graph is shown in Fig. 6.13(c).

High-Pass Filters

To design a high-pass filter, we take the desired frequency response to be

$$H_d(e^{j\omega}) = \begin{cases} e^{-j\omega\tau} & \omega_c \le |\omega| \le \pi \\ 0 & \text{otherwise} \end{cases} \tag{6.36}$$

where $\tau = (N - 1)/2$. The impulse-response coefficients, from Eqs. (1.29) and (6.36), are

$$h_d(n) = \frac{1}{2\pi}\left(\int_{-\pi}^{-\omega_c} e^{j\omega(n-\tau)}\, d\omega + \int_{\omega_c}^{\pi} e^{j\omega(n-\tau)}\, d\omega\right)$$

$$= \frac{1}{\pi(n-\tau)}[\sin(n-\tau)\pi - \sin(n-\tau)\omega_c] \qquad n \ne \tau$$

If N is odd, then τ is an integer and

$$h_d(\tau) = h_d\left(\frac{N-1}{2}\right) = \frac{1}{\pi}(\pi - \omega_c) = 1 - \frac{\omega_c}{\pi} \tag{6.37}$$

For example, if $N = 7$ and $\omega_c = 2$ rad/s, then $\tau = 3$ and

$$\begin{aligned} h_d(n) &= -\frac{\sin 2(n-3)}{\pi(n-3)} \qquad n \ne 3 \\ h_d(3) &= 1 - 2/\pi = 0.36338 \end{aligned} \tag{6.38}$$

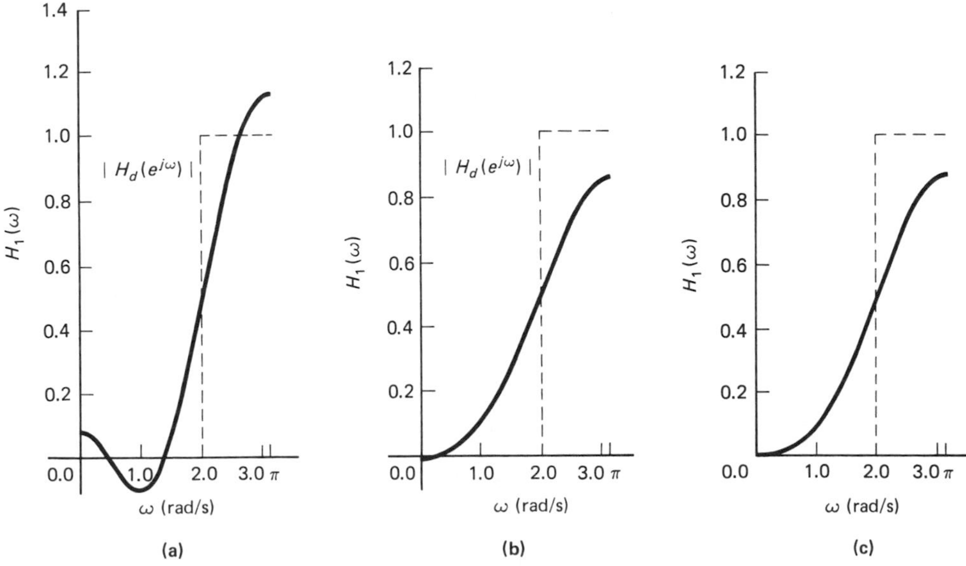

Figure 6.14 Graphs of the high-pass function $H_1(\omega)$ for (a) a rectangular window, (b) a Hann window, and (c) a Hamming window, all for $M = 3$. The desired responses are shown by the dashed lines.

The coefficients for $n \neq 3$ are

$$\begin{aligned} h_d(0) &= h_d(6) = 0.02965 \\ h_d(1) &= h_d(5) = 0.12045 \\ h_d(2) &= h_d(4) = -0.28944 \end{aligned} \tag{6.39}$$

By using a rectangular window, the pseudomagnitude is

$$H_1(\omega) = 0.36338 - 0.57888 \cos \omega + 0.2409 \cos 2\omega + 0.0593 \cos 3\omega \tag{6.40}$$

For a Hann window, the filter coefficients are obtained from Eqs. (6.27), (6.32), (6.38), and (6.39) with the result that

$$H_1(\omega) = 0.36338 - 0.43416 \cos \omega + 0.06022 \cos 2\omega \tag{6.41}$$

By using the window coefficients of Eq. (6.34) for the Hamming window, the pseudomagnitude response is

$$H_1(\omega) = 0.36338 - 0.44574 \cos \omega + 0.07468 \cos 2\omega + 0.00474 \cos 3\omega \tag{6.42}$$

The graphs of $H_1(\omega)$ for these three cases are shown in Fig. 6.14.

Bandpass Filters

In the case of a bandpass filter, the desired frequency response is

$$H_d(e^{j\omega}) = \begin{cases} e^{-j\omega\tau} & \omega_{c_1} \leq |\omega| \leq \omega_{c_2} < \pi \\ 0 & \text{otherwise} \end{cases} \tag{6.43}$$

where $\tau = (N - 1)/2$. The filter coefficients are given by

$$h_d(n) = \frac{1}{2\pi}\left[\int_{-\omega_{c2}}^{-\omega_{c1}} e^{j(n-\tau)\omega}\, d\omega + \int_{\omega_{c1}}^{\omega_{c2}} e^{j(n-\tau)\omega}\, d\omega\right] \tag{6.44}$$

$$= \frac{1}{(n-\tau)\pi}[\sin (n - \tau)\omega_{c_2} - \sin (n - \tau)\omega_{c_1}] \qquad n \neq \tau \tag{6.45}$$

$$= \frac{1}{\pi}(\omega_{c_2} - \omega_{c_1}) \qquad n = \tau \tag{6.46}$$

For the special case $N = 7$, $\omega_{c_1} = 1$ rad/s and $\omega_{c_2} = 2$ rad/s, we have $\tau = 3$ and

$$\begin{aligned} h_d(0) &= h_d(6) = -0.04462 \\ h_d(1) &= h_d(5) = -0.26517 \\ h_d(2) &= h_d(4) = 0.02159 \\ h_d(3) &= 0.31831 \end{aligned} \tag{6.47}$$

For a rectangular window,

$$H_1(\omega) = 0.31831 + 0.04318 \cos \omega - 0.53034 \cos 2\omega - 0.08924 \cos 3\omega \tag{6.48}$$

for a Hann window,

$$H_1(\omega) = 0.31831 + 0.03238 \cos \omega - 0.13258 \cos 2\omega \tag{6.49}$$

and for a Hamming window,

$$H_1(\omega) = 0.31831 + 0.03325 \cos \omega - 0.16441 \cos 2\omega - 0.00714 \cos 3\omega \tag{6.50}$$

The graphs of $H_1(\omega)$ for each of these cases are shown in Fig. 6.15.

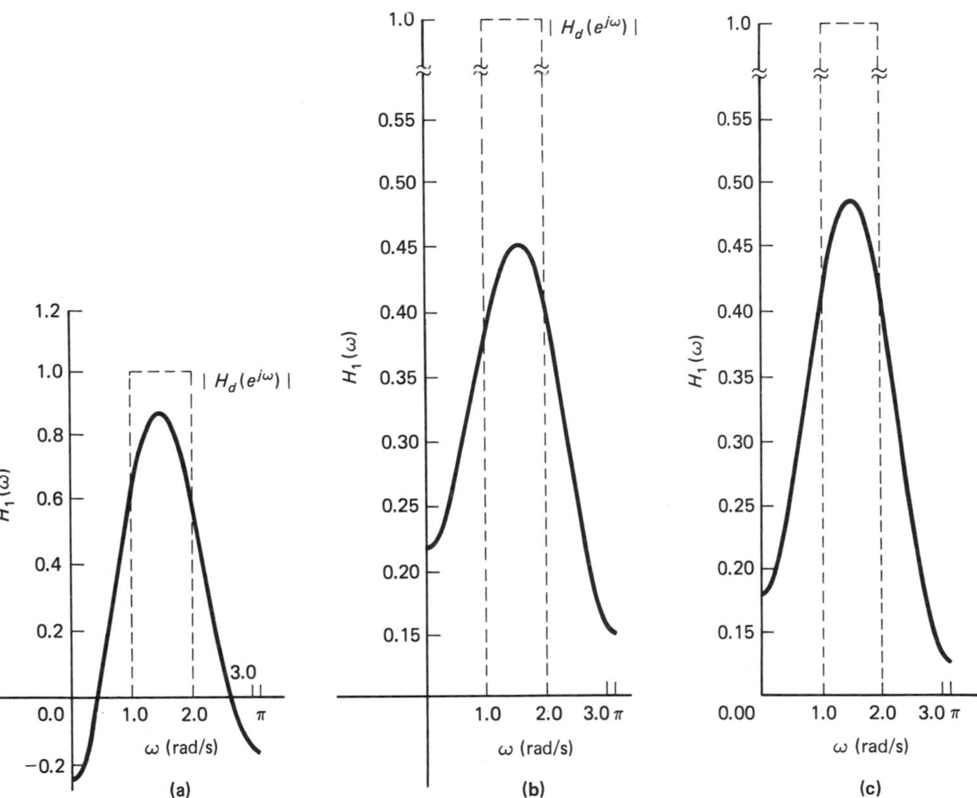

Figure 6.15 Graphs of the bandpass function $H_1(\omega)$ for (a) a rectangular window, (b) a Hann window, and (c) a Hamming window, all for $M = 3$. The desired responses are shown by the dashed lines.

Band-Reject Filters

Finally, for a band-reject filter, the desired frequency response is

$$H_d(e^{j\omega}) = \begin{cases} e^{-j\omega\tau} & 0 \le |\omega| \le \omega_{c_1},\ \omega_{c_2} \le |\omega| \le \pi \\ 0 & \text{otherwise} \end{cases} \tag{6.51}$$

with filter coefficients

$$h_d(n) = \frac{1}{2\pi}\left(\int_{-\pi}^{-\omega_{c2}} e^{j(n-\tau)\omega}\, d\omega + \int_{-\omega_{c1}}^{\omega_{c1}} e^{j(n-\tau)\omega}\, d\omega + \int_{\omega_{c2}}^{\pi} e^{j(n-\tau)\omega}\, d\omega\right)$$

$$= \frac{1}{\pi(n-\tau)}[\sin (n-\tau)\omega_{c_1} - \sin (n-\tau)\omega_{c_2} + \sin (n-\tau)\pi] \qquad n \neq \tau \tag{6.52}$$

$$= \frac{1}{\pi}(\pi - \omega_{c_2} + \omega_{c_1}) \qquad n = \tau \tag{6.53}$$

If we take $N = 7$, $\omega_{c_1} = 1$ rad/s and $\omega_{c_2} = 2$ rad/s, then $\tau = 3$ and

$$h_d(0) = h_d(6) = 0.04462$$
$$h_d(1) = h_d(5) = 0.26517$$
$$h_d(2) = h_d(4) = -0.02159$$
$$h_d(3) = 0.68169$$

The frequency response for the filter obtained using a rectangular window has

$$H_1(\omega) = 0.68169 - 0.04318 \cos \omega + 0.53034 \cos 2\omega + 0.08924 \cos 3\omega \tag{6.54}$$

If a Hann window is used, then

$$H_1(\omega) = 0.618169 - 0.03238 \cos \omega + 0.13258 \cos 2\omega \tag{6.55}$$

The filter resulting from a Hamming window has

$$H_1(\omega) = 0.68169 - 0.03325 \cos \omega + 0.16441 \cos 2\omega + 0.00714 \cos 3\omega \tag{6.56}$$

The graphs of $H_1(\omega)$ for these three cases are shown in Fig. 6.16.

Further Examples

The results obtained for $N = 7$ are useful in providing the reader with examples that can be worked out using a handheld calculator. In Figs. 6.17–6.20 are shown the graphs of $H_1(\omega)$ for the four types of filters when $N = 25$. These latter graphs are much better approximations to the ideal responses than those obtained for $N = 7$. Finally, in Figs. 6.21–6.23 are responses of a low-pass filter using a rectangular window, a bandpass filter using a Hann window, and a low-pass filter using a Hamming window, respectively, all with $N = 1024$. In Fig. 6.21, the Gibbs effect is clearly evident for the rectangular window. In Fig. 6.22, a slight overshoot at $\omega = 1$ and $\omega = 2$ rad/s is evident and the overshoot for the Hamming window in Fig. 6.23 is not noticeable.

A program written by Rabiner, McGonegal and Paul [34, 141] is available for designing windowed filters. The program is capable of designing low-pass, high-pass, bandpass, and band-reject filters using either a rectangular, a Bartlett, a Hamming, a Hann, a generalized Hamming, a Kaiser, or a Chebyshev window.

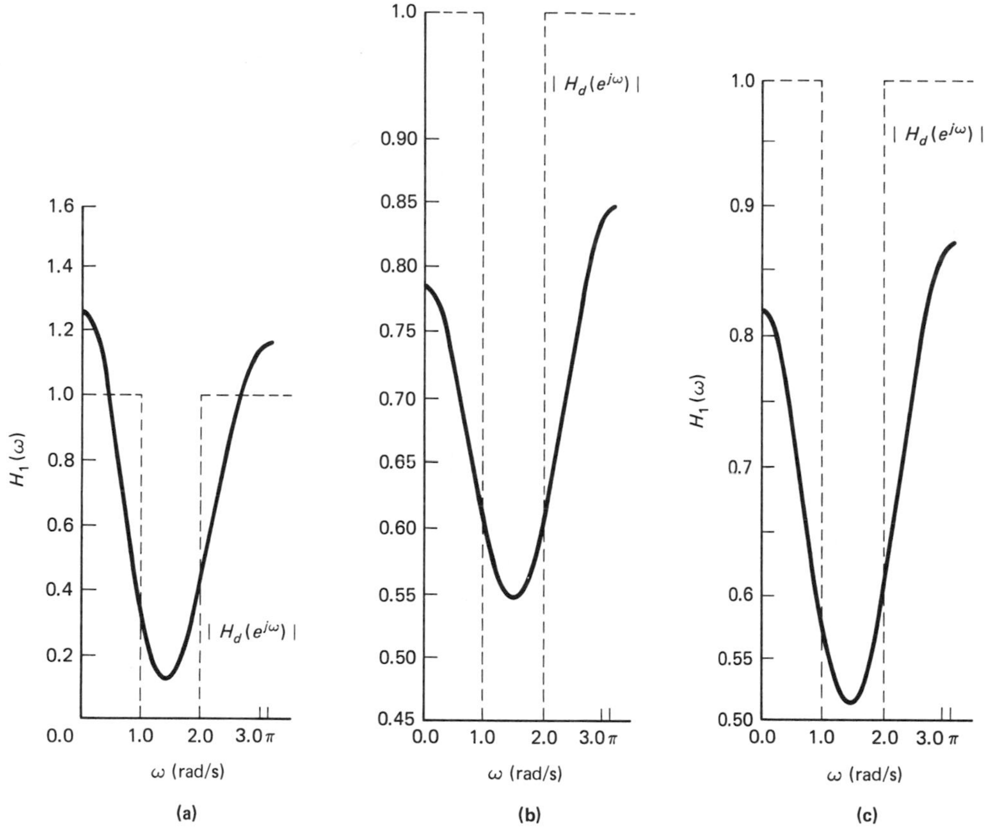

Figure 6.16 Graphs of the band-reject function $H_1(\omega)$ for (a) a rectangular window, (b) a Hann window, and (c) a Hamming window, all for $M = 3$. The desired responses are shown by the dashed lines.

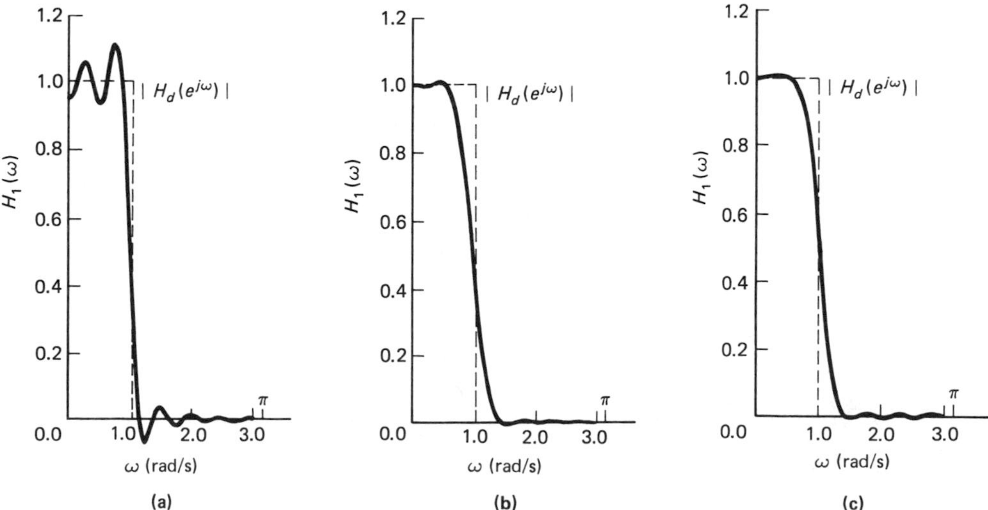

Figure 6.17 Graphs of the low-pass function $H_1(\omega)$ for (a) a rectangular window, (b) a Hann window, and (c) a Hamming window, all for $M = 12$. The desired responses are shown by the dashed lines.

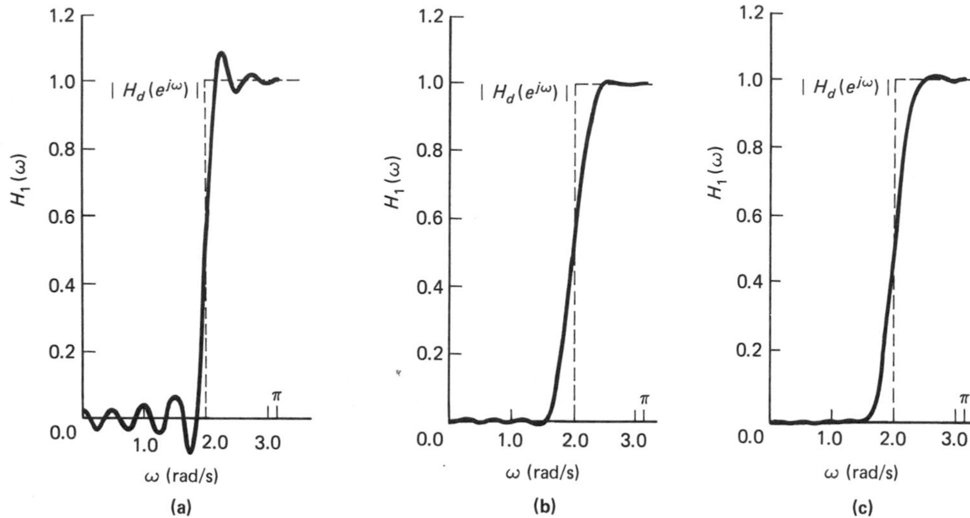

Figure 6.18 Graphs of the high-pass function $H_1(\omega)$ for (a) a rectangular window, (b) a Hann window, and (c) a Hamming window, all for $M = 12$. The desired responses are shown by the dashed lines.

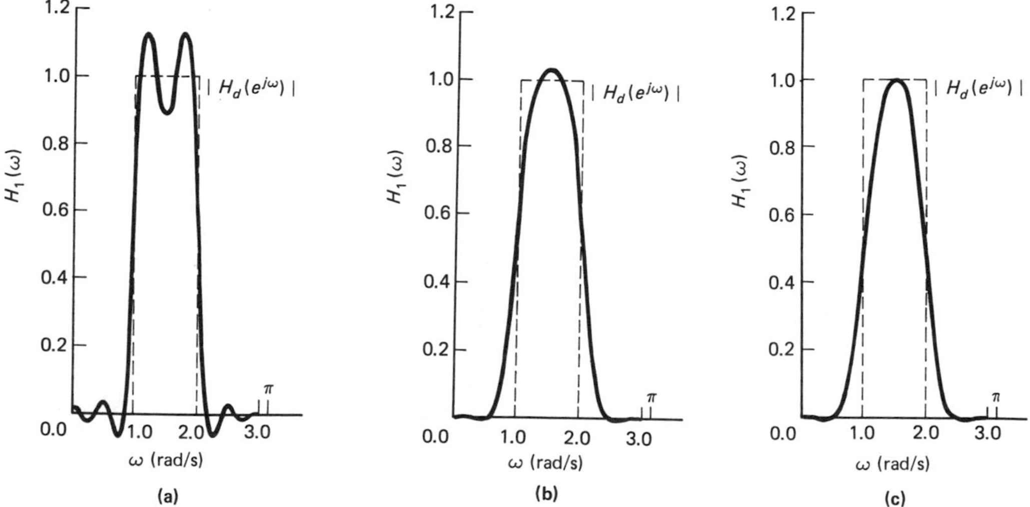

Figure 6.19 Graphs of the bandpass function $H_1(\omega)$ for (a) a rectangular window, (b) a Hann window, and (c) a Hamming window, all for $M = 12$. The desired responses are shown by the dashed lines.

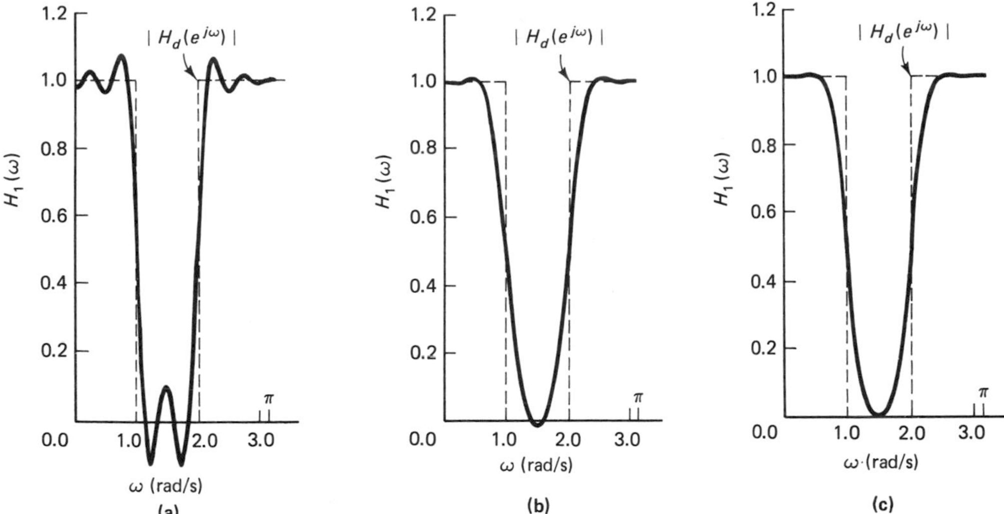

Figure 6.20 Graphs of the band-reject function $H_1(\omega)$ for (a) a rectangular window, (b) a Hann window, and (c) a Hamming window, all for $M = 12$. The desired responses are shown by the dashed lines.

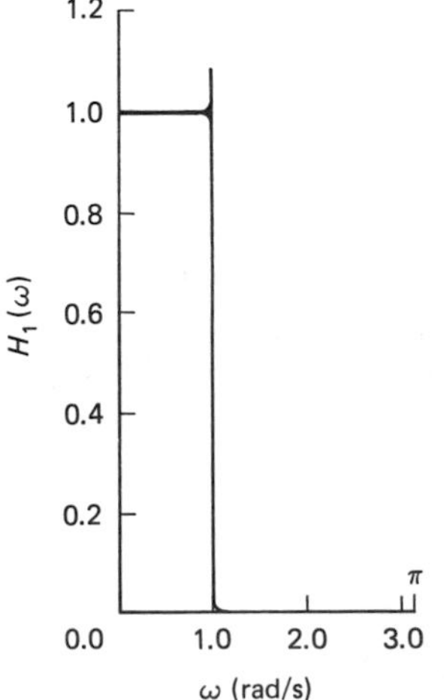

Figure 6.21 Graph of $H_1(\omega)$ for a lowpass filter using a rectangular window with $N = 1024$.

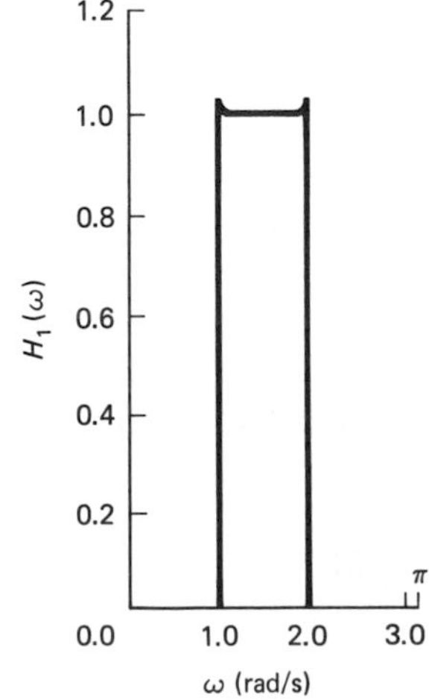

Figure 6.22 Graph of $H_1(\omega)$ for a bandpass filter using a Hann window with $N = 1024$.

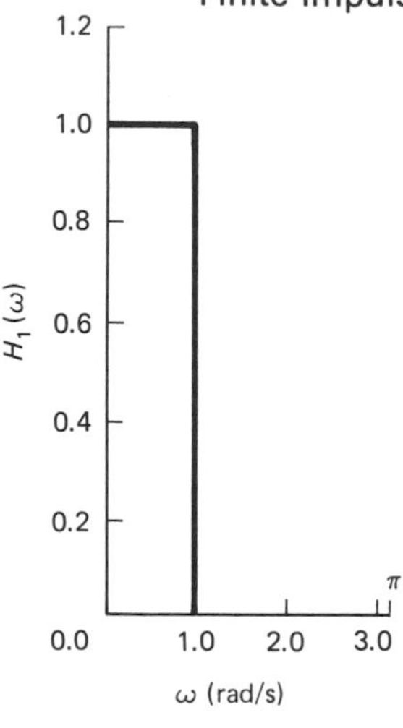

Figure 6.23 Graph of $H_1(\omega)$ for a lowpass filter using a Hamming window with $N = 1024$.

EXERCISES

6.4.1 Design a filter with $H_d(e^{j\omega})$ given in Ex. 6.2.1 using a Hann window with $N = 5$.

6.4.2 Repeat Ex. 6.4.1 using a Hamming window.

6.4.3 Repeat Ex. 6.4.1 for $H_d(e^{j\omega})$ given in Ex. 6.2.3.

6.4.4 Repeat Ex. 6.4.3 using a Hamming window.

6.5 THE KAISER WINDOW

Definition

A good window should be a time-limited function with a Fourier transform that is bandlimited. In the case of continuous-time functions, Slepian and Pollak [162] have shown that a class of functions called *prolate spheroidal wave functions* have the property that they are limited as much as possible in both the time and frequency domains. Kaiser has chosen a class of windows having properties closely approximating those of the prolate spheroidal wave functions. This family of windows, known as the *Kaiser windows* or as the *I_0–sinh windows,* is defined by [55, 87]

$$w_K(\alpha, n) = \frac{I_0\left\{\alpha\left[1 - \left(\frac{2n}{N-1}\right)^2\right]^{1/2}\right\}}{I_0(\alpha)} \qquad -\frac{N-1}{2} \le n \le \frac{N-1}{2} \tag{6.57}$$

$$= 0 \qquad \text{otherwise}$$

where $I_0(x)$ is the modified Bessel function of the first kind of order zero, given by

$$I_0(x) = \sum_{k=0}^{\infty} \left[\frac{(x/2)^k}{k!}\right]^2 \tag{6.58}$$

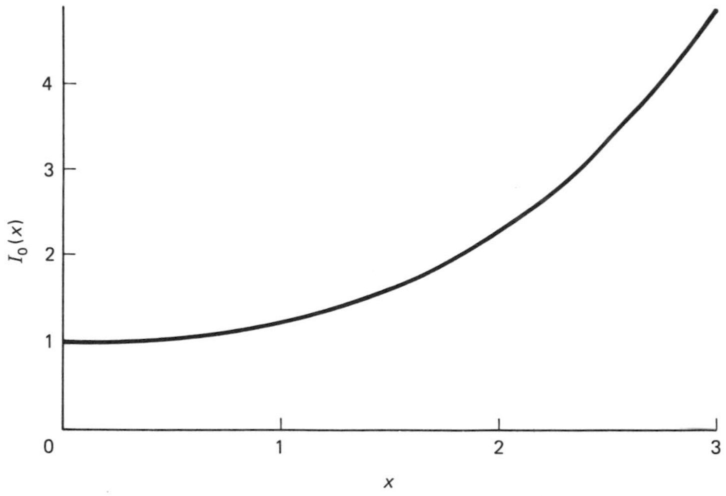

Figure 6.24 Graph of $I_0(x)$.

A graph of $I_0(x)$ is shown in Fig. 6.24. When the parameter α is varied, both the transition bandwidth and the peak ripple in the side lobes change. This gives the filter designer the flexibility to trade off main-lobe width for side-lobe ripple amplitude. Graphs of the Kaiser window for $\alpha = 8.885$ and $N = 7$ and $N = 25$ are shown in Fig. 6.25. From Eq. (6.57), we see that $w_K(\alpha,\ 0) = 1$ and $w_K[(\alpha,\ (N-1)/2)] = I_0(0)/I_0(\alpha) = 1/I_0(\alpha)$. Since $I_0(8.885)$ is very large, as seen from Fig. 6.24, we see that $w_k[8.885,(N - 1)/2] \simeq 0$.

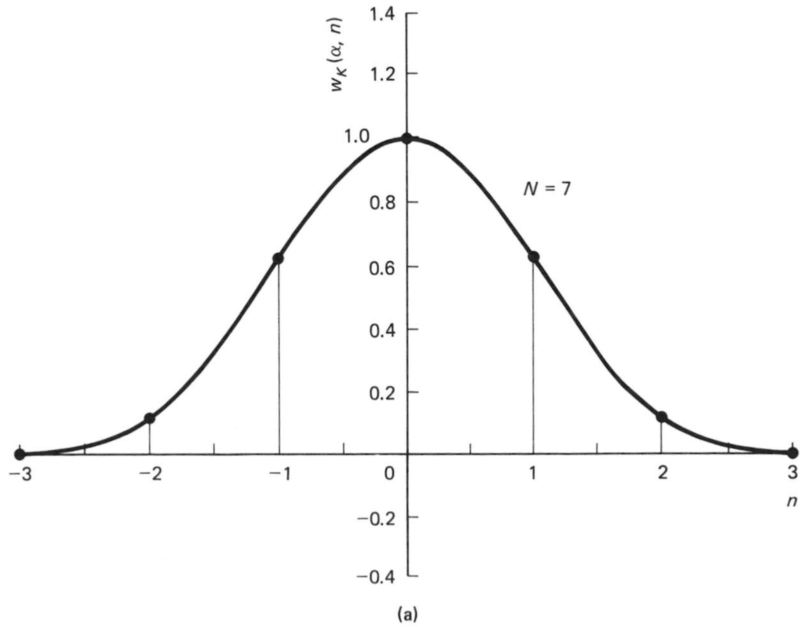

Figure 6.25 Graphs of the Kaiser window $w_K(\alpha, n)$ for $\alpha = 8.885$ and (a) $N = 7$ and (b) $N = 25$.

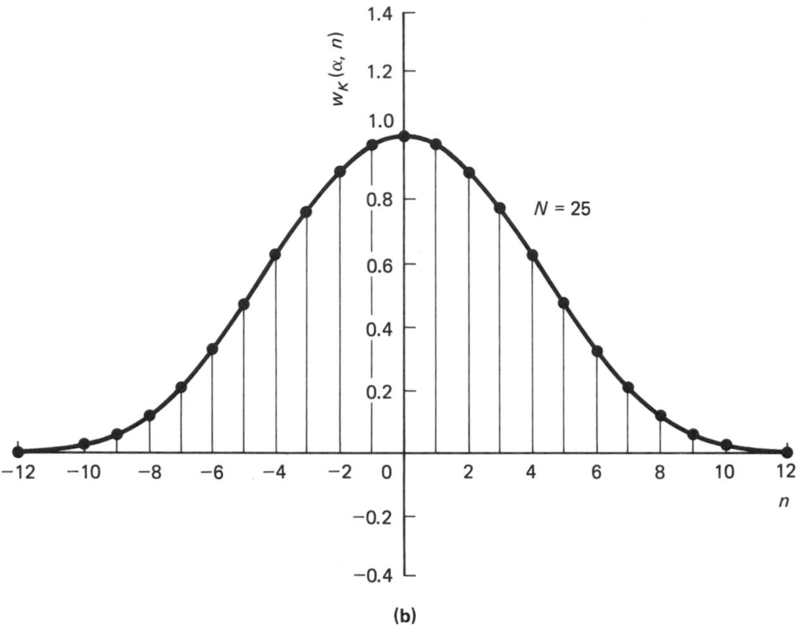

Figure 6.25 *(continued)*

Frequency Response

To get an idea how the frequency response of the Kaiser window behaves, let us consider the continuous-time function

$$w_a(\alpha, t) = \begin{cases} \dfrac{I_0\{\alpha[1 - (t/\tau)^2]^{1/2}\}}{I_0(\alpha)} & |t| \leq \tau \\ 0 & |t| \leq \tau \end{cases} \tag{6.59}$$

The sequence $w_K(\alpha, n)$ can be obtained by sampling $w_a(\alpha, t)$ for $T = 1$ and $\tau = (N - 1)/2$ and so the frequency response $W_K(\alpha, e^{j\omega})$ should be related to $W_a(\alpha, j\Omega)$ for $\omega = \Omega$. The Fourier transform $W_a(\alpha, j\Omega)$, given by Campbell and Foster [17], is

$$W_a(\alpha, j\Omega) = \frac{2}{I_0(\alpha)} \frac{\sin\{\tau[\Omega^2 - (\alpha/\tau)^2]^{1/2}\}}{[\Omega^2 - (\alpha/\tau)^2]^{1/2}} \tag{6.60}$$

and the sampled signal $w(\alpha, n) = w_a(\alpha, nT)$ has the frequency response

$$W(\alpha, e^{j\omega}) = \frac{1}{T} \sum_{k=-\infty}^{\infty} W_a(\alpha, j\omega/T + j2\pi k/T) \tag{6.61}$$

The analog frequency response is sufficiently bandlimited, so that a reasonable approximation is

$$W(\alpha, e^{j\omega}) \approx \frac{1}{T} W_a(\alpha, j\omega/T) \tag{6.62}$$

Assuming for the moment that this approximation is valid for $T = 1$, we obtain

$$W_K(\alpha, e^{j\omega}) \approx W_a(\alpha, j\omega) = \frac{2}{I_0(\alpha)} \frac{\sin \{\tau[\omega^2 - (\alpha/\tau)^2]^{1/2}\}}{[\omega^2 - (\alpha/\tau)^2]^{1/2}} \tag{6.63}$$

We will see later that this is a reasonable approximation.

The function $W_a(\alpha, j\omega)$, which is real, has the value $(2\tau/\alpha) \sinh \alpha/I_0(\alpha)$ when $\omega = 0$ and steadily decreases as ω increases to α/τ. When $\omega > \alpha/\tau$, the behavior of $W_a(\alpha, j\omega)$ is somewhat like that of the function sinc $(\tau u/\pi)$, discussed in Sec. 1.5, where $u = [\omega^2 - (\alpha/\tau)^2]^{1/2}$. Consequently, $W_a(\alpha, j\omega)$ has ripples with decreasing amplitude and is zero when

$$\tau u = \tau[\omega^2 - (\alpha/\tau)^2]^{1/2} = \pm k\pi \qquad k = 1, 2, 3, \ldots$$

or when

$$\omega = \omega_k = \pm[\alpha^2 + k^2\pi^2]^{1/2}/\tau \qquad k = 1, 2, 3, \ldots \tag{6.64}$$

The main lobe occurs between the zeros $\pm\omega_1$ and has width $2\omega_1 = 2(\alpha^2 + \pi^2)^{1/2}/\tau$. The first side lobe, occurring between $u = \pi/\tau$ and $u = 2\pi/\tau$, has a minimum that is approximately $-4\tau/3\pi I_0(\alpha)$, the value of $W_a(\alpha, j\Omega)$ when $u = 3\pi/2\tau$. We can see how the ripples in the side lobes depend upon α from the graph of $I_0(\alpha)$ in Fig. 6.24. From this graph, we see that $I_0(\alpha)$ increases rather rapidly as α increases and therefore the ripples decrease quite rapidly as α is increased.

Graphs of $W_K(\alpha, e^{j\omega})$ are shown in Fig. 6.26 for $N = 7$ and $N = 25$ and various values of α ranging from 3 to 8.885. In particular, when $\alpha = 8.885$, the Kaiser window becomes the Blackman window [55]. Kaiser's weights resemble the Hamming window (raised cosine on a platform) [55], since $w(0) = 1$ and

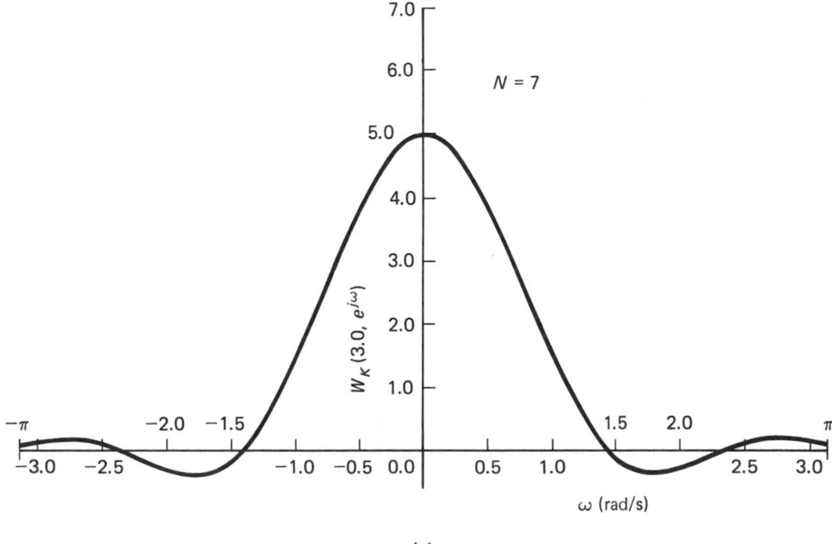

Figure 6.26 Graphs of $W_K(\alpha, e^{j\omega})$ for the Kaiser window for various values of α with $N = 7$ and $N = 25$.

Figure 6.26 (continued)

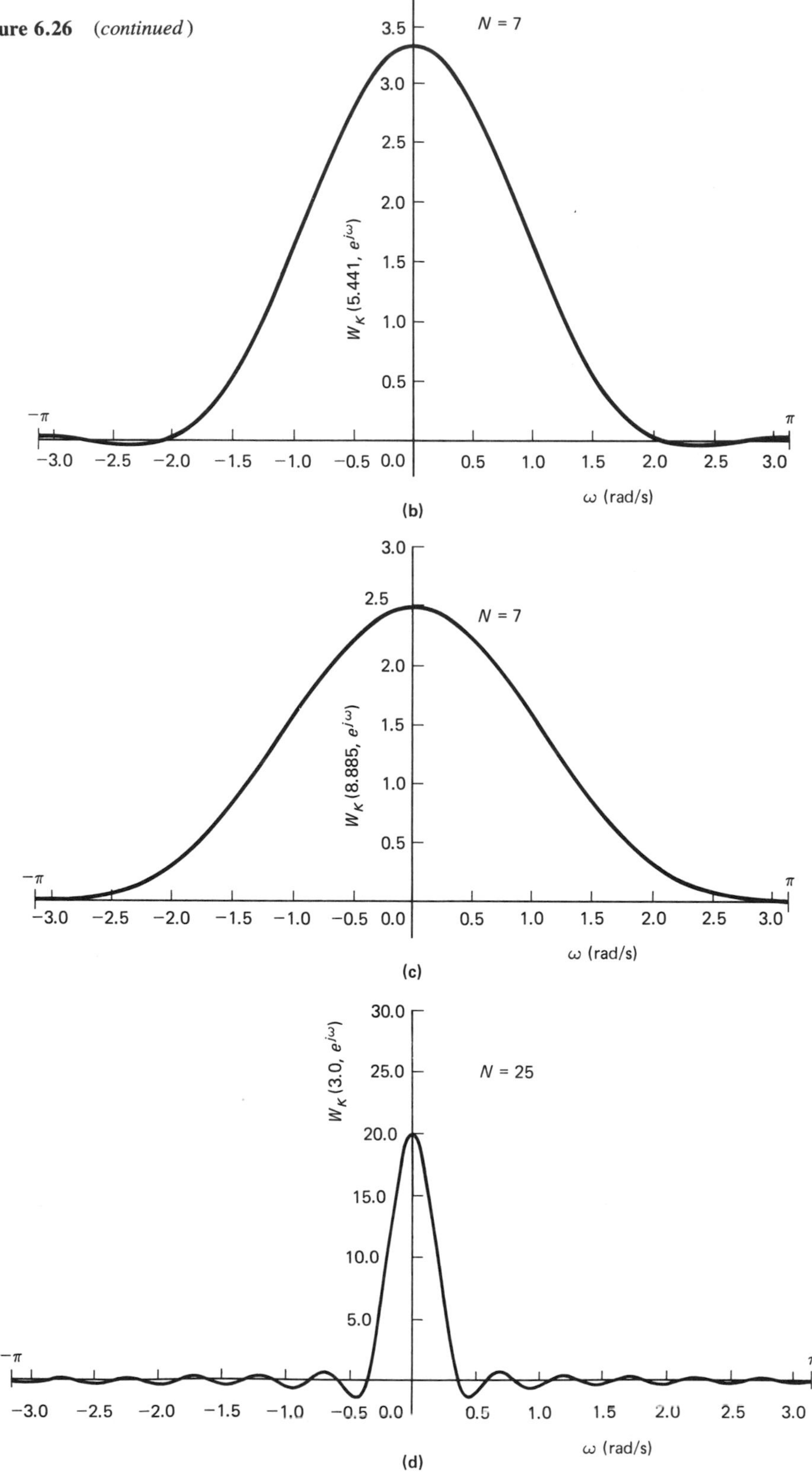

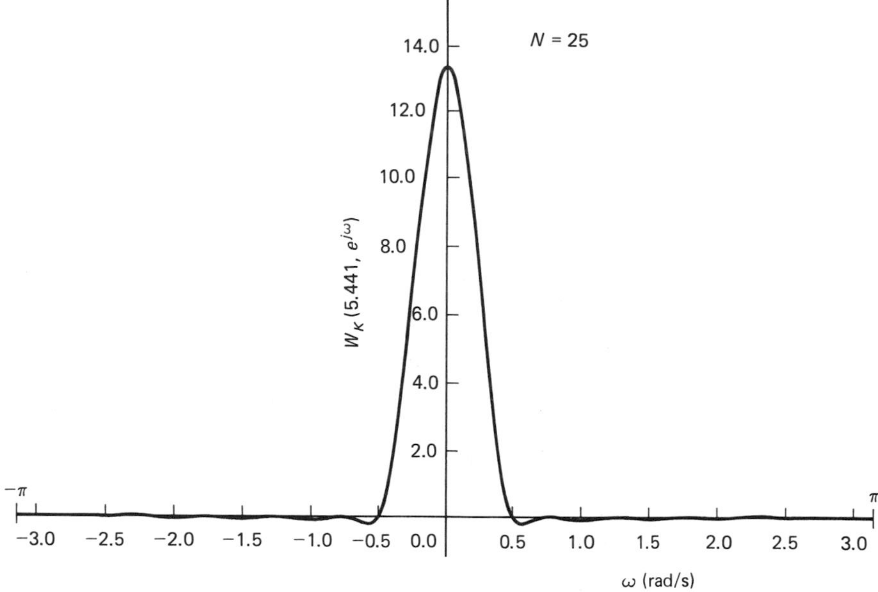

(e)

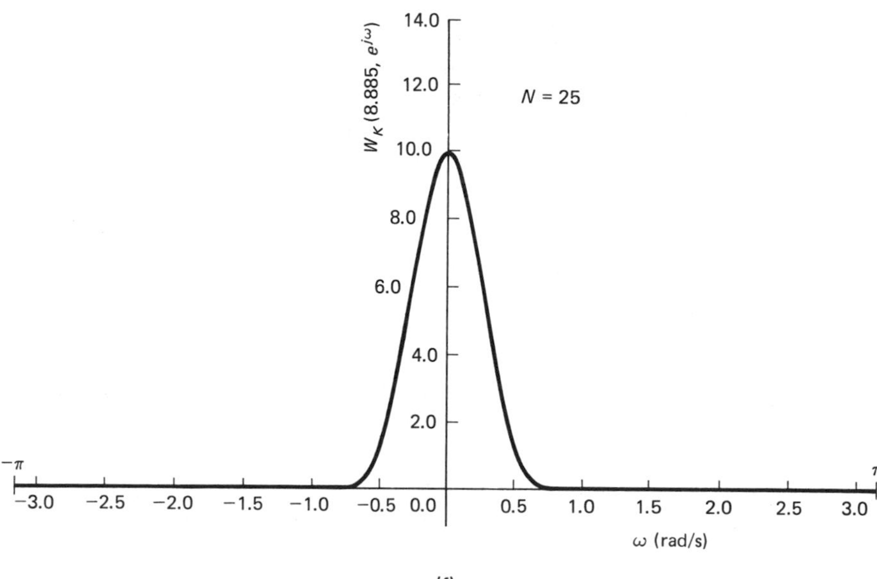

(f)

Figure 6.26 *(continued)*

$w([N-1]/2) = I_0(0)/I_0(\alpha) = 1/I_0(\alpha)$. For $\alpha = 5.4414$, the Kaiser window does not become the Hamming window, but they have the same main-lobe width. However, the windows differ in the side lobes. This is evidenced by the regular rolloff of the successive side lobes of the Kaiser window as compared to the first side lobes being leveled in the Hamming window. This can be seen by comparing the graphs of Fig. 6.26(c) and (e) with those of Fig. 6.8 and Fig. 6.12, respectively. There is very good agreement between the graphs of $W_K(\alpha, e^{j\omega})$ and the approximation obtained using $T = 1$ and $\tau = (N-1)/2$. For example, the values

$$\omega_1 = \frac{2}{N-1}(\alpha^2 + \pi^2)^{1/2}$$

are good approximations to the smallest zeros of $W_k(\alpha, e^{j\omega})$. Similarly, for $\alpha = 8.885$, we note that the ratio of the peak values for $N = 25$ and $N = 7$ is approximately 4. From the expression

$$W_a(\alpha, j0) = (N-1)(\sinh \alpha)/[\alpha I_0(\alpha)]$$

we see that this ratio is $(25-1)/(7-1) = 4$ also.

Filter Design Using a Kaiser Window

The parameter α allows the filter designer the flexibility to trade off width of transition band for peak ripple in the side lobes. Typical values of α are in the range $4 < \alpha < 9$. To illustrate the effect of varying α, we use a Kaiser window to design the low-pass filter considered in Sec. 6.4. The results are shown in Fig. 6.27 for $N = 7$ and 25 for several values of α. These graphs should be compared with those of Figs. 6.13 and 6.17. The envelopes of $h(n)$ for $N = 7$, $\alpha = 1.51$ and $\alpha = 8.885$, and for $N = 25$ and $\alpha = 8.885$ are shown in Fig. 6.28.

Kaiser [87] has developed design procedures using tables and empirical formulas that can be used to obtain the value of α that results in the desired filter characteristics. For discussion purposes, we use the ideal low-pass filter characteristic shown in Fig. 6.29; however, the design procedure also applies to the other types of filters. In the figure, Δf represents an acceptable normalized transition bandwidth in hertz and δ is the amplitude of the allowable ripple. An attenuation A characterizing the ripple width is defined by

$$A = -20 \log_{10} \delta \tag{6.65}$$

and a quantity D is defined by

$$D = (2N)(\Delta f) \tag{6.66}$$

where $2N + 1$ is the filter length. Kaiser found that the parameter α could be computed from the empirical equations

$$\alpha = \begin{cases} 0.1102(A - 8.7) \qquad A > 50 \\ 0.5842(A - 21)^{0.4} + 0.07886(A - 21) \qquad 21 < A < 50 \\ 0 \qquad A \leq 21 \end{cases} \tag{6.67}$$

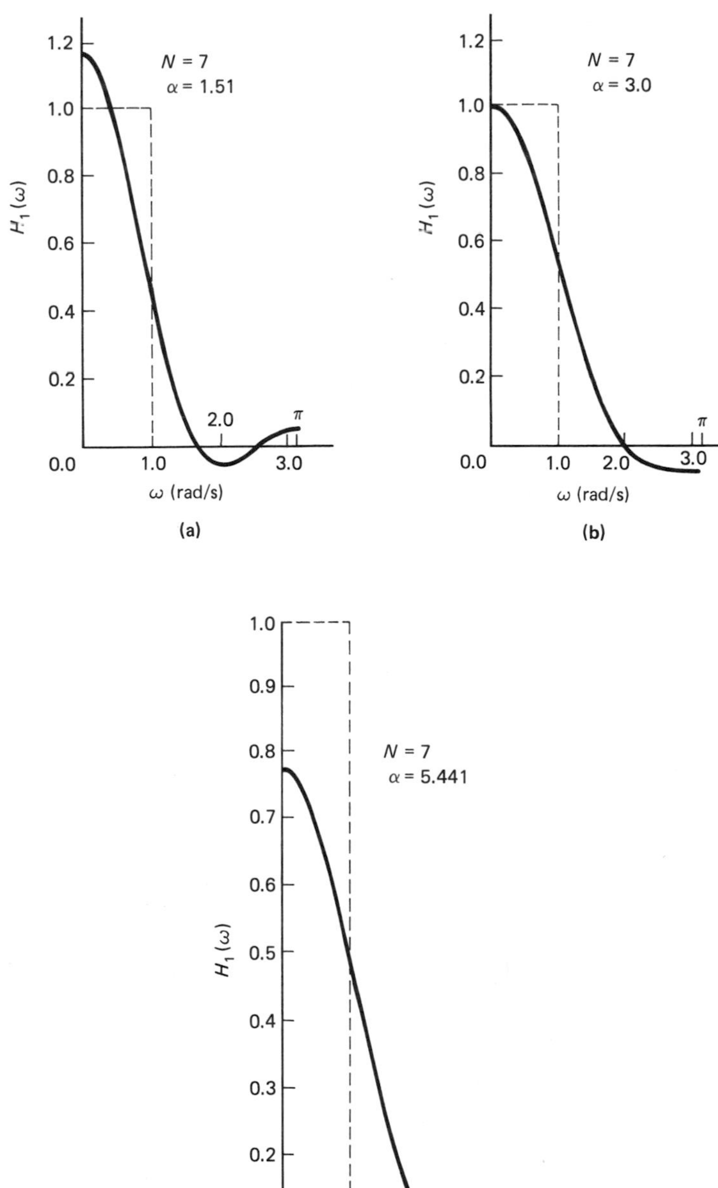

Figure 6.27 Graphs of $H_1(\omega)$ for low-pass filters using the Kaiser window for various values of α with $N = 7$ and $N = 25$. The desired responses are shown by the dashed lines.

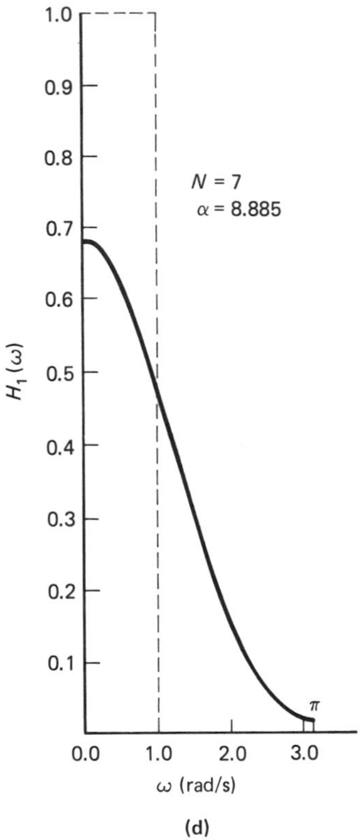

(d)

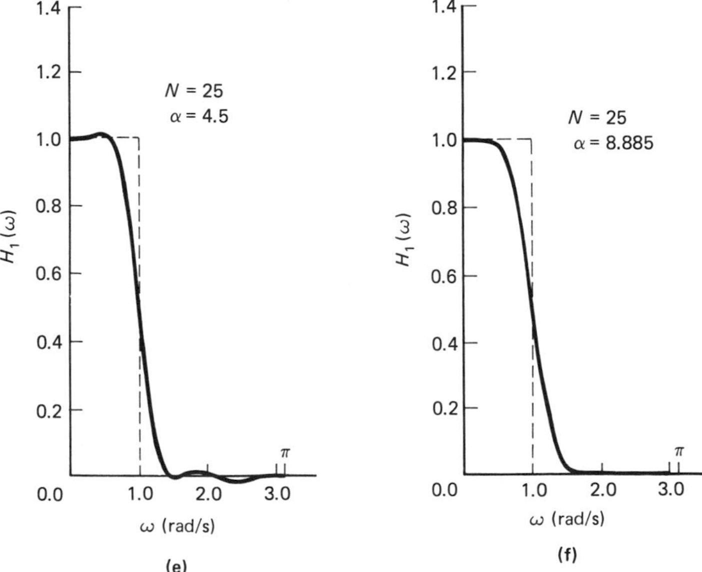

Figure 6.27 *(continued)*

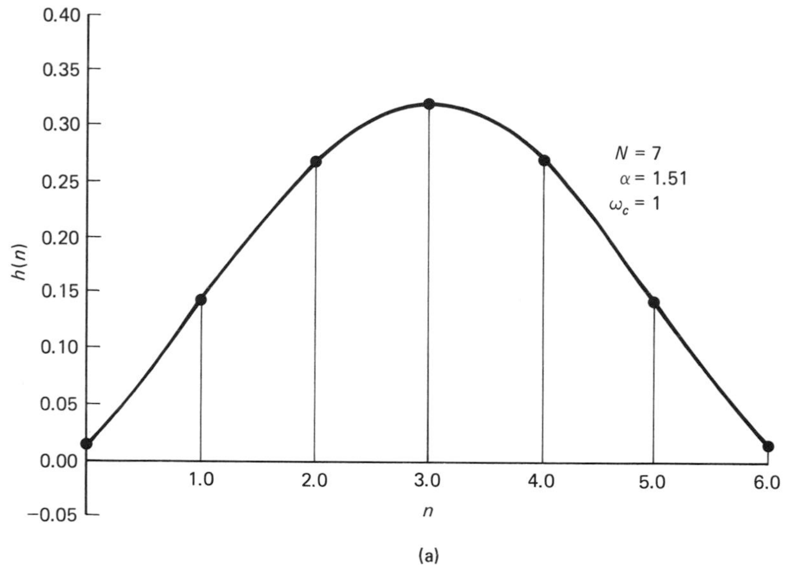

(a)

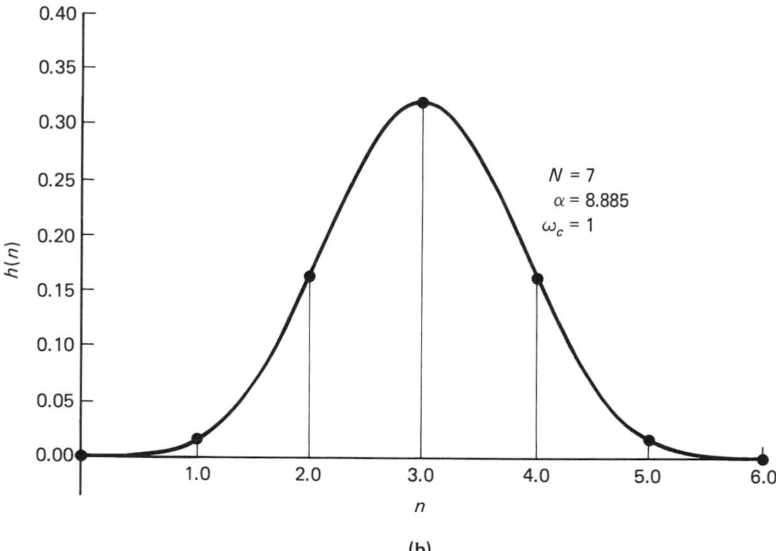

(b)

Figure 6.28 Envelope of the impulse response for low-pass filters using a Kaiser window with (a) $N = 7$ and $\alpha = 1.51$, (b) $N = 7$ and $\alpha = 8.885$, and (c) $N = 25$ and $\alpha = 8.885$. The cutoff frequency $\omega_c = 1$ rad/s.

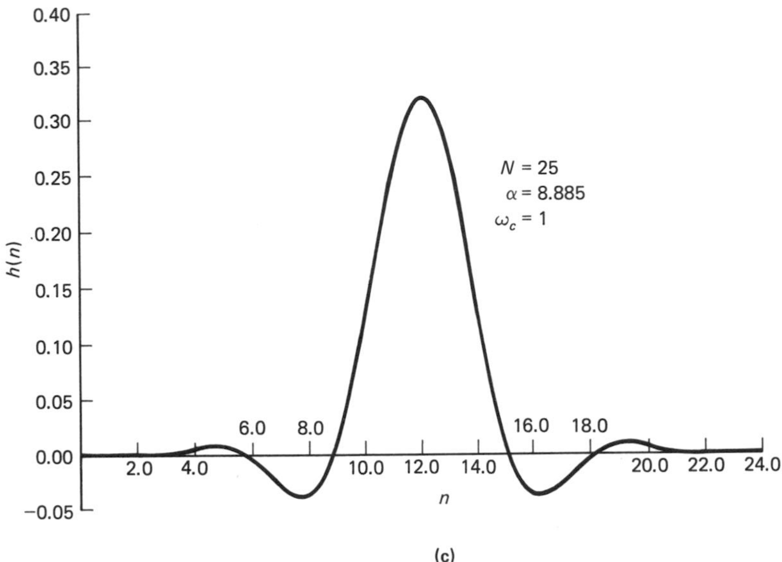

(c)

Figure 6.28 *(continued)*

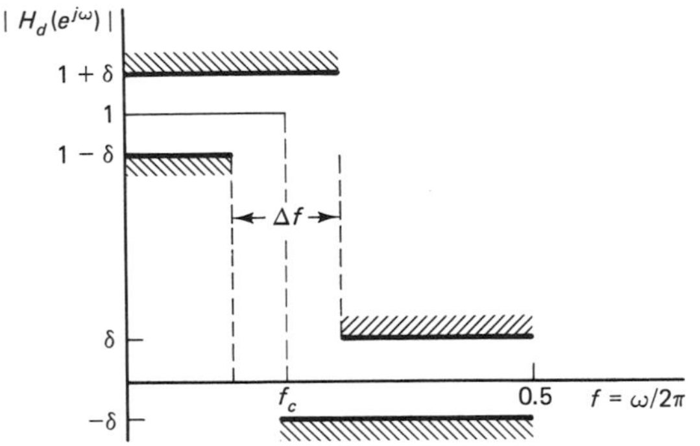

Figure 6.29 Ideal filter amplitude response.

and D could be determined from

$$D = \begin{cases} \dfrac{A - 7.95}{14.36} & A > 21 \\ 0.9222 & A < 21 \end{cases} \tag{6.68}$$

The value of N is obtained by solving Eq. (6.66) with D given by Eq. (6.68). The result is

$$N = \begin{cases} \dfrac{A - 7.95}{28.72(\Delta f)} & A > 21 \\ \dfrac{0.4611}{\Delta f} & A < 21 \end{cases} \tag{6.69}$$

Design Example Using Kaiser's Equations

To illustrate the procedure, let us take $\delta = 0.05$. First, from Eq. (6.65), we have

$$A = -20 \log_{10} (0.05) = 26.02$$

Then the value of α is found from the second of Eq. (6.67) to be

$$\alpha = 0.5842(5.02)^{0.4} + 0.07886(5.02) = 1.5098$$

If we desire a transition width $\Delta f = 0.1$ Hz, then from the first of Eqs. (6.69)

$$N = \frac{26.02 - 7.95}{28.72(0.25)} = 2.517$$

Taking $N = 3$, we see that the filter length is $2N + 1 = 7$. The filter coefficients $h(n)$ can be obtained from the graph in Fig. 6.28(a). The graph of the frequency response for the filter with $\omega_c = 1$ rad/s is that shown in Fig. 6.27(a). In most practical cases, Δf would be much smaller than 0.05 and a larger value of $2N + 1$, the filter length, would be required. It should be noted that if we want a larger value of α, then we should reduce the value of δ representing the allowable ripple.

In addition to Kaiser's paper, the reader might wish to read the discussion of Kaiser windows given by Hamming [55]. A computer program for designing filters using Kaiser windows is given in [34,141].

The Causal Kaiser Window

In the event that we wish to consider a window with nonzero values on the interval $0 \le n \le N - 1$, we simply replace n by $n - (N - 1)/2$ in Eq. (6.57). This corresponds to shifting the sequence to the right by $(N - 1)/2$ units. The resulting window is given by

$$w_K(\alpha, n) = \frac{I_0\left\{\dfrac{2\alpha}{N-1}\left[\left(\dfrac{N-1}{2}\right)^2 - \left(n - \dfrac{N-1}{2}\right)^2\right]^{1/2}\right\}}{I_0(\alpha)} \qquad 0 \le n \le N - 1 \tag{6.70}$$

For convenience, we can define a new parameter $\beta = 2\alpha/(N - 1)$. Equation (6.70) then becomes

$$w'_K(\beta, n) = \frac{I_0\left\{\beta\left[\left(\dfrac{N-1}{2}\right)^2 - \left(n - \dfrac{N-1}{2}\right)^2\right]^{1/2}\right\}}{I_0\left[\beta\left(\dfrac{N-1}{2}\right)\right]} \qquad 0 \le n \le N - 1 \tag{6.71}$$

EXERCISES

6.5.1 Show that when $\alpha = 0$, a Kaiser window becomes a rectangular window.

6.5.2 Estimate the values of $w_K(\alpha, n)$ from the graph of Fig. 6.24 for $N = 7$

and $\alpha = 2$. These values can be obtained also from [1], where values of $e^{-x}I_0(x)$ are tabulated.

6.5.3 Determine the values of α and N for a Kaiser window if $\delta = 0.01$ and $\Delta f = 0.1$ Hz are the desired low-pass filter parameters.

6.6 FREQUENCY-SAMPLING TECHNIQUE

Design Procedure Using Frequency Sampling

Another method that is used for designing FIR filters is that of *frequency sampling*. A set of samples is determined from a desired frequency response and is identified as discrete Fourier-transform (DFT) coefficients. The filter coefficients are then determined as the inverse discrete Fourier-transform (IDFT) of this set of samples.

The set of sample points that are used in this procedure can be obtained by sampling a desired frequency response $H_d(e^{j\omega})$ at N points ω_k, $k = 0, 1, \ldots, N - 1$, uniformly spaced around the unit circle. For a *Type 1 design,* the frequency samples are chosen to be

$$\omega_k = \frac{2\pi k}{N} \qquad k = 0, 1, \ldots, N - 1$$

Sampling the desired frequency response at these frequencies, we obtain

$$\begin{aligned} \tilde{H}(k) &= H_d(e^{j\omega})\big|_{\omega=\omega_k} && k = 0, 1, \ldots, N - 1 \\ &= H_d(e^{j2\pi k/N}) && k = 0, 1, \ldots, N - 1 \end{aligned} \tag{6.72}$$

If we now consider this set of points as DFT samples, then we can use the IDFT to compute the coefficients $h(n)$ from

$$h(n) = \frac{1}{N}\sum_{k=0}^{N-1} \tilde{H}(k)e^{j2\pi nk/N} \qquad n = 0, 1, \ldots, N - 1 \tag{6.73}$$

In order for these numbers to be the impulse-response coefficients of a FIR filter, they must all be real. This can happen if all the complex terms appear in complex conjugate pairs. This suggests that the terms can be matched by comparing the exponentials. The term $\tilde{H}(k)e^{j2\pi nk/N}$ should be matched by the term that has the exponential $e^{-j2\pi nk/N}$ as a factor. The matching terms then are $\tilde{H}(k)e^{j2\pi nk/N}$ and $\tilde{H}(N - k)e^{j2\pi n(N-k)/N}$ since $2\pi n(N - k)/N = 2\pi n - 2\pi nk/N$. These terms are complex conjugates if $\tilde{H}(0)$ is real and

$$\tilde{H}(N - k) = \tilde{H}^*(k) \qquad k = 1, 2, \ldots, (N - 1)/2 \tag{6.74}$$

for N odd, and if

$$\begin{aligned} \tilde{H}(N - k) &= \tilde{H}^*(k) \qquad k = 1, 2, \ldots, N/2 - 1 \\ \tilde{H}(N/2) &= 0 \end{aligned} \tag{6.75}$$

for N even. The procedure using the frequency samples amd DFT coefficients of Eq. (6.72) is referred to as a *Type 1 design*.

Assuming that $H_d(e^{j\omega})$ is chosen so the requirements of Eq. (6.74) or (6.75) for N odd or even, respectively, are satisfied, the filter coefficients can then be written

$$h(n) = \frac{1}{N}\left\{\tilde{H}(0) + 2\sum_{k=1}^{(N-1)/2} \text{Re}\,[\tilde{H}(k)e^{j2\pi nk/N}]\right\} \qquad N \text{ odd} \tag{6.76}$$

and

$$h(n) = \frac{1}{N}\left\{\tilde{H}(0) + 2\sum_{k=1}^{(N/2)-1} \text{Re}\,[\tilde{H}(k)e^{j2\pi nk/N}]\right\} \qquad N \text{ even} \tag{6.77}$$

Once the filter coefficients $h(n)$ have been determined, the system function of the filter is given by

$$H(z) = \sum_{n=0}^{N-1} h(n)z^{-n}$$

To see how $H(e^{j\omega})$ and $H_d(e^{j\omega})$ are related, we note that

$$H(e^{j2\pi k/N}) = \sum_{n=0}^{N-1} h(n)e^{-j2\pi kn/N}$$

This sum, as seen by Eq. (1.59), is simply the DFT $\tilde{H}(k)$, and so, using Eq. (6.72), we obtain

$$H(e^{j2\pi k/N}) = \tilde{H}(k) = H_d(e^{j2\pi k/N})$$

Consequently, the values of $H(e^{j\omega})$ and $H_d(e^{j\omega})$ coincide when $\omega = 2\pi k/N$. At other values of ω, we can think of $H(e^{j\omega})$ as an interpolation of the sampled desired frequency response.

If the frequency response $H(e^{j\omega})$ is not suitable, then we might try varying $H_d(e^{j\omega})$ and/or N. It is possible to improve the approximation that $H(e^{j\omega})$ provides for $H_d(e^{j\omega})$ by making some of the frequency samples $\tilde{H}(k)$ unconstrained. Their values are determined then so as to optimize the frequency response [136].

Design Example

As an example of the frequency-sampling technique, let us consider the design of a low-pass filter with cutoff frequency $\omega_c = \pi/2$. Since we are interested frequently in linear phase, we take the ideal response to be

$$H_d(e^{j\omega}) = \begin{cases} e^{-j(N-1)\omega/2} & 0 \le \omega \le \pi/2 \\ 0 & \pi/2 \le \omega \le \pi \end{cases} \tag{6.78}$$

The ideal magnitude and phase responses with samples for $N = 17$ are shown in Fig. 6.30. We have added π rad to the phase at $\omega = \pi/2$; we could have subtracted π rad just as well.

The DFT samples, which are obtained from $H_d(e^{j\omega})$, are given by

$$\tilde{H}(k) = H_d(e^{j2\pi k/17}) = \begin{cases} e^{-j16\pi k/17} & 0 \le k \le 4 \\ 0 & 5 \le k \le 12 \\ e^{-j16\pi(k-17)/17} & 13 \le k \le 16 \end{cases}$$

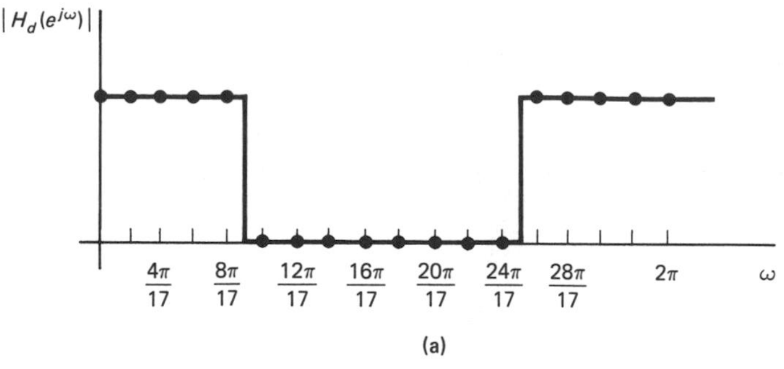

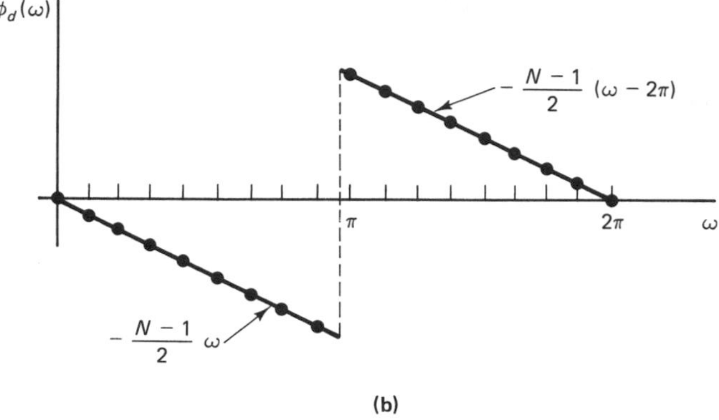

Figure 6.30 Magnitude and phase responses for $H_d(e^{j\omega})$ of Eq. (6.78).

These samples satisfy Eq. (6.74), as can be seen easily from Fig. 6.30, since $\tilde{H}(16) = \tilde{H}^*(1)$, $\tilde{H}(15) = \tilde{H}^*(2)$, and so forth. The coefficients $h(n)$ given by Eq. (6.73) are then real and can be determined from Eq. (6.76), which becomes

$$\begin{aligned}17h(n) &= \tilde{H}(0) + 2\sum_{k=1}^{8} \text{Re}\,[\tilde{H}(k)e^{j2\pi nk/17}] \\ &= 1 + 2\sum_{k=1}^{4} \text{Re}\,[e^{-j2\pi k(8-n)/17}] \\ &= 1 + 2\sum_{k=1}^{4} \cos\frac{2\pi k(8-n)}{17}\end{aligned}$$

The sum involving the complex exponentials can be summed as a geometric series or we may use Ex. 1.1.4. The result is

$$h(n) = \frac{1}{17}\frac{\sin[9\pi(8-n)/17]}{\sin[\pi(8-n)/17]} \qquad n = 0, 1, 2, \ldots, 16 \tag{6.79}$$

The reader is asked to verify this result in Prob. 6.20. The frequency response $H(e^{j\omega})$ is then found from

$$\begin{aligned}H(e^{j\omega}) &= H(z)\big|_{z=e^{j\omega}} \\ &= \sum_{n=0}^{16} h(n)z^{-n}\big|_{z=e^{j\omega}}\end{aligned}$$

We leave its computation to the reader (see Prob. 6.21).

General Expression for the System Function

We can determine the general expression for $H(z)$ in terms of $\tilde{H}(k)$ by using Eq. (6.73) or (6.76) for N odd and Eq. (6.77) for N even as the expression for $h(n)$. By using Eq. (6.73) the general expression

$$H(z) = \sum_{n=0}^{N-1} h(n)z^{-n} \tag{6.80}$$

becomes

$$H(z) = \sum_{n=0}^{N-1} h(n)z^{-n} = \sum_{n=0}^{N-1} \left(\frac{1}{N}\sum_{k=0}^{N-1} \tilde{H}(k)e^{j2\pi kn/N}\right)z^{-n} \tag{6.81}$$

If we now interchange the order of the summations, we can write

$$H(z) = \frac{1}{N}\sum_{k=0}^{N-1} \tilde{H}(k)\left(\sum_{n=0}^{N-1} e^{j2\pi kn/N}z^{-n}\right)$$

The sum in the parentheses is a finite geometric series, and from Eq. (1.5) its value is

$$\begin{aligned}\sum_{n=0}^{N-1} (e^{j2\pi k/N}z^{-1})^n &= \frac{1 - (e^{j2\pi k/N}z^{-1})^N}{1 - e^{j2\pi k/N}z^{-1}} \\ &= \frac{1 - z^{-N}}{1 - e^{j2\pi k/N}z^{-1}}\end{aligned}$$

Substituting this value into the expression for $H(z)$ gives

$$H(z) = \frac{1 - z^{-N}}{N}\sum_{k=0}^{N-1} \frac{\tilde{H}(k)}{1 - e^{j2\pi k/N}z^{-1}} \tag{6.82}$$

This corresponds to a recursive realization of $H(z)$, whereas the expression of Eq. (6.80) corresponds to a nonrecursive realization.

The expression of Eq. (6.82) for $H(z)$ is actually a polynomial since $1 - e^{j2\pi k/N}z^{-1}$ is a factor of $1 - z^{-N} = 1 - e^{j2\pi kN/N}z^{-N}$ for each value of k. Consequently, $H(z)$ is a sum of polynomials in z^{-1} of degree $N - 1$.

Frequency Response

The frequency response of the FIR filter can be obtained by setting $z = e^{j\omega}$ in Eq. (6.82), resulting in the expression

$$\begin{aligned}H(e^{j\omega}) &= \frac{1 - e^{-j\omega N}}{N}\sum_{k=0}^{N-1} \frac{\tilde{H}(k)}{1 - e^{j2\pi k/N}e^{-j\omega}} \\ &= \frac{e^{-j\omega N/2}}{N}\sum_{k=0}^{N-1} \frac{\tilde{H}(k)(e^{j\omega N/2} - e^{-j\omega N/2})}{e^{-j(\omega/2 - k\pi/N)}(e^{j(\omega/2-k\pi/N)} - e^{-j(\omega/2-k\pi/N)})} \\ &= \frac{e^{-j\omega(N-1)/2}}{N}\sum_{k=0}^{N-1} \frac{\tilde{H}(k)e^{-j\pi k/N}\sin \omega N/2}{\sin (\omega/2 - k\pi/N)}\end{aligned} \tag{6.83}$$

Noting that

$$\sin (\omega N/2 - k\pi) = (-1)^k \sin \omega N/2$$

we can write Eq. (6.83) as

$$H(e^{j\omega}) = \frac{e^{-j\omega(N-1)/2}}{N} \sum_{k=0}^{N-1} \tilde{H}(k) \frac{(-1)^k e^{-j\pi k/N} \sin N(\omega/2 - k\pi/N)}{\sin (\omega/2 - k\pi/N)} \tag{6.84}$$

It is of some interest to note that this expression, Eq. (6.84), for $H(e^{j\omega})$ can be expressed in terms of the Chebyshev polynomials of the second kind, defined by [78]

$$U_n(x) = \frac{\sin (n+1)\theta}{\sin \theta} \qquad n = 0, 1, 2, \ldots \tag{6.85}$$

where $\theta = \text{arc cos } x$. Making the identification

$$\theta_k = \omega/2 - k\pi/N = \text{arc cos } x_k$$

and noting that

$$x_k = \cos \theta_k = \cos (\omega/2 - k\pi/N) \tag{6.86}$$

we obtain

$$H(e^{j\omega}) = \frac{e^{-j\omega(N-1)/2}}{N} \sum_{k=0}^{N-1} (-1)^k e^{-j\pi k/N} \tilde{H}(k) U_{N-1}(x_k) \tag{6.87}$$

As seen from Eq. (6.86), x_k is a function of k, N, and ω.

Type 2 Design

The set of frequency samples indicated by Eq. (6.72) includes the frequency $\omega = 0$. In some cases, it may be desirable to omit this frequency and use the set of frequencies given by $\omega_k = 2\pi(2k + 1)/2N$, $k = 0, 1, \ldots, N - 1$. In this case, the DFT coefficients are defined by

$$\tilde{H}(k) = H_d(e^{j\pi(2k+1)/N}) \qquad k = 0, 1, \ldots, N - 1 \tag{6.88}$$

Using these frequency samples as DFT coefficients results in a *Type 2 design*.

We will give the important Type 2 design results, but the development is left to the problems. The condition that $h(n)$ be real is, for N odd,

$$\begin{aligned} \tilde{H}(N - k - 1) &= \tilde{H}^*(k) \qquad k = 0, 1, \ldots, (N-1)/2 - 1 \\ \tilde{H}\left(\frac{N-1}{2}\right) &= 0 \end{aligned} \tag{6.89}$$

and for N even,

$$\tilde{H}(N - k - 1) = \tilde{H}^*(k) \qquad k = 0, 1, \ldots, N/2 - 1 \tag{6.90}$$

When these conditions are satisfied, the filter coefficients can be found from

$$Nh(n) = 2 \sum_{k=0}^{(N-3)/2} \text{Re}\,[\tilde{H}(k)e^{jn\pi(2k+1)/N}] \qquad N \text{ odd} \tag{6.91}$$

or

$$Nh(n) = 2 \sum_{k=0}^{(N/2)-1} \text{Re}\,[\tilde{H}(k)e^{jn\pi(2k+1)/N}] \qquad N \text{ even} \tag{6.92}$$

If it is convenient, we can interchange the operations of summation and taking the real part.

The transfer function $H(z)$ can be written

$$H(z) = \frac{1 + z^{-N}}{N} \sum_{k=0}^{N-1} \frac{\tilde{H}(k)}{1 - e^{j\pi(2k+1)/N}z^{-1}} \tag{6.93}$$

and the resulting frequency response is

$$H(e^{j\omega}) = \frac{e^{-j\omega(N-1)/2}}{N} \sum_{k=0}^{N-1} \frac{\tilde{H}(k)e^{-j\pi(k+\frac{1}{2})/N} \cos \omega N/2}{j \sin \left[\dfrac{\omega}{2} - \dfrac{\pi}{N}\left(k + \dfrac{1}{2}\right)\right]} \tag{6.94}$$

For more information concerning the frequency-sampling design technique, including development of the Type 2 results, the reader should consult Oppenheim and Schafer [126], Rabiner and Gold [136], and the papers by Rabiner and Schafer [143, 145].

EXERCISES

6.6.1 Determine the filter coefficients $h(n)$ obtained by sampling $H_d(e^{j\omega})$ given in Eq. (6.78) for $N = 7$. Obtain the frequency response $H(e^{j\omega})$ and compare it with $H_d(e^{j\omega})$.

6.6.2 Use the frequency samples given by Eq. (6.88) and determine the filter coefficients $h(n)$ if $H_d(e^{j\omega})$ is given in Eq. (6.78) and $N = 7$. Obtain the frequency response $H(e^{j\omega})$ and compare it with $H_d(e^{j\omega})$ and $H(e^{j\omega})$ obtained in Ex. 6.6.1.

6.7 OPTIMAL FIR FILTERS

Description of the Chebyshev-Type Approximation

As we have seen in Chapter 5, optimal analog filters are those that had equal ripples in both the passband and the stopband. This is also true for linear-phase FIR digital filters. The design procedure for these optimal filters is based on a minimax or Chebyshev-type approximation. The technique involves the determination of a weighted error function based on the desired response and the general form of the response function. The coefficients in the response function are then determined so as to minimize the maximum error that occurs.

We are concerned with obtaining the impulse-response coefficients $h(n)$, but we use the expressions for $H(e^{j\omega})$ given in Eqs. (2.85) and (2.96) for linear-phase

FIR filters. In Ex. 6.7.1, the reader is asked to show that in both of these cases, we can write

$$H(e^{j\omega}) = e^{-j\omega(N-1)/2}H_1(\omega) \tag{6.95}$$

where $H_1(\omega)$ is the pseudomagnitude given by

$$H_1(\omega) = Q(\omega)P(\omega) \tag{6.96}$$

The quantities $Q(\omega)$ and $P(\omega)$ are described in Table 6.1 for the cases N even and N odd.

TABLE 6.1

N	$Q(\omega)$	$P(\omega)$
Odd	1	$\sum_{n=0}^{(N-1)/2} a(n) \cos n\omega$
Even	$\cos \omega/2$	$\sum_{n=0}^{(N/2)-1} \tilde{b}(n) \cos n\omega$

In the case of a low-pass filter, the desired function that is to be approximated by $H_1(\omega)$ is given by

$$\hat{H}_1(\omega) = \begin{cases} 1 & 0 \le \omega \le \omega_p \\ 0 & \omega_s \le \omega \le \pi \end{cases} \tag{6.97}$$

where $0 \le \omega \le \omega_p$ is the passband, and $\omega_s \le \omega \le \pi$ is the stopband. The weighted error function is defined by

$$E(\omega) = W(\omega)[\hat{H}_1(\omega) - H_1(\omega)] \tag{6.98}$$

where the weighting function $W(\omega)$ is

$$W(\omega) = \begin{cases} 1/K & 0 \le \omega \le \omega_p \\ 1 & \omega_s \le \omega \le \pi \end{cases} \tag{6.99}$$

According to the theory of Chebyshev approximation, the minimum of the maximum absolute error $|E(\omega)|$ over the pass- and stopbands, $0 \le \omega \le \omega_p$ and $\omega_s \le \omega \le \pi$, respectively, is achieved when $E(\omega)$ exhibits at least $N_0 + 2$ *alternations*. Denoting $E_{\max}$ as the maximum value of $|E(\omega)|$ over the pass- and stopbands, this means that there is at least $N_0 + 2$ frequencies in this range such that

$$E(\omega_i) = -E(\omega_{i-1}) = \pm E_{\max} \qquad i = 1, 2, \ldots, N_0 + 1 \tag{6.100}$$

where $\omega_0 < \omega_1 < \cdots < \omega_{N_0+1}$. The value of N_0 is the upper limit on the summation describing $P(\omega)$ in Table 6.1. That is, for N odd, we have $N_0 = (N-1)/2$, and for N even, $N_0 = N/2 - 1$.

To see how $H_1(\omega)$ behaves in the optimum case, let us determine its value when $E(\omega)$ has its maximum absolute value. From Eq. (6.98), we find that

$$H_1(\omega) = \hat{H}_1(\omega) - E(\omega)/W(\omega) \tag{6.101}$$

Now assuming that $\max|E(\omega)| = \delta_2$ and that $E(\omega_i) = \delta_2$ for $0 \le \omega_i \le \omega_p$, we have

$$\begin{aligned} H_1(\omega_i) &= \hat{H}_1(\omega_i) - E(\omega_i)/W(\omega_i) \\ &= 1 - K\delta_2 \end{aligned}$$

For convenience, we take $K = \delta_1/\delta_2$. It then follows that

$$H_1(\omega_i) = 1 - \delta_1$$

For $\omega_{i+1} > \omega_i$, $0 \le \omega_{i+1} \le \omega_p$, we know that $E(\omega_{i+1}) = -\delta_2$, and, consequently,

$$H_1(\omega_{i+1}) = 1 + \delta_1$$

Since ω_i and ω_{i+1} are typical consecutive frequencies at which $E(\omega)$ alternates in the passband, we can see that $H_1(\omega)$ varies between $1 - \delta_1$ and $1 + \delta_1$ in the passband. This is illustrated in Fig. 6.31.

Next let ω_i and ω_{i+1} be consecutive extrema of $E(\omega)$ in the stopband, $\omega_s \le \omega \le \pi$. Assuming the same values of $E(\omega)$ as before, we see from Eq. (6.101) that

$$H_1(\omega_i) = -E(\omega_i) = -\delta_2$$

$$H_1(\omega_{i+1}) = -E(\omega_{i+1}) = \delta_2$$

In the stopband, $H_1(\omega)$ varies from $-\delta_2$ to δ_2, as shown in Fig. 6.31.

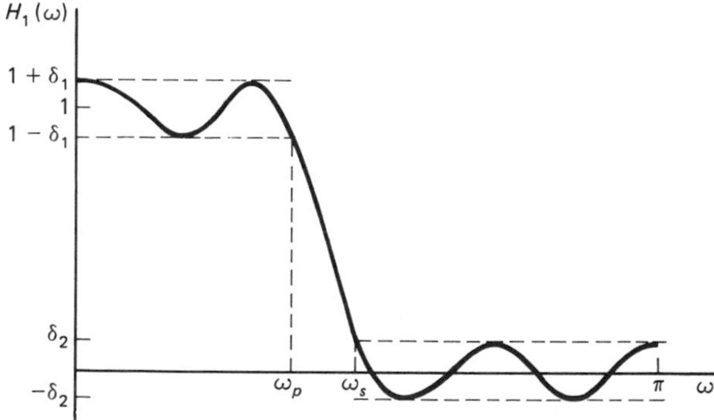

Figure 6.31 Equiripple approximation for $H_1(\omega)$.

Relative Extrema of $H_1(\omega)$

It can be seen from the preceding discussion and Fig. 6.31 that the Chebyshev-approximation procedure results in an equiripple approximation for $H_1(\omega)$. This response is similar to that of the elliptic analog filter, which is optimum for analog filters. The maximum deviation from the ideal filter response for the example is δ_1 in the passband and δ_2 in the stopband.

If the number of relative extrema of $H_1(\omega)$ in the passband is N_p and the number in the stopband is N_s, then for $H_1(\omega)$ shown in Fig. 6.31, we have $N_p = 3$ and $N_s = 4$. The total number of relative extrema is denoted by N_e, with $N_e = N_p + N_s$. For this example, $N_e = 7$. We also note that there are four alternations in the passband and five in the stopband, corresponding to the values of ω for which $H_1(\omega)$ is $1 + \delta_1$, $1 - \delta_1$, δ_2, or $-\delta_2$.

The two parameters N_p and N_s are arbitrary except that their sum is fixed by the value of N_0, which determines the number of relative extrema of $H_1(\omega)$. To see what this relation is, let us first consider the case where N is odd. In this case, we have, from Table 6.1,

$$H_1(\omega) = \sum_{n=0}^{(N-1)/2} a(n) \cos n\omega$$

with $N_0 = (N - 1)/2$. The relative extrema are the zeros of $dH_1/d\omega$. Before computing this derivative, it is convenient to first express $H_1(\omega)$ as a polynomial in $\cos \omega$. This can be achieved using the relation [78]

$$\begin{aligned} \cos n\omega &= \frac{n}{2} \sum_{k=0}^{[n/2]} \frac{(-1)^k (n-k-1)!\, 2^{n-2k}}{k!(n-2k)!} (\cos \omega)^{n-2k} \\ &= \sum_{m=0}^{n} \alpha_{mn} \cos^m \omega \end{aligned} \tag{6.102}$$

where $[n/2]$ is the largest integer less than or equal to $n/2$. The coefficients α_{mn} can be determined by comparing these two expressions. (The reader is asked to develop these relations in Prob. 6.41.) Substituting this value for $\cos n\omega$ into $H_1(\omega)$, we obtain the desired form

$$\begin{aligned} H_1(\omega) &= \sum_{n=0}^{(N-1)/2} a(n) \left(\sum_{m=0}^{n} \alpha_{mn} \cos^m \omega \right) \\ &= \sum_{n=0}^{(N-1)/2} \bar{a}(n) \cos^n \omega \end{aligned} \tag{6.103}$$

where $\bar{a}(n)$ is related to $a(n)$ and α_{mn}. Finally, we obtain from Eq. (6.103) the derivative

$$\begin{aligned} \frac{dH_1(\omega)}{d\omega} &= -\sin \omega \sum_{n=1}^{(N-1)/2} n\bar{a}(n) \cos^{n-1} \omega \\ &= -\sin \omega \sum_{n=0}^{(N-3)/2} (n+1)\bar{a}(n+1) \cos^n \omega \end{aligned}$$

The zeros of this expression that occur in the interval $0 \leq \omega \leq \pi$ are solutions of the two equations

$$\sin \omega = 0 \tag{6.104}$$

$$\sum_{n=0}^{(N-3)/2} (n+1)\bar{a}(n+1) \cos^n \omega = 0 \tag{6.105}$$

Roots of Eq. (6.104) are $\omega = 0$ and $\omega = \pi$. Now the expression in Eq. (6.105) is a polynomial in $\cos \omega$ of degree $(N - 3)/2$ and consequently there are at most $(N - 3)/2$ values of ω in $0 \leq \omega \leq \pi$ that satisfy Eq. (6.105). Consequently, there are at most $(N - 3)/2 + 2 = (N + 1)/2$ relative extrema of $H_1(\omega)$ in $0 \leq \omega \leq \pi$. The number of these relative extrema for which $H_1(\omega)$ is $1 + \delta_1$, $1 - \delta_1$, $+\delta_2$, or $-\delta_2$ is N_e, and from the above discussion, it follows that

$$N_e \leq \frac{N+1}{2} = N_0 + 1$$

From $\hat{H}_1(\omega)$ in Eq. (6.97), $E(\omega)$ in Eq. (6.98), and $W(\omega)$ given in Eq. (6.99), we see that $H_1(\omega)$ and $E(\omega)$ have relative extrema at the same points.

Equiripple and Extraripple FIR Filters

We now relate the information concerning the extrema of $H_1(\omega)$ to the alternations of $E(\omega)$. In order to have an optimal filter, we must have at least $N_0 + 2$ alternations. Possible alternations can occur at the $N_e = N_0 + 1$ relative extrema of $H_1(\omega)$. There are always two alternations at ω_p and ω_s, so that the number can be $N_0 + 3$. Consequently, the unique optimum approximation for the low-pass response has either $N_0 + 2$ or $N_0 + 3$ alternations of the error function $E(\omega)$. The number $N_0 + 3$ occurs when there is an alternation at both $\omega = 0$ and $\omega = \pi$. If there is an alternation at only one of these frequencies, then we have $N_0 + 2$ alternations. The filters having $N_0 + 3$ alternations are referred to as *extraripple filters* and those having $N_0 + 2$ alternations are *equiripple filters*. Thc pseudomagnitude in Fig. 6.31 is that of an extraripple filter.

In the case where N is even, it can be shown that $N_e \leq N/2$. This is considered in Prob. 6.42.

Examples of Extraripple Pseudomagnitudes

To illustrate the ideas considered so far in this section, let us consider the case where $N = 7$. For the extraripple case, we have $N_e = (N + 1)/2 = 4$ and $1 \leq N_p \leq N_e - 1 = 3$. The last relation follows since there must be at least one passband extremum and one stopband extremum. There are three possible cases for which these constraints can be satisfied. They are the cases where (a) $N_p = 1$, (b) $N_p = 2$, and (c) $N_p = 3$. In each case, $N_s = N_e - N_p$. These three cases are illustrated in Fig. 6.32.

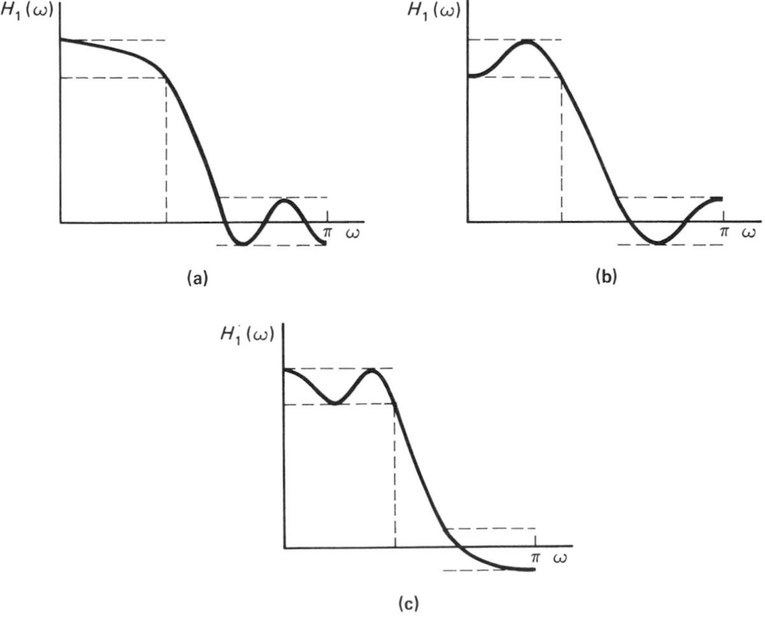

Figure 6.32 Extraripple filters for $N = 7$ and (a) $N_p = 1$, (b) $N_p = 2$, and (c) $N_p = 3$.

Description of a Computer Program

The parameters that can be used to describe $H_1(\omega)$ are N_0, N, δ_1, δ_2, K, ω_p, ω_s, N_p, and N_s. There are various procedures that have been developed in which certain of these parameters are specified and the others are variable [60, 62, 63, 106, 131, 132, 138, 140]. Some of these methods are considered in the problems. Perhaps the most widely used method is that of McClellan, Parks and Rabiner [107]. Their pro-

```
FINITE IMPULSE RESPONSE (FIR) LINEAR PHASE DIGITAL FILTER DESIGN
                    REMEZ EXCHANGE ALGORITHM

                       BANDPASS FILTER

                      FILTER LENGTH = 35

                       IMPULSE RESPONSE
              H( 1) =  0.12984430E-02 = H( 35)
              H( 2) = -0.16478480E-02 = H( 34)
              H( 3) = -0.38898340E-02 = H( 33)
              H( 4) = -0.51338300E-02 = H( 32)
              H( 5) = -0.29354840E-02 = H( 31)
              H( 6) =  0.34807880E-02 = H( 30)
              H( 7) =  0.11580970E-01 = H( 29)
              H( 8) =  0.15774610E-01 = H( 28)
              H( 9) =  0.10304960E-01 = H( 27)
              H(10) = -0.63291630E-02 = H( 26)
              H(11) = -0.28041150E-01 = H( 25)
              H(12) = -0.41671420E-01 = H( 24)
              H(13) = -0.32293580E-01 = H( 23)
              H(14) =  0.87236090E-02 = H( 22)
              H(15) =  0.77156900E-01 = H( 21)
              H(16) =  0.15502690E 00 = H( 20)
              H(17) =  0.21681860E 00 = H( 19)
              H(18) =  0.24033690E 00 = H( 18)

                       BAND  1          BAND  2          BAND
 LOWER BAND EDGE     0.000000000     0.160000000
 UPPER BAND EDGE     0.079999980     0.500000000
 DESIRED VALUE       1.000000000     0.000000000
 WEIGHTING           1.000000000     1.000000000
 DEVIATION           0.003215526     0.003215526
 DEVIATION IN DB     0.027878920   -49.854930000

 EXTREMAL FREQUENCIES
   0.0000000   0.0277777   0.0520833   0.0729165   0.0800000
   0.1600000   0.1669444   0.1860416   0.2120832   0.2381247
   0.2659024   0.2954162   0.3231938   0.3527076   0.3822214
   0.4117352   0.4412490   0.4707627   0.5000000
```

Figure 6.33 Design data for an optimal FIR low-pass filter.

gram is available [34] and is capable of designing a wide variety of optimal FIR filters including low-pass, high-pass, bandpass, and band-reject filters, as well as differentiators and Hilbert transformers. Filters can have as many as 10 bands. The input parameters for their program are the filter length N, the band edges in hertz, and the weight in each band. They use the Remez exchange algorithm to obtain the correct extrema frequencies. The output data provided by the program are the impulse-response coefficients $h(n)$, the extremal frequencies in hertz of $H_1(\omega)$, and the deviations.

Design Example

We now give an example of a filter design using this program. The filter to be designed is a low-pass filter with $N = 35$. The desired function that is to be approximated is given by

$$\hat{H}_1(\omega) = \begin{cases} 1 & 0 \le f \le 0.08 \\ 0 & 0.16 \le f \le 0.5 \end{cases}$$

and the weighting function is

$$W(\omega) = \begin{cases} 1 & 0 \le f \le 0.08 \\ 1 & 0.16 \le f \le 0.5 \end{cases}$$

where f is the frequency in hertz with $\omega = 2\pi f$. The filter coefficients $h(n)$, the passband and stopband deviations δ_1 and δ_2, respectively, and the extremal frequencies in hertz are shown in the computer printout of Fig. 6.33. From these data, we see that $\delta_1 = \delta_2 = 0.003215526$.

EXERCISES

6.7.1 **(a)** Starting with Eq. (2.86) for N odd,

$$H_1(\omega) = h\left(\frac{N-1}{2}\right) + 2 \sum_{n=0}^{(N-3)/2} h(n) \cos \left(\frac{N-1}{2} - n\right)\omega$$

substitute $(N - 1)/2 - n = k$ and obtain the relation

$$H_1(\omega) = \sum_{n=0}^{(N-1)/2} a(n) \cos n\omega$$

given in Table 6.1.

(b) Starting with Eq. (2.96) for N even,

$$H_1(\omega) = 2 \sum_{n=0}^{(N/2)-1} h(n) \cos \left(\frac{N-1}{2} - n\right)\omega$$

first make the substitution $N/2 - n = k$, and then use the trigonometric identity

$$2 \cos \omega/2 \cos n\omega = \cos (n + \tfrac{1}{2})\omega + \cos (n - \tfrac{1}{2})\omega$$

to obtain

$$2\sum_{n=0}^{(N/2)-1} \tilde{b}(n)\cos\frac{\omega}{2}\cos n\omega$$

$$= \tilde{b}(0) + \sum_{n=1}^{(N/2)-1} [\tilde{b}(n) + \tilde{b}(n-1)]\cos(n-\tfrac{1}{2})\omega$$

$$+ \tilde{b}\left(\frac{N}{2}-1\right)\cos\frac{N-1}{2}\omega$$

$$= 2\sum_{n=1}^{N/2} b(n)\cos(n-\tfrac{1}{2})\omega$$

Now determine the relations between $\tilde{b}(n)$ and $b(n)$. This yields the relation given in Table 6.1 for N even.

Answers:

(a) $a(0) = h\left(\frac{N-1}{2}\right)$

$a(n) = 2h\left(\frac{N-1}{2} - n\right) \qquad n = 1, 2, 3, \ldots, \frac{N-1}{2}$

(b) $b(n) = h\left(\frac{N}{2} - n\right) \qquad n = 1, 2, \ldots, \frac{N}{2}$

$\tilde{b}\left(\frac{N}{2} - 1\right) = 2b\left(\frac{N}{2}\right)$

$\tilde{b}(n-1) = 2b(n) - \tilde{b}(n) \qquad n = N/2 - 1, \ldots, 3, 2$

$\tilde{b}(0) = b(1) - \tfrac{1}{2}\tilde{b}(1)$

6.7.2 Determine the possible cases for $H_1(\omega)$ for extraripple filters for $N = 9$. Draw the graphs. (Hint: See Fig. 6.32.)

6.8 ANALYTICAL DESIGN OF OPTIMAL FILTERS

Determination of $H_1(\omega)$ for $N_p = 1$

In general, analytic solutions of the optimal FIR filter problem are not available. However, for the particular cases when $h(n)$ is symmetric and there is one passband or one stopband extremum, analytical solutions have been found. The cases when N is odd and $N_p = 1$ or $N_s = 1$ have been treated by Rabiner and Gold [136]. An analytic solution when N is even and $N_p = 1$ has been given by Rabiner and Herrmann [138].

We first consider the case where N is odd and $N_p = 1$. In the extraripple case, we have $N_e = (N + 1)/2$, and since $N_p = 1$, it follows that $N_s = (N - 1)/2$. Since the frequency function $H_1(\omega)$ is equiripple in the stopband, it seems possible that $H_1(\omega)$ could be related to a Chebyshev polynomial $C_m(x)$ of the first kind, defined in Eq. (5.36). Since the ripple magnitude is δ_2, we consider the relation

$$H_1(\omega) = \delta_2 C_m(x) \tag{6.106}$$

where x is related to ω. Since $|C_m(x)| \leq 1$ for $|x| \leq 1$, we require that this interval correspond to the filter stopband, $\omega_s \leq \omega \leq \pi$. Since by Eq. (6.102), we see that $H_1(\omega)$ can be expressed as a polynomial in cos ω, then an appropriate relation between x and ω is of the form

$$x = a \cos \omega + b \tag{6.107}$$

where the parameters a and b are to be determined.

The value of m is determined so that $H_1(\omega)$ has the proper number of relative extrema. Each value of x such that $C_m(x) = \pm 1$ corresponds to a relative extremum of $H_1(\omega)$. Since there are $m - 1$ relative extrema of $C_m(x)$ and since $|C_m(\pm 1)| = 1$, it follows that there are $m + 1$ values of x for which $C_m(x) = \pm 1$. However, one of these values corresponds to the stopband frequency ω_s. Therefore, we require that $m = N_s = (N - 1)/2$.

To see how x and ω, $H_1(\omega)$ and $C_m(x)$, $m = (N - 1)/2$, are related, we consider the case $N = 7$. The graph of $C_3(x)$ is shown in Fig. 6.34, and the corresponding graph of $H_1(\omega)$ is shown in Fig. 6.35. From these graphs, we see that $x = x_0$ corresponds to $\omega = 0$, $x = x_p$ to $\omega = \omega_p$, $x = 1$ to $\omega = \omega_s$, and $x = -1$ to $\omega = \pi$. The values of x_0 and x_p are seen from the graph to be determined by

$$\begin{aligned} H_1(0) &= 1 + \delta_1 = \delta_2 C_m(x_0) \\ H_1(\omega_p) &= 1 - \delta_1 = \delta_2 C_m(x_p) \end{aligned} \tag{6.108}$$

These values are given explicitly by

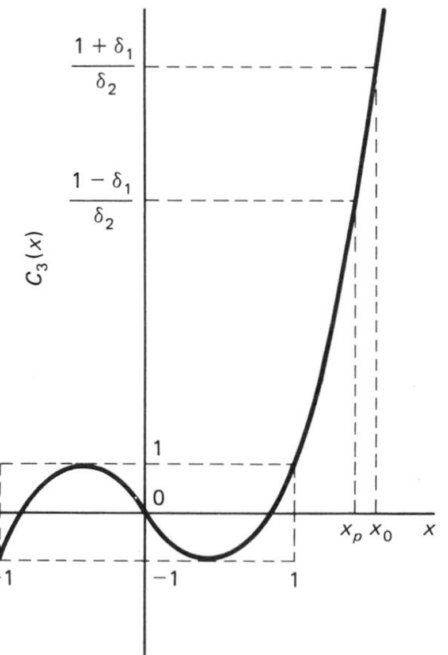

Figure 6.34 Graph of $C_3(x)$.

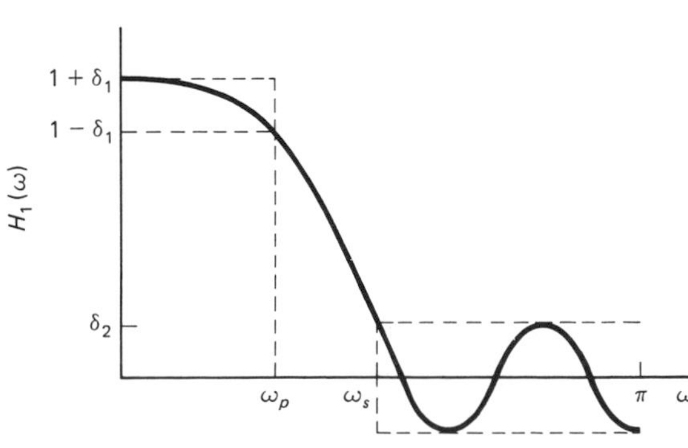

Figure 6.35 Graph of $H_1(\omega)$ corresponding to $C_3(x)$.

$$x_0 = \cosh\left[\frac{1}{m}\cosh^{-1}\left(\frac{1+\delta_1}{\delta_2}\right)\right] \tag{6.109}$$

$$x_p = \cosh\left[\frac{1}{m}\cosh^{-1}\left(\frac{1-\delta_1}{\delta_2}\right)\right] \tag{6.110}$$

We can determine the parameters a and b in Eq.(6.107), which relates x and ω, by taking $x = x_0$ and $x = -1$. Since the corresponding values are $\omega = 0$ and $\omega = \pi$, these relations become

$$x_0 = a + b$$

$$-1 = -a + b$$

and their solution is

$$a = \tfrac{1}{2}(x_0 + 1) \qquad b = \tfrac{1}{2}(x_0 - 1) = a - 1 \tag{6.111}$$

By using these values of a and b, Eq. (6.107) can be written

$$x = \tfrac{1}{2}(x_0 + 1)\cos\omega + \tfrac{1}{2}(x_0 - 1) \tag{6.112}$$

Solving this equation for ω, we obtain

$$\omega = \cos^{-1}\frac{2x - x_0 + 1}{x_0 + 1}\ \text{rad/s} \tag{6.113}$$

The passband and stopband frequencies result when $x = x_p$ and $x = 1$ and therefore are given by

$$\omega_p = \cos^{-1}\frac{2x_p - x_0 + 1}{x_0 + 1}\ \text{rad/s} \tag{6.114}$$

$$\omega_s = \cos^{-1}\frac{3 - x_0}{x_0 + 1}\ \text{rad/s} \tag{6.115}$$

The transition width is $\omega_s - \omega_p$.

Example

As an example of this design procedure, let us take $N = 7$ and $\delta_1 = \delta_2 = 0.1$. First, we see that $m = (N - 1)/2 = 3$. Then by Eqs. (6.109) and (6.110), we have

$$x_0 = \cosh\left(\tfrac{1}{3}\cosh^{-1} 1.1/0.1\right) = 1.579$$

$$x_p = \cosh\left(\tfrac{1}{3}\cosh^{-1} 0.9/0.1\right) = 1.500$$

The passband and stopband frequencies are then calculated from Eqs. (6.114) and (6.115), resulting in

$$\omega_p = \cos^{-1}\frac{2(1.5) - 1.579 + 1}{1 + 1.579} = 0.352\ \text{rad/s}$$

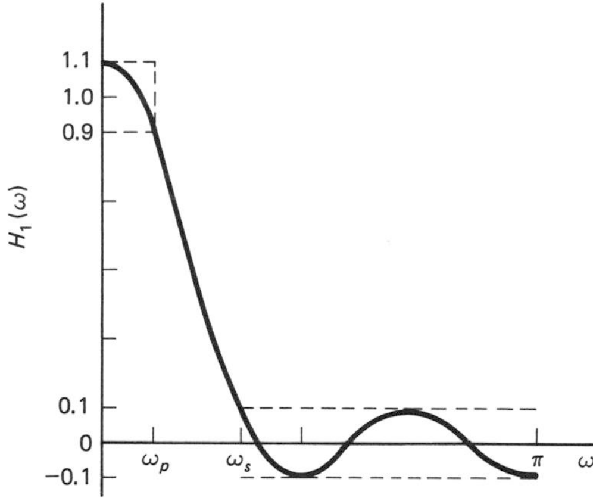

Figure 6.36 A graph of $H_1(\omega)$ for $N = 7$, $N_p = 1$, $\delta_1 = \delta_2 = 0.1$.

$$\omega_s = \cos^{-1} \frac{3 - 1.579}{1 + 1.579} = 0.987 \text{ rad/s}$$

A graph of $H_1(\omega)$ for this example is shown in Fig. 6.36.

To complete the design, we must determine the filter coefficients $h(n)$. This is accomplished by substituting for x in Eq. (6.106) the value given in Eq. (6.112), where $x_0 = 1.579$. For convenience, we use $x = a \cos \omega + b$ and then substitute the correct values for a and b. From Eq. (5.38), we find that

$$\begin{aligned} C_3(x) &= 4x^3 - 3x \\ &= (a \cos \omega + b)[4(a \cos \omega + b)^2 - 3] \\ &= a^3 \cos 3\omega + 6a^2 b \cos 2\omega \\ &\quad + (3a^2 + 12b^2 - 3)\, a \cos \omega \\ &\quad + (6a^2 + 4b^2 - 3)b \end{aligned} \tag{6.116}$$

In obtaining this form for $C_m(x)$, it may be convenient to use the trigonometric identity

$$2 \cos \omega \cos k\omega = \cos (k + 1)\omega + \cos (k - 1)\omega$$

Since, from Eqs. (6.106) and (6.116), we have

$$H_1(\omega) = \delta_2 C_3(x) = \sum_{n=0}^{3} a(n) \cos n\omega$$

and, when coefficients of cos $n\omega$ are matched, we find that

$$\begin{aligned} a(3) &= a^3 \delta_2 = 0.2144 \\ a(2) &= 6a^2 b \delta_2 = 0.2888 \\ a(1) &= (a^2 + 4b^2 - 1)(3a\delta_2) = 0.3861 \\ a(0) &= (6a^2 + 4b^2 - 3)b\delta_2 = 0.2117 \end{aligned}$$

The filter coefficients, given by Ex. 6.7.1, are

$$h(0) = h(6) = \tfrac{1}{2}a(3) = 0.1072$$
$$h(1) = h(5) = \tfrac{1}{2}a(2) = 0.1444$$
$$h(2) = h(4) = \tfrac{1}{2}a(1) = 0.1930$$
$$h(3) = a(0) = 0.2117$$

EXERCISES

6.8.1 Design an extraripple low-pass filter with $N = 3$, $N_p = 1$, $\delta_1 = \delta_2 = 0.1$. Determine ω_p and ω_s and draw the graph for $H_1(\omega)$.

6.8.2 Repeat Ex. 6.8.1 if $\delta_2 = 0.05$ and $N = 7$.

PROBLEMS

6.1 Show that if $h(n) = h(-n)$, the frequency response $H(e^{j\omega})$ is a real function of ω. Obtain the resulting cosine series for $H(e^{j\omega})$.

6.2 For a low-pass filter, the desired response is

$$H_d(e^{j\omega}) = \begin{cases} e^{-j\omega\tau} & -\omega_c \le \omega \le \omega_c < \pi \\ 0 & \omega_c < |\omega| \le \pi \end{cases}$$

Find the filter coefficients $h_d(n)$ and determine τ so that $h_d(n) = h_d(-n)$.
Answer: $\tau = 0$.

6.3 The desired frequency response of a low-pass filter is

$$H_d(e^{j\omega}) = \begin{cases} 1 & -\pi/2 \le \omega \le \pi/2 \\ 0 & \pi/2 < |\omega| < \pi \end{cases}$$

Find $h_d(n)$ and show that it equals $h_d(-n)$. Using the symmetric rectangular window with $M = 3$, find $h(n)$ and show that $h(-n) = h(n)$. Determine $H(e^{j\omega})$ and compare it with $H_d(e^{j\omega})$. Show that $H'(z) = z^{-3}H(z)$ is realizable and realize it with a minimum number of multipliers.

6.4 Using $H_d(e^{j\omega})$ of Prob. 6.3 and the rectangular window for $N = 7$, find $h(n)$ and $H(e^{j\omega})$. Discuss the phase response. Realize $H(z)$ using the minimum number of multipliers. Compare these results with those of Prob. 6.3.

6.5 For $H_d(e^{j\omega})$ in Prob. 6.3, determine $H(e^{j\omega})$ using a symmetric modified rectangular window

$$w(n) = 1 \qquad -2 \le n \le 2$$
$$w(-3) = w(3) = \tfrac{1}{2}$$
$$w(n) = 0 \qquad \text{otherwise}$$

Compare the results with those of Prob. 6.3.

6.6 Repeat Prob. 6.5 using **(a)** a symmetric Hann window and **(b)** a symmetric Hamming window.

6.7 Verify Eq. (6.20) for $W_\alpha(e^{j\omega})$.

6.8 For the desired response

$$H_d(e^{j\omega}) = \begin{cases} e^{-j3\omega} & -3\pi/4 \le \omega \le 3\pi/4 \\ 0 & 3\pi/4 < |\omega| \le \pi \end{cases}$$

determine $H(e^{j\omega})$ for $N = 7$ and compare the responses for **(a)** a rectangular window, **(b)** a modified rectangular window, **(c)** a Hann window, and **(d)** a Hamming window.

6.9 Repeat Prob. 6.8 for the desired response

$$H_d(e^{j\omega}) = \begin{cases} e^{-j3\omega} & -\pi/8 \le \omega \le \pi/8 \\ 0 & \pi/8 < |\omega| \le \pi \end{cases}$$

6.10 Repeat Prob. 6.8 for the desired response

$$H_d(e^{j\omega}) = \begin{cases} 0 & -\omega_c < \omega < \omega_c \\ e^{-j3\omega} & \omega_c \le |\omega| \le \pi \end{cases}$$

if **(a)** $\omega_c = 3\pi/4$ and **(b)** $\omega_c = \pi/4$.

6.11 Repeat Prob. 6.8 for the desired response

$$H_d(e^{j\omega}) = \begin{cases} e^{-j3\omega} & \omega_{c_1} \le |\omega| \le \omega_{c_2} \\ 0 & |\omega| < \omega_{c_1},\ \omega_{c_2} < |\omega| \le \pi \end{cases}$$

if **(a)** $\omega_{c_1} = \pi/4$, $\omega_{c_2} = \pi/2$ and **(b)** $\omega_{c_1} = \pi/2$, $\omega_{c_2} = 3\pi/4$.

6.12 Repeat Prob. 6.8 for the desired response

$$H_d(e^{j\omega}) = \begin{cases} e^{-j3\omega} & -\omega_{c_1} \le \omega \le \omega_{c_1},\ \omega_{c_2} \le |\omega| \le \pi \\ 0 & \omega_{c_1} < |\omega| < \omega_{c_2} \end{cases}$$

if **(a)** $\omega_{c_1} = \pi/4$, $\omega_{c_2} = 3\pi/4$ and **(b)** $\omega_{c_1} = \pi/2$, $\omega_{c_2} = 3\pi/4$.

6.13 Determine $W(e^{j\omega})$ for the Bartlett window of Eq. (6.23). Draw its graph for $M = 3$.

6.14 Determine $W(e^{j\omega})$ for the symmetric (shifted) Blackman window obtained from Eq. (6.24) with n replaced by $n + (N - 1)/2$.

6.15 Determine $H(e^{j\omega})$ using a Bartlett window for the desired responses given in Probs. 6.8–6.12.

6.16 Repeat Prob. 6.15 using a Blackman window.

6.17 The frequency response of a Dolph–Chebyshev or Chebyshev window shown in the figure is described by the three parameters N, δ_p, and ΔF (in Hz). Each of these parameters can be determined from the other two from the equations [34, 141]

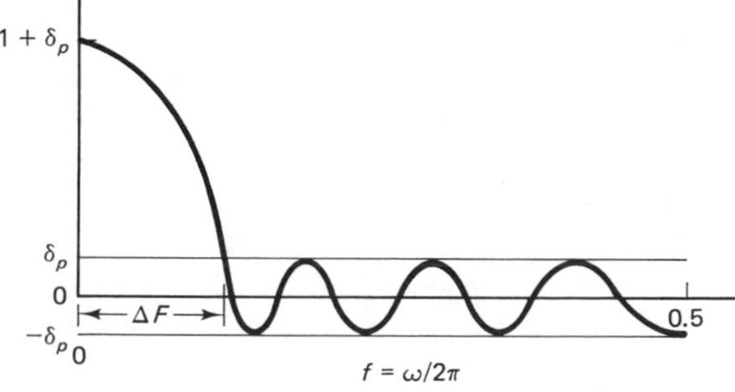

Problem 6.17

$$N \geq 1 + \frac{\cosh^{-1}\left(\dfrac{1+\delta_p}{\delta_p}\right)}{\cosh^{-1}\left(\dfrac{1}{\cos \pi\Delta F}\right)}$$

$$\Delta F = \frac{1}{\pi}\cos^{-1}\frac{1}{\cosh\left[\left(\dfrac{1}{N-1}\right)\cosh^{-1}\left(\dfrac{1+\delta_p}{\delta_p}\right)\right]}$$

$$\delta_p = \frac{1}{\cosh\left[(N-1)\cosh^{-1}\left(\dfrac{1}{\cos \pi\Delta F}\right)\right] - 1}$$

(a) If $N = 55$ and $\delta_p = 0.001$, determine ΔF.
(b) If $\delta_p = 0.001$ and $\Delta F = 0.05$ Hz, determine N. If the window is to be used for a high-pass or bandpass filter, N must be odd.

6.18 Draw the graphs of the functions described by the relations

(a) $W(e^{j\omega}) = \delta_p C_{N-1}(x_0 \cos \omega/2)$, N even

$$x_0 = \cosh\left[\frac{1}{N-1}\cosh^{-1}\left(\frac{1+\delta_p}{\delta_p}\right)\right]$$

(b) $W(e^{j\omega}) = \delta_p C_{(N-1)/2}(x)$, N odd

$$x = \frac{(1-x_0)\cos\omega + x_0\cos\omega_s - 1}{\cos\omega_s - 1}$$

$$x_0 = \cosh\left[\frac{2}{N-1}\cosh^{-1}\left(\frac{1+\delta_p}{\delta_p}\right)\right]$$

Compare these graphs with that given in Prob. 6.17. For each, identify ω_s on the graph and determine its relation to δ_p. Determine the value of $W(e^{j0})$ in each case.

6.19 The Dolph–Chebyshev window, as discussed by Helms [58], has the defining relation

$$W(e^{j\omega}) = \frac{\cos[P\cos^{-1}(z_0\cos\omega/2)]}{\cosh(P\cosh^{-1} z_0)}$$

where P is assumed to be a positive even integer. Express this function in terms of Chebyshev polynomials of the first kind. Determine the frequency ω at which $W(e^{j\omega})$ first drops to the value $1/\cosh(P\cosh^{-1} z_0)$. For $P = 8$, determine the zeros of $W(e^{j\omega})$ and draw its graph. Compare the parameters given here with those in Prob. 6.17.

6.20 Verify Eq. (6.79).

6.21 Obtain $H(e^{j\omega})$ for the filter coefficients $h(n)$ given in Eq. (6.79).

6.22 Show that Eqs. (6.89) and (6.90) result in real coefficients $h(n)$.

6.23 Derive Eqs. (6.91) and (6.92).

6.24 Design an extraripple low-pass filter with $N_p = 1$, $\delta_1 = 0.1$, $\delta_2 = 0.2$ and **(a)** $N = 3$, **(b)** $N = 5$, and **(c)** $N = 7$. Determine ω_p and ω_s and give a realization using the minimum number of multipliers in each case.

6.25 Repeat Prob. 6.24 if $\delta_2 = 0.05$.

6.26 Assume an optimal filter with parameter values N', N_p', N_s', ω_p', ω_s', δ_1', and δ_2'. The frequency function for the filter has the form

$$H_2(\omega') = \sum_{n=0}^{m} a'(n)\cos n\omega'$$

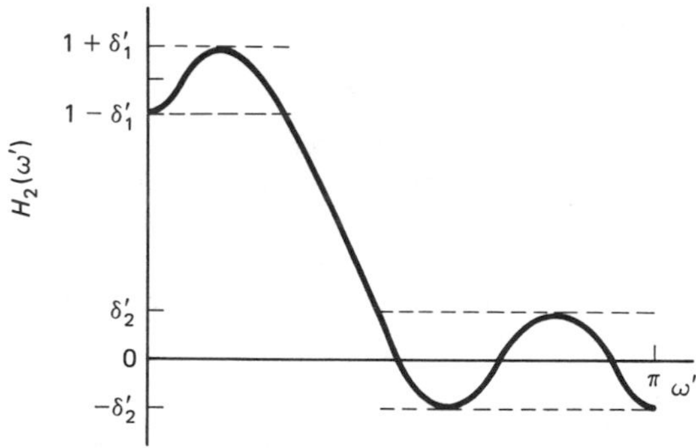

Problem 6.26

where $m = (N - 1)/2$ for N odd. A typical graph is shown in the figure. Make the frequency transformation $\omega' = \pi - \omega$. Determine $H_2(\pi - \omega)$, draw its graph, and show that it is a high-pass filter. To obtain a low-pass filter, define

$$H_1(\omega) = 1 - H_2(\pi - \omega) = \sum_{n=0}^{m} a(n) \cos n\omega$$

and draw its graph. If the new filter has parameters N, N_p, N_s, ω_p, ω_s, δ_1, and δ_2, show that $N = N'$, $N_p = N_s'$, $N_s = N_p'$, $\omega_p = \pi - \omega_s'$, $\omega_s = \pi - \omega_p'$, $\delta_1 = \delta_2'$, and $\delta_2 = \delta_1'$. Also show that $a(0) = 1 - a'(0)$ and $a(n) = (-1)^{n+1}a'(n)$, $n = 1, 2, \ldots, m = (N - 1)/2$.

6.27 It should be noted from Prob. 6.26 that if we want to design an optimal filter with $N_s = 1$, we first design an optimal filter with $N_p' = 1$, $\delta_1' = \delta_2$, and $\delta_2' = \delta_1$. Use this procedure to design a filter with $N = 5$, $N_s = 1$, and $\delta_1 = \delta_2 = 0.1$. Give a direct-form realization for the filter using a minimum number of multipliers.

6.28 Repeat Prob. 6.27 with $\delta_1 = 0.2$ and $\delta_2 = 0.1$.

6.29 Repeat Prob. 6.27 with $N = 7$.

6.30 Using the results of Prob. 6.26, we can design a high-pass filter. If $H_1(\omega)$ is an optimal low-pass filter, then $H_1(\pi - \omega)$ is an optimal high-pass filter with the same value of N_p and N_s as the corresponding low-pass filter. Determine ω_p and ω_s for the high-pass filter in terms of the corresponding low-pass parameters. Use this technique to design a high-pass filter with $N = 5$, $N_p = 1$, $\delta_1 = 0.1$, and $\delta_2 = 0.2$. Determine the filter coefficients and draw the frequency function $H_1(\omega)$ for the high-pass filter.

6.31 Devise a procedure to design an optimal high-pass filter for $N_s = 1$ using low-pass design information. Use the results to design an optimal high-pass filter with $N = 7$, $N_s = 1$, $\delta_1 = 0.2$, and $\delta_2 = 0.1$.

6.32 Use the procedure of Prob. 6.30 to design an optimal high-pass filter with $N = 7$, $N_p = 1$, and $\delta_1 = \delta_2 = 0.1$.

6.33 Repeat Prob. 6.31 for $N = 5$.

6.34 For N even, the optimal filter has the frequency function

$$H_1(\omega) = \cos \omega/2 \sum_{n=0}^{m} \tilde{b}(n) \cos n\omega$$

where $m = N/2 - 1$. Set $\xi = \omega/2$ and follow the procedure for N odd to obtain

$$H_1(\omega) = \cos \xi \sum_{n=0}^{m} \bar{b}(n) \cos^{2n} \xi$$

where $\bar{b}(n)$ is related to $\tilde{b}(n)$. Show that

$$H_1(\omega) = \delta_2 C_{N-1}(x)$$

where $x = a \cos \omega/2 + b$, yields the correct number of relative extrema N_s if $N_p = 1$ and the interval is $0 \leq x \leq 1$. The graph for $C_3(x)$ for the case $N = 4$ is shown in the figure. Determine x_0 and x_p, which are defined exactly as for the case where N is odd. The parameters a and b are determined by requiring that $x = 0$ correspond to $\omega = \pi$ and $x = x_0$ correspond to $\omega = 0$. Determine the relation between x and ω and the frequencies ω_p and ω_s. Finally, show that

$$\begin{aligned} H_1(\omega) &= \delta_2 C_{N-1}(x_0 \cos \omega/2) \\ &= \cos \omega/2 \sum_{n=0}^{m} \tilde{b}(n) \cos n\omega \end{aligned}$$

where $m = N/2 - 1$.

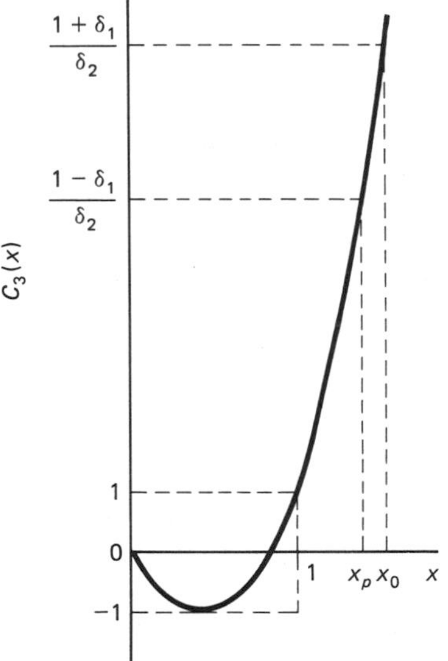

Problem 6.34

6.35 Using the results of Prob. 6.34, design an optimal filter with $N = 4$, $N_p = 1$, $\delta_1 = 0.1$, and $\delta_2 = 0.2$. Realize the filter using a direct-form realization with a minimum number of multipliers.

6.36 For an extraripple low-pass filter with $N_p = 1$, show that the stopband relative extrema, occurring when $C_m(x) = \pm 1$, are given by

$$C_m(x) = 1 \quad : \quad x = \cos 2k\pi/m \qquad 0 < k < (N-1)/4$$

$$C_m(x) = -1 \quad : \quad x = \cos (2k-1)\pi/m \qquad 0 < k < (N+1)/4$$

Apply these results to Probs. 6.24 and 6.35.

6.37 Repeat Probs. 6.35 and 6.36 for $\delta_1 = \delta_2 = 0.1$.

6.38 Repeat Probs. 6.35 and 6.36 for **(a)** $N = 6$ and **(b)** $N = 2$.

6.39 Use the methods given in Sec. 6.8 and Prob. 6.34 to obtain the expressions for $W(e^{j\omega})$ given in Prob. 6.18.

6.40 For $N = 7$, draw the graph of $H_1(\omega)$ for an extraripple FIR filter if $N_p = 2$. Determine the equality constraints on $H_1(\omega)$ at the extremal frequencies and at ω_p and ω_s. If $\delta_1 = \delta_2 = 0.1$, show that the filter coefficients are given by

$$a(0) = 0.5 \qquad a(1) = 0.624$$

$$a(2) = 0 \qquad a(3) = -0.724$$

and that the extremal and cutoff frequencies are

$$0,\ 0.256\pi,\ 0.744\pi,\ \pi$$

$$\omega_p = 0.391\pi \qquad \omega_s = 0.609\pi$$

The set of equations involved can be solved by elimination and also by the use of the Fletcher–Powell algorithm by which the sum of the squares of left-hand sides of the homogeneous equations is minimized. Methods of this type for designing optimal filters were considered by Herrmann [60] and Hofstetter, Oppenheim, and Siegel [63].

6.41 Determine α_{mn} in Eq. (6.102).

6.42 Show that for N even, the number of relative extrema of $H_1(\omega)$ satisfies the inequality $N_e \leq N/2$.

7

DISCRETE FOURIER TRANSFORMS

7.1 DEFINITIONS

In Sec. 1.6, we developed the discrete Fourier transform (DFT) for a periodic sequence and its inverse discrete Fourier transform (IDFT). If we introduce the quantity $W_N = \exp(-j2\pi/N)$, then the DFT of Eq. (1.59) and the IDFT of Eq. (1.60) for the periodic sequences $\tilde{x}(n)$ and $\tilde{X}(k)$ of period N can be written

$$\tilde{X}(k) = \sum_{n=0}^{N-1} \tilde{x}(n) W_N^{nk} \qquad k = 0, 1, \ldots, N-1 \tag{7.1}$$

$$\tilde{x}(n) = \frac{1}{N} \sum_{k=0}^{N-1} \tilde{X}(k) W_N^{-nk} \qquad n = 0, 1, \ldots, N-1 \tag{7.2}$$

Frequently, we refer to these quantities as the DFT and IDFT for $0 \le k \le N - 1$ and $0 \le n \le N - 1$, which is one period of the sequences. Since there are N values of $\tilde{x}(n)$ involved in Eq. (7.1), we refer to it as an *N-point* DFT.

We also saw in Sec. 1.6 how the DFT could be used in computing approximations of the Fourier transform of a continuous-time function. Later in Sec. 6.6, we used the DFT in the design of FIR filters.

In this chapter, we develop some of the more important properties of the DFT.

For notational convenience, we represent the operations of taking the DFT and IDFT of sequences by the expressions

$$\{\tilde{X}(k)\} = \text{DFT}\,\{\tilde{x}(n)\}$$
$$\{\tilde{x}(n)\} = \text{IDFT}\,\{\tilde{X}(k)\}$$

Usually, we omit the braces on the left side of these equations. Later, we show how the DFT can be used in computing convolution sums and how this technique can be extended to FIR filter computations. Clearly, there are many important applications that utilize the DFT and, therefore, it is desirable to develop algorithms that allow its rapid computation. Such algorithms are known as fast Fourier transforms (FFTs) and are the subject of the next chapter.

EXERCISES

7.1.1 Compute the DFT of the sequence $\tilde{x}(n)$ for $N = 5$ if $\tilde{x}(0) = \tilde{x}(4) = 2$, $\tilde{x}(1) = \tilde{x}(3) = 1$, $\tilde{x}(2) = 0$.

7.1.2 Compute the DFT of the sequence defined by $\tilde{x}(n) = (-1)^n$ for
(a) $N = 3$
(b) $N = 4$
(c) N odd
(d) N even

7.1.3 Compute the DFT for $N = 4$ if
(a) $\tilde{x}(n) = 1,\ 0 \le n \le 3$
(b) $\tilde{x}(0) = -\tilde{x}(3) = 2,\ \tilde{x}(1) = -\tilde{x}(2) = 1$

7.1.4 Let $\tilde{x}(n)$ be a periodic sequence of period N. We can determine the value of $\tilde{x}(m)$ in terms of $\tilde{x}(n)$, $0 \le n \le N-1$, by using *congruences modulo N*. Two integers n and m are said to be *congruent modulo N* or *congruent mod N,* where N is an integer called the *modulus,* if

$$m - n = kN$$

where k is some integer. This condition is usually written

$$m \equiv n(\text{mod } N)$$

and we say m is congruent to n, or m and n are congruent modulo N. If we define the set of integers $Z_N = \{0, 1, 2, \ldots, N-1\}$, then each integer m is congruent to a unique integer in Z_N. If $m \equiv n(\text{mod } N)$ for $0 \le n \le N-1$, then n is the representative in Z_N of m. If $N = 7$, find the representatives in Z_7 of the integers $m = 8$, 25, 127, and -37. Note that n is the remainder obtained by dividing m by N if $m > 0$ and is the remainder plus N if $m < 0$. For example, if we divide -37 by 7, we take the quotient to be -5 and the remainder to be -2.

7.2 PROPERTIES OF THE DFT

There are certain properties of the DFT that are of fundamental importance to its usefulness in signal processing. We develop some of these properties in this section and give other properties, leaving their proofs to the problems at the end of the chapter.

Linearity

The first property to be considered is linearity. Let $\tilde{x}_1(n)$ and $\tilde{x}_2(n)$ be two periodic sequences of period N. Then the sequence

$$\tilde{x}_3(n) = a_1\tilde{x}_1(n) + a_2\tilde{x}_2(n) \tag{7.3}$$

where a_1 and a_2 are constants, is also periodic with period N and has as its DFT

$$\tilde{X}_3(k) = \text{DFT}\,\{\tilde{x}_3(n)\} = \sum_{n=0}^{N-1} [a_1\tilde{x}_1(n) + a_2\tilde{x}_2(n)]W_N^{nk}$$

Writing this expression as two sums, we see that

$$\begin{aligned}\tilde{X}_3(k) &= a_1 \sum_{n=0}^{N-1} \tilde{x}_1(n)W_N^{nk} + a_2 \sum_{n=0}^{N-1} \tilde{x}_2(n)W_N^{nk} \\ &= a_1\tilde{X}_1(k) + a_2\tilde{X}_2(k)\end{aligned}$$

or

$$\text{DFT}\,\{a_1\tilde{x}_1(n) + a_2\tilde{x}_2(n)\} = a_1\,\text{DFT}\,\{\tilde{x}_1(n)\} + a_2\,\text{DFT}\,\{\tilde{x}_2(n)\} \tag{7.4}$$

and, consequently, the DFT is a linear operation. In a similar manner, it can be shown that the IDFT is also linear.

Shift Property

The next property of interest is that of shifting a sequence. Let $\tilde{x}_1(n) = \tilde{x}(n - m)$, the sequence obtained by shifting the sequence $\tilde{x}(n)$ to the right m units. This shifted sequence has the DFT given by

$$\begin{aligned}\tilde{X}_1(k) = \text{DFT}\,\{\tilde{x}_1(n)\} &= \sum_{n=0}^{N-1} \tilde{x}_1(n)W_N^{nk} \\ &= \sum_{n=0}^{N-1} \tilde{x}(n - m)W_N^{nk}\end{aligned}$$

Now setting $n - m = r$ in the sum, we obtain

$$\begin{aligned}\tilde{X}_1(k) &= \sum_{r=-m}^{N-1-m} \tilde{x}(r)W_N^{(r+m)k} \\ &= W_N^{mk} \sum_{r=-m}^{N-1-m} \tilde{x}(r)W_N^{rk}\end{aligned}$$

Since $\tilde{x}(r)W_N^{rk}$ is periodic of period N also, it follows that the values obtained in the previous sum are exactly the same as when $r = 0, 1, \ldots, N-1$. Using these limits in the sum results in

$$\begin{aligned}\tilde{X}_1(k) &= W_N^{mk} \sum_{r=0}^{N-1} \tilde{x}(r)W_N^{rk} \\ &= W_N^{mk}\tilde{X}(k)\end{aligned}$$

or

$$\text{DFT}\,\{\tilde{x}(n-m)\} = W_N^{mk}\,\text{DFT}\,\{\tilde{x}(n)\} \tag{7.5}$$

We see from this result that shifting the sequence $\tilde{x}(n)$ to the right m units corresponds to multiplying the DFT $\tilde{X}(k)$ by W_N^{mk}. A similar result can be obtained when the IDFT coefficients $\tilde{X}(k)$ are shifted. If $\tilde{X}_1(k) = \tilde{X}(k-m)$, then $\tilde{x}_1(n) = W_N^{-mn}\tilde{x}(n)$ or IDFT $\{\tilde{X}(k-m)\} = W_N^{-mn}$ IDFT $\{\tilde{X}(k)\}$.

Condition that $\tilde{x}(n)$ Be Real

In many applications, we can have $\tilde{X}(k)$ given and require that the resulting coefficients $\tilde{x}(n)$ be real. This was the case when frequency sampling was used to design FIR filters in Sec. 6.6. The constraint that yields a real IDFT sequence $\tilde{x}(n)$ is that $\tilde{X}(k) = \tilde{X}^*(-k)$ (* means complex conjugate) or because of the periodicity property, $\tilde{X}(k) = \tilde{X}^*(N-k)$. It should be noted that this requires that $\tilde{X}(0)$ be real since $\tilde{X}(0) = \tilde{X}^*(N) = \tilde{X}^*(0)$ because of the periodicity.

Symmetry Properties

The DFT has numerous symmetry properties. For example, the DFT of the sequence $\tilde{x}(-n)$ is $\tilde{X}(-k)$ and the DFT of $x^*(n)$ is $\tilde{X}^*(-k)$. These and other properties are similar to those of the Fourier transform considered in Chapter 1 and, consequently, we leave their development to Ex. 7.2.4.

EXERCISES

7.2.1 Compute the DFT of the sequence $\tilde{x}(0) = 3$, $\tilde{x}(1) = 2$, $\tilde{x}(2) = 0$, and $\tilde{x}(3) = -1$ if $N = 4$. Show that this is the sum of the DFTs in Ex. 7.1.3.

7.2.2 Compute the DFT of the sequence $\tilde{x}(0) = -2$, $\tilde{x}(1) = 2$, $\tilde{x}(2) = 1$, and $\tilde{x}(3) = -1$ if $N = 4$. Compare the result with that of Ex. 7.1.3(b).

7.2.3 Define $\tilde{X}(k) = e^{-j0.8k\pi}$, $k = 0, 1, 2$, and $\tilde{X}(k) = e^{j0.8\pi(5-k)}$, $k = 3, 4$, for $N = 5$. Show that $\tilde{X}(k) = \tilde{X}^*(5-k)$ and compute the IDFT.

7.2.4 Show that

(a) DFT $\{\tilde{x}(-n)\} = \tilde{X}(-k)$

(b) DFT $\{\tilde{x}^*(n)\} = \tilde{X}^*(-k)$

7.3 CIRCULAR CONVOLUTION

Periodic Convolution

In Chapter 6, we found that the Fourier transform of the product of two sequences was given by the complex convolution of their Fourier transforms. This relation was given in Eq. (6.12). In an analogous way, we can compute the DFT of a product of two periodic sequences. Since the DFT and the IDFT are so similar, the procedures for calculating the DFT or IDFT of a product are essentially the same. In this section, we develop the IDFT expression and leave the determination of the DFT relation to Ex. 7.3.3.

Suppose, we have two sequences $\tilde{x}_1(n)$ and $\tilde{x}_2(n)$, both periodic of period N, whose DFTs are

$$\tilde{X}_1(k) = \sum_{m=0}^{N-1} \tilde{x}_1(m) W_N^{mk} \tag{7.6}$$

$$\tilde{X}_2(k) = \sum_{r=0}^{N-1} \tilde{x}_2(r) W_N^{rk} \tag{7.7}$$

We define a new sequence $\tilde{x}_3(n)$ with DFT given by the product

$$\tilde{X}_3(k) = \tilde{X}_1(k)\tilde{X}_2(k) \tag{7.8}$$

The sequence $\tilde{x}_3(n)$ is then determined from the IDFT relation $\tilde{x}_3(n) = \text{IDFT}\,\{\tilde{X}_3(k)\}$, which is equivalent to

$$\begin{aligned} N\tilde{x}_3(n) &= \sum_{k=0}^{N-1} \tilde{X}_3(k) W_N^{-nk} \\ &= \sum_{k=0}^{N-1} \tilde{X}_1(k)\tilde{X}_2(k) W_N^{-nk} \end{aligned}$$

Replacing $\tilde{X}_1(k)$ and $\tilde{X}_2(k)$ by their values from Eqs. (7.6) and (7.7) and rearranging the resulting sums, we obtain

$$N\tilde{x}_3(n) = \sum_{m=0}^{N-1} \tilde{x}_1(m) \sum_{r=0}^{N-1} \tilde{x}_2(r) \left[\sum_{k=0}^{N-1} W_N^{-(n-m-r)k} \right]$$

Since [see Eq. (1.57)]

$$\sum_{k=0}^{N-1} e^{j2\pi k(n-m-r)/N} = \begin{cases} N & \text{for } r = n - m + pN \\ 0 & \text{otherwise} \end{cases}$$

where p is an integer, the above expression for $N\tilde{x}_3(n)$ simplifies to

$$\begin{aligned} \tilde{x}_3(n) &= \sum_{m=0}^{N-1} \tilde{x}_1(m)\tilde{x}_2(n - m + pN) \\ &= \sum_{m=0}^{N-1} \tilde{x}_1(m)\tilde{x}_2(n - m) \end{aligned} \tag{7.9}$$

The last expression is a result of the periodicity of $\tilde{x}_2(n)$.

There is some similarity between Eq. (7.9) and the convolution sum. The main difference is that the sum in Eq. (7.9) is taken over only one period. Another difference is that $\tilde{x}_3(n)$ is periodic of period N since the sequences $\tilde{x}_1(n)$ and $\tilde{x}_2(n)$ are both periodic of period N. For this reason, Eq. (7.9) is called a *periodic convolution*.

Example

As an example of a periodic convolution, let us consider the sequences $\tilde{x}_1(n) = \tilde{x}_2(n)$ defined by

$$\tilde{x}_1(0) = 1 \qquad \tilde{x}_1(1) = 2 \qquad \tilde{x}_1(2) = 2 \qquad \tilde{x}_1(3) = 1 \qquad N = 4 \tag{7.10}$$

The graph of $\tilde{x}_1(n)$ is shown in Figure 7.1(a). The periodic convolution is then given by Eq. (7.9), and for $n = 0$, we have

$$\begin{aligned}\tilde{x}_3(0) &= \tilde{x}_1(0)\tilde{x}_2(0) + \tilde{x}_1(1)\tilde{x}_2(-1) + \tilde{x}_1(2)\tilde{x}_2(-2) + \tilde{x}_1(3)\tilde{x}_2(-3)\\ &= \tilde{x}_1(0)\tilde{x}_1(0) + \tilde{x}_1(1)\tilde{x}_1(3) + \tilde{x}_1(2)\tilde{x}_1(2) + \tilde{x}_1(3)\tilde{x}_1(1)\\ &= (1)^2 + (2)(1) + (2)^2 + (1)(2)\\ &= 9\end{aligned}$$

Similarly, we find that $\tilde{x}_3(1) = 8$, $\tilde{x}_3(2) = 9$, and $\tilde{x}_3(3) = 10$. The value of $\tilde{x}_3(n)$ can be computed quite readily using the graphs of $\tilde{x}_1(m)$ and $\tilde{x}_2(n - m)$, which are shown

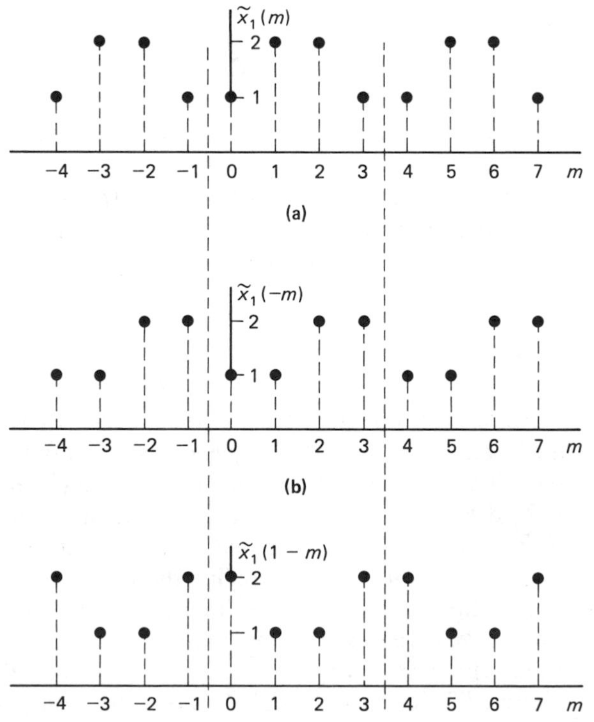

Figure 7.1 Graphs of (a) $\tilde{x}_1(m)$, (b) $\tilde{x}_1(-m)$, and (c) $\tilde{x}_1(1 - m)$.

in Fig. 7.1 for $n = 0, 1$. We simply multiply the corresponding values of $\tilde{x}_1(m)$ and $\tilde{x}_1(n - m)$ and sum these products.

Use of Tabular Arrays

A representation that can be used more easily in computing the periodic convolution is Table 7.1, which is for the example just considered. In general, the table contains the values of $\tilde{x}_1(m)$ and $\tilde{x}_2(n - m)$, which are needed for the computations. The first row contains the values of $\tilde{x}_1(m)$, $-(N - 1) \leq m \leq N - 1$. Each succeeding row gives the values of $\tilde{x}_2(n - m)$ for $n = 0, 1, 2, \ldots, N - 1$ and the appropriate values of m. The values of $\tilde{x}_2(n - m)$ for $n \neq 0$ are obtained by shifting the values of $\tilde{x}_2(-m)$ to the right n places. Since $N - 1$ is the largest value of n to be considered, we shift the values of $\tilde{x}_2(-m)$ to the right $N - 1$ places to obtain the last row in the table. All the products in the periodic convolution are formed by multiplying the proper entries in the columns for $0 \leq m \leq N - 1$. The entries in the row for $\tilde{x}_2(-m)$ are given for $-(N - 1) \leq m \leq N - 1$ because these values eventually must be shifted to the right $N - 1$ places to obtain the entries for $\tilde{x}_2(N - 1 - m)$.

TABLE 7.1 DISPLAY OF VALUES FOR COMPUTING THE PERIODIC CONVOLUTION OF $\tilde{x}_1(n)$ and $\tilde{x}_2(n) = \tilde{x}_1(n)$

m	-3	-2	-1	0	1	2	3
$\tilde{x}_1(m)$	2	2	1	1	2	2	1
$\tilde{x}_2(-m) = \tilde{x}_2(4 - m)$	1	2	2	1	1	2	2
$\tilde{x}_2(1 - m)$		1	2	2	1	1	2
$\tilde{x}_2(2 - m)$			1	2	2	1	1
$\tilde{x}_2(3 - m)$				1	2	2	1

Table 7.1 contains the information for the example where $\tilde{x}_1(n) = \tilde{x}_2(n)$ and $\tilde{x}_1(n)$ is given by Eq. (7.10). The reader should verify that the procedure just outlined using this table agrees with the values of $\tilde{x}_3(n)$ obtained using the summation for the periodic convolution.

Use of Circular Arrays

For the convolution of Eq. (7.9), we can imagine that the two sequences are displayed on two circles with the same diameter. The sequence $\tilde{x}_2(n - m)$ is time-reversed with respect to the sequence $\tilde{x}_1(m)$. The circle having the values $\tilde{x}_1(m)$ displayed on its circumference is held fixed and the circle with the values $\tilde{x}_2(n - m)$ displayed on its circumference is rotated so that $\tilde{x}_1(m)$ and $\tilde{x}_2(n - m)$ are aligned. The values in the convolution are then obtained by multiplying the values on one circle by the corresponding values on the other circle. This interpretation has caused this convolution sum to be given the name *circular convolution* also.

Actually, we can use a single circle with the values of $\tilde{x}_1(m)$ on the outside and $\tilde{x}_2(n - m)$ on the inside of the circle. To see how the values should be placed on the circle, consider the convolution for $n = 0$, which is

$$\tilde{x}_3(0) = \sum_{m=0}^{N-1} \tilde{x}_1(m)\tilde{x}_2(-m)$$

Consequently, when the proper values of $\tilde{x}_1$ and $\tilde{x}_2$ are placed on the circle in Fig. 7.2(a), we see that the value $\tilde{x}_1(m)$ matches $\tilde{x}_2(-m)$. Since $\tilde{x}_2(n)$ is periodic, it follows that $\tilde{x}_2(-m) = \tilde{x}_2(N - m)$ and, in particular, $\tilde{x}_2[-(N - 1)] = \tilde{x}_2[N - (N - 1)] = \tilde{x}_2(1)$. Continuing this procedure in the clockwise direction about the circle, we then encounter $\tilde{x}_2(2)$, $\tilde{x}_2(3)$, and so forth. It then follows that the values on the outside corresponding to the sequence $\tilde{x}_1(m)$ are $\tilde{x}_1(0)$, $\tilde{x}_1(1)$, and so forth counterclockwise, and the values on the inside corresponding to the sequence $\tilde{x}_2(-m)$ are $\tilde{x}_2(0)$, $\tilde{x}_2(1)$, and so forth in the clockwise direction. This is shown in Fig. 7.2(a). The value of $\tilde{x}_3(0)$ is obtained by multiplying the matching values of $\tilde{x}_1$ and $\tilde{x}_2$ on the circle and summing these products.

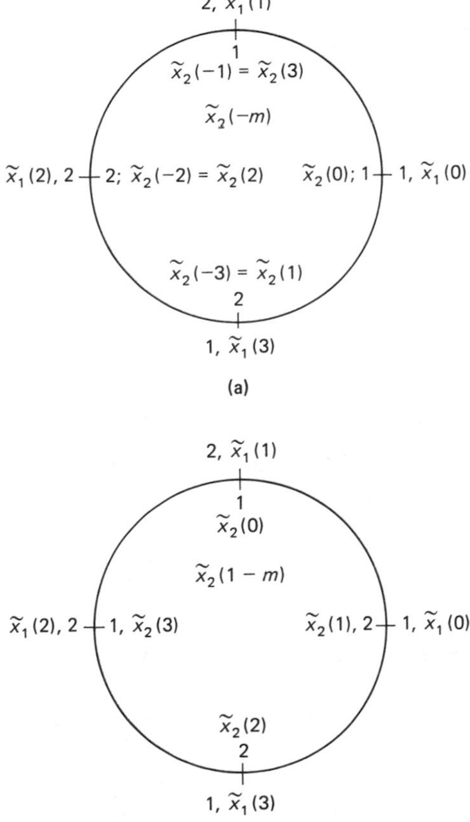

Figure 7.2 Circular representation for computing the convolution for (a) $n = 0$ and (b) $n = 1$.

The remaining values of $\tilde{x}_3(n)$ are obtained by rotating the values of $\tilde{x}_2$ on the inside of the circle and then computing the sum of the products of the matching values of $\tilde{x}_1$ and $\tilde{x}_2$. In the case of $\tilde{x}_3(1)$, we must match $\tilde{x}_1(m)$ and $\tilde{x}_2(1 - m) = \tilde{x}_2(N - m + 1)$. In other words, we match $\tilde{x}_1(0)$ with $\tilde{x}_2(1)$, $\tilde{x}_1(1)$ with $\tilde{x}_2(0)$, $\tilde{x}_1(2)$ with $\tilde{x}_2(-1) = \tilde{x}_2(3)$, and $\tilde{x}_1(3)$ with $\tilde{x}_2(-2) = \tilde{x}_2(2)$. This amounts to adding one to the argument of each term $\tilde{x}_2(-m)$ and corresponds to rotating the values of $\tilde{x}_2$ on the inside of the circle in Fig. 7.2(a) one position *counterclockwise*. This is shown in Fig. 7.2(b). Summing the products of the aligned values gives

$$\tilde{x}_3(1) = (1)(2) + (2)(1) + (2)(1) + (1)(2) = 8$$

To obtain $\tilde{x}_3(2)$, the values inside the circle in Fig. 7.2(a) are rotated two positions counterclockwise and for $\tilde{x}_3(3)$, three positions.

It should be noted that the arguments of $\tilde{x}_1(m)$ and $\tilde{x}_2(n - m)$ should have the sum n. If we use positive arguments, then the sum should be either n or $n + N$.

Stockham's Methods

An alternate way to determine the circular convolution of $\tilde{x}_1(n)$ and $\tilde{x}_2(n)$ is to first calculate the DFTs $\tilde{X}_1(k)$ and $\tilde{X}_2(k)$, form their product, and then compute the IDFT of $\tilde{X}_3(k) = \tilde{X}_1(k)\tilde{X}_2(k)$. This procedure is summarized in the relation

$$\sum_{m=0}^{N-1} \tilde{x}_1(m)\tilde{x}_2(n - m) = \text{IDFT}\,\{\tilde{X}_1(k)\tilde{X}_2(k)\} \qquad n = 0, 1, \ldots, N - 1 \tag{7.11}$$

This technique is feasible whenever the DFTs and the IDFT involved can be computed economically.

To illustrate this procedure, we use Eq. (7.11) to calculate the circular convolution of $\tilde{x}_1(n)$ and $\tilde{x}_2(n)$, where $\tilde{x}_1(n) = \tilde{x}_2(n)$ is given in Eq. (7.10). First, we note that

$$\begin{aligned} \tilde{X}_1(k) &= \sum_{n=0}^{3} \tilde{x}_1(n)W_4^{nk} \\ &= 1 + e^{-j3k\pi/2} + 2(e^{-jk\pi/2} + e^{-jk\pi}) \\ &= (1 + j^k)[1 + 2(-1)^k] \end{aligned}$$

and so

$$\tilde{X}_1(0) = 6 \qquad \tilde{X}_1(1) = -1 - j1, \qquad \tilde{X}_1(2) = 0 \qquad \tilde{X}_1(3) = -1 + j1$$

Since, for our example, we have

$$\tilde{X}_3(k) = \tilde{X}_1(k)\tilde{X}_2(k) = [\tilde{X}_1(k)]^2$$

it follows that

$$\begin{aligned} 4\tilde{x}_3(n) &= \text{IDFT}\,\{\tilde{X}_3(k)\} = \sum_{k=0}^{3} [\tilde{X}_1(k)]^2 W_4^{-nk} \\ &= (6)^2 + (-1 - j1)^2 e^{j\pi n/2} + (0)^2 e^{j\pi n} + (-1 + j1)^2 e^{j3\pi n/2} \\ &= 36 + 2j^{n+1}[1 - (-1)^n] \end{aligned}$$

The resulting values are

$$\tilde{x}_3(0) = 9 \qquad \tilde{x}_3(1) = 8 \qquad \tilde{x}_3(2) = 9 \qquad \tilde{x}_3(3) = 10$$

which are the same as those obtained by evaluating the convolution sum of Eq. (7.9) directly.

This alternate technique for evaluating the circular convolution is due to Stockham [171]. He was also able to apply this technique to the evaluation of linear convolutions by decomposing them into smaller circular convolutions. His overlap–add method is presented in the next section. His overlap–save method, also developed independently by Helms [59], is considered in Prob. 7.26. Helms also used these techniques for solving linear difference equations of the form of Eq. (2.29). These methods are also presented in Oppenheim and Schafer [126], Rabiner and Gold [136], and in a chapter in Gold and Rader written by Stockham [172].

EXERCISES

7.3.1 Given the two sequences $\tilde{x}_1(n)$ and $\tilde{x}_2(n)$ of length $N = 4$ defined by

$$\tilde{x}_1(0) = 1 \qquad \tilde{x}_1(1) = 2 \qquad \tilde{x}_1(2) = 2 \qquad \tilde{x}_1(3) = 1$$

$$\tilde{x}_2(0) = 2 \qquad \tilde{x}_2(1) = 1 \qquad \tilde{x}_2(2) = 1 \qquad \tilde{x}_2(3) = 2$$

determine the periodic convolution of $\tilde{x}_1$ and $\tilde{x}_2$ using
(a) graphs such as Fig. 7.1
(b) a table such as Table 7.1
(c) circular representations such as Fig. 7.2

7.3.2 For the sequences of Ex. 7.3.1, compute
(a) $\tilde{X}_3(k) = \tilde{X}_1(k)\tilde{X}_2(k)$
(b) the periodic convolution $\tilde{x}_3(n)$ using the IDFT

7.3.3 Calculate the DFT of the product of the two periodic sequences $\tilde{x}_1(n)$ and $\tilde{x}_2(n)$, each having period N.

7.4 LINEAR CONVOLUTION

In this section, we see how we can use the circular convolution to compute the linear convolution of two sequences when one of the sequences has finite duration. The other sequence can have finite or infinite duration.

Two Finite-Duration Sequences

We consider first the linear convolution when both sequences have finite duration. This result is then used to calculate the linear convolution when one of the sequences has infinite duration.

Let $x_1(n)$ have duration N_1 and $x_2(n)$ have duration N_2. If $x_3(n)$ denotes their linear convolution, then

$$x_3(n) = \sum_{m=0}^{n} x_1(m)x_2(n-m) \tag{7.12}$$

The sequence $x_3(n)$ also has a finite duration, which is $N_1 + N_2 - 1$. When the graph of $x_2(-m)$ is shifted $(N_2 - 1) + (N_1 - 1) + 1 = N_1 + N_2 - 1$ units to the right, there is no overlap of nonzero terms of this graph and that of $x_1(m)$. Consequently, it follows that $x_3(n) = 0$ for $n \geq N_1 + N_2 - 1$.

To illustrate this idea, consider the sequences $x_1(n)$ having duration $N_1 = 7$ and $x_2(n)$ having duration $N_2 = 5$ and defined by

$$x_1(n) = \begin{cases} 1 & 0 \le n \le 6 \\ 0 & \text{otherwise} \end{cases}$$

$$x_2(n) = \begin{cases} (-1)^n & 0 \le n \le 4 \\ 0 & \text{otherwise} \end{cases}$$

Their graphs are shown in Figs. 7.3(a) and (b). From the graph of $x_2(-n)$ shown in Fig. 7.3(c) and the graph of $x_1(n)$, we can see that $x_1(n) * x_2(n) = 0$ for $n \ge N_1 + N_2 - 1 = 7 + 5 - 1 = 11$. The sequence $x_2(11 - n)$ is shown in Fig. 7.3(d).

For convenience, we make both $x_1(n)$ and $x_2(n)$ have duration $N \ge N_1 + N_2 - 1$ by filling in zeros. Then $x_1(n)$, $x_2(n)$, and $x_3(n)$ are sequences with the same finite duration. The graphs of $x_1(n)$ and $x_2(n)$ that result after filling in zeros to make the duration $N = 11$ for both sequences are shown in Fig. 7.3(e) and (f).

Since $x_2(n - m) = 0$ for $m > n$, the linear convolution of Eq. (7.12) of two finite-duration sequences can be written

$$x_3(n) = \sum_{m=0}^{N-1} x_1(m)x_2(n - m) \qquad 0 \le n \le N - 1 \tag{7.13}$$

where $x_3(n)$ has finite duration N. This expression can now be thought of as a circular convolution, and it can be evaluated directly or by using Stockham's method and DFTs. We recall that if we use Stockham's method, the procedure is to first find $\tilde{X}_1(k)$ and $\tilde{X}_2(k)$ and then compute

$$\tilde{X}_3(k) = \tilde{X}_1(k)\tilde{X}_2(k) \qquad 0 \le k \le N - 1 \tag{7.14}$$

The linear convolution $x_3(n)$ is then the IDFT of $\tilde{X}_3(k)$, given by

$$x_3(n) = \frac{1}{N}\sum_{k=0}^{N-1} \tilde{X}_3(k)W_N^{-nk} \qquad 0 \le n \le N - 1 \tag{7.15}$$

As an example, let us take $x_1(n) = x_2(n)$, where $x_1(0) = 1$, $x_1(1) = 2$, and $x_1(n) = 0$ otherwise. We have $N_1 = N_2 = 2$, so that we can take $N = N_1 + N_2 - 1 = 3$. Using these values, we obtain

$$\tilde{X}_1(k) = \tilde{X}_2(k) = 1 + 2e^{-j2\pi k/3} \qquad k = 0, 1, 2$$

and

$$\tilde{X}_3(k) = \tilde{X}_1(k)\tilde{X}_2(k) = (1 + 2e^{-j2\pi k/3})^2 \qquad k = 0, 1, 2$$

Then, using the IDFT, we have

$$\begin{aligned} x_3(n) &= \tfrac{1}{3}\sum_{k=0}^{2} \tilde{X}_3(k)W_3^{-nk} \\ &= \tfrac{1}{3}[9 + (1 + 2e^{-j2\pi/3})^2 e^{j2\pi n/3} + (1 + 2e^{-j4\pi/3})^2 e^{j4\pi n/3}] \qquad n = 0, 1, 2 \end{aligned}$$

The resulting terms in the convolution sum are determined to be

$$x_3(0) = 1 \qquad x_3(1) = 4 \qquad x_3(2) = 4 \qquad x_3(n) = 0 \text{ for } n > 2$$

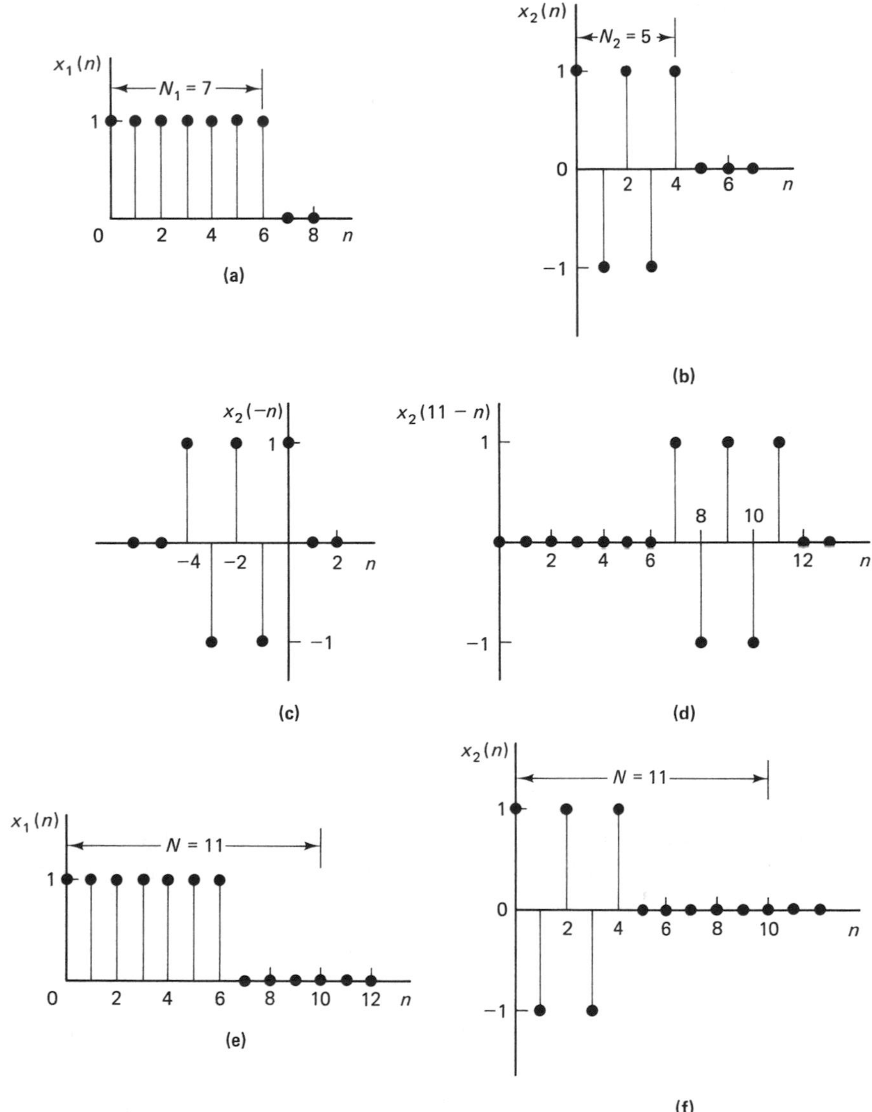

Figure 7.3 Graphs of (a) $x_1(n)$, (b) $x_2(n)$, (c) $x_2(-n)$, and (d) $x_2(11 - n)$. Graphs of (e) $x_1(n)$, and (f) $x_2(n)$ that result after filling in zeros to make the sequence lengths $N = N_1 + N_2 - 1 = 11$.

The reader is asked to verify these results using Eq. (7.15) directly and to show that these results agree with those obtained using the linear convolution of Eq. (7.12).

Convolution When One Sequence Has Infinite Duration

In many applications, we need to convolve a finite-duration sequence with an infinite-duration sequence. For example, this situation arises when the input sequence $x(n)$ to a FIR filter has infinite duration. In order to utilize the DFT, we segment the sequence $x(n)$ of infinite duration into segments of length L. If we then define new sequences $x_k(n)$ by

$$x_k(n) = \begin{cases} x(n) & kL \leq n \leq (k+1)L - 1 \\ 0 & \text{otherwise} \end{cases} \tag{7.16}$$

we can write for a general sequence

$$x(n) = \sum_{k=-\infty}^{\infty} x_k(n)$$

For a causal sequence, we have $x_k(n) = 0$ for $k < 0$ and then

$$x(n) = \sum_{k=0}^{\infty} x_k(n)$$

The response $y(n)$, which is the convolution of $h(n)$ and $x(n)$, then is given by

$$\begin{aligned} y(n) &= h(n) * x(n) = h(n) * \sum_{k=-\infty}^{\infty} x_k(n) \\ &= \sum_{k=-\infty}^{\infty} h(n) * x_k(n) \qquad (7.17) \\ &= \sum_{k=-\infty}^{\infty} y_k(n) \end{aligned}$$

where $y_k(n) = h(n) * x_k(n)$, a convolution of two finite-duration sequences. Denoting the duration of $h(n)$ by M, we see that each $y_k(n)$ is a finite-duration sequence of duration $L + M - 1$ and can be evaluated using a $(L + M - 1)$-point DFT.

We now consider each convolution $y_k(n)$, seeing how it can be evaluated and how these convolutions are combined to yield $y(n)$. The first convolution sum in Eq. (7.17) for a causal input is

$$\begin{aligned} y_0(n) &= h(n) * x_0(n) \\ &= \sum_{m=0}^{n} h(n-m)x_0(m) \end{aligned}$$

As we saw earlier in this section, by filling in zeros, we can write this sum in the form of a $(L + M - 1)$-point DFT:

$$y_0(n) = \sum_{m=0}^{L+M-2} h(n-m)x_0(m)$$

where, in particular, we have from Eq. (7.16)

$$x_0(n) = \begin{cases} x(n) & 0 \leq n \leq L - 1 \\ 0 & \text{otherwise} \end{cases}$$

The next convolution to be evaluated is given by

$$y_1(n) = \sum_{m=0}^{n} h(n-m)x_1(m) \tag{7.18}$$

From our previous discussion concerning the convolution of finite-duration sequences, we know that $y_1(n)$ has finite duration. Since $x_1(n) = 0$ for $n < L$ and $n > 2L - 1$, it can be seen from the graphs of $h(-m)$ and $x_1(m)$ that $y_1(n) = 0$ for $n < L$ and $n > 2L + M - 2$ and is given by the summation

$$y_1(n) = \sum_{m'=L}^{2L+M-2} h(n - m')x_1(m')$$

If we change the index of this summation by setting $m' = m + L$, we obtain

$$y_1(n) = \sum_{m=0}^{L+M-2} h(n - m - L)x_1(m + L) \tag{7.19}$$

This expression can be evaluated as the circular convolution of the two sequences $x_1'(n) = x_1(n + L)$ and $h'(n) = h(n - L)$.

Another way to view the response $y_1(n)$ given in Eq. (7.19) is to consider it as the response of a linear time-invariant system. Let the response to $x_1'(n) = x_1(n + L)$ be $y_1'(n)$, where

$$y_1'(n) = \sum_{m=0}^{n} h(n - m)x_1'(m) \tag{7.20}$$

If we shift the input sequence $x_1'(m)$ to the right by L units, then the response is shifted also and, therefore,

$$y_1'(n - L) = \sum_{m=0}^{n} h(n - m)x_1'(m - L)$$

From the definition of $x_1'(n)$, we see that $x_1'(m - L) = x_1(m + L - L) = x_1(m)$. Therefore, $y_1'(n - L)$ is the system response due to $x_1(n)$ and, consequently, we must have $y_1'(n - L) = y_1(n)$.

To see how the computation proceeds for the calculation of $y_1(n)$, we note that $x_1'(n) = x_1(n + L)$, which is obtained by shifting $x_1(n)$ to the left by L units. Since $x_1'(0) = x_1(L)$, the first possible nonzero value of $x_1(n)$, we see that $x_1'(n) = 0$ for $n < 0$ and $n > L$. Once we have obtained $x_1'(n)$ by shifting $x_1(n)$ to the left, we then calculate $y_1'(n)$ from Eq. (7.20), which can be considered as a circular convolution. Once $y_1'(n)$ is known, we can then obtain $y_1(n)$ by shifting $y_1'(n)$ to the right by L units. In summary, we shift $x_1(n)$ to the left L units, compute the circular convolution of this sequence and $h(n)$, and then shift the resulting sequence L units to the right to obtain $y_1(n)$.

The procedure for calculating $y_k(n)$ for $k > 1$ is essentially the same as that given before for calculating $y_1(n)$. The procedure is to shift $x_k(n)$ to the left kL units, compute the circular convolution of this shifted sequence and $h(n)$, and then shift the resulting sequence kL units to the right to obtain $y_k(n)$.

Any two consecutive sequences $y_k(n)$ and $y_{k+1}(n)$ overlap by $M - 1$ points. For example, $y_0(n)$ is nonzero for $0 \leq n \leq L + M - 2$ and $y_1(n)$ is nonzero for $L \leq n \leq 2L + M - 2$. The last $M - 1$ points of $y_0(n)$ overlap the first $M - 1$ points of $y_1(n)$. Since we can compute $y_k(n)$ and $y_{k+1}(n)$ by first shifting $x_k(n)$ and $x_{k+1}(n)$ to the origin and then shifting $y_k'(n)$ by kL units to the right and $y_{k+1}'(n)$ by $(k + 1)L$ units to the right, we see that they overlap by $M - 1$ points in the same way as did $y_0(n)$ and $y_1(n)$.

Since consecutive sequences $y_k(n)$ and $y_{k+1}(n)$ overlap, this procedure for obtaining the output by adding the outputs produced by the segmented inputs, as indicated by Eq. (7.17), is frequently referred to as the *overlap-add method.*

As a very simple example to illustrate this method, let

$$h(n) = \begin{cases} 1 & n = 0, 1 \\ 0 & \text{otherwise} \end{cases} \tag{7.21}$$

and

$$x(n) = u(n)$$

Taking $L = 3$ and noting that $M = 2$, we have

$$x_k(n) = \begin{cases} 1 & 3k \leq n \leq 3(k+1) - 1 = 3k + 2 \\ 0 & \text{otherwise} \end{cases} \tag{7.22}$$

For this simple example, we have $y_0(0) = y_0(3) = 1$, $y_0(1) = y_0(2) = 2$, and $y_0(n) = 0$, otherwise. These values can be obtained quite easily from Table 7.2.

TABLE 7.2 DISPLAY OF VALUES FOR STEP RESPONSE OF THE SYSTEM OF EQ. (7.21).

m	-3	-2	-1	0	1	2	3
$x_0(m)$				1	1	1	0
$h(-m)$	0	0	1	1	0	0	0
$h(1-m)$		0	0	1	1	0	0
$h(2-m)$			0	0	1	1	0
$h(3-m)$				0	0	1	1

Since $x_k(n) = x_0(n - 3k)$, it follows that $y_k(n) = y_0(n - 3k)$. The graphs of $y_0(n)$, $y_1(n)$, and $y_2(n)$ are shown in Figure 7.4. Since $M = 2$, the overlap in consecutive outputs is one point. In the graph, this is indicated by vertical dashed lines. For $n \leq 8$, it is seen that $y(n) = y_0(n) + y_1(n) + y_2(n)$. For example, $y(3) = 1 + 1 + 0 = 2$, $y(4) = 0 + 2 + 0 = 2$, and $y(6) = 0 + 1 + 1 = 2$. The reader should verify that these are the values that are obtained from the convolution $h(n) * x(n)$.

As another example of the overlap-add method, let us determine the response of the filter described by Eq. (7.21) to the input $x(n) = (\frac{1}{2})^n u(n)$. Table 7.3 is used for computing $y_0(n)$ and $y_1(n)$. Using the input values $x_0(m)$ and the values of $h(n - m)$, we obtain

$$y_0(0) = 1 \qquad y_0(1) = \tfrac{3}{2} \qquad y_0(2) = \tfrac{3}{4} \qquad y_0(3) = \tfrac{1}{4} \qquad y_0(n) = 0 \text{ for } n > 3$$

Next using the values $x_1'(m) = x_1(m + 3)$, we obtain

$$y_1'(0) = \tfrac{1}{8} \qquad y_1'(1) = \tfrac{3}{16} \qquad y_1'(2) = \tfrac{3}{32} \qquad y_1'(3) = \tfrac{1}{32} \qquad y_1'(n) = 0 \text{ for } n > 3$$

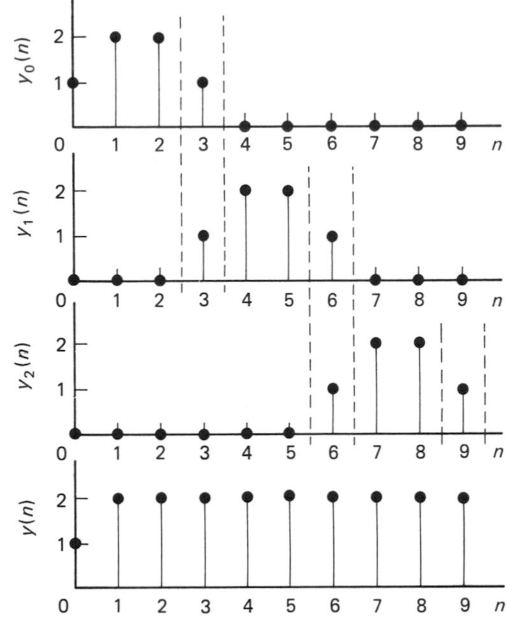

Figure 7.4 Graphs of $y_0(n)$, $y_1(n)$, $y_2(n)$, and $y(n)$ for the system described by Eq. (7.21) with input $u(n)$.

TABLE 7.3 A TABLE FOR COMPUTING $y_0(n)$ AND $y_1(n)$ FOR THE SYSTEM OF EQ. (7.21) FOR THE INPUT $x(n) = (\frac{1}{2})^n u(n)$

m	-3	-2	-1	0	1	2	3
$x_0(m)$				1	$\frac{1}{2}$	$\frac{1}{4}$	0
$x_1(m+3)$				$\frac{1}{8}$	$\frac{1}{16}$	$\frac{1}{32}$	0
$h(-m)$	0	0	1	1	0	0	0
$h(1-m)$		0	0	1	1	0	0
$h(2-m)$			0	0	1	1	0
$h(3-m)$				0	0	1	1

Since $y_1(n) = y_1'(n-3)$, it follows that

$$y_1(0) = y_1(1) = y_1(2) = 0 \qquad y_1(3) = \tfrac{1}{8} \qquad y_1(4) = \tfrac{3}{16}$$

$$y_1(5) = \tfrac{3}{32} \qquad y_1(6) = \tfrac{1}{32} \qquad y_1(n) = 0 \text{ for } n > 6$$

The values of $y(n)$ for $0 \leq n \leq 5$ can be obtained from $y(n) = y_0(n) + y_1(n)$.

Another method for computing the linear convolution, called the *overlap-save method,* is treated in Probs. 7.26–7.28.

It should be clear that evaluation of the linear convolution using the DFT is not at all amenable to hand calculations. It is probably not clear that this would be a useful procedure using a computer. However, as we have already mentioned, there are efficient means for calculating the DFT and so this method could be feasible. Such procedures for efficiently computing the DFT are considered in the next chapter.

EXERCISES

7.4.1 Given the two sequences defined by

$$x_1(n) = \begin{cases} 1 & 0 \le n \le 1 \\ 0 & \text{otherwise} \end{cases}$$

$$x_2(n) = \begin{cases} 1 & 0 \le n \le 2 \\ 0 & \text{otherwise} \end{cases}$$

fill in zeros to obtain a common sequence duration as small as possible. Then compute the linear convolution using Eq. (7.9).

7.4.2 Compute the linear convolution of the two sequences in Ex. 7.4.1 using Eq. (7.11).

7.4.3 Use the *overlap-add method* to find the step response of a system with

$$h(0) = 2 \qquad h(1) = 1 \qquad h(n) = 0 \text{ otherwise}$$

PROBLEMS

7.1 Compute DFT $\{\tilde{x}(n)\}$ if one period of $\tilde{x}(n)$ is given by
(a) $\{1, 1, -1, -1\}$
(b) $\{1, j, -1, -j\}$
(c) a^n, $n = 0, 1, 2, \ldots, N-1$

7.2 Compute DFT $\{\tilde{x}(n)\}$ if one period of $\tilde{x}(n)$ is given by
(a) e^{an}, $n = 0, 1, 2, \ldots, N-1$
(b) $\{1, 1, 1, 1, 0, 0, 0, 0\}$
(c) an, $n = 0, 1, 2, \ldots, N-1$

7.3 **(a)** Compute the N-point DFT of the sequence $\tilde{x}(n) = 1$.
(b) For $N = 5$, compute DFT $\{1, 1, 1, 0, 0\}$ and compare the results with DFT $\{1, 1, 1\}$ for $N = 3$.

7.4 Compute DFT $\{\tilde{x}(n)\}$ using the linearity property if the period of $\tilde{x}(n)$ is N:
(a) $\cos an$
(b) $\sin an$
(c) $\cosh an$
(d) $\sinh an$
(e) $b^n \cos an$

7.5 Compute DFT $\{1, 0, 1, 0, 1\}$ using linearity and the results of Ex. 7.1.2 and Prob. 7.3.

7.6 Compute DFT $\{-1, 1, 1, -1\}$ directly from the definition and by applying the shift property to the result of Prob. 7.1(a).

7.7 Show that

$$N \sum_{n=0}^{N-1} |\tilde{x}(n)|^2 = \sum_{k=0}^{N-1} |\tilde{X}(k)|^2$$

7.8 Obtain a relation that gives the IDFT as a DFT by showing that $\tilde{x}(n)$ is the DFT of some sequence.

7.9 Show that IDFT $\{\tilde{X}(k-m)\} = W_N^{-mn}$ DFT $\{\tilde{X}(k)\}$.

7.10 Compute DFT $\{1, 1, 1, 1\}$ using Eq. (7.1). Verify that $\tilde{X}(k) = 0$ for $k \neq 0$ by using the shift property.

7.11 Compute DFT $\{(-1)^n\}$ for a general value of N. Using the shift property, determine DFT $\{(-1)^{n+1}\}$.

7.12 Show that DFT $\{W_N^{-kn}\tilde{x}(n)\} = \tilde{X}(m-k)$. Illustrate this property with $\tilde{x}(n)$ given in Prob. 7.1(c).

7.13 Derive the following results:

(a) DFT $\{x(n)y(n)\} = \frac{1}{N}\sum_{k=0}^{N-1} X(k)Y(m-k)$

(b) DFT $\{x(n)\cos 2\pi kn/N\} = \frac{1}{2}[X(m-k) + X(m+k)]$

(c) DFT $\left\{x(n)\sin\frac{2\pi kn}{N}\right\} = \frac{1}{2j}[X(m-k) - X(m+k)]$

(d) DFT $\{x(n)\}$ is real and even if $x(n)$ is real and even

(e) DFT $\{x(n)\}$ is purely imaginary and odd if $x(n)$ is real and odd

(f) DFT $\{e^{jm\omega n}\} = N\delta(k-m)$, $0 \leq m \leq N-1$, if $\omega = 2\pi/N$.

7.14 Show that

$$\text{DFT}\left\{\sum_{m=0}^{N-1} x(m)y(m-n)\right\} = X(k)Y^*(k)$$

Use this result to obtain DFT $\{\sum_{m=0}^{N-1} x(m)y(m+n)\}$.

7.15 From the z-transform

$$H(z) = \sum_{n=-\infty}^{\infty} h(n)z^{-n}$$

determine the "DFT coefficients"

$$\tilde{H}(k) = H(z)\big|_{z=e^{j2\pi k/N}} = \sum_{n=-\infty}^{\infty} h(n)e^{-j2\pi nk/N}$$

If $\tilde{h}(n) = \text{IDFT}\{\tilde{H}(k)\}$, show that

$$\tilde{h}(n) = \sum_{r=-\infty}^{\infty} h(n-rN)$$

This process involves sampling in the frequency domain and results in aliasing in the time domain. Apply these results to the sequence $x(n) = 2^{-n}u(n)$ for

(a) $N = 5$

(b) $N = 10$

7.16 Apply the results of Prob. 7.15 to the function

$$H(z) = \sum_{n=0}^{\infty} a^n z^{-n} \qquad |a| < 1$$

7.17 Given the two sequences $\tilde{x}_1(n) = a^n$ and $\tilde{x}_2(n) = b^n$ of length 4, determine the circular convolution of $\tilde{x}_1$ and $\tilde{x}_2$ using

(a) graphs as in Fig. 7.1

(b) a table such as Table 7.1

(c) circular representations such as Fig. 7.2.

7.18 For the sequences $\tilde{x}_1$ and $\tilde{x}_2$ of Prob. 7.17, compute $\tilde{X}_3(k) = \tilde{X}_1(k)\tilde{X}_2(k)$ and $\tilde{x}_3(n) =$ IDFT $\{\tilde{X}_3(k)\}$.

7.19 Repeat Prob. 7.17 for

(a) $\tilde{x}_1(n) = 1$ and $\tilde{x}_2(n) = (-1)^n$

(b) $\tilde{x}_1(n) = 1$ and $\tilde{x}_2(n) = 2^{-n}$

7.20 Repeat Prob. 7.18 for the sequences given in Prob. 7.19.

7.21 Consider the sequences defined by

$$x_1(n) = \begin{cases} 1 & 0 \le n \le 2 \\ 0 & \text{otherwise} \end{cases}$$

$$x_2(n) = \begin{cases} 1 & 0 \le n \le 2 \\ 0 & \text{otherwise} \end{cases}$$

(a) Compute the linear convolution of these sequences directly from Eq. (7.12).

(b) Compute the linear convolution by filling in zeros and using the DFT to compute the circular convolution.

7.22 Repeat Prob. 7.21 if

$$x_2(n) = \begin{cases} 2^{-n} & 0 \le n \le 3 \\ 0 & \text{otherwise} \end{cases}$$

7.23 Repeat Prob. 7.21 if

(a) $x_2(n) = \begin{cases} -1 & 0 \le n \le 2 \\ 0 & \text{otherwise} \end{cases}$

(b) $x_1(n) = \begin{cases} 1 & 0 \le n \le 1 \\ 0 & \text{otherwise} \end{cases}$

$x_2(n) = \begin{cases} 1 & 0 \le n \le 4 \\ 0 & \text{otherwise} \end{cases}$

7.24 Use the overlap-add method to find the step response of a filter with $h(n) = 2^{-n}[u(n) - u(n-3)]$ and $L = 3$.

7.25 Find the step response of the low-pass filter described by Eqs. (6.25) and (6.26) if

(a) a rectangular window is used

(b) a Hann window is used

Take $N = 5$, $\tau = (N-1)/2$, $L = 3$, and use the overlap-add method.

7.26 Another method for computing the response of a FIR filter due to an infinite-duration sequence is the *overlap-save method* [171, 126]. Let the duration of $h(n)$ be M. The input sequence $x(n)$ is divided into segments of length L, so that successive segments have $M - 1$ points in common. If the input segments are denoted by $x_0(n)$, $x_1(n)$, $x_2(n)$, . . . , where

$$x_k(n) = \begin{cases} x[n + k(L - M + 1)] & 0 \le n \le L - 1 \\ 0 & \text{otherwise} \end{cases}$$

and

$$y_k(n) = \sum_{m=0}^{L-1} h(n-m)x_k(m)$$

show that $y_k(n)$ and $y_{k+1}(n)$ overlap by $M - 1$ points. Determine a procedure whereby $y(n) = h(n) * x(n)$ can be determined from $y_k(n)$, $k = 0, 1, 2, \ldots$. It should be noted that the first $M - 1$ points of $y_k(n)$ are incorrect and should be discarded. Use this procedure to find the response to $x(n)$ if

$$h(n) = \begin{cases} 1 & 0 \le n \le 2 = M - 1 \\ 0 & \text{otherwise} \end{cases}$$

$$x(n) = \begin{cases} 1 & n \text{ even} \\ \frac{1}{2} & n \text{ odd} \\ 0 & n < 0 \end{cases}$$

and $L = 5$.

7.27 Solve Prob. 7.24 using the overlap-save method with $L = 5$.

7.28 Solve Prob. 7.25 using the overlap-save method with $L = 5$.

7.29 Define the vectors

$$\mathbf{x} = [x(0) \quad x(1) \quad \cdots \quad x(N-1)]^T$$

and

$$\mathbf{X} = [X(0) \quad X(1) \quad \cdots \quad X(N-1)]^T.$$

Then the DFT relation of Eq. (7.1) can be written in the matrix form

$$\mathbf{X} = \mathbf{F}\mathbf{x}$$

Determine the matrices $\mathbf{F}$ and $\mathbf{F}^{-1}$.

7.30 A permutation of the indices $(0, 1, \ldots, N-1)$ is a one-to-one mapping P: $(0, 1, 2, \ldots) \rightarrow (a, b, c, \ldots)$ which defines the mappings $0 \rightarrow a$, $1 \rightarrow b$, $2 \rightarrow c$, etc. A permutation can be represented by a permutation matrix that is obtained by interchanging rows of the identity matrix. For example, the permutation P: $(0, 1, 2, 3) \rightarrow (3, 2, 0, 1)$ has the matrix representation

$$\mathbf{P} = \begin{bmatrix} 0 & 0 & 0 & 1 \\ 0 & 0 & 1 & 0 \\ 1 & 0 & 0 & 0 \\ 0 & 1 & 0 & 0 \end{bmatrix}$$

If this transformation is applied to the indices of $\mathbf{x}$, show that

$$\mathbf{x}_p = [x(3) \quad x(2) \quad x(0) \quad x(1)]^T = \mathbf{P}[x(0) \quad x(1) \quad x(2) \quad x(3)]^T = \mathbf{P}\mathbf{x}$$

Determine the permutation matrix for the mapping P: $(0, 1, 2, 3) \rightarrow (2, 0, 1, 3)$. (Probs. 7.30–7.33 are based on [89].) The procedure involved here is referred to as *scrambling*.

7.31 Let X_p be the DFT of the permuted sequence represented in matrix form by $\mathbf{P}\mathbf{x}$. If we write $\mathbf{X}_p = \mathbf{T}\mathbf{X}$, determine the relation between the transformation matrix $\mathbf{T}$ and the permutation matrix $\mathbf{P}$. Apply this result to the permutations of Prob. 7.30 and find $\mathbf{T}$ in terms of $\mathbf{P}$ and the matrix $\mathbf{F}$ introduced in Prob. 7.29.

7.32 A new subclass of permutations can be defined by the mapping

$$s \equiv k_1 r \pmod N \qquad r = 1, 2, \ldots, N-1$$

which is a modulo-N equivalence. We have

$$a \equiv b \pmod N$$

if and only if $a - b$ is divisible by N. In this case, we say that a is congruent to b modulo N. For example, if $N = 3$, then the numbers $4, 7, 10, \ldots, 3n + 1, \ldots$ are all congruent to 1 modulo 3. (For a further discussion of modulo representations, see Ex.

7.1.4.) The greatest common divisor of two positive integers k_1 and N is denoted by (k_1, N). The two integers are *relatively prime* if $(k_1, N) = 1$, that is, their only common divisor is 1. In the previous mapping $r \to s$, we select k_1 so that k_1 and N are relatively prime. For example, if $N = 8$, then k_1 could be any of the numbers 3, 5, 7. The permutation determined by this procedure is known as a *uniform* permutation with $\mathbf{P} = \mathbf{U}$. If $k_1 = 3$, determine the permutation matrix $\mathbf{U}$, which is described by the table that follows. Determine the permutation U. It should be noted that $0 \to 0$.

r	1	2	3	4	5	6	7
$3r$	3	6	9	12	15	18	21
$s \equiv 3r \pmod 8$	3	6	1	4	7	2	5

7.33 Repeat Prob. 7.32 if $N = 8$ and $k_1 = 5$.

7.34 Let $(p, N) = 1$. For $N = 16$ and $p = 13$, determine the set $pn \pmod N$, $n = 0, 1, 2, \ldots, N - 1$, and the permutation matrix $\mathbf{U}$. Express the transform relation

$$X_P(k) = \sum_{n=0}^{N-1} x[pn \pmod N)]W_N^{nk} \qquad k = 0, 1, 2, \ldots, N-1$$

in matrix form for $N = 16$ and $p = 13$. Also determine this relation for the general case.

7.35 The descrambling of a $\boldsymbol{U}$ permutation can be achieved by another (inverse) $\boldsymbol{U}$ permutation. This is accomplished by taking

$$k_1 k_2 \equiv 1 \pmod N$$

$$r \equiv k_2 s \pmod N$$

Then $k_2 = k_1^{-1}$, the inverse of k_1 in Z_N, and $k_1 = k_2^{-1}$.

(a) In Prob. 7.32, if $k_1 = 3$, show that $k_2 = 3$ and, therefore, the inverse $\mathbf{U}$ matrix is the same as the $\mathbf{U}$ matrix.

(b) Determine the inverse $\mathbf{U}$ matrix if $k_1 = 5$ and $N = 8$.

(c) Repeat part (b) if $N = 16$ and $k_1 = 3$.

7.36 From number theory, it is known that any positive integer M has the unique prime power factorization [33]

$$M = p_1^{r_1} p_2^{r_2} \cdots p_s^{r_s} = q_1 q_2 \cdots q_s$$

where $q_i = p_i^{r_i}$ and each p_i is a prime. The number of positive integers (including 1) less than M and relatively prime to M is called the *totient function* or *Euler's ϕ-function* of M, denoted by $\phi(M)$, and is given by

$$\phi(M) = M \prod_{i=1}^{s} (1 - p_i^{-1}) = \prod_{i=1}^{s} q_i(1 - p_i^{-1}) = \prod_{i=1}^{s} \phi(q_i)$$

For example, $\phi(2) = 2(1 - \frac{1}{2}) = 1$. Euler showed that $x^{\phi(M)} \equiv 1 \pmod M$. If M is a prime, then $\phi(M) = M - 1$ and so $x^{M-1} \equiv 1 \pmod M$. If $\alpha^K \equiv 1 \pmod M$, then α is a *root of unity* of order K if K is the least such integer. If $K = \phi(M)$, then α is called a *primitive root*. The number of primitive roots is known to be $\phi[\phi(M)]$. If p is a prime, there exists a primitive root q in $Z_p = \{0, 1, 2, \ldots, p-1\}$ such that $q^n \pmod p$, $n = 1, 2, \ldots, p - 1$, is a permutation of $\{1, 2, \ldots, p-1\}$.

(a) Illustrate this for $p = 5$. Show that $q = 3$ generates the nonzero elements of Z_5 and determine the permutation matrix $\mathbf{U}$.

(b) For $p = 11$, determine the primitive roots and for each one determine $\mathbf{U}$.

7.37 If p is a prime, show that a p-point DFT is essentially a circular convolution. Use the transformations

$$k \longrightarrow g^k(\text{mod } p) \qquad n \longrightarrow g^{-n}(\text{mod } p)$$

to accomplish this, where $g^{-1}(\text{mod } p)$ is the inverse of the element g, $gg^{-1} \equiv 1(\text{mod } p)$. Illustrate this concept for $p = 5$ and $g = 2$. Write the resulting expression for the DFT in matrix form. Here g is a primitive root in Z_p. The term $x(0)$ should be excluded from the summation in making this transformation since $0 \rightarrow 0$.

7.38 Suppose $x(n)$ is a finite-duration sequence of length N. Evaluate $X(z_k)$, where

$$z_k = AW^{-k} \qquad k = 0, 1, \ldots, M-1$$

and

$$A = A_0 e^{j\theta_0}$$

$$W = W_0 e^{-j\phi_0}$$

where A_0 and W_0 are positive real numbers. Using the identity [11]

$$nk = \tfrac{1}{2}[n^2 + k^2 - (k-n)^2]$$

determine $g(n)$ so that

$$X(z_k) = W^{k^2/2} \sum_{n=0}^{N-1} g(n) W^{-(k-n)^2/2} \qquad k = 0, 1, \ldots, M-1$$

which is a circular convolution. This expression is known as the *chirp z-transform* (CZT). Obtain a representation of the CZT using a linear system with complex impulse respone $h(n) = W^{-n^2/2}$. (For further information concerning the CZT, see [126, 136, 146].)

7.39 The *discrete Hartley transform* (DHT) [13, 92, 100, 163, 189] is defined by

$$\text{DHT}\,\{x(n)\} = H(k) = \sum_{n=0}^{N-1} x(n)\,\text{cas}(2\pi kn/N) \qquad 0 \le k \le N-1$$

where $\text{cas}\,\theta = \cos\theta + \sin\theta$.

(a) Show that

$$\text{cas}\,\theta_1\,\text{cas}\,\theta_2 = \cos(\theta_1 - \theta_2) + \sin(\theta_1 - \theta_2)$$

(b) Using the results of Ex. 1.1.4, show that

$$\sum_{k=0}^{N-1} \text{cas}\,(2\pi kn/N)\,\text{cas}\,(2\pi km/N) = N\delta_{nm} \qquad 0 \le n, m \le N-1$$

(c) Using the result of part (b), obtain the relation for the *inverse discrete Hartley transform* (IDHT):

$$x(n) = \text{IDHT}\,\{H(k)\} = \frac{1}{N}\sum_{k=0}^{N-1} X(k)\,\text{cas}\left(\frac{2\pi kn}{N}\right) \qquad 0 \le n \le N-1$$

7.40 Derive the following relation between the discrete Hartley transform DHT and the DFT:

$$\text{DHT}\,\{x(n)\} = \text{Re DFT}\,\{x(n)\} - \text{Im DFT}\,\{x(n)\}$$

7.41 Show that

(a) $\text{Re DFT}\,\{x(n)\} = \tfrac{1}{2}[\text{DHT}\,\{x(N-n)\} + \text{DHT}\,\{x(n)\}]$

(b) Im DFT $\{x(n)\} = \frac{1}{2}[\text{DHT}\{x(N-n)\} - \text{DHT}\{x(n)\}]$
(c) DHT $\{x(n+M)\} = H(k)\cos(2\pi Mk/N) - H(N-k)\sin(2\pi Mk/N)$

7.42 For the sequence $\{1, 2, 3, 4\}$, compute
(a) the DHT from the definition in Prob. 7.39
(b) the DFT and verify the result of Prob. 7.40
(c) Compute DHT $\{1, 0, 2, 0, 3, 0, 4, 0\}$ and compare the results with those of part (a).
(d) Determine the relation between DHT $\{a, b, c, d\}$ and DHT $\{a, 0, b, 0, c, 0, d, 0\}$.

7.43 Define the even and odd parts of the Hartley transform by

$$H_e(k) = \sum_{n=0}^{N-1} x(n)\cos(2\pi kn/N)$$

$$H_o(k) = \sum_{n=0}^{N-1} x(n)\sin(2\pi kn/N) = \sum_{n=1}^{N-1} x(n)\sin(2\pi kn/N)$$

respectively.
(a) Show that $H_e(-k) = H_e(k)$ and $H_o(-k) = -H_o(k)$. Note that $H(k)$ is periodic with period N.
(b) Show that

$$\text{DFT}\{x(n)\} = H_e(k) - jH_o(k)$$

(c) Use the result of part (b) and the result of Prob. 7.42(a) to obtain DFT $\{1, 2, 3, 4\}$.

7.44 Take $N = 2M + 1$ for the Hartley transform and let $x(n) = x_e(n) + x_o(n)$, where $x_e(n)$ is an even function of n, and $x_o(n)$ is an odd function of n. Show that

$$\text{DHT}\{x_e(n)\} = H_e(k) \qquad \text{DHT}\{x_o(n)\} = H_o(k)$$

7.45 Let $N = 2^m$ be the transform length. The *discrete Walsh transform* $W(k)$ of the sequence $\{x(n)\}$ is defined by [3, 46, 53, 190]

$$W(k) = \frac{1}{N}\sum_{n=0}^{N-1} x(n)g(n, k) \qquad k = 0, 1, 2, \ldots, N-1$$

where the kernel $g(n, k)$ is given by the product

$$g(n, k) = \prod_{i=0}^{m-1}(-1)^{b_i(n)b_{m-i-1}(k)}$$
$$= (-1)^{\sum_{i=0}^{m-1} b_i(n)b_{m-i-1}(k)}$$

where $b_i(n)$ is the ith bit in the binary representation of n.
(a) For $N = 4$, let $g(n, k)$ be the entry in the table that follows. Complete the table using + for $g(n, k) = +1$ and − for $g(n, k) = -1$.
(b) Use the table obtained in part (a) and obtain $W(k)$, $k = 0, 1, 2, 3$, in terms of $x(n)$, $n = 0, 1, 2, 3$.
(c) Obtain the Walsh transform of $\{1, 2, 3, 4\}$.

k \ n	00	01	10	11
00				
01				
10				
11				

7.46 Let k' be the integer obtained from k by bit reversal.

(a) Show that for $b_i(n)$, the ith bit of n,

$$b_{m-i-1}(k) = b_i(k')$$

and, therefore,

$$\sum_{i=0}^{m-1} b_i(n)b_{m-i-1}(k) = \sum_{i=0}^{m-1} b_i(n)b_i(k')$$

where $N = 2^m$ is the transform length.

(b) Let K be the number of times that the ith bits of n and k' are both one, referred to as the number of matched ones. Show that

$$\left[\sum_{i=0}^{m-1} b_i(n)b_i(k')\right](\text{mod } 2) \equiv \begin{cases} 0 & K \text{ even} \\ 1 & K \text{ odd} \end{cases}$$

$$g(n, k) = \begin{cases} +1 & K \text{ even} \\ -1 & K \text{ odd} \end{cases}$$

Verify the values of $g(n, k)$ for the Walsh transform given in the following table. In each row, the value of k' is shown also.

		n			
k	k'	00	01	10	11
		00	00	00	00
00	00	+	+	+	+
		10	10	10	10
01	10	+	+	−	−
		01	01	01	01
10	01	+	−	+	−
		11	11	11	11
11	11	+	−	−	+

For example, when $k = 01$, we compare $k' = 10$ with 00, 01, 10, 11, the values of n. For $k' = 10$ and $n = 00$, there are no matches of ones and $g(0, 1) = +1$; for $k' = 10$ and $n = 11$, there is one match of ones and so $g(3, 1) = -1$.

(c) Show that the procedure in part (b) is equivalent to sliding the number k', the bit reversal of k, along the row in the table and counting the number of matching ones in k' and n.

7.47 **(a)** Construct the table for determining the Walsh transform kernel $g(n, k)$ for $N = 8$.

(b) Determine the Walsh transform for the sequence $\{1, 0, 2, 0, 3, 0, 4, 0\}$.

7.48 The *inverse discrete Walsh transform* $x(n)$ of the transformed sequence $W(k)$ is given by

$$x(n) = \sum_{k=0}^{N-1} W(k)h(n, k) \qquad n = 0, 1, 2, \ldots, N-1$$

where $h(n, k) = g(n, k)$, given in Prob. 7.45. Find $x(n)$ if $W(k) = \{\frac{5}{2}, -1, -\frac{1}{2}, 0\}$. The answer should be the sequence considered in Prob. 7.45(c).

7.49 The *discrete Hadamard transform* $H(k)$ is defined by [3, 53]

$$H(k) = \sum_{n=0}^{N-1} x(n)g(n, k)$$

where the Hadamard kernel $g(n, k)$ is defined by

$$Ng(n, k) = (-1)^{s(n, k)}$$

with

$$s(n, k) = \sum_{n=0}^{m-1} b_i(n)b_i(k) \qquad N = 2^m$$

where $b_i(n)$ is the ith bit in the binary representation of n. Determine the entries in the table for $g(n, k)$ for **(a)** $N = 2$ and **(b)** $N = 4$. Let n identify the column and k identify the row as for the Walsh transform in Prob. 7.45. **(c)** Show that $g(n, k) = g(k, n)$. Therefore, the matrix $\mathbf{H} = [g(n, k)]$ is symmetric, that is, $\mathbf{H} = \mathbf{H}^T$. **(d)** If $\mathbf{H}_N = N\mathbf{H}$, where N is the transform length, show that

$$\mathbf{H}_4 = \begin{bmatrix} \mathbf{H}_2 & \mathbf{H}_2 \\ \mathbf{H}_2 & -\mathbf{H}_2 \end{bmatrix}$$

(e) Show that $\mathbf{H}^2 = \mathbf{I}$, the identity matrix, so that $\mathbf{H}^{-1} = \mathbf{H}^T$.

7.50 Obtain the Hadamard transform of the sequence $\{1, 2, 3, 4\}$. Compare the results with those of Prob. 7.45 for the Walsh transform. The number of sign changes along a column (or row since $\mathbf{H} = \mathbf{H}^T$) of the Hadamard matrix $\mathbf{H}_N$ is often called the *sequency* of that column.

(a) Show that the sequencies of the eight columns (or rows) of $\mathbf{H}_8$ are 0, 7, 3, 4, 1, 6, 2, 5, respectively.

(b) It is desired to arrange the values of k so that the sequency increases as a function of k, that is, the columns are rearranged so that the sequencies are 0, 1, 2, 3, 4, 5, 6, 7, respectively. Show that the *ordered* Hadamard kernel given by

$$Ng(n, k) = (-1)^{\sum_{i=0}^{m-1} b_i(n)p_i(k)}$$

where

$$p_0(k) = b_{m-1}(k)$$

$$p_1(k) = b_{m-1}(k) + b_{m-2}(k)$$

$$p_2(k) = b_{m-2}(k) + b_{m-3}(k)$$

$$\cdots\cdots\cdots\cdots\cdots\cdots$$

$$p_{m-1}(k) = b_1(k) + b_0(k)$$

provides the desired sequency ordering. The arithmetic is performed modulo 2.

(c) Show that this is equivalent to making a transformation from k to k'' described by

$$k: 0\ 4\ 6\ 2\ 3\ 7\ 5\ 1$$

$$k'': 0\ 1\ 2\ 3\ 4\ 5\ 6\ 7$$

where the numbers in the same column are matched. Determine the matrix $\mathbf{H}_8''$ for the transform using the values 0, 4, 6, 2, 3, 7, 5, 1 for the row values. Show that

$$\mathbf{H}_8'' = \begin{bmatrix} 1 & 0 & 0 & 0 & 0 & 0 & 0 & 0 \\ 0 & 0 & 0 & 0 & 1 & 0 & 0 & 0 \\ 0 & 0 & 0 & 0 & 0 & 0 & 1 & 0 \\ 0 & 0 & 1 & 0 & 0 & 0 & 0 & 0 \\ 0 & 0 & 0 & 1 & 0 & 0 & 0 & 0 \\ 0 & 0 & 0 & 0 & 0 & 0 & 0 & 1 \\ 0 & 0 & 0 & 0 & 0 & 1 & 0 & 0 \\ 0 & 1 & 0 & 0 & 0 & 0 & 0 & 0 \end{bmatrix} \mathbf{H}_8 = \mathbf{PH}_8$$

where $\mathbf{P}$ is a permutation matrix.

7.51 Let the Hadamard matrix $\mathbf{H}_{2N}$ be partitioned as

$$\mathbf{H}_{2N} = \begin{bmatrix} \mathbf{A} & \mathbf{B} \\ \mathbf{C} & \mathbf{D} \end{bmatrix}$$

where each of the submatrices is of order N. For each submatrix the $(m - 1)$st bits involved are

	$b_{m-1}(k)$	$b_{m-1}(n)$
A	0	0
B	0	1
C	1	0
D	1	1

(a) Show that for **A**, **B**, and **C**, we have $b_{m-1}(n)b_{m-1}(k) = 0$ and, therefore, for these entries

$$\sum_{i=0}^{m-1} b_i(n)b_i(k) = \sum_{i=0}^{m-2} b_i(n)b_i(k)$$

For the submatrix **D**, show that $b_{m-1}(n)b_{m-1}(k) = 1$ and so

$$\sum_{i=0}^{m-1} b_i(n)b_i(k) = 1 + \sum_{i=0}^{m-2} b_i(n)b_i(k)$$

$$(-1)^{\sum_{i=0}^{m-1} b_i(n)b_i(k)} = -(-1)^{\sum_{i=0}^{m-2} b_i(n)b_i(k)}$$

(b) For $N = 2$, show that

$$\mathbf{H}_4 = \begin{bmatrix} \mathbf{H}_2 & \mathbf{H}_2 \\ \mathbf{H}_2 & -\mathbf{H}_2 \end{bmatrix}$$

(c) For $N = 4$, show that

$$\mathbf{H}_8 = \begin{bmatrix} \mathbf{H}_4 & \mathbf{H}_4 \\ \mathbf{H}_4 & -\mathbf{H}_4 \end{bmatrix}$$

(d) Show that, in general,

$$\mathbf{H}_{2N} = \begin{bmatrix} \mathbf{H}_N & \mathbf{H}_N \\ \mathbf{H}_N & -\mathbf{H}_N \end{bmatrix}$$

7.52 The *inverse Hadamard transform* is given by

$$x(n) = \sum_{k=0}^{N-1} H(k)h(n,k)$$

where the inverse Hadamard kernel $h(n,k)$ is given by

$$h(n,k) = Ng(n,k)$$

Compute the inverse Hadamard transform of the sequence $\{\frac{5}{2}, -\frac{1}{2}, -1, 0\}$. Compare the result with the sequence in Prob. 7.50.

7.53 The *discrete cosine transform* (DCT) of the sequence $x(n)$ is defined by [53]

$$\text{DCT}\{x(n)\} = C(k) = \left(\frac{2}{N}\right)^{1/2} \sum_{n=0}^{N-1} x(n) \cos\frac{(2n+1)k\pi}{2N} \qquad k = 0, 1, 2, \ldots, N-1$$

(a) Show that the *inverse discrete cosine transform* (IDCT) is given by

$$\text{IDCT}\{c(k)\} = x(n) = -\frac{1}{N^{1/2}}C(0) + \left(\frac{2}{N}\right)^{1/2} \sum_{k=1}^{N-1} C(k) \cos\frac{(2n+1)k\pi}{2N}$$

$$n = 0, 1, \ldots, N-1$$

(b) Find the DCT of the sequence $\{1, 2, 3, 4\}$.
Other discrete cosine transforms are discussed in [3, 65, 95, 100, 188].

8

FAST FOURIER-TRANSFORM ALGORITHMS

8.1 INTRODUCTION

As we saw in Sec. 1.6, Fourier analysis can be done using the discrete Fourier transform (DFT). Direct computation using Eq. (1.59) requires about N^2 complex multiplications. Prior to 1965, computer programs were using up hundreds of hours of machine time using algorithms requiring something proportional to N^2 operations with a proportionality factor that could be reduced by using symmetry properties of the trigonometric functions. Then in 1965, a method of computing the DFT was given by Cooley and Tukey [26] that required approximately $N \log_2 N$ operations when N is a power of 2. Their algorithm is also applicable when N is a composite (it has a divisor other than 1 and N). Methods for computing DFTs requiring a number of operations proportional to $N \log_2 N$ became known as *fast Fourier transforms* (FFTs).

As it so often happens, the Cooley–Tukey algorithm was not entirely new and the doubling algorithm on which it is based dates back to the early 1900s. The reduction in the number of computations was not important for the values of N that were used at that time and the algorithm was apparently overlooked when the computer made computations with large N possible. For interesting historical notes on the FFT, the reader is referred to Cooley, Lewis, and Welch [25]. It is among several papers concerning the FFT that have been collected in [142].

The term FFT usually refers to a class of algorithms for efficiently computing the DFT. Specific algorithms are appropriately named as needed.

The computational efficiency of the FFT becomes apparent when the values of N^2 and $N \log_2 N$ are compared for several values of N. For $N = 32$, we have $N^2 = 1024$ and $N \log_2 N = 160$. When $N = 128$, $N^2 = 16{,}384$ and $N \log_2 N = 896$. For $N = 2048$, we have $N^2 = 4{,}194{,}304$ and $N \log_2 N = 22{,}528$.

In this chapter, we consider two FFT algorithms. They are known as decimation-in-time and decimation-in-frequency algorithms. Complete algorithms are given for the case $N = 8$. The representation is in the form of a signal-flow graph. General computational algorithms are presented for the case when N is a power of 2. An introduction is provided for FFT algorithms when N is a composite integer, one having divisors other than 1 and N.

Numerous computer programs for computing the FFT and convolutions are given in [34]. In addition to these, there are also three programs prepared by Brenner [14].

8.2 DECIMATION-IN-TIME (DIT) ALGORITHM

Computational efficiency in the evaluation of the DFT, given in Eq. (7.1) and repeated here for convenience,

$$X(k) = \sum_{n=0}^{N-1} x(n)W_N^{nk} \tag{8.1}$$

is achieved by decomposing this sum of N terms into sums containing fewer terms. These decomposition schemes give rise to highly efficient algorithms known as fast Fourier transforms (FFTs). These techniques are based on ways in which N can be factored. For the most part, we consider values of N that are integral powers of 2, that is, $N = 2^m$, where m is a positive integer. In many cases, the user is allowed to choose the value of N to be used so that this restriction is not a serious one.

First Decomposition

The first decomposition to be considered is known as *decimation in time* (DIT). The terms in the sum are separated into two sets: one set contains the terms for which n is even, and the other set contains the terms for which n is odd. Using this even and odd separation for n, we can write the DFT in the form

$$\begin{aligned} X(k) &= \sum_{n \text{ even}} x(n)W_N^{nk} + \sum_{n \text{ odd}} x(n)W_N^{nk} \\ &= \sum_{n=0}^{(N/2)-1} x(2n)W_N^{2nk} + \sum_{n=0}^{(N/2)-1} x(2n+1)W_N^{(2n+1)k} \end{aligned}$$

Since N is an even integer, we can note that

$$W_N^2 = \exp\,[-2j(2\pi/N)] = \exp\,[-j(2\pi)/(N/2)] = W_{N/2}$$

Using this expression, we can then write

$$X(k) = \sum_{n=0}^{(N/2)-1} x(2n)W_{N/2}^{nk} + W_N^k \sum_{n=0}^{(N/2)-1} x(2n+1)W_{N/2}^{nk} \tag{8.2}$$

Each of the sums in the expression is of the same form as the DFT in Eq. (8.1) except we have $N/2$ instead of N. Since there are N terms in Eq. (8.1), it is referred to as an N-point DFT. Each of the sums in Eq. (8.2) is then an $(N/2)$-point DFT.

Let us denote the N-point DFT of Eq. (8.2) by $X_m(k)$ and the $(N/2)$-point DFTs by $X_{m-1,1}(k)$ and $X_{m-1,2}(k)$, where

$$X_{m-1,1}(k) = \sum_{n=0}^{(N/2)-1} x(2n)W_{N/2}^{nk} \tag{8.3}$$

$$X_{m-1,2}(k) = \sum_{n=0}^{(N/2)-1} x(2n+1)W_{N/2}^{nk} \tag{8.4}$$

In terms of these newly defined quantities, we can write Eq. (8.2) as

$$X_m(k) = X_{m-1,1}(k) + W_N^k X_{m-1,2}(k) \tag{8.5}$$

The N points $X_{m-1,p}(k)$, $p = 1, 2$, are separated into two sets with each set representing $N/2$-point DFTs. From our previous discussion in Chapter 7, we know that the N-point DFT is periodic with period N. Therefore, the $(N/2)$-point DFTs have period $N/2$, that is,

$$X_{m-1,p}(k + N/2) = X_{m-1,p}(k)$$

As we have seen previously in Sec. 4.2, a useful representation of a system of equations is that of a signal-flow graph. Such a representation for Eqs. (8.3)–(8.5) is shown in Fig. 8.1. It should be noted that there are two incoming signals for each

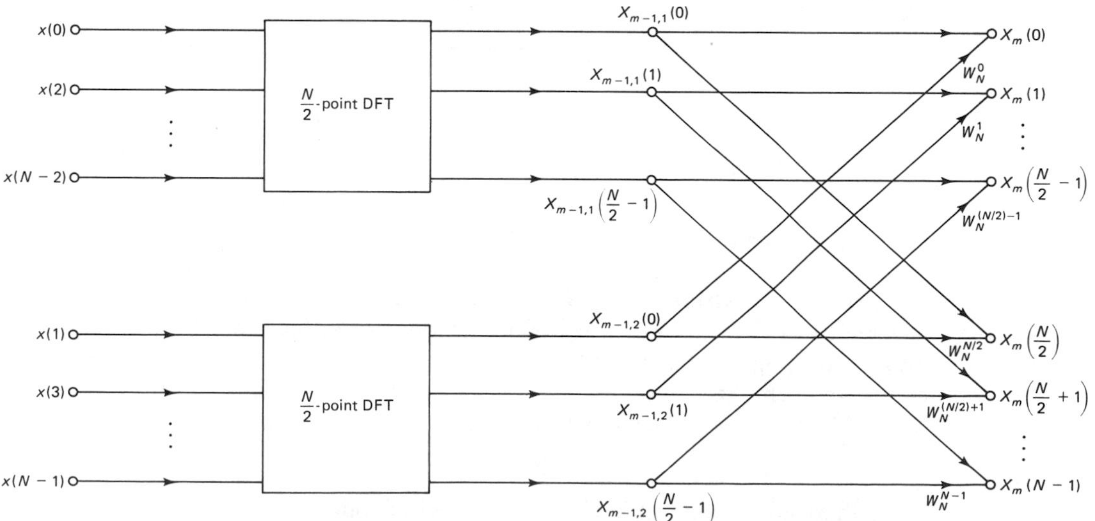

Figure 8.1 Signal-flow graph for the decomposition represented by Eqs. (8.3)–(8.5).

node $X_m(k)$ and two signals leaving each of the nodes $X_{m-1,1}(k)$ and $X_{m-1,2}(k)$. The values of $X_m(k)$ are divided into two sets. The set for which $0 \le k \le (N/2) - 1$ appears in the upper part of the diagram and the set for which $N/2 \le k \le N - 1$ appears in the lower part. Also, the values of $X_{m-1,1}(k)$ appear as the outputs of the upper $(N/2)$-point DFT and the values of $X_{m-1,2}(k)$ appear as the outputs of the lower $(N/2)$-point DFT. Equation (8.5) can be used directly when $0 \le k \le (N/2) - 1$, but for $N/2 \le k \le N - 1$, it should be modified to account for the periodicity of the $(N/2)$-point DFTs $X_{m-1,1}(k)$ and $X_{m-1,2}(k)$. The resulting modification is

$$\begin{aligned} X_m(k + N/2) &= X_{m-1,1}(k + N/2) + W_N^{k+N/2} X_{m-1,2}(k + N/2) \\ &= X_{m-1,1}(k) - W_N^k X_{m-1,2}(k) \end{aligned} \tag{8.6}$$

$$k = 0, 1, 2, \ldots, (N/2) - 1$$

since $(W_N)^{N/2} = -1$. The decomposition represented by Eqs. (8.3)–(8.6) is shown in Fig. 8.2 for the special case $N = 8$.

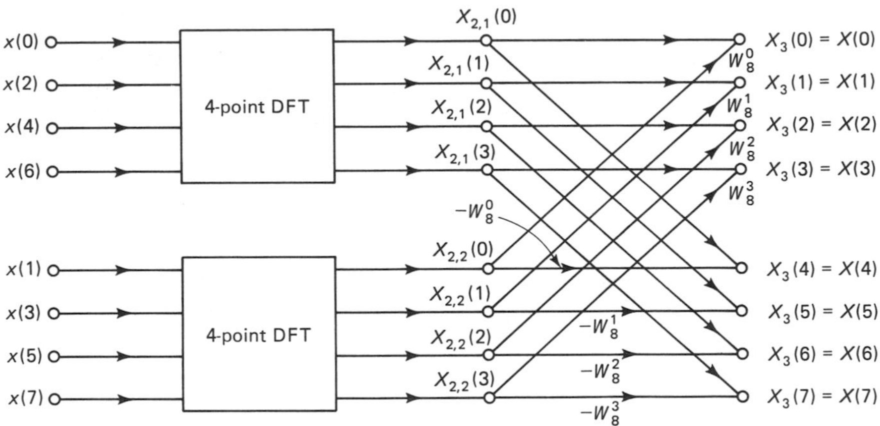

Figure 8.2 Signal-flow graph representing Eqs. (8.5) and (8.6) for $N = 8$.

Number of Computations

Before proceeding further with the decomposition of the N-point DFT, let us compare the number of computations required for Eq. (8.1) and the number required for Eqs. (8.3)–8.6). Using Eq. (8.1) requires N complex multiplications for each value of k and, therefore, a total of N^2. For each value of k, there are $N - 1$ complex additions, making a total of $N(N - 1)$ or about N^2. Using Eqs. (8.3)–(8.6), the decomposed form, requires $(N/2)^2$ complex multiplications in each sum $X_{m-1,1}(k)$ and $X_{m-1,2}(k)$ and N multiplications to compute $W_N^k X_{m-1,2}(k)$, making a total of

$$2(N/2)^2 + N = (N^2/2) + N \tag{8.7}$$

Since there are N complex additions involved in computing $X(k) = X_m(k)$ from Eqs. (8.3)–(8.6), we see that the total number of additions using the decomposed sum is

$$2(N/2)[(N/2) - 1] + N = N^2/2$$

Comparing the number of multiplications just given, we see that fewer multiplications are needed using the decomposed form of Eqs. (8.3)–(8.6) whenever

$$(N^2/2) + N < N^2$$

This inequality is satisfied whenever $N > 2$.

Continuation of the Decomposition

To continue the decomposition of the N-point DFT, we decompose each $(N/2)$-point DFT $X_{m-1,p}(k)$, $p = 1, 2$. First, considering $X_{m-1,1}(k)$, we can separate the values of $x(2n)$ into two sets, one for n even and one for n odd. This decomposition results in

$$\begin{aligned} X_{m-1,1}(k) &= \sum_{n=0}^{(N/4)-1} x(4n)W_{N/2}^{2nk} + \sum_{n=0}^{(N/4)-1} x(4n+2)W_{N/2}^{(2n+1)k} \\ &= \sum_{n=0}^{(N/4)-1} x(4n)W_{N/4}^{nk} + W_{N/2}^{k} \sum_{n=0}^{(N/4)-1} x(4n+2)W_{N/4}^{nk} \end{aligned}$$

Defining the new $(N/4)$-point DFTs,

$$X_{m-2,1}(k) = \sum_{n=0}^{(N/4)-1} x(4n)W_{N/4}^{nk} \tag{8.8}$$

$$X_{m-2,2}(k) = \sum_{n=0}^{(N/4)-1} x(4n+2)W_{N/4}^{nk} \tag{8.9}$$

we can write

$$X_{m-1,1}(k) = X_{m-2,1}(k) + W_N^{2k}X_{m-2,2}(k) \qquad k = 0, 1, 2, 3, \ldots, (N/4) - 1 \tag{8.10}$$

Actually, Eq. (8.10) is valid for $0 \le k \le (N/2) - 1$, but we can use the periodicity of the $(N/4)$-point DFTs to obtain

$$X_{m-1,1}(k + N/4) = X_{m-2,1}(k) - W_N^{2k}X_{m-2,2}(k) \qquad k = 0, 1, 2, \ldots, (N/4) - 1 \tag{8.11}$$

In a similar fashion, it can be shown that (see Prob. 8.1)

$$\begin{aligned} X_{m-1,2}(k) &= X_{m-2,3}(k) + W_N^{2k}X_{m-2,4}(k) \\ X_{m-1,2}(k + N/4) &= X_{m-2,3}(k) - W_N^{2k}X_{m-2,4}(k) \end{aligned} \tag{8.12}$$

are valid for $0 \le k \le (N/4) - 1$, where

$$X_{m-2,3}(k) = \sum_{n=0}^{(N/4)-1} x(4n+1)W_{N/4}^{nk} \tag{8.13}$$

$$X_{m-2,4}(k) = \sum_{n=0}^{(N/4)-1} x(4n+3)W_{N/4}^{nk} \tag{8.14}$$

The decomposition of the N-point DFT that has been accomplished in terms of $(N/4)$-point DFTs and described by Eqs. (8.3)–(8.6) and (8.10)–(8.14) is shown in Fig. 8.3. For the special case $N = 8$, the $(N/4)$-point DFTs are simply 2-point

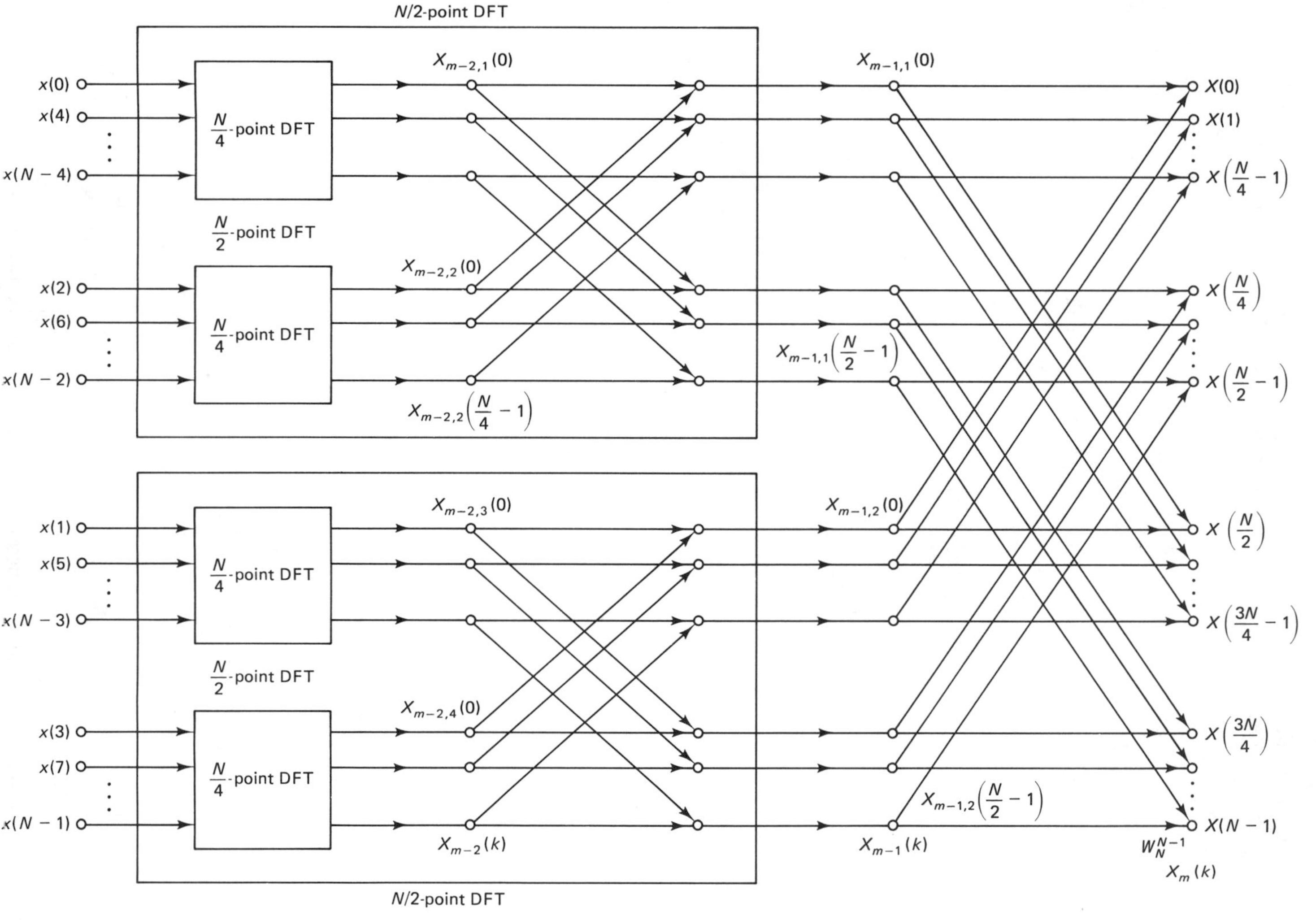

Figure 8.3 Decomposition of the N-point DFT in terms of $(N/4)$-point DFTs. (Branch gains have been omitted.)

DFTs and the decomposition is complete. The signal-flow graph for this special case is shown in Fig. 8.4. The reader is asked to write the relations for this case in Prob. 8.2 and to verify this signal-flow graph.

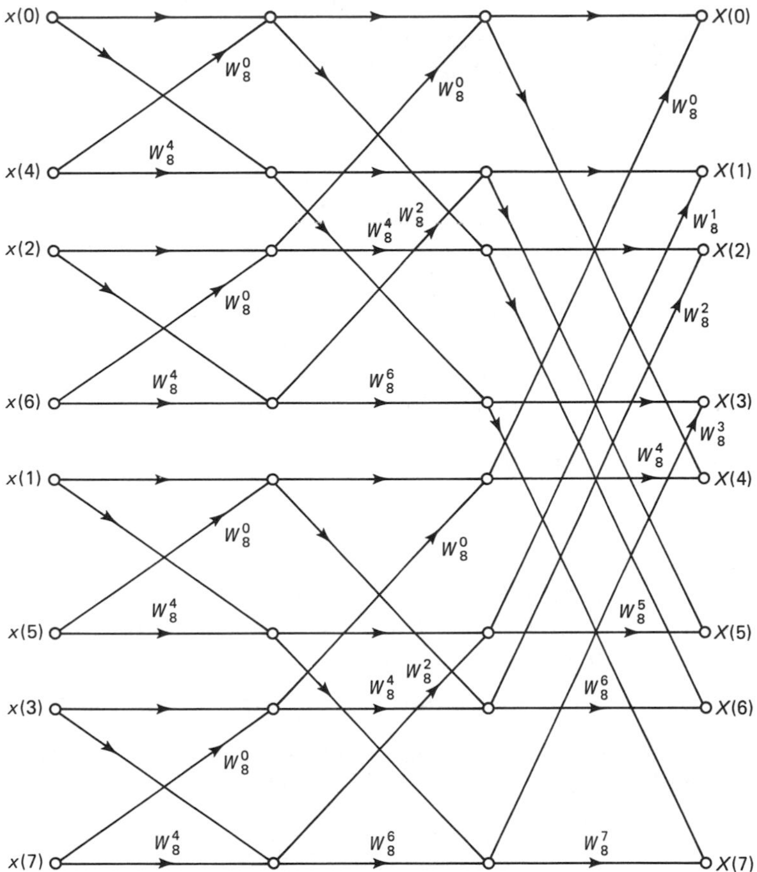

Figure 8.4 Decimation-in-time decomposition of an 8-point DFT.

It should be noted that at each stage of the decomposition of the N-point DFT, there are sets of N values to be computed. As the decomposition proceeds, these sets of values are subdivided into twice as many subsets as in the previous decomposition. To see how the decomposition evolves, we denote the sets obtained by $X_m(k)$, $X_{m-1}(k)$, $X_{m-2}(k)$, . . . , $X_0(k)$, where $X_m(k)$ is the set of DFT outputs $X(k)$, and $X_0(k)$ is the set of input values $x(n)$ rearranged in a prescribed order. Then in the first decomposition, the set $X_m(k)$ is determined from the two sets $X_{m-1,1}(k)$ and $X_{m-1,2}(k)$, which are the subsets making up the set $X_{m-1}(k)$. These calculations are described by Eqs. (8.5) and (8.6), where $X_{m-1}(k)$ is determined by Eqs. (8.3) and (8.4) as outputs of two $(N/2)$-point DFTs. This decomposition is represented by the signal-flow graph of Fig. 8.1. In the next decomposition, the values $X_{m-1}(k)$ are determined from the four sets $X_{m-2,1}(k)$, $X_{m-2,2}(k)$, $X_{m-2,3}(k)$, and $X_{m-2,4}(k)$, which together make up the set $X_{m-2}(k)$. Each of these four sets is the output of an $(N/4)$-point DFT as described by Eqs. (8.8), (8.9), (8.13), and (8.14). The elements of the set $X_{m-1}(k)$ are related to the elements of the set $X_{m-2}(k)$ as described by Eqs. (8.10)–(8.12). The relations obtained thus far for the decomposition of the N-point DFT are illustrated by the signal-flow graph of Fig. 8.3. In this figure, it is shown how each of the $(N/2)$-point DFTs are replaced by two $(N/4)$-point DFTs. The branch gains have been left off.

We can generalize the relations given previously for the first two decompositions. It can be shown that

$$\begin{aligned} X_{m-r,p}(k) &= X_{m-r-1,2p-1}(k) + W_N^{2^r k} X_{m-r-1,2p}(k) \\ X_{m-r,p}(k + N/2^{r+1}) &= X_{m-r-1,2p-1}(k) - W_N^{2^r k} X_{m-r-1,2p}(k) \end{aligned} \tag{8.15}$$

where $0 \le k \le (N/2^{r+1}) - 1$, $0 \le r \le m - 1$, and $1 \le p \le 2^r$. In particular, we define $X_{m,1}(k) = X(k)$, which occurs when $r = 0$. The value of r denotes the stage in the decomposition proceeding from $X_{m,1}(k) = X_m(k) = X(k)$. The set of values of $X_{m-r,p}(k)$ given in Eq. (8.15) is also denoted by $X_{m-r}(k)$.

Number of Multiplications

We now consider the number of multiplications required to calculate the N-point DFT as the decomposition proceeds. After the decomposition into $(N/4)$-point DFTs, we see that $2(N/4)^2 + N/2$ multiplications are required to calculate either of $X_{m-1,p}(k)$, $p = 1, 2$, given in Eqs. (8.10)–(8.12). Once these values are known, N multiplications are required to determine $X(k) = X_m(k)$, making a total of

$$2[2(N/4)^2 + N/2] + N = (N^2/4) + 2N \tag{8.16}$$

Continuing this procedure, we find that when we decompose the N-point DFT to the point where we require the computation of $(N/2^r)$-point DFTs, the number of complex multiplications is

$$N^2/2^r + rN \tag{8.17}$$

Note the generalization that Eq. (8.17) is of Eqs. (8.16) and (8.7). For the case where $N = 2^m$ and the decomposition is carried out completely as shown in Fig. 8.4 for $N = 8$, we see that the number of multiplications is simply mN. This is true since there are only 2-point DFTs to evaluate and the multiplications are those involving powers of W_N, which are included in the second term of Eq. (8.17). This can also be seen by observing that in each stage, the only multiplications are those involving W_N^s for some value of s and that there are N of these multiplications. Since there are m stages, we require $mN = N \log_2 N$ multiplications. Since $N \log_2 N < N^2 + N$ for $N \ge 2$ and $\log_2 N \ll N + 1$ (much less than) for large N, there is a considerable saving in the number of complex multiplications. The same is also true for the number of complex additions. Since complex multiplication is made up of real multiplication and real addition, the decimation-in-time FFT algorithm is a very efficient means for computing the DFT when $N = 2^m$.

Example

The signal-flow graph for the special case $N = 4$ is shown in Fig. 8.5. (The reader is asked to obtain this result in Ex. 8.2.1.) We take $x(n) = (-1)^n$ and compute its DFT using Fig. 8.5. Carrying out the computations as indicated by the signal-flow graph, we obtain $X(0) = X(1) = X(3) = 0$, and $X(2) = 4$. This can be verified directly from Eq. (8.1) by noting that

$$X(k) = \sum_{n=0}^{3} (-1)^n W_4^{nk} = \sum_{n=0}^{3} (-W_4^k)^n$$

$$= \frac{1 - (-W_4^k)^4}{1 - (-W_4^k)} = 0 \text{ for } k \neq 2$$

$$X(2) = 1 - W_4^2 + W_4^4 - W_4^6 = 4$$

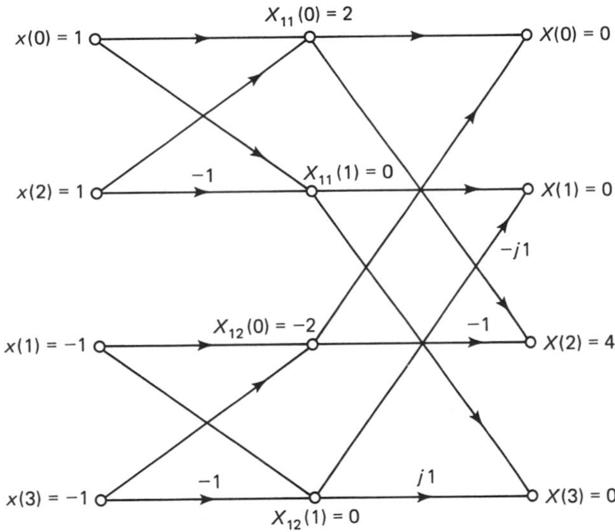

Figure 8.5 Signal-flow graph for the FFT when $N = 4$.

EXERCISES

8.2.1 Write the equations for the decimation-in-time algorithm for $N = 4$, and verify the signal-flow graph shown in Fig. 8.5.

8.2.2 Use the signal-flow graph of Fig. 8.5 to compute DFT $\{1\}$ for $N = 4$. Check the answer by using Eq. (8.1).

8.2.3 Compute the DFTs in Ex. 7.1.3 using the results of Ex. 8.2.1.

8.3 COMPUTATIONAL EFFICIENCY

The signal-flow graph of Fig. 8.4 describes a computational algorithm for the DFT for $N = 8$. In addition, it suggests a useful method for storing the original data and also the results of the intermediate computations. At each stage in the computation, we see that a set of N complex numbers is transformed into another set of N complex numbers. This procedure is repeated for each of the m stages.

The computations that are used at each stage are described by Eq. (8.15) and can be represented by the signal-flow graph shown in Fig. 8.6. The computation described here is referred to as a *butterfly computation*. The set of points $X_{m-r,p}(k)$ and $X_{m-r,p}(k + N/2^{r+1})$ is denoted by $X_{m-r}(k)$ and the set of points $X_{m-r-1,2p-1}(k)$ and $X_{m-r-1,2p}(k)$ is denoted by $X_{m-r-1}(k)$ as mentioned previously. The value of r is the stage number starting with $r = 0$ with the set of values $X_m(k) = X(k)$.

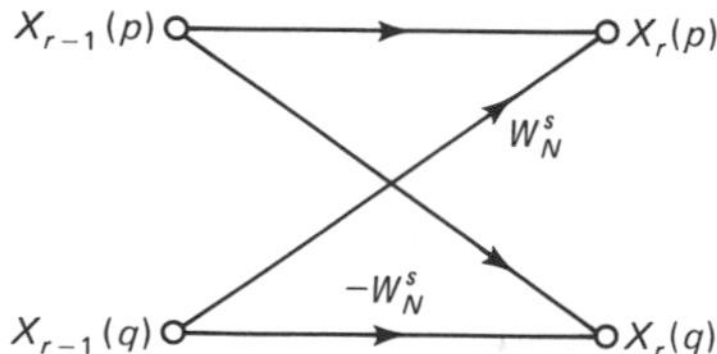

Figure 8.6 Butterfly representing Eq. (8.15) and used in the computations of the rth stage.

For the case $N = 8$, the sets of values that are to be determined at each stage are $X_0(k)$, $X_1(k)$, $X_2(k)$, and $X_3(k) = X(k)$. The set $X_0(k)$ is simply the set $\{x(n)\}$ rearranged. In particular, we have

$$\{X_0(k)\} = \{x(0), x(4), x(2), x(6), x(1), x(3), x(5), x(7)\}$$

$$\{X_1(k)\} = \{X_{11}(0), X_{11}(1), X_{12}(0), X_{12}(1), X_{13}(0), X_{13}(1), X_{14}(0), X_{14}(1)\}$$

$$\{X_2(k)\} = \{X_{21}(0), X_{21}(1), X_{21}(2), X_{21}(3), X_{22}(0), X_{22}(1), X_{22}(2), X_{22}(3)\}$$

$$\{X_3(k)\} = \{X(0), X(1), X(2), X(3), X(4), X(5), X(6), X(7)\}$$

We can simplify the butterfly computations represented by the signal-flow graph of Fig. 8.6 by introducing a new node denoted by $W_N^{2rk} X_{m-r,2p}(k)$, as shown in the butterfly in Fig. 8.7. This butterfly has one less multiplication than that of Fig. 8.6. There are $N/2$ of these butterflies per stage, and, therefore, the number of multiplications is reduced from N to $N/2$ per stage, so that $(N/2) \log_2 N$ multiplications are required for computing the FFT using the simplified butterfly of Fig. 8.7.

It can be seen from the nature of the computational procedure that only one complex array of N storage registers is needed to carry out the computations if the values of $X_{m-r,p}(k)$ and $X_{m-r,p}(k + N/2^{r+1})$ determined from Eq. (8.15) are stored in the same registers that $X_{m-r-1,2p-1}(k)$ and $X_{m-r-1,2p}(k)$ were stored in. (Since we are

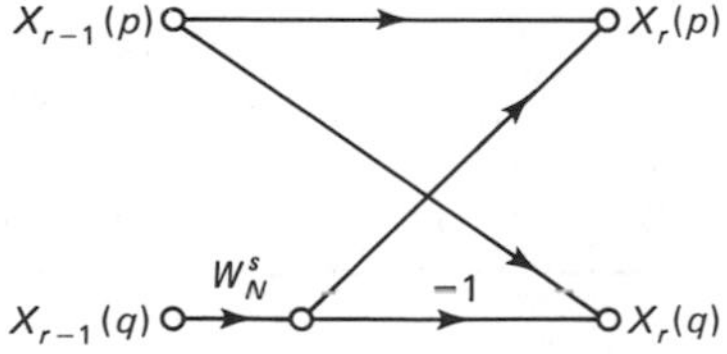

Figure 8.7 Simplified butterfly representing Eq. (8.15).

dealing with complex numbers, each complex number requires two storage locations.) A computation of this form is commonly called an *in-place computation.*

In order to do the computations described in this algorithm in place, it is necessary that the input data be stored in the nonsequential order that is listed as $X_0(k)$. This ordering of the input data is in *bit-reversed* order. By this we mean that if for $X_0(k)$, the integer k is expressed in binary representation, then $X_0(k) = x(k')$, where k' is the integer obtained from k by reversing the order of the bits in the binary representation. For the case $N = 8$, when the arguments are expressed in binary form, we have

$$X_0(0) = X_0(000) = x(000) = x(0)$$

$$X_0(1) = X_0(001) = x(100) = x(4)$$

$$X_0(2) = X_0(010) = x(010) = x(2)$$

$$X_0(3) = X_0(011) = x(110) = x(6)$$

$$X_0(4) = X_0(100) = x(001) = x(1)$$

$$X_0(5) = X_0(101) = x(101) = x(5)$$

$$X_0(6) = X_0(110) = x(011) = x(3)$$

$$X_0(7) = X_0(111) = x(111) = x(7)$$

where the binary representation of k *is* determined by

$$k = a_2 2^2 + a_1 2^1 + a_0 2^0 = (a_2 a_1 a_0)_2$$

and the coefficients a_0, a_1, and a_2 are either 0 or 1.

EXERCISES

8.3.1 Draw the signal-flow graph for $N = 4$ for the decimation-in-time algorithm using the butterfly of Fig. 8.7.

8.3.2 Repeat Ex. 8.2.2 using the algorithm of Ex. 8.3.1.

8.3.3 Repeat Ex. 8.3.1 for $N = 8$.

8.3.4 Show that the ordering of the input data is in bit-reversed order for **(a)** $N = 4$ and **(b)** $N = 16$ for decimation in time.

8.4 DECIMATION-IN-FREQUENCY (DIF) ALGORITHM

The next algorithm that we consider for efficiently computing the DFT is known as *decimation in frequency* (DIF). Assuming as before that $N = 2^m$, we again separate the index set into two sets. In this case, the first set consists of the integers $0 \leq n \leq N/2 - 1$, and the second set consists of the integers $N/2 \leq n \leq N - 1$. In terms of these two sets, we can write

$$X(k) = \sum_{n=0}^{(N/2)-1} x(n)W_N^{nk} + \sum_{n'=N/2}^{N-1} x(n')W_N^{n'k}$$

In the second sum, letting $n' = n + N/2$ and noting that $W_N^{N/2} = -1$, we obtain

$$\begin{aligned} X(k) &= \sum_{n=0}^{(N/2)-1} x(n)W_N^{nk} + (-1)^k \sum_{n=0}^{(N/2)-1} x(n + N/2)W_N^{nk} \\ &= \sum_{n=0}^{(N/2)-1} [x(n) + (-1)^k x(n + N/2)]W_N^{nk} \end{aligned} \tag{8.18}$$

To proceed with the decomposition, we separate the values of k into the sets of even values and odd values. We thus have from Eq. (8.18)

$$X(2k) = \sum_{n=0}^{(N/2)-1} x_{11}(n)W_{N/2}^{nk} \tag{8.19}$$

$$X(2k + 1) = \sum_{n=0}^{(N/2)-1} x_{12}(n)W_N^n W_{N/2}^{nk} \tag{8.20}$$

where $W_N^2 = W_{N/2}$, as we saw before, and

$$x_{11}(n) = x(n) + x(n + N/2) \qquad n = 0, 1, \ldots, (N/2) - 1 \tag{8.21}$$

$$x_{12}(n) = x(n) - x(n + N/2) \qquad n = 0, 1, \ldots, (N/2) - 1 \tag{8.22}$$

Each of the sums in Eqs. (8.19) and (8.20) are $(N/2)$-point DFTs. The computations indicated by Eqs. (8.19)–(8.22) are represented by the signal-flow graph of Fig. 8.8.

The next step in the decomposition is to write Eqs. (8.19) and (8.20) as two sums, just as we did in the previous decomposition of $X(k)$. Once this is done, we then form the sums $X(4k)$ and $X(4k + 2)$ from the decomposition of Eq. (8.19) and the sums $X(4k + 1)$ and $X(4k + 3)$ from the decomposition of Eq. (8.20). To illustrate the procedure, we consider the case $N = 8$.

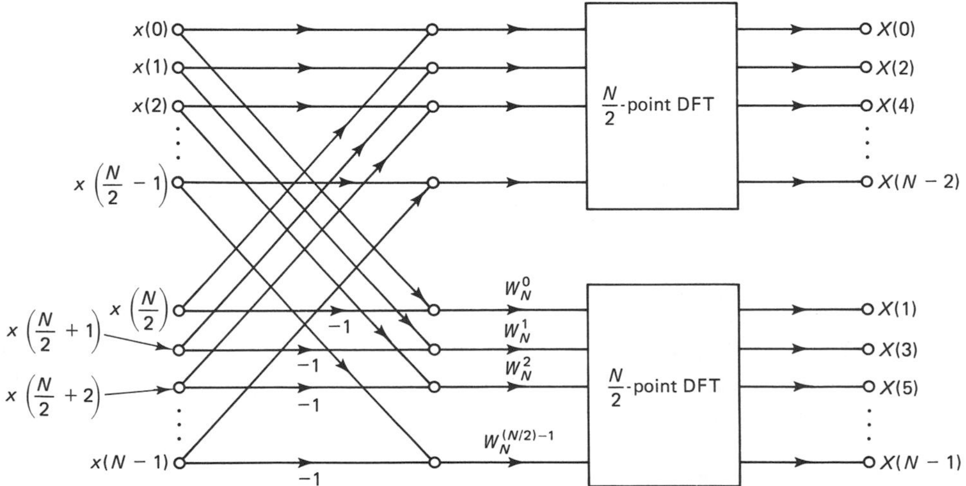

Figure 8.8 Signal-flow graph for the first decomposition of the DFT using decimation in frequency.

For $N = 8$, Eqs. (8.19) and (8.20) have the forms

$$X(2k) = \sum_{n=0}^{3} x_{11}(n)W_4^{nk} \qquad k = 0, 1, 2, 3 \tag{8.23}$$

$$X(2k+1) = \sum_{n=0}^{3} x_{12}(n)W_8^n W_4^{nk} \qquad k = 0, 1, 2, 3 \tag{8.24}$$

Following the previous procedure whereby we separated the index sets into smaller integers and larger integers, we can write

$$\begin{aligned} X(2k) &= \sum_{n=0}^{1} x_{11}(n)W_4^{nk} + \sum_{n=2}^{3} x_{11}(n)W_4^{nk} \\ &= \sum_{n=0}^{1} [x_{11}(n) + (-1)^k x_{11}(n+2)]W_4^{nk} \end{aligned} \tag{8.25}$$

and

$$\begin{aligned} X(2k+1) &= \sum_{n=0}^{1} x_{12}(n)W_8^n W_4^{nk} + \sum_{n=2}^{3} x_{12}(n)W_8^n W_4^{nk} \\ &= \sum_{n=0}^{1} [x_{12}(n)W_8^n + (-1)^k x_{12}(n+2)W_8^{n+2}]W_4^{nk} \end{aligned} \tag{8.26}$$

If we separate $X(2k)$ into the two cases $X(4k)$ and $X(4k+2)$, the result is

$$X(4k) = \sum_{n=0}^{1} x_{21}(n)W_2^{nk} \qquad k = 0, 1 \tag{8.27}$$

$$X(4k+2) = \sum_{n=0}^{1} x_{22}(n)W_4^n W_2^{nk} \qquad k = 0, 1 \tag{8.28}$$

where

$$x_{21}(n) = x_{11}(n) + x_{11}(n+2) \qquad n = 0, 1 \tag{8.29}$$

$$x_{22}(n) = x_{11}(n) - x_{11}(n+2), \qquad n = 0, 1 \tag{8.30}$$

Following the same previous procedure given, we can decompose Eq. (8.26) into

$$X(4k+1) = \sum_{n=0}^{1} x_{23}(n)W_2^{nk} \qquad k = 0, 1 \tag{8.31}$$

$$X(4k+3) = \sum_{n=0}^{1} x_{24}(n)W_4^n W_2^{nk} \qquad k = 0, 1 \tag{8.32}$$

where

$$x_{23}(n) = x_{12}(n)W_8^n + x_{12}(n+2)W_8^{n+2} \qquad n = 0, 1 \tag{8.33}$$

$$x_{24}(n) = x_{12}(n)W_8^n - x_{12}(n+2)W_8^{n+2} \qquad n = 0, 1 \tag{8.34}$$

We see that Eqs. (8.27), (8.28), (8.31), and (8.32) are all 2-point DFTs. The computations involved for these two decompositions are represented by the signal-flow graph of Fig. 8.9 for the case $N = 8$.

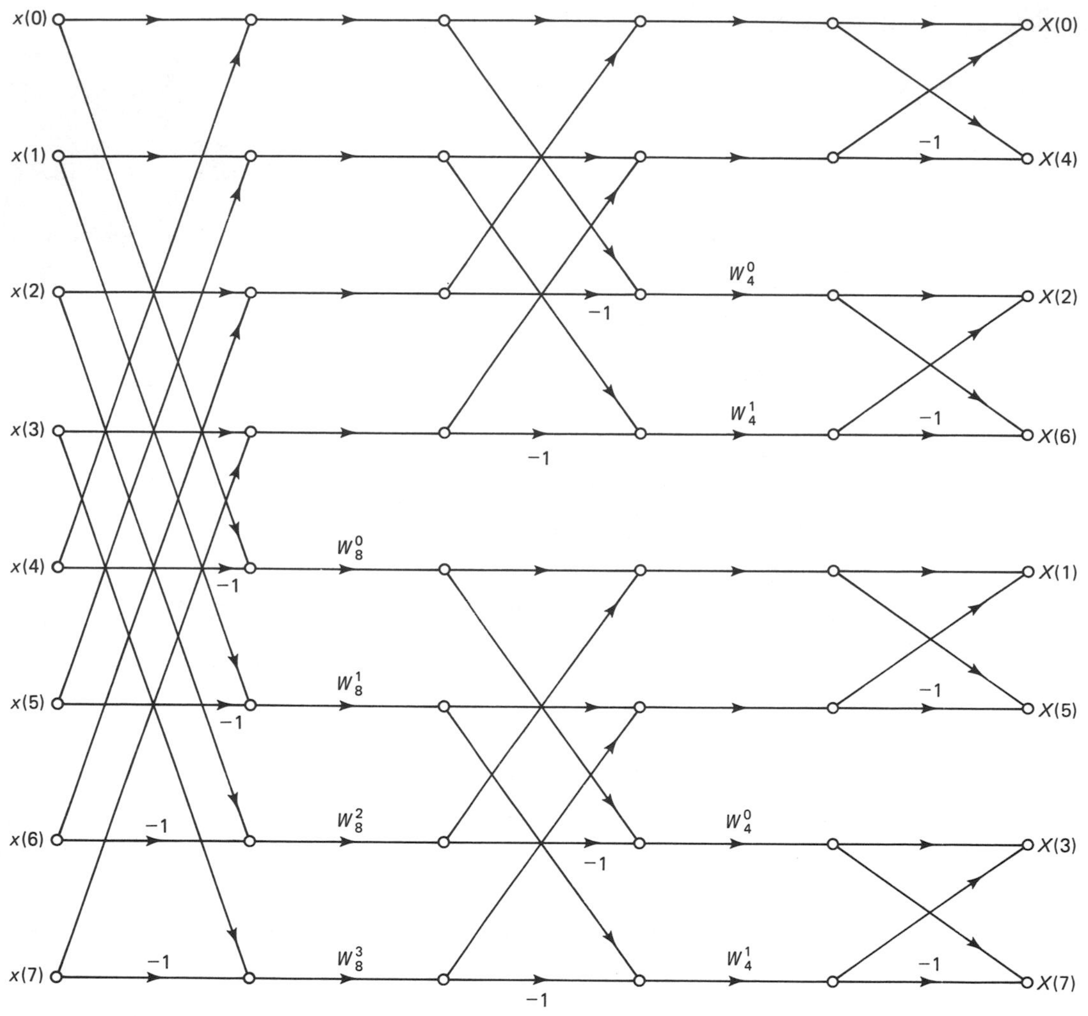

Figure 8.9 Signal-flow graph for the decimation-in-frequency FFT algorithm for $N = 8$.

If we go through the process of counting the number of operations, we find that this number is the same as for the decimation-in-time FFT algorithm.

The computations required for this algorithm can be divided into m stages, with each set of computations represented by the butterfly of Fig. 8.10. These butterflies are clearly evident in Fig. 8.9 for $N = 8$, except the last set of multipliers W_N^0 have been omitted for convenience. The equations corresponding to this butterfly are

$$\begin{aligned} X_r(p) &= X_{r-1}(p) + X_{r-1}(q) \\ X_r(q) &= [X_{r-1}(p) - X_{r-1}(q)]W_N^s \end{aligned} \tag{8.35}$$

These butterflies differ from those of the decimation-in-time algorithm.

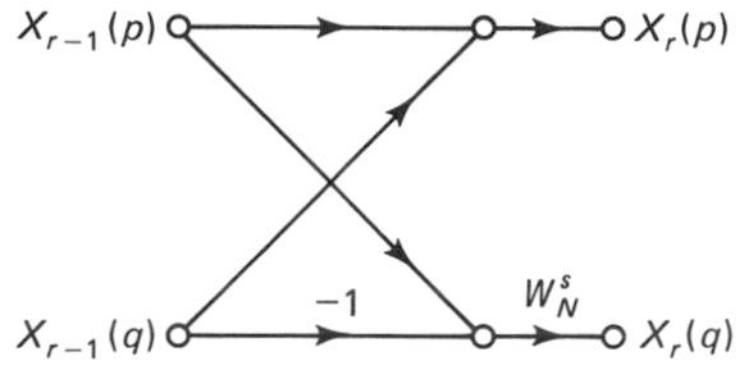

Figure 8.10 A butterfly for the decimation-in-frequency FFT algorithm.

Since the computation at each stage can be represented by a butterfly, the signal-flow graph representing the FFT algorithm can be thought of as an in-place computation. For decimation in frequency, it should be noted that the input sequence is in normal order and the output sequence is in bit-reversed order.

The DIF signal-flow graphs can be obtained by transposing the DIT signal-flow graphs (see Ex. 4.2.5). The reader is asked to show this for $N = 4$ and $N = 8$ in Ex. 8.4.4.

EXERCISES

8.4.1 Write the equations and draw the signal-flow graph for the decimation-in-frequency algorithm for $N = 4$.

8.4.2 Compute the DFTs of Ex. 7.1.3 using the results of Ex. 8.4.1.

8.4.3 Show that the output data is in bit-reversed order for the decimation-in-frequency algorithm for $N = 8$.

8.4.4 Show that if the DIT signal-flow graphs of Figs. 8.4, 8.5, and 8.7 are transposed (see Ex. 4.2.5), the corresponding DIF signal-flow graphs are obtained. For example, the input $x(0)$ is replaced by the output $X(0)$, $x(4)$ by $X(1)$, and so forth in Fig. 8.4.

8.5 DECOMPOSITION FOR N A COMPOSITE NUMBER

In previous sections, we considered two FFT algorithms, decimation-in-time and decimation-in-frequency, for the case where N is a power of 2, that is, $N = 2^m$. These algorithms are highly efficient and are easily represented by butterflies that correspond essentially to 2-point DFTs. Therefore, it is advantageous whenever possible to deal with sequences of length 2^m and this can be done in many cases. However, there are cases where this is not possible and it is necessary to develop algorithms for the case where N is a composite number and can be written as a product of factors such as

$$N = p_1 p_2 \cdots p_m \tag{8.36}$$

In this section, we give a decimation-in-time decomposition and leave decimation in frequency to the exercises at the end of this section and the problems at the end of the chapter.

To start the decomposition, we let $N_1 = p_2 \cdots p_m$, so that $N = p_1 N_1$. If we choose to use decimation in time, then we can separate the input sequence $x(n)$ into

p_1 subsequences of N_1 elements each. This can be accomplished by choosing the p_1 sequences to be

Sequence 1: $x(0), x(p_1), x(2p_1), \ldots, x[(N_1 - 1)p_1]$
Sequence 2: $x(1), x(p_1 + 1), x(2p_1 + 1), \ldots, x[(N_1 - 1)p_1 + 1]$

$\vdots$

Sequence p_1: $x(p_1 - 1), x(2p_1 - 1), \ldots, x(N - 1)$

These sequences can be arranged in the p_1 by N_1 two-dimensional array

$$\begin{array}{ccccccc}
0 & p_1 & 2p_1 & \cdots & rp_1 & \cdots & (N_1 - 1)p_1 \\
1 & p_1 + 1 & 2p_1 + 1 & \cdots & rp_1 + 1 & \cdots & (N_1 - 1)p_1 + 1 \\
2 & p_1 + 2 & 2p_1 + 2 & \cdots & rp_1 + 2 & \cdots & (N_1 - 1)p_1 + 2 \\
\vdots & \vdots & \vdots & & \vdots & & \vdots \\
s & p_1 + s & 2p_1 + s & \cdots & rp_1 + s & \cdots & (N_1 - 1)p_1 + s \\
\vdots & \vdots & \vdots & & \vdots & & \vdots \\
p_1 - 1 & 2p_1 - 1 & 3p_1 - 1 & \cdots & (r + 1)p_1 - 1 & \cdots & N_1 p_1 - 1
\end{array} \tag{8.37}$$

Each element in the array is the sum of rp_1 in the same column in the first row and s in the same row of the first column for $r = 0, 1, 2, \cdots, N_1 - 1$ and $s = 0, 1, 2, \ldots, p_1 - 1$. For example, if $N = 6 = (3)(2)$, then we can take $p_1 = 3$ and $N_1 = 2$. The three sequences of length 2 are

Sequence 1: $x(0), x(3)$
Sequence 2: $x(1), x(4)$
Sequence 3: $x(2), x(5)$

The corresponding array is

$$\begin{array}{cc}
0 & 3 \\
1 & 4 \\
2 & 5
\end{array}$$

Once we have separated the input sequence into p_1 subsequences, the DFT can be written

$$\begin{aligned}
X(k) = \sum_{r=0}^{N_1-1} x(rp_1)W_N^{rp_1k} &+ \sum_{r=0}^{N_1-1} x(rp_1 + 1)W_N^{(rp_1+1)k} \\
&+ \cdots + \sum_{r=0}^{N_1-1} x(rp_1 + p_1 - 1)W_N^{(rp_1+p_1-1)k}
\end{aligned} \tag{8.38}$$

In this sum, it should be noted that the first sum corresponds to the entries of the first row of the previous two-dimensional array, the second sum corresponds to the second row, and so forth. A typical term in this sum has the form

$$\sum_{r=0}^{N_1-1} x(rp_1 + s)W_N^{(rp_1+s)k} = W_N^{sk} \sum_{r=0}^{N_1-1} x(rp_1 + s)W_{N_1}^{rk} \qquad s = 0, 1, \cdots, p_1 - 1 \tag{8.39}$$

since $W_N^{p_1} = W_{N/p_1} = W_{N_1}$. The sum in the right-hand term of this equation is simply the N_1-point DFT of the sequence $x(rp_1 + s)$, and it follows that $X(k)$ in Eq. (8.38) is a weighted sum of p_1 N_1-point DFTs. Using Eq. (8.39), we can write Eq. (8.38) in the more compact form

$$X(k) = \sum_{s=0}^{p_1-1} W_N^{sk} \sum_{r=0}^{N_1-1} x(rp_1 + s)W_{N_1}^{rk}$$

To proceed with the decomposition, we let $N_1 = p_2N_2$ and decompose each of the N_1-point DFTs appearing in the simplified form of $X(k)$ given in the last equation. Each of these sequences of length N_1 is decomposed into p_2 subsequences of length N_2. For example, the set of points $\{rp_1\}$ involved in the first sum of Eq. (8.38) decomposes into the following sets, where $0 \le s \le N_2 - 1$:

$$\{sp_1p_2\}, \{sp_1p_2 + p_1\}, \{sp_1p_2 + 2p_1\}, \ldots, \{sp_1p_2 + (p_2 - 1)p_1\} \tag{8.40}$$

Note that each argument is a multiple of p_1 and the remainder obtained when the argument is divided by p_1p_2 is a multiple of p_1 less than p_2p_1. The sets of points involved in the decomposition of the other sums are obtained by replacing sp_1p_2 in Eq. (8.40) first by $sp_1p_2 + 1$, then by $sp_1p_2 + 2$, and so forth until finally sp_1p_2 is replaced by $sp_1p_2 + p_1 - 1$.

The sets in Eq. (8.40), representing the decomposition of the set $\{rp_1\}$, the first row of Eq. (8.37), can be arranged in the following array:

$$\begin{matrix} 0 & p_1p_2 & \cdots & sp_1p_2 & \cdots & (N_2 - 1)p_1p_2 \\ p_1 & p_1p_2 + p_1 & \cdots & sp_1p_2 + p_1 & \cdots & (N_2 - 1)p_1p_2 + p_1 \\ 2p_1 & p_1p_2 + 2p_1 & \cdots & sp_1p_2 + 2p_1 & \cdots & (N_2 - 1)p_1p_2 + 2p_1 \\ \vdots & \vdots & & \vdots & & \vdots \\ (p_2 - 1)p_1 & p_1p_2 + (p_2 - 1)p_1 & \cdots & sp_1p_2 + (p_2 - 1)p_1 & \cdots & (N_1 - 1)p_1 \end{matrix} \tag{8.41}$$

The arrays for the subsequent rows of Eq. (8.37) are obtained by adding one to each element of the array of Eq. (8.41) to obtain the array for the second row of Eq. (8.37), by adding one to each element of the second row array to obtain the array for the third row of Eq. (8.37), and so forth.

Rather than continue with the general case, suppose we take $N = 12 = 3 \cdot 2 \cdot 2$. The sequences resulting from the first decomposition are for $p_1 = 3$ and $N_1 = 4$:

Sequence 1: $x(0), x(3), x(6), x(9)$
Sequence 2: $x(1), x(4), x(7), x(10)$
Sequence 3: $x(2), x(5), x(8), x(11)$

The corresponding 3×4 array of these sequences is

$$\begin{matrix} 0 & 3 & 6 & 9 \\ 1 & 4 & 7 & 10 \\ 2 & 5 & 8 & 11 \end{matrix} \tag{8.42}$$

Note that the second and third rows are obtained by adding one to each element in the preceding row. This can be seen in general in Eq. (8.37). The first decomposition of $X(k)$ according to Eqs. (8.38) and (8.39) has the form

$$X(k) = \sum_{r=0}^{3} x(3r)W_4^{rk} + W_{12}^{k}\sum_{r=0}^{3} x(3r+1)W_4^{rk} + W_{12}^{2k}\sum_{r=0}^{3} x(3k+2)W_4^{rk}$$

The next step in the decomposition requires that each of the previous three sequences be separated into two subsequences of length two. Since $p_1 p_2 = 6$, the first sequence decomposes into the two sequences $\{x(6s)\}$ and $\{x(6s+3)\}$, $s = 0, 1$, as indicated by Eq. (8.39). The second sequence decomposes into the two subsequences $\{x(6s+1)\}$ and $\{x(6s+3+1)\} = \{x(6s+4)\}$, and the third sequence decomposes into the two subsequences $\{x(6s+2)\}$ and $\{x(6s+5)\}$ for $s = 0, 1$. These six subsequences are

$$\begin{matrix} x(0), x(6) & \text{and} & x(3), x(9) \\ x(1), x(7) & \text{and} & x(4), x(10) \\ x(2), x(8) & \text{and} & x(5), x(11) \end{matrix}$$

Note that in each of these sequences, the arguments differ by $p_1 p_2 = 6$.

The first row of the 3×4 array given in Eq. (8.42) is decomposed into two sequences: the first sequence is $\{sp_1 p_2 = 6s\}$ and the second is $\{sp_1 p_2 + 1 = 6s + 1\}$ for $s = 0, 1$. The two sequences for the second row are obtained by adding one to each element of the sequences for the first row. The two sequences for the third row are obtained by adding one to each element of the sequences for the second row. These results are

$$\begin{matrix} 0 & 3 & 6 & 9 & \longrightarrow & 0 & 6 & \text{and} & 3 & 9 \\ 1 & 4 & 7 & 10 & \longrightarrow & 1 & 7 & \text{and} & 4 & 10 \\ 2 & 5 & 8 & 11 & \longrightarrow & 2 & 8 & \text{and} & 5 & 11 \end{matrix}$$

This second decomposition results in

$$\begin{aligned} X(k) &= \sum_{r=0}^{1} x(6r)W_4^{2rk} + \sum_{r=0}^{1} x(6r+3)W_4^{(2r+1)k} \\ &\quad + W_{12}^{k}\left[\sum_{r=0}^{1} x(6r+1)W_4^{2rk} + \sum_{r=0}^{1} x(6r+4)W_4^{(2r+1)k}\right] \\ &\quad + W_{12}^{2k}\left[\sum_{r=0}^{1} x(6r+2)W_4^{2rk} + \sum_{r=0}^{1} x(6r+5)W_4^{(2r+1)k}\right] \end{aligned}$$

Noting that $W_4^2 = W_2$, we can write

$$X(k) = \sum_{r=0}^{1} x(6r)W_2^{rk} + W_4^k \sum_{r=0}^{1} x(6r+3)W_2^{rk}$$
$$+ W_{12}^k \sum_{r=0}^{1} x(6r+1)W_2^{rk} + W_{12}^k W_4^k \sum_{r=0}^{1} x(6r+4)W_2^{rk} \quad (8.43)$$
$$+ W_6^k \sum_{r=0}^{1} x(6r+2)W_2^{rk} + W_6^k W_4^k \sum_{r=0}^{1} x(6r+5)W_2^{rk}$$

which is a weighted sum of six 2-point DFTs. The reader is asked to construct a signal-flow graph for this algorithm in Ex. 8.5.2.

In our discussion in this section, we have given only a basic introduction. Some of the problems provide added information about this case, including decimation in frequency. For a more detailed discussion of FFT algorithms, the reader is referred to the references [11, 45, 120, 142]. An algorithm for computing the mixed-radix FFT (N a composite number) has been obtained and programmed by Singleton [161].

EXERCISES

8.5.1 **(a)** Complete the decomposition of the DFT for $N = 6 = 3 \cdot 2$ and draw a signal-flow graph.
(b) Repeat part (a) for $N = 2 \cdot 3$.

8.5.2 Construct the signal-flow graph for the decomposition of the DFT for $N = 12$, which is represented by Eqs. (8.36), (8.38), and (8.43).

8.5.3 **(a)** Develop a decimation-in-frequency algorithm for decomposing the DFT for $N = 6 = 3 \cdot 2$. First, write $X(k)$ as two sums of three terms each (determined by $p_1 = 3$). Then separate the values of k into sets of even and odd values (determined by $p_2 = 2$). Use the definitions $x_{11}(n) = x(n) + x(n+3)$ and $x_{12}(n) = x(n) - x(n+3)$ and draw a signal-flow graph.
(b) Repeat part (a) for $N = 2 \cdot 3$. Write the DFT as three sums of two terms each. The subsets for n are $\{0, 1\}$, $\{2, 3\}$, and $\{4, 5\}$. The subsets for k are $\{3r\}$, $\{3r + 1\}$, and $\{3r + 2\}$ for $r = 0, 1$.

PROBLEMS

8.1 Carry out the computations in Eqs. (8.12)–(8.14).

8.2 Write the relations for the decimation-in-time algorithm for $N = 8$ and obtain the signal-flow graph of Fig. 8.4.

8.3 Use the results of Ex. 8.2.1 to compute the DFTs of the following sequences, where $N = 4$:
(a) $x(n) = 2^n$
(b) $x(n) = 2^{-n}$

8.4 Repeat Prob. 8.3 for the following sequences:
(a) $x(n) = \sin n\pi/2$
(b) $x(n) = \cos n\pi/2$
(c) $x(n) = \cos n\pi/4$

8.5 Repeat Prob. 8.3 using the results of Ex. 8.4.1.

8.6 Repeat Prob. 8.5 for the sequences given in Prob. 8.4.

8.7 Develop a decimation-in-time FFT algorithm using 4-point DFTs for the case $N = 4^{\nu}$. Compare the number of multiplications with the algorithm using 2-point DFTs with $N = 2^{2\nu}$.

8.8 Repeat Prob. 8.7 for decimation in frequency.

8.9 Develop a decimation-in-time FFT algorithm for $N = 8$ using a 4-point DFT and a 2-point DFT. Compare the number of multiplications with the algorithm using only 2-point DFTs.

8.10 Use the results of Prob. 8.9 to compute the DFT of the sequence $x(n) = 2^n$.

8.11 Develop a decimation-in-time algorithm for evaluating the DFT for $N = 9 = 3 \cdot 3$.

8.12 Repeat Prob. 8.11 using decimation in frequency.

8.13 Determine the arguments of the sequences that arise using decimation in time to evaluate a 70-point DFT using $N = 7 \cdot 5 \cdot 2$.

8.14 Define the sequence

$$y_k(n) = \sum_{r=0}^{N-1} x(r) W_N^{-k(n-r)}$$

Show that $X(k) = y_k(n)|_{n=N}$. Taking the impulse response as W_N^{-kn}, with k a parameter, construct a block diagram or signal-flow graph for the system having $y_k(n)$ as output and $x(n)$ as input. Show that the system function can be written

$$H_k(z) = \frac{1 - W_N^k z^{-1}}{1 - 2\cos(2\pi k/N)z^{-1} + z^{-2}}$$

Construct a signal-flow graph for this system function. This method is called the *Goertzel algorithm* [50].

8.15 Let the IDFT be written in the matrix form

$$\mathbf{x} = \frac{1}{N}\mathbf{T}\mathbf{X}$$

The rsth element of $\mathbf{T}$ is W_N^{-rs}. Let $\hat{\mathbf{T}}$ be the matrix with rs as the rsth element. Show that if we reduce the exponents of W_N^{-1} in $\mathbf{T}$ by modulo N, we obtain

$$\hat{\mathbf{T}} = \begin{bmatrix} 0 & 0 & 0 & \cdots & 0 \\ 0 & 1 & 2 & \cdots & N-1 \\ 0 & 2 & 4 & \cdots & 2(N-1) \\ 0 & 3 & 6 & \cdots & 3(N-1) \\ \vdots & \vdots & \vdots & & \vdots \\ 0 & N-1 & 2(N-1) & \cdots & (N-1)^2 \end{bmatrix}$$

Write $\hat{\mathbf{T}}$ for $N = 8$. If we permute the rows of $\mathbf{T}$ or $\hat{\mathbf{T}}$, we can write $\mathbf{T}' = \mathbf{Q}\mathbf{T}$. Determine $\mathbf{Q}$ and $\mathbf{T}'$ for $N = 8$ if $\mathbf{T}'$ represents the bit-reversed representation. For $N = 8$, show that

$$\mathbf{T}'_N = \begin{bmatrix} \mathbf{T}'_{N/2} & \mathbf{\Phi} \\ \mathbf{\Phi} & \mathbf{T}'_{N/2} \end{bmatrix} \begin{bmatrix} \mathbf{I} & \mathbf{\Phi} \\ \mathbf{\Phi} & \mathbf{K} \end{bmatrix} \begin{bmatrix} \mathbf{I} & \mathbf{I} \\ \mathbf{I} & -\mathbf{I} \end{bmatrix}$$

where $\mathbf{K}$ is a diagonal matrix with diagonal elements 0, 1, 2, 3, . . . , and $\mathbf{\Phi}$ is a null matrix (has elements that are zero). The FFT algorithm can be obtained if this process is carried out to completion [133].

8.16 In Prob. 7.37, we saw that the p-point DFT for p a prime could be written

$$F[g^k(\text{mod } p)] = \sum_{n=1}^{p-1} f[g^n(\text{mod } p)]W_p^{g^{k-n}} + f(0) \qquad k = 1, 2, \ldots, p-1$$

with $F(0) = \Sigma_{n=0}^{p-1} f(n)$. It can be shown that $g^{p-1} = g^0 = 1$ when p is a prime and so we can write [149]

$$F[g^k(\text{mod } p)] - f(0) = \sum_{n=0}^{N-2} f[g^n(\text{mod } p)]W_p^{g^{k-n}}$$

Since p is a prime, it follows that $p - 1$ is a composite number for $p > 3$. Using FFT techniques, construct a signal-flow graph for computing the DFT of the sequence $f(n)$ when

(a) $N = p = 7$

(b) $N = p = 17$

8.17 Consider the class of transforms defined by [2, 33, 75, 108, 174]

$$X(k) = \sum_{n=0}^{N-1} x(n)\ \alpha^{nk} \qquad k = 0, 1, 2, \ldots, N-1$$

The DFT is the special case arising when $\alpha = W_N$. This transform is known as a *number theoretic transform* (NTT) when x, X, and α are all integers in $Z_M = \{0, 1, 2, \ldots, M-1\}$, where α is a root of unity of order N and modulo M arithmetic is used.

(a) For $N = 4$, show that $\mathbf{X} = \mathbf{Fx}$, where

$$\mathbf{F} = \begin{bmatrix} 1 & 1 & 1 & 1 \\ 1 & \alpha & \alpha^2 & \alpha^3 \\ 1 & \alpha^2 & \alpha^4 & \alpha^6 \\ 1 & \alpha^3 & \alpha^6 & \alpha^9 \end{bmatrix}$$

(b) Obtain a signal-flow graph representing the transform.

(c) Obtain the signal-flow graph for a decimation-in-time algorithm.

8.18 The *Fermat number transform* (FNT) is obtained when $M = F_m$, the mth Fermat number defined by [2, 75, 96, 104, 108]

$$F_m = 2^b + 1 \qquad b = 2^m \qquad m = 0, 1, 2, \ldots$$

is used in the transform of Prob. 8.17. The first five Fermat numbers, F_0, F_1, F_2, F_3, and F_4, are primes. For these primes, the transform length is a divisor of 2^b with the maximum length $N_{\max} = 2^b$. The order of the primitive roots in Z_M is 2^b and there are $\phi[\phi(M)]$ of these (see Prob. 7.36). For $m = 0, 1, 2, 3, 4$, we have $M = 2^b + 1$ and $\phi(M) = M - 1 = 2^b$.

(a) For $m = 1$, show that $M = 5$ and that the primitive roots are 2 and 3.

(b) For each of these primitive roots, obtain the matrix $\mathbf{F}$ in Prob. 8.17.

(c) Using $\alpha = 2$, find the FNT of $\{1, 2, 3, 4\}$.

(d) Use the decimation-in-time algorithm developed in Prob. 8.17 to compute the FNT of $\{1, 2, 3, 4\}$.

8.19 **(a)** For $m = 2$, show that $M = F_2 = 17$ and that the primitive roots are 3, 5, 6, 7, 10, 11, 12, and 14.

(b) For $\alpha = 3$, show that $\{3^k(\text{mod } 17)\}$, $k = 0, 1, 2, \ldots, M - 2 = 15$ is the set of nonzero elements of Z_{17}.

(c) Determine the matrix equation $\mathbf{X} = \mathbf{Fx}$ for $\alpha = 3$.

8.20 The *inverse Fermat number transform* (IFNT) is given by

$$x(n) = N^{-1} \sum_{k=0}^{N-1} X(k)\alpha^{-nk} \qquad n = 0, 1, 2, \ldots, N - 1$$

(a) Determine the IFNT for $m = 1$ and $\alpha = 2$ and $\alpha = 3$.

(b) For $\alpha = 2$, find the IFNT of $\{0, 4, 3, 2\}$.

8.21 Show that $\alpha = 3$ has order $N = 2^b$ for the FNT defined in Prob. 8.18 when $m = 1, 2$, and 3. Obtain a decimation-in-time (DIT) algorithm for $m = 2$.

8.22 The *Rader transform* is obtained from the FNT when $\alpha = 2$. Show that $N = 2b$ for $m = 1$, 2, and 3. Develop a decimation-in-time algorithm for $m = 2$ and draw the signal-flow graph. Calculate the FNT of $\{1, 2, 3, 4\}$ using the latter signal-flow graph.

8.23 For $N = 8$, obtain a signal-flow graph for the decimation-in-time (DIT) algorithm for the Walsh transform defined in Prob. 7.45. Identify the butterfly used in the computations.

8.24 Develop a decimation-in-time (DIT) algorithm for the discrete Hartley transform given in Prob. 7.39 for

(a) $N = 4$

(b) $N = 8$

(c) Draw a signal-flow graph for the computation.

(d) Use the signal-flow graph for $N = 4$ to find DHT $\{1, 2, 3, 4\}$ obtained in Prob. 7.42.

8.25 **(a)** Derive DHT $\{x(n + 1)\}$ in terms of $H(k) = \text{DHT}\{x(n)\}$, where $x(n + N) = x(n)$.

(b) If

$$\{H(0), H(1), \ldots, H(N-1)\} = \text{DHT}\{x(0), x(1), \ldots, x(N-1)\}$$

show that

$$\text{DHT}\{x(0), 0, x(1), 0, \ldots, 0, x(N-1), 0\}$$
$$= \{H(0), H(1), \ldots, H(N-1), H(0), H(1), \ldots, H(N-1)\}$$

(c) If

$$\text{DHT}\{x(0), x(2), x(4), \ldots, x(N-2)\} = \{H_e(0), H_e(1), \ldots, H_e(N/2 - 1)\}$$
$$\text{DHT}\{x(1), x(3), \ldots, x(N-1)\} = \{H_o(0), H_o(1), \ldots, H_o(N/2 - 1)\}$$
$$H_{Ne}(k) = \{H_e(0), H_e(1), \ldots, H_e(N/2 - 1), H_e(0), H_e(1), \ldots, H_e(N/2 - 1)\}$$
$$H_{No}(k) = \{H_o(0), H_o(1), \ldots, H_o(N/2 - 1), H_o(0), H_o(1), \ldots, H_o(N/2 - 1)\}$$

show that

$$H(k) = \text{DHT}\{x(0), x(1), \ldots, x(N-1)\}$$
$$= H_{Ne}(k) + H_{No}(k) \cos 2k\pi/N + H_{No}(N - k) \sin 2k\pi/N$$

Use these relations to develop the first decomposition for a DIT algorithm for the discrete Hartley transform [13, 163].

(d) Use the relations in part (c) to obtain the signal-flow graph for a decimation-in-time (DIT) algorithm for the DHT if $N = 8$.

8.26 The *Mersenne transform* is given by [2, 75, 148, 156]

$$X(k) = \sum_{n=0}^{p-1} x(n)2^{nk} \qquad k = 0, 1, 2, \ldots, p-1$$

where p is a prime, and the arithmetic is modulo $M_p = 2^p - 1$, a *Mersenne number*. Frequently, p is chosen so that M_p is a prime, called a Mersenne prime. It is known that among the 55 prime numbers between 2 and 257 inclusive, there are 12 primes, namely

$$p = 2, 3, 5, 7, 13, 17, 19, 31, 61, 89, 107, 127$$

for which M_p is prime. In mid-1983, the largest known Mersenne prime had 25,962 digits and was the 28th Mersenne prime M_{86243}. The interest in Mersenne primes was spurred by Euclid's result that if $2^n - 1$ is a prime (n must be prime), then $2^{n-1}(2^n - 1)$ is a *perfect* number, a number which equals the sum of its proper divisors (includes 1 but not the number itself). In other words, *euclidean* perfect numbers have a Mersenne prime as a factor. (For example, $6 = 1 + 2 + 3$ is the first perfect number [156].) In the following, take $M_p = M_3$.

(a) Determine the Mersenne transform.

(b) Calculate the transform of the sequence $\{1, 0, 2, 0, 1, 0, 2\}$.

(c) Use the result of Prob. 8.16 to obtain a fast Mersenne transform algorithm.

(d) Using the algorithm of part (c), calculate the transform of the sequence of part (b).

REFERENCES

1. Abramowitz, M., and I. A. Stegun (eds.). *Handbook of Mathematical Functions*. Washington, DC: National Bureau of Standards, 1964.
2. Agarwal, R. C., and C. S. Burrus. "Number Theoretic Transforms To Implement Fast Digital Convolution." *Proc. IEEE* 63 (April 1975): 550–560.
3. Ahmed, N. and K. R. Rao. *Orthogonal Transforms for Digital Signal Processing*. New York: Springer-Verlag, 1975.
4. Anderson, B. D. O., and E. I. Jury. "Stability Test for Two-Dimensional Recursive Filters." *IEEE Trans. Audio Electroacoust*. AU-21(4) (August 1973): 366–372.
5. Andrews, H. C. "Digital Image Processing." *IEEE Spectrum* 16 (April 1978): 38–49.
6. Andrews, H. C., and B. R. Hunt. *Digital Image Restoration*. Englewood Cliffs, NJ: Prentice Hall, 1977.
7. Andrews, H. C., A. G. Tescher, and R. P. Kruger. "Image Processing by Digital Computer." *IEEE Spectrum* 9 (July 1972): 20–32.
8. Antoniou, A. *Digital Filters: Analysis and Design*. New York: McGraw-Hill, 1979.
9. Blackman, R. B., and J. W. Tukey. *The Measurement of Power Spectra*. New York: Dover, 1958.
10. Blahut, R. E. *Fast Algorithms for Digital Signal Processing*. Reading, MA: Addison-Wesley, 1985.
11. Bluestein, L. J. "A Linear Filtering Approach to the Computation of Discrete Fourier Transform." *IEEE Trans. Audio Electroacoust*. AU-18 (December 1970): 451–455.
12. Bose, N. K. *Digital Filters: Theory and Applications*. New York: Elsevier, 1985.

13. BRACEWELL, R. N. "The Fast Hartley Transform." *Proc. IEEE* 72 (August 1984): 1010–1018.

14. BRENNER, N. M. "Three Fortran Programs That Perform the Cooley–Tukey Fourier Transform." Tech. Note 1967–2. Lexington, MA: MIT Lincoln Laboratory, July 1967.

15. BROGAN W. L. *Modern Control Theory*. 2d ed. Englewood Cliffs, NJ: Prentice Hall, 1985.

16. CADZOW, J. A. *Discrete-Time Systems: An Introduction with Interdisciplinary Applications*. Englewood Cliffs, NJ: Prentice Hall, 1973.

17. CAMPBELL, G. A., and R. M. FOSTER. *Fourier Integrals for Practical Applications*. New York: Van Nostrand, 1948.

18. CARLSON, A. B., and D. G. GISSER. *Electrical Engineering—Concepts and Applications*. Reading, MA: Addison-Wesley, 1981.

19. CASTLEMAN, K. R. *Digital Image Processing*. Englewood Cliffs, NJ: Prentice-Hall, 1979.

20. CHEN, C. T. *One-Dimensional Digital Signal Processing*. New York: Marcel Dekker, 1979.

21. CHILDERS, D. G., and A. E. DURLING. *Digital Filtering and Signal Processing*. New York: West Publishing, 1975.

22. CHUA, L. O., and P. M. LIN. *Computer-Aided Analysis of Electronic Circuits*. Englewood Cliffs, NJ: Prentice Hall, 1975.

23. CHURCHILL, R. V., and J. W. BROWN. *Complex Variables and Applications*. 3d ed. New York: McGraw-Hill, 1974.

24. CONSTANTINIDES, A. G. "Spectral Transformations for Digital Filters." *Proc. IEE* 117(8) (August 1970): 1585–1590.

25. COOLEY, J. W., P. A. W. LEWIS, and P. D. WELCH. "Historical Notes on the Fast Fourier Transform." *IEEE Trans. Audio Electroacoust*. AU-15 (June 1967): 76–79.

26. COOLEY, J. W., and J. W. TUKEY. "An Algorithm for the Machine Calculation of Complex Fourier Series." *Math. Computation* 19 (1965): 297–301.

27. CROCHIERE, R. E. "Digital Ladder Structures and Coefficient Sensitivity." *IEEE Trans. Audio Electroacoust*. AU-20 (October 1972) 240–246.

28. CROCHIERE, R. E. "Digital Network Theory and Its Application to the Analysis and Design of Digital Filters." Ph. D. thesis, Department of Electrical Engineering, MIT, 1974.

29. DANIELS, R. W. *Approximation Methods for the Design of Passive, Active, and Digital Filters*. New York: McGraw-Hill, 1974.

30. DECZKY, A. G. "Synthesis of Recursive Digital Filters Using the Minimum-p Error Criterion." *IEEE Trans. Audio Electroacoust*. AU-20 (October 1972): 257–263.

31. DELCARO, L. G., and G. L. SICURANZA. "Design of Two-Dimensional Recursive Digital Filters." *IEEE Trans. Acoust., Speech, Signal Processing* ASSP-25 (December 1977): 577–578.

32. DE MOIVRE, A. *Miscellanea Analytica de Seriebus et Quadraturis*. London, 1730.

33. DICKSON, L. E. *History of the Theory of Numbers*. Vol. 1. Washington, DC: Carnegie Institute, 1919.

34. DSP Committee IEEE ASSP (eds.). *Programs for Digital Signal Processing*. New York: IEEE Press, 1979.

35. DSP Committee IEEE ASSP (eds.). *Selected Papers in Digital Signal Processing,* Vol. II. New York: IEEE Press, 1976.

36. Director, S. W. "Survey of Circuit-Oriented Optimization Techniques." *IEEE Trans. Circuit Theory* CT-18 (January 1971): 3–10.

37. Dorf, R. C. *Modern Control Systems.* 4th ed. Reading, MA: Addison-Wesley, 1986.

38. Fettweis, A. "A Simple Design of Maximally Flat Delay Digital Filters." *IEEE Trans. Audio Electroacoust.* AU-20 (June 1972): 112–114.

39. Fettweis, A. "Some Principles of Designing Digital Filters Imitating Classical Filter Structures." *IEEE Trans. Circuit Theory* CT-18 (March 1971): 314–316.

40. Flanagan, J. L., et al. "Synthetic Voices for Computers." *IEEE Spectrum* 7 (January 1970): 22–45.

41. Fletcher, R., and M. J. D. Powell. "A Rapidly Convergent Descent Method for Minimization." *Comp. J.* 6(2) (July 1963): 163–168.

42. Frank, D. A., B. L. Hornsby, and R. D. Toles. "Structural Filtering Design and Structural Coupling Ground Testing on the AFTI/F-16." Paper #83-2280-CP, *AIAA Guidance and Control Conference,* Gatlinburg, TN, 1983.

43. Franklin, G. F., and J. D. Powell. *Digital Control of Dynamic Systems.* Reading, MA: Addison-Wesley, 1981.

44. Garvin, W. W. *Introduction to Linear Programming.* New York: McGraw-Hill, 1960.

45. Gentleman, W. M., and G. Sande. "Fast Fourier Transforms—For Fun and Profit," in AFIPS, *Proc. Fall Joint Computer Conf.* (1966): 563–578.

46. Gerheim, A. P., and J. W. Stoughton. "Further Results in Walsh Domain Filtering." *IEEE Trans. Acoust., Speech, Signal Processing* ASSP-35 (March 1987): 394–399.

47. Ghausi, M. S., and K. R. Laker. *Modern Filter Design, Active RC and Switched Capacitor.* Englewood Cliffs, NJ: Prentice Hall, 1981.

48. Gibbs, A. J. "On the Frequency-Domain Responses of Causal Digital Filters." Ph.D. thesis, University of Wisconsin, 1969.

49. Gibbs, A. J. "The Design of Digital Filters." *Australian Telecomm. Res.* 4(1) (1970): 29–34.

50. Goertzel, G. "An Algorithm for the Evaluation of Finite Trigonometric Series." *Amer. Math. Monthly* 65 (January 1958): 34–35.

51. Gold, B., and C. M. Rader. *Digital Processing of Signals.* New York: McGraw-Hill, 1969.

52. Gold, B., and C. M. Rader. "Effect of Quantization Noise in Digital Filters," in *1966 Spring Joint Conf., AFIPS Proc.,* Vol. 28, 213–219. Washington, DC: Spartan, 1966.

53. Gonzalez, R. C., and P. Wintz. *Digital Image Processing,* 2d ed. Reading, MA: Addison-Wesley, 1986.

54. Gray, A. H., Jr., and J. D. Markel. "Digital Lattice and Ladder Filter Synthesis." *IEEE Trans. Audio Electroacoust.* AU-21 (December 1973): 491–500.

55. Hamming, R. W. *Digital Filters.* Englewood Cliffs, NJ: Prentice Hall, 1977.

56. Hamming, R. W. *Digital Filters.* 2d ed. Englewoods Cliffs, NJ: Prentice Hall, 1983.

57. Harris, F. J. "On the Use of Windows for Harmonic Analysis with the Discrete Fourier Transform." *Proc. IEEE* 66, (January 1978): 51–83.

58. Helms, H. D. "Digital Filters with Equiripple or Minimax Responses." *IEEE Trans. Audio Electroacoust.* AU-19 (March 1971): 87–94.

59. Helms, H. D. "Fast Fourier Transform Method of Computing Difference Equations and Simulating Filters." *IEEE Trans. Audio Electroacoust.* AU-15 (June 1967): 85–90.

60. Herrmann, O. "On the Design of Nonrecursive Digital Filters with Linear Phase." *Electron Lett.* 6(11) (May 1970): 328–329.

61. Herrmann, O., L. R. Rabiner, and D. S. K. Chan. "Practical Design Rules for Optimum Finite Impulse Response Low-Pass Digital Filters." *Bell Syst. Tech. J.* 52 (July--August 1973): 769–799.

62. Hersey, H. S., D. W. Tufts, and J. T. Lewis. "Interactive Minimax Design of Linear Phase Nonrecursive Digital Filters Subject to Upper and Lower Function Constraints." *IEEE Trans. Audio Electroacoust.* AU-20 (June 1972): 171–173.

63. Hofstetter, E., A. V. Oppenheim, and J. Siegel. "A New Technique for the Design of Nonrecursive Digital Filters," *Proc. 5th Annual Princeton Conf. Inform., Sci., Syst.* (1971): 64–72.

64. Hosticka, B. J., R. W. Brodersen, and P. R. Gray. "MOS Sampled Data Recursive Filters Using Switched Capacitor Integrators." *IEEE J. of Solid-State Circuits* SC-12 (December 1977): 600–608.

65. Hou, H. S. "A Fast Recursive Algorithm for Computing the Discrete Cosine Transform." *IEEE Trans. Acoust., Speech, Signal Processing* ASSP-35 (October 1987): 1455–1461.

66. Hu, J. V., and L. R. Rabiner. "Design Techniques for Two-Dimensional Digital Filters." *IEEE Trans. Audio Electroacoust* AU-20 (October 1972): 249–257.

67. Huang, T. S. "Stability of Two-Dimensional Recursive Filters." *IEEE Trans. Audio Electroacoust.* AU-20 (June 1972): 158–163.

68. Huang, T. S. "Two-Dimensional Windows." *IEEE Trans. Audio Electroacoust.* AU-20 (March 1972): 88–89.

69. Hunt, B. R. "Digital Image Processing." *Proc. IEEE* 63(4) (April 1975): 693–708.

70. Hurewicz, W. "Filters and Servo Systems with Pulsed Data," in H. M. James, N. B. Nichols, and R. S. Phillips (eds.), *Theory of Servomechanisms*. Radiation Laboratory Series, Vol. 25. New York: McGraw-Hill, 1947.

71. Hwang, S. Y. "Realization of Canonical Digital Networks." *IEEE Trans. Acoust., Speech, Signal Processing* ASSP-22(1) (February 1974): 27–39.

72. *Intersil Data Acquisition Handbook: A Technical Guide to A/D and D/A Converters and Their Applications*. Cupertino, CA: Intersil, Inc., 1980.

73. Jackson, L. B. *Digital Filters and Signal Processing*. Hingham, MA: Kluwer Academic Publishers, 1985.

74. Jackson, L. B. "Roundoff-Noise Analysis for Fixed-Point Digital Filters Realized in Cascade or Parallel Form," *IEEE Trans. on Audio Electroacoust.* AU-18 (June 1970): 107–122.

75. Jenkins, W. K., and B. J. Leon. "The Use of Residue Number Systems in the Design of Finite Impulse Response Digital Filters." *IEEE Trans. on Circuits and Syst.* CAS-24(4) (April 1977): 191–201.

76. Jing, Z. "A New Method for Digital All-Pass Filter Design." *IEEE Trans. Acoust., Speech, Signal Processing* ASSP-35 (November 1987): 1557–1564.

77. Johnson, D. E. *Introduction to Filter Theory*. Englewood Cliffs, NJ: Prentice Hall, 1976.

78. JOHNSON, D. E. and J. R. JOHNSON. *Mathematical Methods in Engineering and Physics*. Englewood Cliffs, NJ: Prentice Hall, 1982.

79. JOHNSON, D. E., J. R. JOHNSON, and H. P. MOORE. *A Handbook of Active Filters*. Englewood Cliffs, NJ: Prentice Hall, 1980.

80. JOHNSON, J. R., and D. E. JOHNSON. *Linear Systems Analysis*. Malabar, FL: Robert E. Krieger Publishing, 1981.

81. JONG, M. T. *Methods of Discrete Signal and System Analysis*. New York: McGraw-Hill, 1982.

82. JORDAN, C. *Calculus of Finite Differences, 1939*. Reprint. New York: Chelsea, 1960.

83. JURY, E. I. *Sampled-Data Control Systems*. New York: Wiley, 1958.

84. JURY, E. I. *Theory and Application of the Z-Transform Method*. New York: Wiley, 1964.

85. KAISER, J. F. "Design Methods for Sampled-Data Filters." *Proc. 1st Allerton Conf. Circuit System Theory* (November 1963): 221–236.

86. KAISER, J. F. "Digital Filters," Chapter 7 in F. F. Kuo and J. F. Kaiser (eds.), *System Analysis by Digital Computer*. New York: Wiley, 1966.

87. KAISER, J. F. "Nonrecursive Digital Filter Design Using The I_0–Sinh Window Function." *Proc. IEEE International Symp. on Circuits and Systems*. San Francisco (April 1974): 20–23.

88. KAISER, J. F. "On the Fast Generation of Equally Spaced Values of the Gaussian Function A exp $(-at * t)$." *IEEE Trans. Acoust., Speech, Signal Processing*. ASSP-35 (October 1987): 1480–1481.

89. KAK, S. C., and N. S. JAYANT. "On Speech Encryption Using Waveform Scrambling." *Bell Syst. Tech. J.* 56 (May–June 1977): 781–808.

90. KALMAN, R. E. "A New Approach to Linear Filtering and Prediction Problems." *J. Basic Eng. (ASME Trans.)* 82D (1960): 35–45.

91. KUO, B. C. *Automatic Control Systems*. 5th ed. Englewood Cliffs, NJ: Prentice Hall, 1987.

92. KWONG, C. P., and K. P. SHIU. "Structured Fast Hartley Transform Algorithms." *IEEE Trans. Acoust., Speech, Signal Processing* ASSP-34 (August 1986): 1000–1002.

93. LAPLACE, P. S. *Théorie Analytic des probabilités, pt. I: Du Calcul des fonctions génératrices*. Paris, 1812.

94. LEA, W. A., (ed.). *Trends in Speech Recognition*. Englewood Cliffs, NJ: Prentice Hall, 1980.

95. LEE, B. G. "A New Algorithm to Compute the Discrete Cosine Transform." *IEEE Trans. Acoust., Speech, Signal Processing* ASSP-32 (December 1984): 1243–1245.

96. LEIBOWITZ, L. M. "A Simplified Binary Arithmetic for the Fermat Number Transform." *IEEE Trans. on Acoustics, Speech, Signal Processing* ASSP-24(5) (October 1976): 356–359.

97. LINDSEY, W. C., and C. M. CHIE. "A Survey of Digital Phase-Locked Loops." *Proc. IEEE* 69 (April 1981): 410–431.

98. LIU, B. "Effect of Finite Word Length on the Accuracy of Digital Filters—A Review." *IEEE Trans. Circuit Theory* CT-18 (November 1971): 670–677.

99. LIU, B., and T. KANEKO. "Error Analysis of Digital Filters Realized with Floating-Point Arithmetic." *Proc. IEEE* 57 (October 1969): 1735–1747.

100. MALVAR, H. S. "Fast Computation of the Discrete Cosine Transform and the Discrete Hartley Transform." *IEEE Trans. Acoust., Speech, Signal Processing* ASSP-35 (October 1987): 1484–1485.

101. MASON, S. J. "Feedback Theory—Further Properties of Signal Flow Graphs." *Proc. IRE* 44 (July 1956): 920–926.

102. MASON, S. J. "Feedback Theory—Some Properties of Signal Flow Graphs." *Proc. IRE* 41 (September 1953): 1144–1156.

103. MATHEWS, J. D., J. K. BREAKALL, and G. K. KARAWAS. "The Discrete Prolate Spheroidal Filter as a Digital Signal Processing Tool." *IEEE Trans. Acoust., Speech, Signal Processing* ASSP-33 (December 1985): 1471–1478.

104. MCCLELLAN, J. H. "Hardware Realization of a Fermat Number Transform." *IEEE Trans. on Acoustics, Speech, Signal Processing* ASSP-24(3) (June 1976): 216–225.

105. MCCLELLAN, J. H. "The Design of Two Dimensional Filters by Transformations." *Proc. 7th Annual Princeton Conf. on Inform. Sci. and Syst.* (March 1973): 247–251.

106. MCCLELLAN, J. H. and T. W. PARKS. "A Unified Approach to the Design of Optimum FIR Linear Phase Digital Filters." *IEEE Trans. on Circuit Theory* CT-20 (November 1973): 697–701.

107. MCCLELLAN, J. H., T. W. PARKS, and L. R. RABINER. "A Computer Program for Designing Optimum FIR Linear Phase Digital Filters." *IEEE Trans. Audio Electroacoust.* AU-21 (December 1973): 506–526.

108. MCCLELLAN, J. H. and C. M. RADER. *Number Theory in Digital Signal Processing.* Englewood Cliffs, NJ: Prentice Hall, 1979.

109. MECKLENBRÄUKER, W. F. G., R. M. MERSEREAU. "McClellan Transformations for Two-Dimensional Digital-Filtering: II—Implementation." *IEEE Trans. Circuits and Systems* CAS-23 (July 1976): 414–422.

110. MERSEREAU, R. M., and D. E. DUDGEON. "Two Dimensional Digital Filtering." *Proc. IEEE* 63 (April 1975): 610–623.

111. MERSEREAU, R. M., W. F. G. MECKLENBRÄUKER, and T. F. QUATIERI, JR. "McClellan Transformations for Two-Dimensional Digital Filtering: I—Design." *IEEE Trans. Circuits and Systems* CAS-23 (July 1976): 405–413.

112. MITRA, S. K., and M. P. EKSTROM (eds.). *Two-Dimensional Digital Signal Processing.* Stroudsburg, PA: Dowden, Hutchinson and Ross, 1978.

113. MITRA, S. K., and K. HIRANO. "Digital All-Pass Networks." *IEEE Trans. on Circuits and Syst.* CAS-21(5) (September 1974): 688–700.

114. MITRA, S. K., D. C. HUEY, and R. J. SHERWOOD. "New Methods of Digital Ladder Realization." *IEEE Trans. on Audio Electroacoust.* AU-21(6) (December 1973): 485–491.

115. MITRA, S. K., A. D. SAGAR, and N. A. PENDERGRASS. "Realizations of Two-Dimensional Recursive Digital Filters." *IEEE Trans. Circuits and Systems* CAS-22 (March 1975): 177–184.

116. MITRA, S. K., and R. J. SHERWOOD. "Canonical Realizations of Digital Filters Using the Continued Fraction Expansion." *IEEE Trans. on Audio Electroacoust.* AU-20 (August 1972): 185–194.

117. MITRA, S. K., and R. J. SHERWOOD. "Digital Ladder Networks." *IEEE Trans. on Audio Electroacoust.* AU-21(1) (February 1973): 30–36.

118. MORLEY, M. S. *The Linear IC Handbook.* Blue Ridge Summit, PA: Tab Books, 1986.

119. MORSE, P. M., and H. FESHBACH. *Methods of Theoretical Physics, Part I,* New York: McGraw-Hill, 1953.

120. NAKAYAMA, K. "A New Discrete Fourier Transform Algorithm Using Butterfly Structure Fast Convolution." *IEEE Trans. Acoust., Speech, Signal Processing* ASSP-33 (October 1985): 1197–1208.

121. NUTTALL, A. H. "Efficient Evaluation of Polynomials and Exponentials of Polynomials for Equispaced Arguments." *IEEE Trans. Acoust., Speech, Signal Processing* ASSP-35 (October 1987): 1486–1489.

122. NUTTALL, A. H. "Some Windows with Very Good Sidelobe Behavior." *IEEE Trans. on Acoustics, Speech, Signal Processing* ASSP-29(1) (February 1981): 84–91.

123. OGATA, K. *State Space Analysis of Control Systems*. Englewood Cliffs, NJ: Prentice Hall, 1967.

124. OLIVER, B. M., J. R. PIERCE, and C. E. SHANNON. "The Philosophy of PCM." *Proc. IRE* 36(11) (November 1948): 1324–1331.

125. OPPENHEIM, A. V., (ed.). *Applications of Digital Signal Processing*. Englewood Cliffs, NJ: Prentice Hall, 1978.

126. OPPENHEIM, A. V., and R. W. SCHAFER. *Digital Signal Processing*. Englewood Cliffs, NJ: Prentice Hall, 1975.

127. OPPENHEIM, A. V., and R. W. SCHAFER. *Discrete-Time Signal Processing*. Englewood Cliffs, NJ: Prentice Hall, 1989.

128. OPPENHEIM, A. V., R. W. SCHAFER, and T. G. STOCKHAM, JR. "Nonlinear Filtering of Multiplied and Convolved Signals." *Proc. IEEE* 56 (August 1968): 1264–1291.

129. OPPENHEIM, A. V., and C. J. WEINSTEIN. "Effects of Finite Register Length in Digital Filtering and the Fast Fourier Transform." *Proc. IEEE* 60 (August 1972): 957–976.

130. OPPENHEIM, A. V., A. S. WILLSKY, with I. T. YOUNG. *Signals and Systems*. Englewood Cliffs, NJ: Prentice Hall, 1983.

131. PARKS, T. W., and J. H. MCCLELLAN. "A Program for the Design of Linear Phase Finite Impulse Response Filters." *IEEE Trans. Audio Electroacoust.* AU-20 (August 1972): 195–199.

132. PARKS, T. W., and J. H. MCCLELLAN. "Chebyshev Approximation for Nonrecursive Digital Filters with Linear Phase." *IEEE Trans. Circuit Theory* CT-19 (March 1972): 189–194.

133. PEASE, M. C. "An Adaptation of the Fast Fourier Transform for Parallel Processing." *J. ACM* 15 (April 1968): 252–264.

134. PELED, A., and B. LIU. *Digital Signal Processing*. New York: Wiley, 1976.

135. RABINER, L. R. "Approximate Design Relationships for Low-Pass FIR Digital Filters." *IEEE Trans. Audio Electroacoust.* AU-21 (October 1973): 456–460.

136. RABINER, L. R., and B. GOLD. *Theory and Applications of Digital Signal Processing*. Englewood Cliffs, NJ: Prentice Hall, 1975.

137. RABINER, L. R., N. Y. GRAHAM, and H. D. HELMS. "Linear Programming Design of IIR Digital Filters with Arbitrary Magnitude Function." *IEEE Trans. Acoust., Speech, Signal Processing* ASSP-22 (April 1974): 117–123.

138. RABINER, L. R., and O. HERRMANN. "On the Design of Optimal FIR Low Pass Filters with Even Impulse Response." *Trans. on Audio Electroacoust.* AU-21 (August 1973): 329–336.

139. RABINER, L. R., J. F. KAISER, O. HERRMANN, and M. T. DOLAN. "Some Comparisons between FIR and IIR Digital Filters." *Bell Syst. Tech. J.* 53, (February 1974): 305–331.

140. RABINER, L. R., J. H. MCCLELLAN, and T. W. PARKS. "FIR Digital Filter Design Techniques Using Weighted Chebyshev Approximation." *Proc. IEEE* 63 (April 1975): 595–610.

141. RABINER, L. R., C. A. MCGONEGAL, and D. PAUL. "FIR Windowed Filter Design Pro-

gram—WINDOW," in DSP Committee IEEE ASSP (eds.), *Programs for Digital Signal Processing*. New York: IEEE Press, 1979.

142. RABINER, L. R., and C. M. RADER (eds.). *Digital Signal Processing*. New York: IEEE Press, 1972.

143. RABINER, L. R., and R. W. SCHAFER. "Correction to Recursive and Nonrecursive Realizations of Digital Filters Designed by Frequency Sampling Techniques." *IEEE Trans. Audio Electroacoust*. AU-20(1) (March 1972): 104–105.

144. RABINER, L. R., and R. W. SCHAFER. *Digital Processing of Speech Signals*. Englewood Cliffs, NJ: Prentice Hall, 1978.

145. RABINER, L. R., and R. W. SCHAFER. "Recursive and Nonrecursive Realizations of Digital Filters Designed by Frequency Sampling Techniques." *IEEE Trans. Audio Electroacoust*. AU-19(3) (September 1971): 200–217.

146. RABINER, L. R., R. W. SCHAFER, and C. M. RADER. "The Chirp Z-Transform Algorithm." *IEEE Trans. Audio Electroacoust*. AU-17 (June 1969): 86–92.

147. RADER, C. M. "Convolution and Correlation Using Number Theoretical Transforms," Section 6.19, in L. R. Rabiner and B. Gold, *Theory and Application of Digital Signal Processing*. Englewood Cliffs, NJ: Prentice Hall, 1975.

148. RADER, C. M. "Discrete Convolutions via Mersenne Transforms." *IEEE Trans. Comput*. C-21 (December 1972) 1269–1273.

149. RADER, C. M. "Discrete Fourier Transforms When the Number of Data Samples Is Prime." *Proc. IEEE* 56 (June 1968): 1107–1108.

150. RAGAZZINI, J. R., and G. F. FRANKLIN. *Sampled Data Control Systems*. New York: McGraw-Hill, 1958.

151. REESE, R. L., S. M. EVERETT, and E. D. CRAUN. "Origin of the Julian Period: An Application of Congruences and the Chinese Remainder Theorem." *Am. J. Physics* 49(7) (July 1981): 658–661.

152. ROBERTS, R. A., and C. T. MULLIS. *Digital Signal Processing*. Reading, MA: Addison-Wesley, 1987.

153. ROBINSON, E. A., and S. TREITEL. *Geophysical Signal Analysis*. Englewood Cliffs, NJ: Prentice Hall, 1976.

154. ROSENTHAL, L. H., L. R. RABINER, R. W. SCHAFER, P. CUMMISKEY, and J. L. FLANAGAN. "Automatic Voice Response: Interfacing Man with Machine." *IEEE Spectrum* 11 (July 1974): 61–68.

155. SCHAFER, R. W., and L. R. RABINER. "Design of Digital Filter Banks for Speech Analysis." *Bell Syst. Tech. J.* 50(10) (December 1971): 3097–3115.

156. SCHROEDER, M. R. *Number Theory in Science and Communication*. New York: Springer-Verlag, 1984.

157. SCHWARTZ, M., and L. SHAW. *Signal Processing: Discrete Spectral Analysis, Detection and Estimation*. New York: McGraw-Hill, 1975.

158. SHANKS, J. L., S. TREITEL, and J. H. JUSTICE. "Stability and Synthesis of Two-Dimensional Recursive Filters." IEEE *Trans. Audio Electroacoust*. AU-20(3) (June 1972): 115–128.

159. SHEINGOLD, D. H., (ed.) *Analog-Digital Conversion Notes*. Norwood, MA: Analog Devices, 1977.

160. SHERWOOD, B. A. "The Computer Speaks." *IEEE Spectrum* 16 (August 1979): 18–25.

161. SINGLETON, R. C. "An Algorithm for Computing the Mixed Radix Fast Fourier Transform." *IEEE Trans. Audio Electroacoust*. AU-17 (June 1969): 93–103.

162. Slepian, D., and H. O. Pollak. "Prolate Spheroidal Wave Functions, Fourier Analysis and Uncertainty—I and II. *Bell Syst. Tech. J.* 40(1) (January 1961): 43–84.

163. Sorensen, H. V., D. L. Jones, C. S. Burrus, and M. T. Heidman. "On Computing the Discrete Hartley Transform." *IEEE Trans. Acoust., Speech, Signal Processing* ASSP-33 (October 1985): 1231–1238.

164. "Special Issue on Digital Signal Processing." *Proc. IEEE* 63 (April 1975).

165. "Special Issue on Image Processing." *Proc. IEEE* 69 (May 1981).

166. "Special Issue on Man-Machine Communication by Voice." *Proc. IEEE* 64 (April 1976).

167. "Special Issue on Microprocessor Applications." *Proc. IEEE* 66 (February 1978).

168. "Special Issue on Multidimensional Systems." *Proc. IEEE* 65 (June 1977).

169. Stanley, W. D., G. R. Dougherty, and R. Dougherty. *Digital Signal Processing*. 2d ed. Reston, VA: Reston, 1984.

170. Stieglitz, K. "Computer-Aided Design of Recursive Digital Filters." *IEEE Trans. Audio Electroacoust.* AU-18 (June 1970): 123–129.

171. Stockham, T. G., Jr. "High Speed Convolution and Correlation," in *1966 Spring Joint Comput. Conf., AFIPS Proc.,* Vol. 28, 229–233. Washington, DC: Spartan, 1966.

172. Stockham, T. G., Jr. "High Speed Convolution and Correlation with Applications to Digital Filtering," Chapter 7, in B. Gold and C. M. Rader, *Digital Processing of Signals*. New York: McGraw-Hill, 1969.

173. Strum, R. D., and D. E. Kirk. *First Principles of Discrete Systems and Digital Signal Processing*. Reading, MA: Addison-Wesley, 1988.

174. Szabo, N. S., and R. I. Tanaka. *Residue Arithmetic and its Application to Computer Technology*. New York: McGraw-Hill, 1967.

175. Temes, G. C., and D. A. Calahan. "Computer-Aided Network Optimization—The State of the Art." *Proc. IEEE* 55(11) (November 1967): 1832–1863.

176. Thajchayapong, P., and P. Lomtong. "A Maximally Flat Group Delay Recursive Filter with Controllable Magnitude." *IEEE Trans. Circuits and Syst.* CAS-25(1) (January 1978): 51–53.

177. Thiran, J. P. "Equal-Ripple Delay Recursive Digital Filters." *IEEE Trans. Circuit Theory* CT-18 (November 1971): 664–677.

178. Thiran, J. P. "Recursive Digital Filters with Maximally Flat Group Delay." *IEEE Trans. Circuit Theory* CT-18 (November 1971): 659–663.

179. Tou, J. T. *Digital and Sampled-Data Control Systems*. New York: McGraw-Hill, 1959.

180. Tretter, S. A. *Introduction to Discrete-Time Signal Processing*. New York: Wiley, 1976.

181. Van Valkenburg, M. E. *Introduction to Modern Network Synthesis*. New York: Wiley, 1962.

182. Van Valkenburg, M. E., and B. K. Kinariwala. *Linear Circuits*. Englewood Cliffs, NJ: Prentice Hall, 1982.

183. Weinberg, L. *Network Analysis and Synthesis*. New York: McGraw-Hill, 1962.

184. Weinstein, C. J. "Roundoff Noise in Floating Point Fast Fourier Transform Computation." *IEEE Trans. Audio Electroacoust.* AU-17 (September 1969): 209–215.

185. Weinstein, C. J., and A. V. Oppenheim. "A Comparison of Roundoff Noise in Floating Point and Fixed Point Digital Filter Realizations." *Proc. IEEE* 57 (June 1969): 1181–1183.

186. WILLIAMS, C. S. *Designing Digital Filters*. Englewood Cliffs, NJ: Prentice Hall, 1986.

187. WOOD, L. C., and S. TREITEL. "Seismic Signal Processing." *Proc. IEEE* 63 (April 1975): 649–661.

188. YIP, P., and K. R. RAO. "On the Shift Property of DCT's and DST's." *IEEE Trans. Acoust., Speech, Signal Processing* ASSP-35 (March 1987): 404–406.

189. ZAKHOR, A., and A. V. OPPENHEIM. "Quantization Errors in the Computation of the Discrete Hartley Transform." *IEEE Trans. Acoust., Speech, Signal Processing* ASSP-35 (November 1987): 1592–1602.

190. ZAROWSKI, C. J., and M. YUNIK. "Spectral Filtering Using the Fast Walsh Transform." *IEEE Trans. Acoust., Speech, Signal Processing* ASSP-33 (October 1985): 1246–1252.

INDEX

A

B

C

D

E

F

G

H

I

O

P

Q

R

S

T

U

V

W

Z

FIGURE CREDITS

The following are reprinted with permission:

Fig. 1.7: Carlson/Gisser, *Electrical Engineering: Concepts and Applications,* (Reading, MA: Addison-Wesley Publishing Co., Inc., 1981.)

Fig. P.4.37: Gray and Markel, "Digital Lattice and Ladder Filter Synthesis," *IEEE Trans. Audio Electroacoust.*, vol. AU-21 (Dec 1973).

Fig. P.4.32; Fig. 4.11: S. Y. Hwang, "Realization of canonical digital networks," *IEEE Trans. Acoust., Speech, and Signal Processing,* vol. ASSp-22, no. 1 (Feb 1974).

Fig. 1.8: *Intersil Data Acquisition Handbook, A Technical Guide to A/D and D/A Converters and Their Applications* (Cupertino, CA: Intersil, Inc., 1980)

Fig. 5.22: D. E. Johnson, *Introduction to Filter Theory* (Englewood Cliffs, NJ: Prentice-Hall, Inc., 1976).

Figs. 5.1–5.3, 5.6, 5.9–5.10, 5.15–5.16, 5.19; Tables 5.1–5.3: D. E. Johnson/J. R. Johnson/H. P. Moore, *A Handbook of Active Filters,* (Englewood Cliffs, NJ: Prentice-Hall, Inc., 1980)

Fig. 1.13: B. C. Kuo, *Automatic Control Systems,* 5/E. (Englewood Cliffs, NJ: Prentice-Hall, Inc., 1987)

Fig. 1.14; Fig. P.4.35: Lindsay and Chie, "Digital Phase-Locked Loops," *Proceedings of the IEEE* (April 1981).

Fig. 4.27(b), Fig. 4.31, Figs. P.4.24–4.25, 4.27: S. K. Mitra and R. J. Sherwood, "Canonical Realizations of Digital Filters Using the Continued Fraction Expansion," *IEEE Trans. on Audio and Electroacoust.*, vol. AU-20 (August 1972).

Figs. 4.33, 4.35; Fig. P.4.30: S. K. Mitra and R. J. Sherwood, "Digital Ladder Networks," *IEEE Trans. on Audio and Electroacoust.*, vol. AU-21, no. 1 (Feb 1973).

Fig. 4.34: S. K. Mitra, D. C. Huey, and R. J. Sherwood, "New Methods of Digital Ladder Realization," *IEEE Trans. Audio and Electroacoust.* vol. AU-21, no. 6 (Dec 1973).

Fig. 6.2: P. M. Morse and H. Feshbach, *Methods of Theoretical Physics, Part I,* McGraw-Hill Book Company, 1953.

Figs. 8.1, 8.4, 8.6–8.10: A. V. Oppenheim/R. W. Schafer, *Digital Signal Processing*, (Englewood Cliffs, NJ: Prentice-Hall, Inc., 1975)

Fig. Prob. 6.17: Rabiner, L. R., C. A. McGonegal, and D. Paul, "FIR Windowed Filter Design Program— WINDOW," in DSP Committee IEEE ASSP (eds.), *Programs for Digital Signal Processing*. NY IEEE Press, 1979.

Fig. 5.21: L. R. Rabiner/R. W. Schafer, *Digital Processing of Speech Signals,* p. 268. (Englewood Cliffs, NJ: Prentice-Hall, Inc., 1978)

Figs. 1.15–1.16: M. E. Van Valkenburg/B. K. Kinariwala, *Linear Circuits,* (Englewood Cliffs, NJ: Prentice-Hall, Inc., 1982)